Methods for Evaluating Pesticides for Control of Plant Pathogens

Edited by
Kenneth D. Hickey

Prepared jointly by
The American Phytopathological Society
and the
Society of Nematologists

APS PRESS
The American Phytopathological Society

Library of Congress Catalog Card Number: 86-070603
International Standard Book Number: 0-89054-071-3

Reference in this volume to a company name or product name is intended for explicit description only and does not imply approval or recommendation by the U.S. Department of Agriculture, The American Phytopathological Society, or the Society of Nematologists.

Printed in the United States of America

The American Phytopathological Society
3340 Pilot Knob Road
St. Paul, Minnesota 55121, USA

Contents

Part III. Field Test Procedures for Fruit and Nut Crops

Part IV. Field Test Procedures for Vegetable Crops

Part V. Field Test Procedures for Field Crops

Part VI. Field Test Procedures for Ornamentals and Turfgrasses

Part VII. Field Test Procedures for Seed Treatments

Part VIII. Field Test Procedures for Tree Injection

Part IX. Field Test Procedures for Soil Treatments

Part X. Nematicide Test Procedures

Preface

The importance of agrichemicals in worldwide production of food and fiber is well documented and, in some countries, the continued use of agrichemicals is presumed essential to sustaining profitable production levels. In the United States, the development and manufacture of pesticides are limited to about 50 companies that produce about 35,000 pesticide products available for retail sale. Fungicides, bactericides, and nematicides make up less than 11% of the total volume of pesticides used but often require considerably more detailed and lengthy laboratory, greenhouse, and field evaluations than other pesticides before they are released for commercial use. The quality of testing of new chemicals for control of plant diseases has improved considerably since The American Phytopathological Society (APS) published *Methods for Evaluating Plant Fungicides, Nematicides, and Bactericides,* the predecessor of this book, in 1978. The number of scientists evaluating chemicals for control of plant pathogens has declined during the past decade, prompting the developers of new pesticides to search for test sites where consistent, reliable, and efficient evaluations are obtained.

The purpose of this book is to continue the effort to present state-of-the-art methods that will assist scientists in conducting efficient evaluations that will produce reliable results that developers and regulatory agencies can easily interpret and use. The methods described are presented by individuals experienced in the subject matter who have found the methods worthy of trial. They should not be interpreted as standards or practices recommended by APS, the Society of Nematologists (SON), or the Editorial Committee for this book. The book is intended to be a reference for persons interested in evaluating chemicals for control of plant pathogens. Questions concerning details of the methods presented should be directed to the specific author. Individuals who are conducting chemical evaluations are encouraged to publish the results in the annual *Fungicide and Nematicide Tests* volumes published by the New Fungicide and Nematicide Data Committee of APS.

Some of the chapters presented by plant pathologists and nematologists in *Methods for Evaluating Plant Fungicides, Nematicides, and Bactericides* have been revised by their authors and included in this book. In addition, a number of new chapters have been solicited to expand the coverage of the previous methods book. This book is not intended to replace the previous methods book, since some of the chapters in that book are not included here.

We the members of the Editorial Committee for this book (a subcommittee of the APS New Fungicide and Nematicide Data Committee) recognize that test procedures often evolve over considerable periods of time and that a number of individuals may contribute to their development. We acknowledge the contributions of all authors and of their colleagues who helped in the development and evaluation of methods described in this book.

The manuscripts of the chapters in this book were reviewed by three or more persons knowledgeable in the subject matter who consider the procedures adequate for the purpose stated. The chapters published here were selected on the basis of reliability and relevance of subject matter and thoroughness of procedure under the supervision and advice of the APS New Fungicide and Nematicide Data Committee, the APS Chemical Control Committee, and the SON Editorial Committee. The SON Editorial Committee solicited and edited the manuscripts on nematicide test procedures. L. R. Krusberg made major efforts in editing the chapters on nematicide testing. In addition, all manuscripts were reviewed by two colleagues of the authors, K. D. Hickey, and one additional referee. We express our appreciation and thanks to the following individuals who served as reviewers of manuscripts for this publication:

P. B. Adams
C. W. Averre
P. A. Backman
P. H. Dernoeden
C. R. Drake
A. W. Engelhard
L. B. Forer
K. D. Hickey
L. R. Krusberg
R. C. Lambe
J. W. Lorbeer
D. E. Mathre
D. J. Mitchell
W. C. Nesmith
A. O. Paulus
P. M. Phipps
P. B. Shoemaker
R. N. Sonoda
M. Szkolnik
J. W. Travis
K. S. Yoder
E. I. Zehr

The technical editing services of Margaret E. Shaffer are much appreciated.

Editorial Committee:
K. D. Hickey, Chairman
C. W. Averre
A. W. Engelhard
W. C. Nesmith

Fungicide and Nematicide Testing and Pesticide Registrations

JOSEPH E. ELSON and WILLIAM L. BIEHN, Interregional Research Project No. 4, Cook College, Rutgers University, New Brunswick, NJ 08903

Results of field trials are useful in judging the effectiveness of pesticides under various conditions. In the United States, the Environmental Protection Agency (EPA) must register effective agents before they can be used commercially. Registration is an expensive, time-consuming process, which pesticide companies find difficult to justify economically except in the case of major commodities.

Interregional Research Project No. 4, a National Agricultural Program for Clearance of Pesticides, Biorationals, and Animal Drugs for Minor or Specialty Uses (IR-4), was initiated to aid in the development of data for establishing tolerances and assisting in the registration of pesticides for minor or specialty uses. IR-4 interacts with all the land-grant universities plus many other universities, state and federal agencies, industry, and the EPA. It is responsible for accumulating and evaluating available data; determining whether additional efficacy, phytotoxicity, and environmental data are needed; seeking out resources for producing these data where they are lacking; advising industry of available information; and eliciting the cooperation of industry in label registration.

The EPA requires certain data to establish tolerances and to register product labels. Researchers should consider, whenever they can, registration requirements when planning fungicide and nematicide trials. If researchers take extra time to make careful notes of phytotoxic effects or collect samples for residue analysis, delays in registration may be prevented, and thus a product may be made available a season earlier than it would have been if additional studies had been required.

Even though the EPA no longer requires performance data to support label claims, chemical companies still require efficacy and phytotoxicity data to support their product registration. Thus, researchers who conduct field evaluations of pesticides to determine efficacy rates have a unique opportunity: they can compile useful field data that will aid the registration process.

To expedite registration, the chemical companies or the EPA (or both) must have complete information from representative growing areas about a product's use. Listed below are types of data that are needed.

1. Identity of the host, disease, and pathogen. Knowing the name of the crop, including the crop variety, the accepted common name of the disease, and the name of the pathogen that is to be controlled is absolutely essential. Such information is basic for labeling products for disease control.

2. Plot size. A description of plot size provides a basis for evaluating the precision of the test.

3. Number of replicates. Four replicates are usually sufficient to obtain statistically sound data.

4. Application timing. The actual dates of application must be recorded. To assist in label preparation, the state of plant growth (prebloom, bloom, first true leaf, etc.) should also be recorded where appropriate.

5. Dosage rate and dilution rate. The dosage and dilution rates that are used must be representative of those suggested for labeling; also, rates must be used that are twice the maximum dosage to be recommended. These rates enables the EPA to set residue tolerances with a safety margin and to determine potential phytotoxic reactions. Researchers should record the dosage rates as the amount of active ingredient per area (such as kilograms per hectare) and the spray volume per area. If the material is banded, band width, row spacing, and the amount of active ingredient per linear measure (such as kilograms per 500 m of row) should be indicated.

6. Formulation of the material. Data must be obtained from plots treated with the same formulation as the one to be registered, since variations in carriers, emulsifiers, and other inert ingredients may affect the performance of a product.

7. Method of application. The method of application, such as directed spray to the base of the plant, foliar application, slurry seed treatment, or dip, should be specified, along with such information as the type of sprayer, the type of nozzle, and pressure.

8. Efficacy. Researchers should indicate the prevalence of disease and the level of control compared to such levels in untreated plots and, if possible, compared to a commercial standard.

9. Yields. Where appropriate, yield data should be taken to verify the usefulness of the product. Qualitative and quantitative data are recommended.

10. Phytotoxicity. Type, amount, and duration of injury to the host plant should be recorded. Researchers should indicate whether injury is considered commercially acceptable. The effect of temporary damage on final yields is of concern to reviewers.

11. Environmental conditions. Information on local temperature and precipitation (or irrigation) should be made available.

12. Soil type. The soil type should be identified in the case of soil-applied pesticides. Such information as soil moisture, percentage of organic matter, and pH should be recorded, if it is appropriate.

13. Additional information. The amount and time of application of pesticides other than the test product must also be recorded to assist in the evaluation of the data.

For pesticides that are to be used on food crops, the

EPA requires that a tolerance be established for each pesticide to be used on each commodity. Collection of samples for residue analysis, although cumbersome for researchers, is advisable if the pesticide may be needed for commercial use on a particular commodity. Proper planning and coordination with an industry technical representative or an IR-4 field coordinator can expedite this aspect of the field testing program and add materially to the overall value of the plot work.

The time for taking samples varies with individual crops, but normally they are taken at the earliest stage of maturity. Plots should be sampled individually, and samples should not be combined. Usually, they should be stored frozen before analysis. Because residue requirements vary according to crop and chemical, specific harvest and handling instructions should be obtained from the analytical laboratory.

Because the requirements of pesticide registration are complex, any data that researchers can provide to IR-4 are appreciated. Consideration of the registration requirements at the time of testing of fungicides and nematicides may make an effective pesticide available for use sooner than it would otherwise have been.

Any questions about how researchers can assist IR-4 should be directed to the state IR-4 liaison representative, the IR-4 regional coordinator, or either of the authors.

Methods for Safe Handling, Storage, and Disposal of Pesticides Used in Research and Demonstration

MICHAEL J. WEAVER, Department of Plant Pathology and Chemical, Drug and Pesticide Unit, Virginia Polytechnic Institute and State University, Blacksburg 24061

The safe handling, storage, and disposal of pesticides by professionals, students, technicians, and farm workers are extremely important for successful agricultural research and demonstration programs. The consequences of misuse of pesticides, including environmental contamination, spills, and poisonings, can be very embarrassing and expensive to both industrial and institutional establishments. Improper use of crop protection chemicals and experimental chemicals can destroy an otherwise well-organized and respected program.

Therefore, before implementation of a research or demonstration program involving the use of pesticides for experimental or general pest control, training of all participants in the proper and safe handling of these chemicals is of the utmost importance.

HANDLING REGISTERED PESTICIDES

The majority of pesticides handled in research and demonstration work are registered compounds. Some of these are used for plot maintenance. Others are used for testing on sites or pests that are not approved in their current registrations, to obtain efficacy, phytotoxicity, and residue data for proposed amendments to those registrations. Registered pesticides are relatively safe if they are used according to the directions on the label and if proper precautions are taken in their handling, storage, and disposal. Following the directions on the label is extremely critical to prevent residues from exceeding federal tolerance levels, to safeguard people and the environment, and to provide adequate pest control.

Registered products are sometimes referred to as Section 3 compounds, since their registrations fall under that section of the Federal Insecticide, Fungicide, and Rodenticide Act (FIFRA) and its amendments. Similar jargon is used for products with special local need registrations (section 24Cs) and emergency exemptions (Section 18s). The federal agency responsible for enforcement of FIFRA is the U.S. Environmental Protection Agency (EPA). In addition, each state has laws governing pesticide use and agencies charged with their enforcement.

The EPA classifies registered pesticides into two groups, general-use and restricted-use pesticides. General-use pesticides, which form the larger of the two classes, are defined by the EPA as products that, when used according to label directions, do not pose a risk of unreasonable adverse effects on people or the environment. According to FIFRA, as amended by the Federal Environmental Protection Control Act of 1972, applicators do not have to be trained or certified to use these products. However, requirements may vary from state to state, and some states have established regulations that are more stringent than the federal ones.

Restricted-use pesticides, the second group of registered products, are defined by the EPA as those that pose an unreasonable risk to people or the environment unless great care is taken in their handling and storage. Special training and certification are required for individuals who purchase or handle these products. The list of restricted-use pesticides varies from state to state, since some states restrict many more products than the EPA does. To determine which products are restricted, one should contact the state pesticide regulatory agency or the state university's pesticide coordinator.

HANDLING PESTICIDES AT A STATE LAND-GRANT UNIVERSITY

Because the use of pesticides by any government agency or private company is scrutinized more highly than use by private landowners, safety must be of the highest priority.

A university experiment station is a good example of a place where this is true. Land-grant universities set an example to the public through research and subsequent demonstration by their extension divisions. In addition, the teaching mission, through its educational efforts, sets an example to future leaders and agriculturalists. Improper use of agricultural chemicals, lack of respect for the law, poor methodology, and the bad image that results from these abuses cannot be tolerated by the public.

Pesticide safety at an experiment station should involve a well-coordinated training and safety program. Typically, such a program starts with the development of a comprehensive pesticide policy, which sets down guidelines under which the program and its personnel are to operate. It should include guidelines for worker health and safety, public information, storage and disposal, transportation, contractual agreements with outside groups, education, emergency procedures, and other safeguards to protect the public, the environment, and personnel. (Suggestions for these guidelines are spelled out in more detail below.) This comprehensive policy should be put together by a committee, which is responsible for updating it. The guidelines should be reviewed by a general counsel and published in a booklet for distribution to all workers involved with pesticides and pest management.

The next phase of a training and safety program is implementation of the comprehensive policy. The total program should be reviewed to determine where inadequacies exist and where to start improvements. If a need exists, knowledgeable personnel should be drafted to operate the program.

Worker health and safety. A worker health and safety program should be implemented for anyone working with pesticides, whether in the field, in greenhouses, or in laboratories. The program should include a policy on who is permitted to handle pesticides. Annual physical examinations should be given to all personnel who are exposed to pesticides in their work. Students and other cooperators should not be overlooked. Special attention should be given to personnel who apply the more toxic compounds. Blood cholinesterase tests should be administered to those who work with organophosphate or *n*-alkyl carbamate pesticides. Tests should be administered to establish baseline cholinesterase levels and then repeated at appropriate intervals throughout the period of exposure to monitor changes in these levels in order to prevent chronic poisoning.

The selection, stocking, and availability of proper protective clothing and equipment is an important function of a worker health and safety program. Each worker should be fitted with his or her own respirator, gas mask, gloves, eye protection, spray suits, and other necessary clothing and equipment. The right respirator must be found to fit the facial shape of an individual; respirators should not just be distributed to workers without being fitted. Workers should be encouraged to shave facial hair that might interfere with the proper seal of masks and respirators. Respirators and gas masks, especially, should be of the kind approved for the particular type of exposure. Extra (approved) canisters and filters should be supplied, and workers should be instructed on the proper intervals for changing them and how to detect problems. Probably the ideal occasion for fitting workers with protective gear and instructing them on its proper use is individual interviews, in which health histories and job duties can be evaluated by qualified health and safety specialists.

Training of all workers in the proper and safe handling of pesticides is an essential function of a worker health and safety program. Training should include instruction in the reading of labels; use of protective equipment and clothing; calibration of spraying equipment; proper mixing of pesticides; methods for protecting the environment; handling of spills, poisonings, and other emergencies; and procedures for cleaning up, storing, and disposing of pesticides. A good training program should also provide general knowledge of pesticides, the law, pest identification, application equipment, and pesticide safety.

Excellent training of this type is available in every state through the Pesticide Applicator Training Program, which in most states is coordinated by the Cooperative Extension Service. The cost should be minimal, since in each state personnel are already available who can easily train workers in all aspects of safety and proper use of pesticides. State pesticide coordinators, state pesticide applicator training coordinators, and state pesticide regulatory staff can provide more information about this program. Other courses, offered by industry and special groups, could also be very helpful.

Public information. The dissemination of information about pest control to the public is another area of concern in a pesticide program. Pest control recommendations published annually by the university's extension division are particularly important. A public information policy should regulate who is responsible for these recommendations and how they are to be used by personnel. Recommendations dealing with food and feed commodities should be updated at least annually, and those dealing with nonfood commodities should be updated at least every three years.

Storage. Pesticide storage often causes problems. Hazards associated with storage include accumulation of old, unwanted, and unused chemicals; the danger of fire; contamination of soil, water, and air by leakage and spills; the potential for unauthorized entry and use; and other unsafe conditions created by inadequate storage facilities. Policies should establish a routine and procedures for preventing these problems. Recommendations for the proper storage of pesticides are provided below.

Disposal. The safe disposal of unused pesticides and their containers is a major problem, which must be addressed using the most modern methods available. Two major federal statutes govern the disposal of pesticide wastes: FIFRA and the Resource, Conservation and Recovery Act (RCRA). However, other federal, state, and local laws also apply to this activity. For homeowners and farmers, who generate small amounts of wastes, the pesticide label (in most cases) spells out procedures for the disposal of containers and left-over pesticides. The EPA provides a simple set of guidelines for disposal for these generators of small quantities of wastes.

For generators of large quantities of wastes, such as experiment stations or other research facilities, disposal is complicated and usually is not handled at the site because of lack of technology, funds, or space and out of concern for public safety. In most cases, disposal is handled by contracting with a professional waste management firm. However, institutions do have choices in the methods by which they deal with the problem.

Most institutions collect pesticides and other chemical wastes at a common point and pay a waste contractor to package and haul them to a special treatment facility or disposal site. Each chemical must be properly identified before the wastes can be hauled away for disposal. Identification is often a big problem because of the deterioration of labels or containers or because unrecognizable names have been used to label a product (this is often the case with experimental compounds).

Another method of dealing with hazardous-waste disposal is on-site destruction of wastes by high-temperature incineration. This method has many problems, however, which have limited its extensive use.

In the United States, over 300 types of incinerators are marketed by over 50 companies. Of the 1.7 million metric tons of hazardous wastes destroyed in over 2,300 incinerators, industrial boilers, and kilns in 1984, only a small amount was pesticide wastes; most of the wastes destroyed were industrial chemicals, oils, and other commercial compounds. Only five major commercial incinerators in the nation can destroy pesticide wastes, which therefore have to be hauled long distances for disposal.

The EPA has tested very few pesticides to determine whether they can be completely destroyed by incineration. Those tested thus far include DDT, aldrin, picloram, malathion, toxaphene, atrazine, captan, zineb, mirex, Agent Orange with the dioxin contaminant (2,4,5-T and 2,4-D mixture), pentachlorophenol (PCP), chlordecone (Kepone), and chlordane. Test results indicated that all of these pesticides could be almost completely (over 99.99%) destroyed, with the exception of zineb. The zinc metal component of zineb produced a zinc oxide effluent, which could be eliminated if the metal was stripped and recovered before incineration. Incineration of pesticide wastes that are completely organic apparently causes little problem.

Few university experiment stations have access to or own on-site incinerators, for obvious reasons. Such facilities are undesirable to operate because of the great expense of capital outlay and operation and the complex regulatory environment, in addition to the lack of information about incineration. A number of local, state, and federal agencies monitor and certify these facilities, and numerous regulations charge these agencies with enforcement. The major federal laws alone include RCRA, the Toxic Substance Control Act, and the Clean Air Act.

In Iowa, California, and Virginia, biological methods are being utilized to dispose of water used to wash and flush spray equipment, left-over diluted pesticides, and runoff around filling stations. Pits or beds are lined with impervious materials (concrete, clay, or plastic) and stratified with soil and gravel. Pesticides are deposited in the pits either directly or through pipes buried in the beds. Soil microorganisms then use the pesticide wastes as a carbon source. Several pits have been covered by an enclosure similar to a greenhouse to prevent entry and to encourage microbial action and evaporation. Few problems have been reported with these pits, but installation of devices to monitor or detect leakage, volatilization, and concentration is advisable. In addition, a long-term plan should be developed to address the problem of what to do when a pit is no longer operational.

A recently proposed solution to the problem of hazardous-waste disposal is the exchange of wastes, but the method is really not new. The general philosophy here is that "one man's trash is another man's treasure." Programs and services are being offered to give away, sell, or exchange wastes to prevent their disposal by conventional means and to promote conservation of resources by recycling. One of the first options that should be explored to dispose of any pesticide is finding someone else who can use the material in a proper and legal manner. Arrangements have been made with some manufacturers to take back their products if they cannot be used for legal or other reasons. The suspension, ban, or cancellation of many products has resulted in the development of restitution policies by the EPA. These policies should encourage owners of unwanted or unusable pesticides to return them to manufacturers or distributors.

Several other methods of disposal are under investigation or being utilized in a test phase by several institutions. These include physical treatment, chemical treatment, other biological methods, and land application.

A set of the EPA's recommendations for disposal, taken from federal regulations, is given below. New RCRA regulations, affecting most generators of pesticide wastes, were implemented in 1986. One should contact the regional EPA office or the state agency for hazardous-waste management with questions about these regulations or disposal procedures.

Transportation. Guidelines for the transportation of pesticides, mixed and in concentrated form, from one location to another are an important concern of a pesticide program. Policy must be set to regulate the proper methods for transporting these materials in service vehicles. Special compartments for holding chemicals should be constructed in these vehicles to prevent spillage or escape of fumes in the event of an accident or jarring from a quick turn or stop. The use of automobiles and vans that are not retrofitted with a special compartment to transport chemicals is not a good policy, because lack of a separate cargo compartment poses a greater risk of exposure.

The selection of the least hazardous routes through uncongested areas, or avoidance of public roadways altogether, should also be investigated. In many cases, spraying rigs can be moved from one field to another without crossing or using public thoroughfares. In other situations, storage of chemicals near a testing site might be an alternative to transportation. A spill on a highway can be a serious problem, but most spills can be avoided if steps are taken to properly transport pesticide concentrates and mixtures.

Contractual agreements with outside groups. A policy should govern employees who contract with industry or other funding sources to test chemicals. Procedures should be designed to safeguard employees from exposure to pesticides, especially those for which toxicological tests are incomplete or results are unavailable. Another part of this policy, which is also required with experimental-use pesticides, is the establishment of a guarantee that all left-over chemicals be returned to the group from which they were sent. A contract with outside groups should include this and other pertinent guarantees. Contractors should provide technical data and safety data sheets to guide investigators in the selection of protective equipment and in the choice of safety precautions to protect themselves, other workers, the public, and the environment. This subject is addressed in more detail below.

Education, emergency procedures, and other safeguards. Finally, policies should be developed to protect the public, workers, and the environment. These guidelines should address problems associated with emergencies, public opinion, and long-term effects of testing facilities on the environment.

Employees should be educated in how to handle poisonings, spills, fires, and other conditions that might cause problems with storage and handling of pesticides. In case of a spill into water, an especially sensitive situation, state water control authorities and emergency personnel must be notified. If pesticides are spilled into navigable waters, the EPA must be notified under requirements of the Clean Water Act. The National Response Center, a facility of the U.S. Coast Guard (a division of the U.S. Department of Transportation), must be contacted at 800-424-8802 to comply with these regulations.

Personnel and procedures should be selected to handle the public in the event of problems with testing facilities. A representative should be chosen from the management to answer questions the public or news media might ask about these problems. Information presented to the public should not come from more than one source within a group, to avoid giving varying stories or leaving the impression of incompetence. The truth should not be hidden, and often the best approach is a straightforward one. Initial statements should be well thought out and researched before they are released, since they leave a lasting impression, which might be impossible to change.

Other policies are probably required, according to unique situations associated with local conditions, for an overall strategy to operate a good testing and demonstration program.

HANDLING EXPERIMENTAL-USE PRODUCTS

Besides registered products, experimental-use compounds constitute a second group of pesticides handled in research and demonstration. Experimental-use compounds are regulated by the EPA, by authority of FIFRA and the Code of Federal Regulations 40 Part 172. Under this last set of regulations, the EPA requires an application for an experimental-use permit (EUP) from anyone who desires to accumulate the information needed to register a pesticide under Section 3, for a pesticide that has not previously been registered or for a use of a registered pesticide if that use has not previously been approved in its registration. Applications are submitted in triplicate to the EPA, Registration Division, Office of Pesticide Programs, Washington, DC 20460. Experimental-use products may not be sold or distributed except to participants in the testing or their cooperators under provisions of the permit.

An EUP is not required for the testing of chemicals in a laboratory or greenhouse or in limited field trials to determine their pesticidal properties, toxicity, or other properties if the person conducting the test does not expect to receive benefit in pest control by such use. Under these conditions, such chemicals are not considered pesticides according to the regulations.

Limitations on these exemptions include testing on a cumulative area of not more than 10 acres of land and not more than 1 surface-acre of water for use of a particular substance or mixture of substances against a particular pest. Food or feed crops resulting from such tests must be destroyed unless they are to be fed to experimental animals. In addition, the applicant must provide a method to handle any chemicals left over from testing, temporary tolerances for raw agricultural commodities, a days-to-harvest interval, safety information, testing methods and protocols, toxicological data on acute toxicity and chronic toxicity for food crops, and other pertinent data.

Most researchers do not apply for an EUP but are more likely to be cooperators with chemical manufacturers. Whether involved with filing an EUP or cooperating on one, an investigator should know the details and requirements of this process. To file an EUP, an applicant must supply the following information: name and addresses of the applicant, all participants, and cooperators; the registration number of the product, if it is registered; purpose, objectives, and details of the proposed testing program; results of prior testing for phytotoxicity and effects on target and nontarget organisms and on the environment; a proposed method of storage and disposition of any unused experimental-use pesticide and its containers; and, if residue is expected to remain in food or feed, a tolerance (or exemption from one) or a certification that treated materials will be destroyed or disposed of in a manner that poses no threat to people or the environment. Additional requirements for unregistered pesticides include a complete statement of composition for the formulation to be tested, with chemical and physical properties of each active ingredient; data on the rate of breakdown of residues remaining in the environment and in the treated crop and data regarding reentry intervals; and toxicity data relating to the potential for causing injury to users or others who may be exposed, including any epidemiological information regarding human populations. With chemicals that have not previously been registered, the EPA may require additional studies to be conducted during the permit period.

In addition to providing this information, an applicant must comply with labeling requirements similar to those for registering a pesticide under Section 3, except label statements are specific for the EUP. Also, applicants must comply with a surveillance and reporting program and renewal requirements. In some states, a state agency could have authorization to issue EUPs, and applicants direct their requests to that local authority.

This process is obviously complicated, yet researchers who are potential applicants or cooperators should be familiar with it to safeguard themselves when they negotiate and participate in experimental-use projects.

Handling, storage, and disposal of pesticides for experimental use require even greater care than the procedures outlined above for registered compounds. Experimental-use compounds are in most cases at the end of the long process of testing for final registration. However, some may never reach the market, for a variety of reasons.

When a researcher makes an agreement with a chemical company to handle and test an experimental product, a list of testing methods and protocols should accompany the product to the testing site. The researcher deserves as much information as possible from the manufacturer.

According to a source at one chemical company, for experimental-use compounds labeled for food use, most of the toxicological data, except for the results of a few chronic toxicity tests, have been gathered by the time the product reaches the testing site. For compounds labeled for nonfood uses, these chronic toxicity tests may not have already been performed, if they are performed at all. This is more reason for researchers working with these compounds to use extreme caution in their storage, handling, and disposal. Especially great care should be taken to avoid exposure to concentrates in mixing or handling these chemicals.

Procedures for handling, storage, and disposal of experimental-use pesticides, except as mentioned here, should follow those described above for registered pesticides and in the sections to follow.

HANDLING EXPERIMENTAL-USE COMPOUNDS EXEMPTED FROM EXPERIMENTAL-USE PERMITS

As was pointed out above, under certain conditions, no EUP is needed to test chemicals for pesticidal properties. The amount of land to be treated nationally in such a test is limited to less than 10 acres. Obviously, most of this work is done on very small plots. The chemicals tested (which are designated *numbered compounds*, with a company code and number, e.g., CGA-12345, DPX-71624, MO-81467) are usually of technical grade (high concentrations) with very limited toxicological data available for determining long-term effects. Acute toxicity is usually known for these compounds. Even more extreme caution should be used by researchers who work with these chemicals. In most cases, these chemicals have no label or experimental-use number, and the manufacturer may provide only protocols or methods. The same guarantees should be extended by the manufacturer to the researcher as those given for experimental-use compounds regulated under a permit.

STORAGE AND DISPOSAL OF PESTICIDES

The EPA has spelled out recommendations for proper storage and disposal of pesticides and their containers. This information is invaluable to researchers, instructors, and extension workers, since it indicates the philosophy of the lawmakers and the agency responsible for enforcement. The recommendations are an excellent guide for the development of institutional and company pesticide policies. However, one should also take into account any state or local regulations regarding storage and disposal of these chemicals.

The information provided here is taken from the Code of Federal Regulations 40 Parts 165.2 and 165.10. An attempt has been made to condense much of the information, and some of the regulations presented here are paraphrased.

Storage of Pesticides

Recommendations for storage of pesticides and their containers apply to all pesticides, excess pesticides, used empty containers, and containers that contain pesticides. If their uncontrolled release into the environment would cause unreasonable adverse effects on the environment, pesticides and excess pesticides and their containers should be stored only in facilities where due regard has been given to the hazardous nature of the pesticide, site selection, protective enclosures, and operating procedures and where adequate measures are taken to assure personal safety, accident prevention, and detection of potential environmental damages. These storage recommendations should be observed at sites and facilites storing pesticides and excess pesticides and their containers that are classified as highly toxic or moderately toxic and are required to bear the signal words *DANGER*, *POISON*, or *WARNING* or the skull and crossbones symbol on the label. These recommendations are not necessary at facilities storing most pesticides registered for use in the home or garden or pesticides classified as slightly toxic (bearing the word *CAUTION* on the label). All facilities storing pesticides that are or may in the future be covered by an EUP or other special permit should be in conformance with these recommendations. Experimental-use chemicals exempted from an EUP should be treated with even greater care in storage, since not all storage criteria may be available. Presently, these are only recommendations and are not mandated by federal law.

Temporary storage of highly toxic or moderately toxic pesticides for the period immediately before application, and of the quantity required for a single application, may be undertaken by the user at isolated sites and facilities where flooding is unlikely, where provisions are made to prevent unauthorized entry, and where separation from water systems and buildings is sufficient to prevent contamination by runoff, percolation, or wind-blown particles or vapors.

Storage sites. Storage sites should be selected with due regard to the amount, toxicity, and environmental hazard of pesticides and the number and sizes of containers to be handled. Sites should be located where flooding is unlikely and where the texture and structure of the soil and geologic and hydrologic characteristics will prevent the contamination of any water system by runoff or percolation, if such a location is practical. If containment of drainage from the site is warranted, drainage should be contained by natural or artificial barriers or dikes, monitored, and, if it is contaminated, disposed of as an excess pesticide. Consideration should also be given to containing wind-blown pesticide dusts and particles.

Storage facilities. Pesticides should be stored in a dry, well-ventilated, separate room, building, or covered area where fire protection is provided. Where relevant and practical, the following precautions should be taken:

1. The entire storage facility should be secured by a climb-proof fence, and doors and gates should be kept locked to prevent unauthorized entry.
2. Identification signs should be placed on rooms, buildings, and fences to advise of the contents and warn of their hazardous nature, in accordance with suggestions under Fire control, below. In areas where the need exists, signs should be multilingual, in languages spoken in the area and by personnel.
3. All items of movable equipment used for handling pesticides at the storage site that might be used for other purposes should be labeled *contaminated with pesticides* and should not be removed from the site unless they are thoroughly decontaminated.
4. Provision should be made for decontamination of personnel and equipment such as tractors, trucks, tables, measuring devices, and tarpaulin covers. Where feasible, a washbasin, a shower with a delayed-closing pull chain valve, and an eyewash station should be provided for routine and emergency decontamination. All contaminated water should be disposed of as an excess pesticide. Where required, decontamination areas should be paved or lined with impervious materials and should include gutters. Contaminated runoff should be collected and treated as an excess pesticide.

Operational procedures. Pesticide containers should be stored with the label plainly visible. If containers are not in good condition when they are received, the contents should be placed in a suitable container and properly relabeled. If dry excess pesticides are received in paper bags that are damaged, the bag and the contents should be placed in a sound container that can be sealed. Metal or rigid plastic containers should be checked carefully to insure that the lids and bungs are tight. Where relevant and practical, the following provisions should be considered:

Classification and separation

1. Each pesticide formulation should be segregated and stored under a sign containing the name of the formulation. Rigid containers should be stored in an upright position, and all containers should be stored off the ground, in an orderly way, so as to permit ready access and inspection. They should be accumulated in rows or units so that all labels are visible, and with lanes to provide effective access. A complete inventory should be maintained, indicating the number and identity of containers in each storage unit.

2. Excess pesticides and containers should be further segregated according to the method of disposal to ensure that entire shipments of the same class of pesticides are disposed of properly and that accidental mixing of containers of different categories does not occur during the removal operation.

Container inspection and maintenance. Containers should be checked regularly for corrosion and leaks. If these are found, the container should be transferred to a sound, suitable, larger container and be properly labeled. Materials such as adsorptive clay, hydrated lime, and sodium hypochlorite should be kept on hand for use as appropriate for the emergency treatment or detoxification of spills or leaks.

Safety precautions. In addition to precautions specified on the label of the container and in other labeling, rules for personal safety and accident prevention similar to those listed below should be available in areas where personnel congregate:

Accident prevention measures

1. All containers of pesticides should be inspected for leaks before being handled.

2. Containers should not be mishandled so as to create emergencies by carelessness.

3. Unauthorized persons should not be permitted in the storage area.

4. Pesticides should not be stored next to food or feed or other articles intended for consumption by humans or animals.

5. All equipment should be inspected before it is used, and equipment found to be contaminated or mechanically defective should be treated.

Safety measures

1. Food, beverages, tobacco, eating utensils, and smoking equipment should not be stored in the storage or loading areas.

2. Eating, drinking, and smoking or use of tobacco is not allowed in areas where pesticides are present.

3. Unlined rubber gloves and boots, waterproof aprons, and eye protection should be worn by workers handling containers of pesticides.

4. Workers should not put their fingers in their mouths or rub their eyes or exposed skin while working.

5. Workers should wash their hands before eating, smoking, or using the toilet and immediately after loading or transferring pesticides.

6. Persons working regularly with organophosphate and *n*-alkyl carbamate pesticides should have periodic physical examinations, including cholinesterase tests.

Protective clothing and respirators. When handling pesticides that are in concentrated form, workers should wear protective clothing. Contaminated garments should be removed immediately, and extra sets of clean clothing should be maintained nearby.

Particular care should be taken when handling certain pesticides to protect against absorption through the skin and inhalation of fumes. Respirators or gas masks with proper canisters approved for the particular type of exposure, as noted in the label directions, should be used when such pesticides are handled.

Fire control. Where large quantities of pesticides are stored, or where conditions may otherwise warrant, the owner of stored pesticides should inform the local fire department, hospitals, public health officials, and police department in writing of the hazards that such pesticides may present in the event of a fire. A floor plan of the storage area indicating where different pesticide classifications are regularly stored should be provided to the fire department. A guide called *Pre-planning and Guidelines for Handling Agricultural Chemical Fires*, available from the National Agricultural Chemicals Association, is very helpful in developing such a plan. The fire chief should be furnished with the appropriate telephone numbers in order to have access 24 hours a day to the following: 1) the person or persons responsible for the pesticide storage facility; 2) the appropriate state authorities—state pesticide regulatory agency, state emergency operations center, and state environmental protection agency; 3) the U.S. Coast Guard (National Response Center—telephone 800-424-8802); and 4) the Pesticide Safety Team Network of the National Agricultural Chemicals Association (CHEMTREC—telephone 800-424-9300).

Suggestions for fire hazard abatement include the following:

1. The outside of each storage area should be plainly labeled with *DANGER, POISON*, and *PESTICIDE STORAGE* signs, where applicable. The local fire department should be consulted regarding the use of the current hazard signal system of the National Fire Protection Association.

2. A list should be posted, outside the storage area, of the types of chemicals stored therein. The list should be updated to reflect changes in types stored. Another copy of this list should be kept in a secure place away from the storage area and where fire cannot reach it if the storage area should burn. Although computerized inventory systems are good for management, one should not rely on these alone in an emergency. A power, hardware, or software failure could make this information unavailable when it is needed the most.

Suggested fire-fighting procedures include the following:

1. Fire fighters should wear air-supplied breathing apparatus and rubber clothing, avoid breathing or otherwise contacting toxic smoke and fumes, and wash completely as soon as possible after encountering smoke and fumes.

2. Water used in fire fighting should be contained within the storage site's drainage system.

3. Fire fighters should be given cholinesterase tests after fighting a fire involving organophosphate or *n*-alkyl carbamate pesticides, if they have been heavily exposed to the smoke. Baseline cholinesterase tests should be a part of the regular physical examinations for such fire fighters.

4. Persons near such fires should be evacuated if they may come in contact with smoke, fumes, contaminated surfaces, or runoff.

Monitoring. An environmental monitoring system should be considered in the vicinity of storage facilities. Samples from the surrounding groundwater and surface water, wildlife, and plant environment, as appropriate, should be tested in a regular program to assure minimal environmental insult. Analyses should be performed according to the *Official Methods of Analysis of the Association of Official Analytical Chemists* and such other methods and procedures as may be suitable.

Disposal of Pesticides

Recommendations for disposal of pesticides and their containers apply to all pesticides and pesticide-related wastes (and their containers), including those that are or may in the future be registered for general use or restricted use or covered under an experimental-use permit, except single containers disposed of by open-field burial registered for farm and ranch use and home and garden chemicals disposed of in routine municipal refuse after securing in several layers of paper.

Two options should be thoroughly exhausted before other methods of disposal are sought: 1) use of the chemical for its originally intended purposes at the prescribed dosage rates provided these uses are legal under current laws and 2) return to the manufacturer or distributor for potential relabeling, recovery of resources, or reprocessing into other materials. Transportation must be in accordance with regulatiosns of the U.S. Department of Transportation and Interstate Commerce Commission. In some situations, excess pesticides may be suitable for export to countries where their use is currently legal.

EPA recommendations for disposal of pesticides are as follows.

Organic pesticides (except organic mercury, lead, cadmium, and arsenic compounds, which are discussed below) should be disposed of according to the following procedures:

1. The most highly recommended method is incineration in a pesticide incinerator at the specified combination of temperature and dwell time or at such other lower temperature and related dwell time that causes complete destruction of the pesticide. As a minimum, all emissions should be verified to meet the requirements of the Clean Air Act of 1970 relating to gaseous emissions. Any liquids, sludges, or solid residues generated should be disposed of in accordance with all applicable federal, state, and local pollution control requirements. Municipal solid waste incinerators may be used to incinerate excess pesticides or pesticide containers provided they meet the criteria of a pesticide incinerator and precautions are taken to ensure proper operation.

2. If appropriate incineration facilities are not available, organic pesticides may be disposed of by burial in a specially designated landfill. Records to locate such buried pesticides within the landfill site should be maintained.

3. The environmental impact of the soil injection method of pesticide disposal has not been clearly defined nationally, and therefore this disposal method should be undertaken only with specific guidance. It is recommended that advice be requested from the EPA regional administrator in the region where the material will be disposed of before undertaking disposal by this method.

4. Chemical methods and procedures exist that degrade some pesticides to forms that are not hazardous to the environment. However, practical methods are not available for all groups of pesticides. Until a list of such methods is available, it is recommended that advice be requested from an EPA regional administrator.

5. If adequate incineration facilities, specially designated landfill facilities, or other approved procedures are not available, temporary storage of pesticides for disposal should be undertaken. Storage facilities, management procedures, safety precautions, and fire and explosion control procedures should conform to those described under Storage of Pesticides.

6. The effects of subsurface emplacement of liquid by well injection and the fate of injection materials are uncertain with available knowledge and could result in serious environmental damage requiring complex and costly solutions on a long-term basis. Well injection should not be considered for pesticide disposal unless all reasonable alternative measures have been explored and found less satisfactory in terms of environmental protection. The EPA's policy is to oppose well injection where proof does not exist that such methods are safe to people and the environment and that methods of adequate monitoring, testing, treating well failures, and abandonment are made.

Metallo-organic pesticides (except organic mercury, lead, cadmium, or arsenic compounds) should be disposed of according to the following procedures:

1. After first being subjected to an appropriate chemical or physical treatment to recover the heavy metals from the hydrocarbon structure, such compounds should be incinerated in a pesticide incinerator as described above.

2. If appropriate treatment and incineration are not available, such compounds should be buried in a specially designated landfill as described above.

3. Disposal by soil of metallo-organic pesticides should be undertaken only in accordance with the procedures set forth above.

4. Chemical degradation methods and procedures that can be demonstrated to provide safety to public health and the environment should be undertaken only as described above.

5. If adequate disposal methods are not available, the pesticides should be stored according to the procedures given under Storage of Pesticides.

6. Well injection of metallo-organic pesticides should be undertaken only in accordance with the procedures set forth above.

Organic mercury, lead, cadmium, arsenic, and all inorganic pesticides should be disposed of according to the following procedures:

1. The pesticides should be chemically deactivated by conversion to nonhazardous compounds, and the heavy metal resources should be recovered. Methods that are appropriate will be described and classified according to their applicability to the different groups of pesticides. Until a list of practical methods is available, however, each use of such procedures should be undertaken only as noted above.

2. If chemical deactivation facilities are not available, such pesticides should be encapsulated and buried in a specially designated landfill. Records to permit location for retrieval of such materials within the landfill should

be maintained.

3. If none of the above options is available, such pesticides should be placed in suitable containers (if necessary) and temporary storage should be provided until adequate disposal facilities or procedures are available.

EPA also recommends the following procedures for disposal of pesticide containers and residues, according to pesticide groups:

Group I containers. Combustible containers that formerly contained organic or metallo-organic pesticides, except organic mercury, lead, cadmium, or arsenic compounds, should be disposed of in a pesticide incinerator or buried in a specially designated landfill, as noted above, except that small quantities of such containers may be burned in open fields by the user of the pesticide when such open burning is permitted by state and local regulations, or buried singly by the user in open fields with due regard for protection of surface and subsurface water and in accordance with label directions.

Group II containers. Non-combustible containers that formerly contained organic or metallo-organic pesticides, except organic mercury, lead, cadmium, or arsenic compounds, should be triple-rinsed. Containers in good condition may then be returned to the pesticide manufacturer or formulator or to a drum reconditioner for reuse with the same chemical class of pesticide previously contained, provided reuse is legal under currently applicable U.S. Department of Transportation regulations. Other rinsed metal containers should be punctured to facilitate drainage prior to transport to a facility for recycling as scrap metal or for disposal. All rinsed containers may be crushed and disposed of by burial in a sanitary landfill, in conformance with state and local standards, or buried in the field by the user of the pesticide. Unrinsed containers should be disposed of in a specially designated landfill or subjected to incineration in a pesticide incinerator.

Group III containers. Containers (both combustible and noncombustible) that formerly contained organic mercury, lead, cadmium, or arsenic or inorganic pesticides and have been triple-rinsed and punctured to facilitate drainage may be disposed of in a sanitary landfill. Such containers that are not rinsed should be encapsulated and buried in a specially designated landfill.

Residue disposal. Residues and rinse liquids should be added to spray mixtures in the field. If not, they should be disposed of in a manner prescribed for each specific type of pesticide as described above.

CONCLUSION

The information provided here should assist researchers, students, technicians, farm workers, administrators, instructors, extension workers, and others who deal with the problems associated with handling, storage, and disposal of pesticides used for research and demonstration. Much of the information was developed by consulting the federal laws and through contacts with sources in industry, government, and universities. A set of general references is given below for further information on this subject. Because pesticide information changes almost daily, one should consult the appropriate sources before attempting to handle, store, or dispose of any pesticide or pesticide container. These sources include the pesticide label; federal, state, and local statutes and authorities; and the literature. This article was not developed to substitute for the information provided by these sources but is intended to be a general overview of the subject.

BIBLIOGRAPHY

1. Anonymous. 1984. Selected Environmental Law Statutes, 1984 ed. West Publishing Co., St. Paul, MN. 758 pp.
2. Bohmont, B. L. 1983. The New Pesticide User's Guide. Reston Publishing Co., Inc., Reston, VA. 452 pp.
3. Dewey, J. E., McDonald, S. A., Pendleton, R. F., and Smith, W. G. 1978. Pesticide Applicator Training Manual—Core Manual. Cornell University, Ithaca, NY. 136 pp.
4. McDonald, S. A. 1984. Applying Pesticide Correctly: A Guide for Private and Commercial Applicators. North Carolina State University, Raleigh. 128 pp.
5. Morgan, D. P. 1982. Recognition and Management of Pesticide Poisonings, 3rd ed. U.S. Environmental Protection Agency, Washington, DC. 120 pp.
6. National Agricultural Chemicals Association. 1971. Pre-planning and Guidelines for Handling Agricultural Chemical Fires. National Agricultural Chemicals Association, Washington, DC. 8 pp.
7. Swope, A. D., Costas, P. P., Jackson, J. O., and Weitzman, D. J. 1983. Guidelines for the Selection of Chemical Protective Clothing. American Conference of Governmental Industrial Hygienists, Cincinnati, OH. 239 pp.
8. Ware, G. W. 1978. The Pesticide Book. W. H. Freeman & Co., San Francisco, CA. 186 pp.
9. Wereley, L. S., ed. 1984. Pesticides Guide: Registration, Classification and Applications. J. J. Keller & Associates, Inc., Neenah, WI. 1,135 pp.

Use of Statistics in Planning, Data Analysis, and Interpretation of Fungicide and Nematicide Tests

LARRY A. NELSON, Department of Statistics, North Carolina State University, Raleigh 27695-8203

Statistics is the science and art of applying logical principles in experimental design and technique and in the collection and analysis of data to make valid (though uncertain) inductive inferences. Statistical analysis and inference procedures usually include measures of the degree of uncertainty. The proper use of statistics depends on the application of these principles to each of several phases of scientific investigation. Federer (3) listed the following steps in scientific investigation: 1) formulation of questions to be answered and hypotheses to be tested, 2) a critical, logical analysis of the problem or problems to be raised, 3) selection of a procedure for research, 4) selection of suitable measuring instruments and control of the personal equation, 5) a complete analysis of the data and interpretation of the results in light of experimental conditions and hypotheses tested, and 6) preparation of a complete, correct, readable report of the experiment.

Statistics is not a substitute for this process, but it does provide the tools for implementing some of these steps. The steps are not entirely statistical, but all may involve at least some statistical aspects. A statistician can be of assistance in asking questions that sharpen a researcher's understanding and formulation of the problem under investigation (steps 1 and 2). A statistician can also assist in preparing the tabular portion of the research report (step 6). Steps 3 and 5 contain the most highly statistical elements. Unfortunately, many researchers only consult statisticians for matters involved in step 5.

The purpose of this chapter is to review and clarify the basic statistical principles as they apply to each phase of research, so that the reader might, through better understanding of these principles, improve the quality of fungicide and nematicide tests. More reliable testing should facilitate the registration of chemicals and enhance the acceptance of research findings. Some of the suggestions given here cannot be found in textbooks on statistical methods.

This chapter is divided into three sections, corresponding to the important phases in the application of statistics: planning, data and their analysis, and interpretation and reporting of results. Various aspects of steps 1–4 are included in the discussion of planning. The second section, on data and their analysis, deals with aspects of step 5, as does the third section, in the discussion of the interpretation of results. A short discussion of the reporting of results in the third section is designed to help in implementing step 6. A detailed flow chart of the aspects of experimentation to be discussed is given in Fig. 1.

PLANNING

Planning is designed to assure that treatments selected for an experiment most effectively provide relevant comparisons, estimates, or hypothesis tests and that the analysis of data is straightforward. Planning also helps to assure that the size of the experiment is appropriate for a particular situation. Undersized experiments often fail to detect important differences between treatments. Oversized experiments can be costly. Planning also involves choosing an appropriate method of assigning the treatments to the plots, thus assuring unbiased estimates of means, an estimate of their reliability, and valid statistical tests.

Role of statistician. Before testing begins, a researcher should consult a statistician to work out the details of the experimental design. The statistician's most valuable contribution is often made by asking questions that cause the researcher to reexamine all aspects of the problem, including the reasons for conducting the test. The test can then be tailored to the particular experimental setting, to make it efficient and

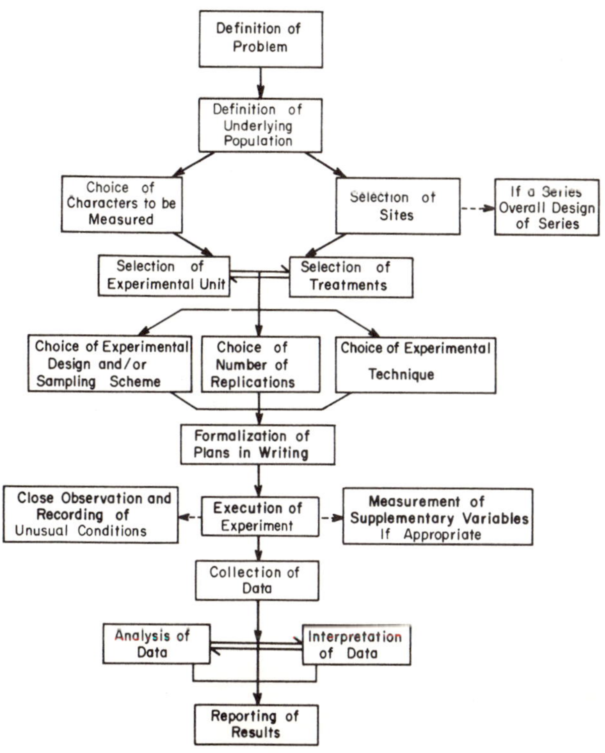

Fig. 1. Flow chart of decisions and operational aspects of experimentation.

ensure that it yields a maximum amount of information with minimum effort and cost.

Steps in planning. The scientific method suggests that planning of research be done in steps. These include defining the objectives of a test, formulating a set of detailed specifications for the test, and determining specifically the method to be used in the analysis of the data.

Defining objectives. Objectives may be questions to be answered, hypotheses to be tested, specifications to be met, or effects to be estimated. Many researchers are overly ambitious and thus set objectives that are too broad and diffuse. On the other hand, the objectives may be so limited that experiments can be condensed into a single test.

A statement of objectives should be clear, concise, and specific. It should define the extent of the population for which generalizations are to be made (e.g., one farm, six counties, poorly drained soils). The results of the research are usually based on a sample (one or several experiments), which should be representative of the population to which the conclusions are to be applied.

Detailing specifications. The detailed specifications or procedures for an experiment should be formalized in writing. The proposed field layout may be shown in a diagram drawn to scale. Some of the more important specifications are the underlying population for which inferences are to be made, characters to be measured, sites, treatments, experimental material, experimental design, replication, experimental technique, design and randomization in a series of tests, supplementary variables, and sampling procedures.

Determining the method of analysis. The procedures for data analysis should be described in writing during the planning stage. Reference works on statistical methods may be cited for details of analyses. An outline of the sources of variation (e.g., blocks, treatments, harvest, error) and their respective degrees of freedom is also useful.

A brief discussion of each important specification listed above follows.

Underlying population. In every experimental situation, some target population exists to which the results apply. The population must be specified in the planning stage to adequately sample the range of its properties and to determine the most appropriate statistical methods for finding the distribution of a given property. For example, a farmer might conduct a test to determine the optimal rate of nematicide for a particular field. The population would consist of all the nematodes (of various species) in that field. A single randomized block experiment in that field would provide the needed information. In contrast, an extension worker might wish to study methods of controlling a particular pathogen throughout a state. To make valid inferences, a series of experiments at different locations, representing the range of conditions under which the pathogen occurs, would need to be conducted.

Characters to be measured. The variables that are submitted to statistical analysis are called characters. Examples are the number of diseased plants per 3 m of row and the proportion of plants affected by the root-knot nematode. In some areas of research, certain standard characters have been established, and information on them is obtained routinely. In other cases, the researcher is not certain about which characters best reflect the effects of the controlled variables. Therefore, a number of them may be measured to select one or more that will be useful for future studies.

The time at which characters are measured is important. In some cases, multiple measurements of characters should be taken throughout the season to adequately sample climatic effects.

Sites. Sites are drawn to represent the area about which inferences are to be made. The final selection of sites is made after careful screening of a group of prospective sites. Often these are chosen with the aim of representing certain environmental regimes. On other occasions, sites may be selected randomly (i.e., each prospective site has an equal chance of being selected for the sample of sites). The area within a site should be as homogeneous as possible in such characteristics as soil type, cropping history, slope, drainage, and fertility.

Treatments. Selection of treatments involves consideration of the number of treatments, their function, factorial arrangement of treatments, the level of factors, and the use of controls.

Number of treatments. The number of treatments is determined to a great extent by the factors (and their levels) that provide information on the problem being studied. A factor is a quantitative or qualitative variable that is under investigation in an experiment as a possible cause of variation. The level of a factor is the magnitude or intensity with which it is brought to bear. An upper limit on the number of treatments may be imposed by limited availability of experimental material or by the number of treatments that give a reasonable block size (e.g., 15 treatments).

Function of treatments. The purpose of one type of experiment is to compare chemicals and spot the winner. Often the applications must be rate-specific to the chemicals. The treatments then consist of package combinations of chemicals and rates. The chemical and rate that perform best may then be recommended. In other experiments, the purpose is to estimate the parameters of the response curve for a single quantitative factor, to explore a response surface for the optimum combination of levels of more than one quantitative factor for obtaining maximum yield, or to find an inflection point in a response curve. In these cases, levels of the factor or factors under investigation should be chosen so that they bracket the major region where response occurs and give precise estimates of model parameters, optima, and inflection points. Equal spacing of the levels of a quantitative factor may or may not be desirable.

Factorial arrangement of treatments. Treatments in factorial experiments are combinations of levels of two or more factors. For example, an experiment to study three levels of fungicide and two methods of application is a 3×2 factorial experiment, with six treatments. One benefit of factorial arrangement is the possibility of estimating the interaction of factors. Another is the added precision, in obtaining means for a factor, gained by averaging over all levels of the other factors. Response surfaces, showing relationships among responses to different treatments, can be fitted to data from factorial experiments with quantitative variables. These surfaces facilitate estimation of optimal rates of the input factors. An example is a growth chamber study in which four rates of fungicide are varied with five rates of herbicide, and growth response is measured. The purpose is to study the interaction of the two variables and draw a contour map for the response.

With many factors at several levels, the number of treatments is large. Consequently, incomplete factorials are used in some cases. Care should be taken in selecting treatment combinations for incomplete factorials to assure that important comparisons are valid. Another way to reduce the number of treatments in experiments is to include some treatments that are factorially arranged and others that are not part of a factorial. The treatments are assigned to the experimental plots with no distinction between those that are factorial and those that are not. For example, the following treatments might be selected: Rate A, Method 1; Rate A, Method 2; Rate B, Method 1; Rate B, Method 2; Rate C, Method 1; and Rate D, Method 2. The first four treatments are factorially arranged; the last two are not.

Level of factors. Whether treatments are factorial or nonfactorial, the rates of quantitative variables and the specific classes of qualitative variables used in treatments must be specified. For the quantitative variables, deciding first on a range of rates that corresponds to the region of response is important. Failure to bracket the region of response could result in an inaccurate picture of the shape of the response curve and could bias estimates of the optimal rate of the input factor. Once the range of rates is established, one must decide how many treatment rates should be included within the range and how they should be spaced. In cases in which a linear response is expected (i.e., a curved response pattern is not likely in the population), only two levels, one at the low end and one at the high end of the range, would be sufficient. Usually the response pattern is not known, and therefore at least one point in the middle of the range, in addition to the endpoints, is necessary to permit a check for curvature. In cases in which the shape of the curve is more complicated (e.g., S-shaped curves), several levels are needed to estimate the number of parameters necessary to describe the curve adequately. The desire to keep the experiment within reasonable size places an upper limit on the number of levels of each factor.

Use of controls. Controls are often needed for a basis of comparison, especially in cases in which treatments are not known to be effective. In factorial experiments in which none of the treatment combinations is an untreated check, a control can be randomized with the factorial treatments. Few situations call for a separate control for each treatment. The added precision gained from such an arrangement is usually offset by the additional experimental material required. Many experiments necessitate several types of controls. For example, in a nematicidal evaluation, if a chemical has both nematicidal and insecticidal properties, three checks should be included—one with nematicide only, one with insecticide only, and one with neither nematicide nor insecticide. An analysis of variance for a 2×2 factorial can be conducted on data for the three checks and the chemical treatment. This factorial arrangement permits testing for the separate effects of insecticide and nematicide and the interaction of insecticide with nematicide. Similar approaches are needed for chemicals that control several pathogens. In one check, none of the pathogens should be controlled, and each other check should allow for control of all but one pathogen. In some cases, several controls are necessary to pinpoint the mechanism of the response to the chemical treatment. In other cases, however, no controls are needed. In fact, in tests involving airborne pathogens, controls may be detrimental, because they supply spores that can contaminate the treated plots. In such tests, omitting the controls and using one of the chemical treatments as a standard for comparison is advisable.

Experimental material. Treatments are applied to experimental units that collectively are called the experimental material. Examples of experimental units are plots of land, trees, petri dishes, and pots of soil in a greenhouse.

Selection of the experimental units to be used in an experiment depends on the purpose of the experiment. If the experimental units (e.g., plots) are to be used as a medium on which treatments are compared, they should be as homogeneous as possible. That is, the factors that might influence response (e.g., plant population, soil fertility and moisture, and size of plants) are nearly constant. On the other hand, if the objective is to evaluate some property of the experimental units themselves (e.g., average size of two different nematodes), less selection and more randomness is needed in the choice of units. Selection may narrow the population to which the results apply.

In the field, experimental units are plots of land. The size and shape of plots must be determined, as well as the need for such items as border rows. Aspects of experimental technique for field plots have been described by Gomez (7), Gomez and Gomez (8), LeClerg et al (14), and Wishart and Sanders (22). Smith (18) reported some quantitative techniques for estimating optimal plot size from uniformity trial data and costs. He found that, for plots between one-fourth and four times optimal size, experiments did not differ greatly in efficiency. Local conditions often call for modifications of the optimal size estimated by his procedure. Mechanical restrictions imposed by equipment and techniques used in the experimentation often dictate the size and shape of plots.

In general, plots in most field experiments should be long and narrow, with the elongation in the direction of the gradient. Plots with this configuration provide minimum error and yet are convenient to handle with row-crop equipment. Plots in fungicide tests for foliar diseases where spores move freely over the borders are an exception. Square plots are preferable in such tests, with only the center portion being used for experimental data. Square plots have proportionately less border than do long, narrow ones. The superiority of square plots over long, narrow ones is greater when dispersal of spores occurs at normal atmospheric turbulence than when it occurs at low turbulence. Van der Plank (21) reported that mutual interference among plots under conditions of low air turbulence can be considerably decreased by increasing the plot size: with low turbulence, increasing the sides of square plots from 1 to 10 m reduced mutual interference by more than half, and increasing them from 10 to 100 m reduced it by more than five sixths. When air turbulence was normal during dispersal of spores, however, increasing the sides of square plots from 1 to 100 m reduced mutual interference among plots by less than one half.

Usually, experience with a particular crop or pest

species in one locality helps a researcher to understand the underlying population and the spatial and temporal distribution of the characters being studied. This experience should be useful in devising a suitable plot and a precise experimental arrangement that can provide unbiased estimates of treatment effects and is operationally convenient. The purpose of the experiment is important in the determination of plot size and shape. Breeding experiments may require plot configurations that are completely different from those used in experiments designed to compare the effects of chemical treatments.

Border areas (or guard rows) are used in cases in which the treatment imposed on one plot is expected to influence the adjacent plot. A correct judgment of the degree of this influence is important. Insufficient border area can allow interplot interference that may cause representational or cryptic error. An example is the movement of disease from unsprayed to sprayed plots. Information on the sprayed plots is not representative of what farmers who use spray would find, because the sprayed plots have been contaminated with disease and yield less than uncontaminated plots would.

If border areas are made larger than they need to be, they may occupy an excessive portion of the experimental area. Border areas that constitute at least 50–60% of the experimental area are not uncommon. Even where the movement of spores threatens mutual interference among plots, the border area can be reduced by keeping only a small range of disease between the best treatment and the worst treatment (21). Minimizing the range of disease implies that untreated plots are omitted and that one chemical is chosen as a standard for comparison.

Shoemaker (17) described three different designs to be used in experiments involving airborne pathogens. In the first design, for screening and ranking fungicides (protectants) under severe disease pressure, unsprayed zones border the sides of each treatment plot, and an unsprayed check treatment is optional. In the second design, protected (sprayed) buffer zones are placed between treatment plots, and the unsprayed check is completely eliminated. Either protectants or eradicants may be used in the treatments. Stricter precautions against interference must be taken if eradicants are included in the test. Interference could still cause problems in the second design if the range in effectiveness of the materials is great. The third design involves only treatments that have been selected from preliminary testing and are candidates for recommendation. They do not differ substantially in their effectiveness to control disease. Plots should be large, and data should be taken from their centers. Unsprayed checks are omitted.

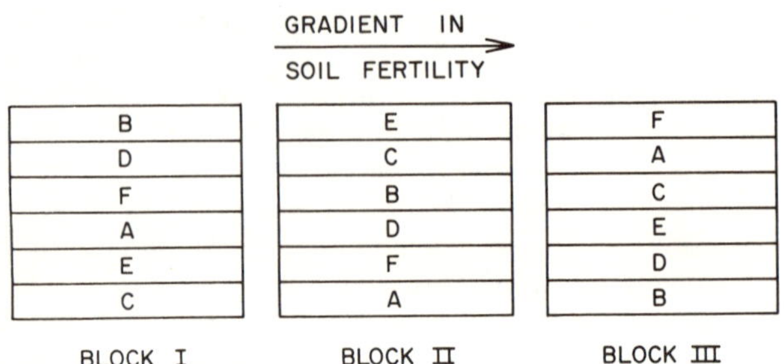

Fig. 2. A randomized complete block design in a field experiment to compare varieties for disease resistance.

Experimental design. Several experimental designs are used extensively in many fields of research. The choice of an appropriate one for a specific situation involves a number of considerations, but the general rule is to keep the design as simple as possible. A balanced design in which all treatment combinations have the same number of observations (usually replications) aids analysis and interpretation. Without this equal replication, analysis can be difficult and the resulting estimates poor. Imbalance is not a problem in the simple designs discussed below if textbooks in experimental design are followed. If the experimental plots are sampled and the number of samples varies from plot to plot, however, the balance is lost and problems result. Missing data (e.g., missing because of environmental causes or mortality) can also cause imbalance.

With information on the variability pattern of the experimental material, the number and nature of treatments, and the experimental techniques employed, a particular design usually emerges as the logical choice. The commonly used designs differ mainly in the way in which treatments are *randomly* assigned to the experimental units. Randomization is the process of assigning treatments to plots in such a way that all treatments have an equal chance of being assigned to a particular plot. It provides assurance that a treatment is not continually favored or handicapped in various replications because of some extraneous source of variation, known or unknown. Randomization, like an insurance policy, protects against disturbances that could occur. Tables of random numbers or computers can be used as sources of random numbers. The commonly used experimental designs help to control known extraneous sources of variation by blocking. The basic difference between these designs is the number of restrictions on randomization, the restrictions being necessary because of the blocking.

In the *randomized complete block design*, which is by far the most popular design, one restriction is placed on randomization. A complete set of treatments is randomized within each block (replication). An example of a randomized block design is shown in Fig. 2. The object is to block in such a way that, although the blocks may differ considerably from one another, the units within each block are relatively uniform. Environmental sources of variation, such as moisture and natural fertility of the soil, are often a basis for blocking. In other cases, each run of a set of operations is a block, and the runs are performed at different times. Nearly square, compact blocks are preferred to long, narrow blocks in the field.

In addition to its precision, an advantage of the randomized complete block design is its simplicity. Treatments are readily assigned to the experimental plots at random, and field layout and analysis of data are simple. Furthermore, missing plot values can readily be estimated. This design accommodates a wide range in numbers of treatments and replications, although practical considerations place limits on both.

The *Latin square design* provides error control (blocking) in two directions. It is useful for high-precision experiments having two to 10 treatments, because the number of replications must equal the number of treatments. For squares having two or three treatments, several squares are usually needed to provide a reliable estimate of experimental error.

The Latin square design is not as simple to randomize as the randomized complete block design, because it has two restrictions on randomization—each row and each column must contain a complete set of treatments (Fig. 3)—and the field layout could be more complex. Also, missing data cause more problems than they do with the randomized complete block design.

The *split-plot design* is used in cases in which 1) the nature of the experimental material or mechanical aspects of the research require differential plot sizes for various factors or 2) more precision is desired for some effects (subplot factors, such as fungicide and interaction) than for others (whole-plot factors, such as crop variety). This design is commonly used in plant science experiments. One use is for perennial experiments, in which years are the subplots. In split-plot experiments, the factor requiring a more sensitive test is usually assigned to the subplot. Two separate randomizations are required, one for the whole plot and one for the subplot treatments within each whole plot (Fig. 4). If sampling is done within subplots, the numbers of samples must be constant to maintain balance.

The split-plot principle can be extended to more than one split (e.g., the split split-plot design). Other variants involve stripping of plots for one factor over plots for a second factor (e.g., the split-block design). An example of a split-block design is shown in Fig. 5. Split-block designs are usually used where mechanical factors prevent the formation of smaller plots within the whole plots (e.g., spraying in one direction, plowing in another). Consultation with a statistician in planning a complex stripped design should assure that the experimental data can be analyzed.

The *completely randomized design*, although simple and flexible with regard to the number of replications within each treatment, does not control error variation through blocking. Therefore, it is not precise enough for most field experiments. It is used, however, for laboratory and greenhouse experiments. An example of a completely randomized design is shown in Fig. 6.

Cochran and Cox (1) have published an excellent detailed description of the designs described above and a discussion of their use. Their text also discusses more complicated designs (e.g., lattice designs for experi-

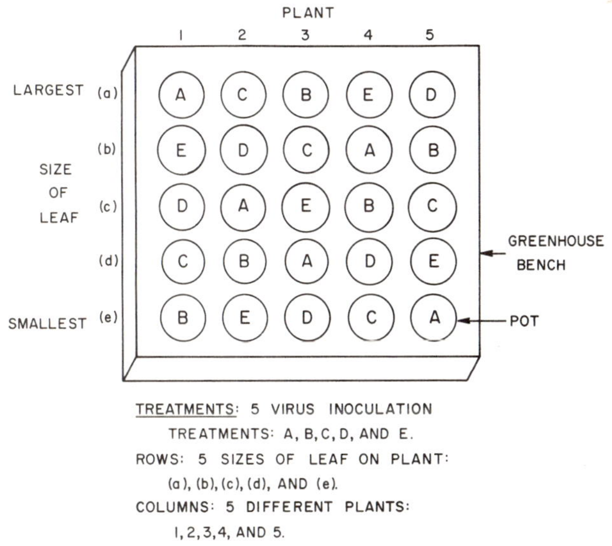

TREATMENTS: 5 VIRUS INOCULATION TREATMENTS: A, B, C, D, AND E.
ROWS: 5 SIZES OF LEAF ON PLANT: (a), (b), (c), (d), AND (e).
COLUMNS: 5 DIFFERENT PLANTS: 1, 2, 3, 4, AND 5.

Fig. 3. A Latin square design in a greenhouse experiment to compare virus inoculation treatments.

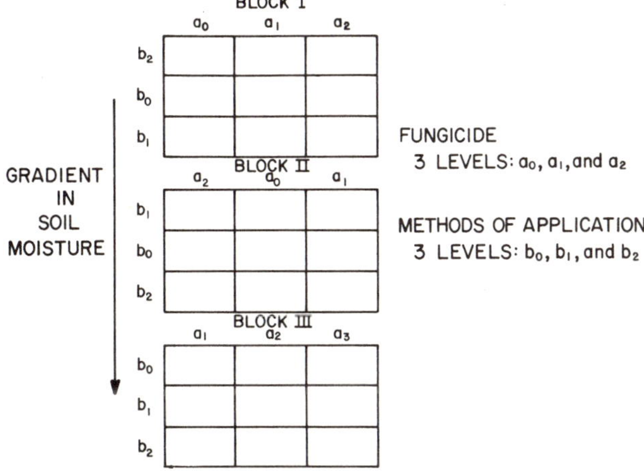

Fig. 5. A split-block design in a field experiment in which three levels of fungicide are stripped across three methods of application.

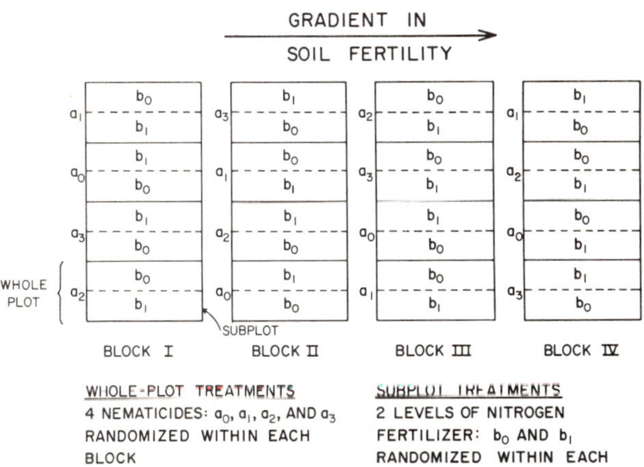

Fig. 4. A split-plot design in a field experiment to compare four nematicides (whole-plot treatments) and two levels of nitrogen fertilizer (subplot treatments).

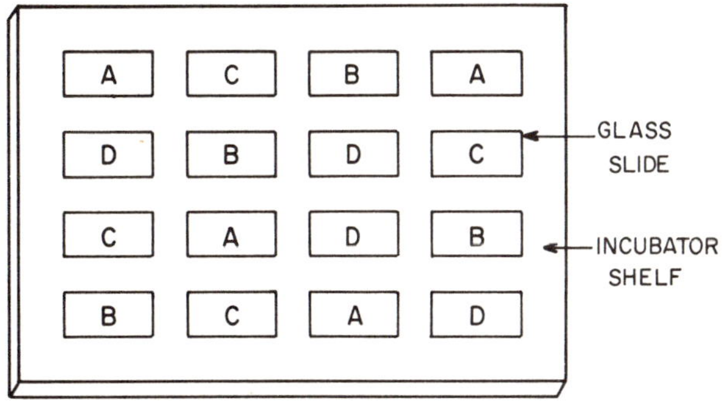

Fig. 6. A completely random design in an experiment with four replications of each of four fungicides assigned to glass slides containing fungus spores.

ments involving large numbers of treatments). Comprehensive bibliographies of experimental design are available (4,5).

Replication. Replication assures that an estimate of experimental error can be determined for a test. Moreover, it is a simple way to increase the precision of estimates of means and the sensitivity of tests of significance. Thus, with replication, one is more apt to detect real differences among means. Beyond a certain number of replications, the benefits of increased precision do not offset the cost of another replication. Table 2.1 of Cochran and Cox (1) can be used as a guide in choosing an appropriate number of replications if the coefficient of variation is available from previous experiments. The required precision, the magnitude of differences to be measured, and the significance level must also be specified. The table is accompanied by examples of its use. The amount of land, material, time, and money available for a study usually limits the number of replications. Single-replication experiments have been used in demonstrations to obtain preliminary indications but are not recommended for precise comparisons of treatments.

Repetition of experiments in several locations and years is one form of replication. Conclusions are usually better founded if based on data replicated over space and time. Interaction of treatments with environmental effects can be detected through repetition of this kind. For example, the response pattern of treatments in locations 1, 2, and 3 may be different from that in locations 4, 5, and 6. The presence of a significant interaction of treatments and sites might suggest that separate recommendations are in order for the two sets of locations.

Experimental technique. Statisticians have made one of their more important contributions to research programs by emphasizing that good technique increases the precision of experiments. Some ways of improving technique are the following:

1. Procedures for conducting various phases of the experiment and a time schedule for their execution should be written out.

2. All personnel who deal with the treatments, plots, and data should be made aware of the various sources of error and the need for good technique.

3. The treatments should be applied uniformly.

4. Sufficient control should be exercised over external influences so that every treatment produces its effect under controlled, comparable conditions. For example, having each of three men harvest an individual replication is better than having the three harvest all replications as a team, and having one person harvest all three replications is still better. If environmental conditions are impossible to control, readings on major environmental variables should be taken and used as covariables.

5. Suitable unbiased measures of the effects of treatments should be devised.

6. Gross errors should be prevented.

Statisticians sometimes use data from previous years to help to find changes in technique that might improve precision. In other cases, they design new experiments specifically to investigate ways of improving technique in a particular experiment. One type, preliminary experiments, is established to focus on limited aspects of technique. Another type involves superimposing a sampling and measurement study on existing experimental plots to facilitate estimation of the relative sizes of different sources of variation (e.g., plant-to-plant and leaf-to-leaf variation compared with plot-to-plot variation).

All of the designs mentioned above assume proper randomization. Completely systematic arrangement of treatments is usually to be avoided. An example of improper randomization is aerial application of fungicide treatments in long strips, such as that shown in Fig. 7. A properly randomized arrangement (although more difficult and expensive to perform) is shown in Fig. 8. Improper randomization results in an invalid test for treatments and a possible underestimation of experimental error.

Randomization can be a problem in split-plot experiments involving growth chambers. For example, if temperature-humidity treatments are whole-plot factors, often not enough chambers are available for replication of these treatments. Changing the temperature-humidity setting for random assignment of these treatments to various chambers is also difficult. As a result, no valid test exists for the effects of temperature and humidity treatments. Multiple tests using different combinations of temperature and humidity in a chamber serve as a source of replication for the whole-plot factor, thus alleviating the problem.

A researcher must exercise judgment in implementing randomization. In some cases, certain possible randomizations are eliminated and, if one of these is obtained in the randomization process, a new set of random numbers is drawn. For example, if the original randomization of three treatments (1, 2, and 3) in four blocks of a randomized complete block design turned out to be 1, 2, 3; 1, 2, 3; 1, 2, 3; 1, 2, 3 for plots in the left-to-right

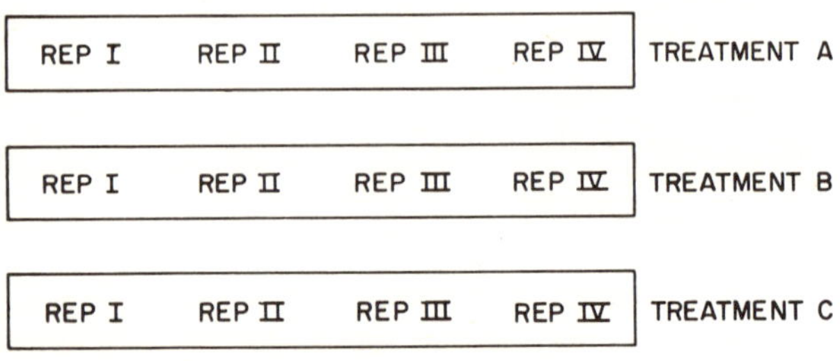

Fig. 7. An improper randomization of fungicide treatments in long strips using aerial application.

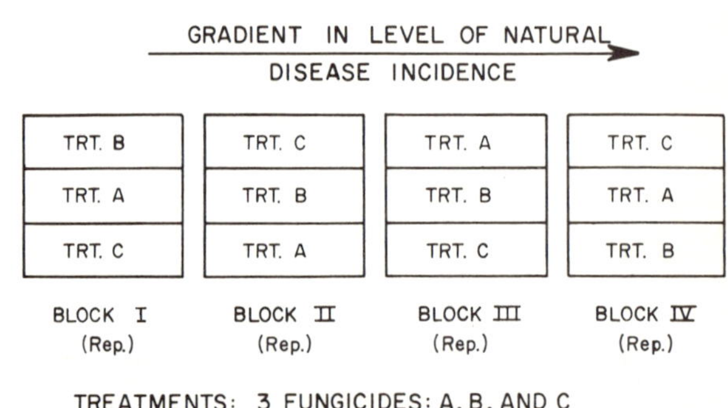

Fig. 8. A properly randomized arrangement for the treatments of Fig. 7.

positions, rerandomizing to eliminate the systematic arrangement of the treatments would seem desirable. In other cases, some systematic arrangement of treatments is intentionally built into the experiment. In certain experiments designed for university agricultural extension programs, the treatments are arranged systematically in one replication to make the best demonstrational trial. For example, fungicide rates might be progressively increased from one side of a replication to the other. In another type of experiment, nematicides from different companies might be placed together in groups within one replication, and treatments are then randomly assigned to plots within the other replications. The increased facility for making visual comparisons of effects offsets the statistical disadvantages of this practice.

Design and randomization in a series of tests. Parallel design (using the same treatments and the same number of replications) and independent randomization of individual tests in a series of experiments facilitate the combined analysis of variance.

For a series of experiments, statistical models, which include all known sources of variation (i.e., those that are deliberately controlled in the experiments and also those that are uncontrolled), have recently been fitted to experimental data. The uncontrolled variables (rainfall, temperature, soil properties) primarily explain site-to-site variability, although they can also account for some intrasite variability. To develop an appropriate model, one must recognize the variables (both controlled and uncontrolled) that affect the response and measure them at an appropriate time during the experiment. An example of the use of multiple linear regression for modeling data involving controlled and uncontrolled variables was given by Laird and Cady (13). Some workers approach the problem using simulation models (systems science), but whether one should draw a sharp distinction between the two approaches is doubtful. In any case, the success of modeling data from a group of experiments representing a range of environmental conditions depends on the ability to specify an appropriate model. With such a model, estimation of the extent of the interaction between environmental and controlled factors should be possible. The model also allows a researcher to synthesize the extant knowledge and understanding of a system, critically analyze the hypotheses about a system's functioning, identify fundamental constraints on a system's functioning, identify mechanisms to which a system's behavior is most sensitive, and identify additional research needs.

Supplementary variables. Throughout the course of an experiment, the researcher should try to determine whether environmental factors not controlled by the design are affecting the results. Readings on these supplementary variables should be recorded for possible use in statistical control. These variables are called covariables. They are used in an analysis of covariance for increasing the precision of an experiment and adjusting treatment means to a constant level of the covariable (e.g., adjusting total plot yield for plant population where stands are uneven).

Some covariables may have to be estimated on a rather subjective scale (e.g., index of wind damage). Even with such subjectivity, precision may be improved by including covariables in the analysis.

Sampling procedures. Sampling is one of the most useful statistical procedures in nematological research. It may be performed to estimate some character for an entire field or an individual plot in an experiment. Measurement of some important character may require a larger plot than does measurement of supplementary characters, and expense and time may be saved if the supplementary characters are measured on only a fraction of the plot. Also, this procedure allows estimation of the relative sizes of experimental and sampling errors, which is useful information for improving techniques for future experiments. For example, for estimating the percentage of plants affected by a disease, only 10 of 100 plants in a plot may be sampled. Yates and Zacopanay (23) studied the loss of information from sampling instead of complete harvesting in a number of experiments on cereals. At a 6% sampling rate, the loss of information was 31.2%, or about one third. A 12% sampling rate caused an information loss of 18%, and an 18% sampling rate caused a loss of 13%.

The objective in sampling is to estimate the value of a parameter that would be obtained if all the individuals in the population were measured. The difference between the sample value and population value is sampling error. The best sampling procedures are those that give a small sampling error.

The unit in which actual measurement of a character is made is called the sampling unit. Some commonly used units are an insect, a 1.5-m segment of a row, a plant, a day (sampling time), and a 2-m^2 area of a plot. A good sampling unit must be easy to identify, easy to measure, and fairly uniform. Although several different units can satisfy these requirements, the best is one that gives the highest precision at low cost. Precise estimates are those with low variance. The cost is usually measured in labor, and low-cost estimates are those that save labor.

Once the sampling unit has been determined, consideration should be given to the population being sampled and how these units are distributed in it. How the samples are taken, as well as their number, depends on the distribution of units in the population. Visual observation or actual preliminary samplings may be used to study the distribution of units in the population.

A sampling plan specifies the sampling unit, sampling design, and sample size in measuring a character. Data from previous or ongoing experiments or sample surveys may be useful in developing an appropriate sampling plan. Also, experiments may be specifically designed for this purpose. Gomez and Gomez (8) discussed at length the use of data from previous or ongoing experiments for the development of sampling plans, taking their examples from the sampling of certain rice characters. They also described experiments for evaluating sampling plans.

A sampling design defines the manner in which sampling units (e.g., segments of a row) are selected from the population (e.g., a plot). Three common sampling designs are discussed here.

In *simple random sampling*, each of the N population items has an equal chance of being selected in the sample of n sampling units. A simple procedure for taking such a sample is to number the items of the population from 1 to N and select, at random, n numbers in the range of 1 to N from a random number table. If nothing is known about the structure of the population other than its size, no sampling design is better than simple random sampling.

In *stratified random sampling*, the population is first divided into s subsections, and then a simple random sample of n sampling units is taken from each subsection. The subsections are referred to as strata.

Stratified sampling increases the precision of estimates if the variation between units of different strata is greater than that between units within the same stratum. Precision is never lost by stratifying, and considerable precision may be gained by it. Stratification is analogous to blocking in experimental design. The number of samples per stratum may be either constant or variable. Nothing is wrong with clusters of variable size as long as the chance of selection in each stratum is known and the estimator can be adjusted accordingly. Oversampling the more variable strata and undersampling the less variable strata are profitable. To maximize precision, the sampling fraction of each stratum should be proportional to the square root of the variance in that stratum. If the cost of an observation differs from one stratum to another, the sampling fraction of each stratum should be proportional to the square root of the ratio of the variance to the cost of an observation in that stratum.

Multistage or *nested sampling* is a sampling design ordinarily used in sample surveys and in many sampling experiments in the biological sciences. For our illustration, we assume that the universe has three tiers, A, B, and C. From each of the n_1 levels of the top tier A (e.g., rows), n_2 random samples of B (e.g., plants) are taken. Within each level of B, n_3 random samples of C (e.g., leaves) are taken. Such a scheme permits estimation of the components of variance caused by the factors at the various tier levels (i.e., rows, plants, and leaves). Considerably more information is available with this method than with a one-stage simple random sampling, and the cost is only a little higher. Using cost and variance estimates for units at the various tier levels, one can calculate the numbers of these units that should be used in future experiments (15).

Size of sample. Several different methods are available to estimate the required size of a sample, depending on the sampling design and several other considerations. One such method was just mentioned, for finding optimal sample sizes of various stages within a multistage sampling scheme. Perhaps the simplest case is estimating a population average from a simple random sample. The investigator must specify both how close the estimate should be to the average and the desired chance of estimating it that closely. For example, assume that the mean number of diseased plants per row of plants is being estimated. The estimate is desired to be within 13 plants of the population average, with 99 chances out of 100 of being that close to the average. In other words, the investigator wants a 99% confidence interval whose half-length is 13 plants. The half-length of a 99% confidence interval is 2.6 times the standard error of the sample average (i.e., $Z = 2.6$, from a table for the normal distribution). The sample size must therefore be fixed so that the standard error equals $13/2.6 = 5$ plants (i.e., $\sigma_{\bar{X}}^2 = 25$ plants). If n is the number of plants in the sample, the variance of the sample average is the population variance times $1/n$. Therefore n must be equal to the population variance divided by 25, i.e., equal to 0.04 times the population variance. The estimate of the population variance is obtained either by previous experience, from a small pilot survey, or by intelligent guesswork. The confidence statement is as follows:

$$\text{Probability}\ (\bar{X} - Z_{0.01}\sigma_{\bar{X}} < \mu < \bar{X} + Z_{0.01}\sigma_{\bar{X}}) = 0.99$$

where $\bar{X}$ is the sample mean, $Z_{0.01}$ is the 1% tabular value of Z, $\sigma_{\bar{X}}$ is the population standard error of the mean, and μ is the population mean. The actual calculations proceed as follows:

$$\text{Let}\quad Z_{0.01}\sigma_{\bar{X}} = 13$$

$$2.6(\sigma/\sqrt{n}) = 13$$

$$(2.6)^2(\sigma^2/n) = 169$$

$$169n = 6.76\sigma^2$$

$$n = 0.04\sigma^2$$

Suppose that, from previous studies in assessing the extent of disease incidence, σ^2 (measured from a number of different rows) is known to be approximately 150. Then the sample size that should be taken is $(0.04)(150) = 6$ rows. The principal problem that can occur is that the estimate of σ^2 may not be accurate.

If one is dealing with attribute data, such as the proportion of leaves that are diseased, only one parameter, P, specifies the population completely. For the preceding example, P is the proportion of diseased leaves in the population. Because P is usually not known, it is estimated by p, the proportion of diseased leaves in the sample. The estimated sampling variance of p is $p(1 - p)/n$, which cannot exceed $1/4n$, and therefore its standard error cannot exceed $\sqrt{(1/4n)}$.

Thus, to estimate a proportion P to within ± 0.05 with 95% confidence, 0.05 must be set equal to two standard errors, and the standard error is therefore 0.025. Its square, the sampling variance, is 0.000625, and this cannot exceed $1/4n$, where n is the sample size. Thus, the sample size must be at most $1/[(4)(0.000625)] = 400$. Such a sample is large enough to achieve the specified aim, whatever the parameter P and whatever the statistic p.

Graybill and Kneebone (9) gave two procedures that are useful in determining sample size based on a knowledge of the coefficient of variation of the data, the acceptable length of the confidence interval expressed as a percentage of the mean, and the risk level of Type I error that one is willing to accept.

Federer (3) discussed a method for determining the number of samples that should be taken in experimental plots. Not only the sample size but also the method of sampling needs to be determined. Plant materials within a plot are often systematically sampled, although the first plant sampled is chosen at random. Thus, every fifth plant in a row of corn might be taken to constitute the sample. This method of sampling provides a representative sample of the entire plot, and it is more convenient and timesaving than sampling completely at random.

THE DATA AND THEIR ANALYSIS

Data

The data reflect not only treatment effects but also variation produced by a number of causes (known and unknown) other than treatment. A careful study of the data and appropriate analysis are necessary to separate the treatment effects, which are important, from the random effects, which are of little interest.

Use of rating scales. Measurement of percentage of disease appears to be a problem, and workers are divided over how it is best accomplished. Many studies have been conducted using disease rating scales with five to 11 classes. Assumptions about the underlying stimulus-response relationship also vary from worker to worker. Some have assumed a linear relationship between disease manifestation and the degree of intensity of the stimulus. Horsfall and Barratt (11) developed an 11-class logarithmic rating scale based on the Weber-Fechner law. This law states that visual acuity is proportional to the logarithm of the intensity of the stimulus. The Horsfall-Barratt procedure has been widely used for many years. However, Hebert (10) recently reported that the assumption that all estimates depending on visual perception obey the Weber-Fechner law is false. He pointed to evidence that a stimulus-response curve is affected by many factors besides the sensitivity of the observer to the stimulus.

Hebert (10) noted that others have used a percent scale developed by James (12) and suggested that an experiment should be conducted to compare the performances of the logarithmic and percent scales. Another alternative is to estimate disease directly, without a scale. Such direct estimation has much to recommend it, because better estimates of experimental error should be available with it than with the scales. Converting to a scale also introduces a strong possibility of making an incorrect assumption about the stimulus-response relationship. If scales are used, they would seem to be better used only for classifying diseased material into categories of disease intensity and not as an aid in estimating the actual percentage of disease.

Accuracy of data. Two problems that continually arise in the handling and reporting of data are carrying more digits than are biologically measurable and accurate and, on the other hand, rounding data too much before analysis. The former can give a false impression that the data were measured with a great deal of accuracy. Chapter 3 of Cochran and Cox (1) contains a good discussion of the number of significant figures that should be carried in data. Rounding errors can cause extremely erroneous results in some regression analyses. In a series of calculations, the general rule is to round only after all the calculations have been performed.

Data variation patterns. Before analysis of variance is performed, variation patterns in the data should be carefully studied to find unusual values, or outliers, that might bias the conclusions about treatment effects. An easy way to identify outliers is to obtain the range in values for the various replications of each treatment. An extremely large or small range for a treatment indicates the presence of at least one outlier for that treatment. In Table 1, the unusually large range of 12 for treatment 3 suggests that the reading of 14 for treatment 3 in block 2 should be checked further. The number 14 would also be identified as unusual if the treatments within each block were ranked according to the numerical value of the data, from highest to lowest. For blocks 1, 3, and 4, the ranking is treatment 1, treatment 2, treatment 3, and treatment 4, whereas that for block 2 is treatment 3, treatment 1, treatment 2, and treatment 4. Rank ordering also serves as a check for the possibility of a block times treatment interaction.

In many data sets, 5–10% of the readings have been estimated to be in error because of, for example, misreading of instruments, copying errors, misplaced decimals, and misidentification of treatment and replication numbers. Other errors are introduced if unusual environmental conditions are present during an experiment and affect its results but are not observed or recorded. The possibility of such errors emphasizes the need to take notes on unusual conditions, such as flooding or wind damage. If such information is available, statistical techniques (e.g., analysis of covariance or missing plot techniques) can be applied if aberrant observations are later identified in the data.

Nonhomogeneity of data. Many sets of data are obviously not homogeneous. Some treatments may have all zero values. Common sense and good judgment should be used in excluding data that have zero variance or whose variance is different from that of the remainder of the data. Sometimes the untreated check should be omitted from the overall analysis of variance, because its variance is higher or lower than that of the other treatments. Biological reasons often exist for expecting this variance to be different. A statistician should be consulted if any question arises about the homogeneity of variance in a data set. If error heterogeneity is not rectified, the analysis of variance may not reflect a true picture of the data.

Percentage data. If the choice arises between conducting an analysis of variance on actual numbers of nematodes controlled and conducting it on the percentage of control, analyzing the actual numbers is usually better. Percentages based on a wide range of denominators are apt to have different variances. A good way to handle the problem is to analyze and report the actual numbers and also include the percentages, calculated from the means of the data, in the table of means. No standard error is reported with the percentages. Percentages may be used in analysis of variance in cases in which they are all based on the same divisor, because in such cases the conversion to percentages is equivalent to data coding.

Data transformation. Analysis of variance requires normality and equality of variances of the data. A slight departure from these requirements need not invalidate the results of analysis of variance, but a drastic departure may require corrective action. This corrective action usually consists of transforming the data from the original scale to a new one. The process is called transformation. Researchers should consult a statistician on questions about the need for data

Table 1. Use of range to inspect data for outliers[a]

Treatment	Block 1	Block 2	Block 3	Block 4	Range
1	4	6	6	10	6
2	3	5	7	9	6
3	2	14	6	8	12
4	1	3	5	8	7

[a] Hypothetical data from a randomized complete block experiment with four treatments and four blocks.

transformation and the choice of an appropriate method. Common methods of normalizing data and equalizing variances are angular or inverse sine transformation, square root transformation, and logarithmic transformation. In angular transformation, the transformed data are angles from 0 to 90°, which correspond to numbers ranging from 0 to 1 in the original scale. In square root transformation, each number in the transformed scale is the square root of a number in the original scale. In logarithmic transformation, the numbers that are analyzed are the logarithms of numbers in the original scale. Theoretical variances have been worked out for data on transformed scales and serve as a basis for comparison to determine whether the appropriate transformation has been made. Steel and Torrie (20) have discussed data transformation in detail.

The disadvantage of using a transformation is that the comparisons are made and reported on a scale (e.g., log or square root) that may not be familiar to readers. Converting the means and standard errors back to the original scale for reporting and comparison is not valid. When transformations are made, a common practice is to perform an analysis of variance on the original data as well as on the transformed data. Over a period of years, I have found that data transformation does not greatly affect the pattern of results of tests of significance, except in data sets having extreme deviations from the statistical assumptions.

Analyses of Data

Several advantages are gained from subjecting data to statistical analyses before reporting them. First, analysis is often necessary to separate treatment effects from random effects. Second, statistical analysis has the effect of reducing the bulk of the data, so that their essential features can be presented concisely. Third, analyses can be performed to calculate standard errors, which give a measure of the precision of the experiment that generated the data.

Statisticians now place more emphasis on estimating effects than on routine performance of tests of hypotheses. For example, the statement that rates overall are significant in a nematicide rate test with treatments of 0, 25, 50, and 75 kg/ha is not particularly informative. Neither is it particularly helpful to know that the mean for 75 kg/ha is not significantly different from that for 50 kg/ha, but the mean for 50 kg/ha is significantly higher than that for 25 kg/ha, which in turn is significantly higher than the check mean. A more informative approach is to choose a realistic response model, such as the quadratic polynomial $\hat{Y} = b_0 + b_1(\text{rate}) + b_{11}(\text{rate})^2$, and find the parameter estimates b_0, b_1, and b_{11} through appropriate statistical techniques. Useful interpretations, such as the anticipated yield without nematicide (i.e., b_0), the initial response to nematicide (i.e., b_1), and the optimal rate of nematicide to apply for maximum yield (i.e., $-b_1/2b_{11}$) would then be available. Another example is the comparison of two chemical fungicides. We are more interested in estimating the magnitude of their mean yield difference than in finding whether this difference is statistically significant at some preselected probability level. Thus, we develop methods that provide unbiased estimates of the magnitude of the difference, because such estimates are the basis for recommendations and economic analysis.

The following analyses are used in nematicide and fungicide tests:

Student's t test. This test is appropriate for a simple comparison of the means of two groups.

Chi-square tests. Applications of chi-square tests in nematological work are numerous. They are used with count data to test independence of factors. They are also used in breeding work to test hypotheses of particular genetic ratios. Chapters 20–22 of Steel and Torrie (20) provide further information on the uses of chi-square tests.

Analyses of variance. Methods of analysis of variance for commonly used designs are reported in several texts (2,3,6,19,20) and should be routine to most researchers. These are probably the most common form of data analysis.

Regression. This is used in survey or observational studies to relate the variation in a dependent variable to variation in one or more independent variables.

Methods of computation. Electronic computers accurately and efficiently perform routine calculations used in statistical analyses. They are invaluable in cases in which the design is complicated or large numbers of characters measured at one or more locations need to be analyzed. Statistical system packages (e.g., Biomedical Package [BMDP], General Statistical Package [GENSTAT], International Mathematical and Statistical Libraries [IMSL], Statistical Analysis System [SAS], and Statistical Package for the Social Sciences [SPSS]), available at many computing centers, can be used with minimum knowledge of programming and limited instructions to the computer. At small facilities, a researcher may need to devote considerable effort to writing computer programs for his or her own needs, and these should be checked for accuracy before they are used routinely. Statistical packages are also becoming available for microcomputers. If a computer is used, the researcher must study the data carefully before the analysis, check for errors in data entry and programming, and understand the analysis that the computer is performing. Too often, data are entered and analyzed by the computer before their patterns have been studied carefully.

Electronic calculators remain a viable option if the data set is small and the analysis straightforward. Analysis by electronic calculators has the advantage of permitting a close scrutiny of the data.

INTERPRETATION AND REPORTING OF RESULTS

Interpretation

Interpretation of experimental results is one of the most important steps in experimental research. Its theoretical aspects are based on the mathematical laws of probability. Experimental observations are limited experiences that are carefully planned in advance and designed to form a secure basis of new knowledge. In the interpretation phase, the results of these planned experiences are extrapolated to the underlying population (e.g., a five-county area). A degree of uncertainty is associated with this induction, but probabilities can be attached to the uncertainties.

Examples are the probabilities associated with F and t statistics, which are calculated in tests of significance.

Interpretation of the results of statistical analyses is also an art. Because textbooks on statistical methods do not treat data interpretation in depth, some individuals have developed an intuitive faculty for it. This would account for the fact that two researchers might interpret the analysis of a set of data in different ways. Whether an intuitive approach or a more structured approach is used, much of the interpretation process involves careful study of variation patterns in the data. The results of statistical analysis and the conclusions should reflect trends seen in the data before the analysis was made.

The actual techniques used in interpretation vary with the purposes of the experiments and the nature of the treatments. Interpretational techniques for a few experimental situations are discussed below.

Treatment relationships. If an experiment is well designed, the treatments have a logical structure that suggests a set of biologically meaningful comparisons. Experiments are often specifically set up to allow comparisons that have the property of orthogonality, which is desirable from a mathematical and statistical point of view. Orthogonality means that the total of the sums of squares of $t-1$ independent comparisons equals the treatment sum of squares with $t-1$ df.

Notwithstanding the statistical and arithmetic advantages of orthogonality, the biological basis for choosing a set of comparisons among treatments should override the desire to keep all comparisons orthogonal. Selection of comparisons that are of interest from a biological point of view can best be accomplished through the cooperative efforts of a researcher and a statistician. Once a set of meaningful comparisons has been defined, the statistician can provide the analysis, including the F tests (or t tests) for these comparisons, and can provide assistance in interpretation. The logic of treatment comparison varies a great deal, depending on the nature of the study.

One logical treatment structure is increasing levels of a quantitative factor, such as nematicide rate. In this case, the analysis would consist of fitting the data to a response curve and presenting the curve with its equation in a figure such as Fig. 9. Fitting of a response curve completes the statistical analysis and permits answers to such questions as "What is the estimated change in Y as X is changed from 10 to 15, and what is the standard error of this estimated change?" or "How much does X have to change to produce in Y a (biologically) meaningful increase of S units, and what is the standard error of this estimated change in X?"

In variety trials, in which the object often is to pick one or more outstanding varieties, the treatment structure is usually not explicit. The outstanding varieties can be picked on the basis of treatment means without the aid of analysis of variance. The question would be "If we pick the top k varieties in the experiment, what is the probability that the top variety in the population is included?" Combinatorial mathematics is involved in determining such a probability.

Other qualitative treatment sets have more structure, which suggests orthogonal comparisons. For example, consider an experiment involving two chemicals, C_1 and C_2, and, for each chemical, two wetting agents with which it can be applied, W_1 or W_2 for C_1 and W_3 or W_4 for C_2. Treatment 1 is C_1 applied with W_1; treatment 2, C_1 with W_2; treatment 3, C_2 with W_3; and treatment 4, C_2 with W_4. Treatment 5 is an untreated control. The logical set of comparisons to be made in an analysis of variance are the control versus the other treatments; C_1 versus C_2; W_1 versus W_2 for C_1; and W_3 versus W_4 for C_2. Each of these comparisons has 1 df. The correct way to interpret the data is to perform F tests or t tests for each of these orthogonal comparisons. Inferences drawn from the t tests are exactly those drawn from the F tests.

Interpretation of factorial experiments. Another common treatment structure is factorial arrangement, in which treatments consist of combinations of all levels of two or more factors. A logical set of comparisons is implied by the treatment structure. However, the particular analysis and interpretation vary according to whether the factors are qualitative or quantitative. In either case, an analysis of variance and testing of main effects and interaction should be among the first steps in the analysis. For treatments consisting of combinations of levels of quantitative factors, estimating a response surface from the data is appropriate. For treatments consisting of combinations of levels of qualitative factors, logical comparisons among levels of each factor and interaction comparisons are appropriate as part of the analysis of variance. (The coefficients in interaction comparisons are the products of the coefficients for the various comparisons.) For treatments consisting of combinations of levels of quantitative and qualitative factors, construction of appropriate comparisons should be possible for each of the factors and interactions.

Some experiments include one or more sets of treatments that are factorially arranged together with one or more additional treatments. An example is the following experiment, with five treatments: a 2 × 2 factorial set of treatments, consisting of fungicide applied at two rates, r_1 and r_2, on a standard variety and a new variety of onion; and a check treatment, which is an untreated control of the standard variety. The logical comparisons among these five treatments are the check versus the other four treatments; the standard variety versus the new variety; r_1 versus r_2; and (standard variety versus new variety) × (r_1 versus r_2). A textbook in statistical methods should be consulted for the methodology of constructing comparisons and

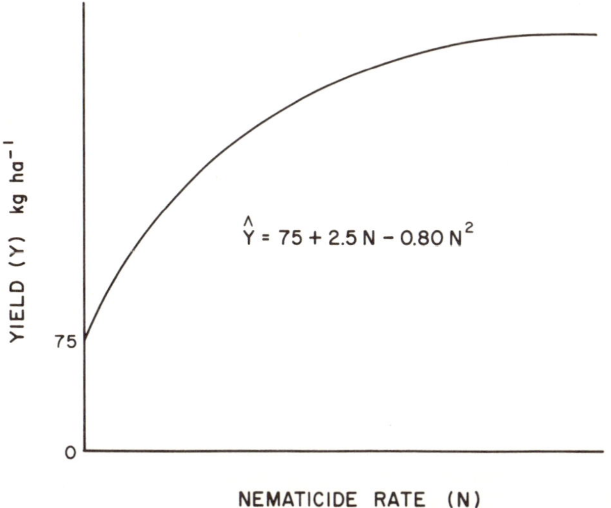

Fig. 9. Response curve fitted to yield data from plots treated with increasing levels of nematicide.

calculating the sum of squares for each comparison. Multiple comparisons among all treatments are not appropriate in this example because specific comparisons are suggested by the factorial structure of four of the five treatments. The power of the tests should be focused upon the particular meaningful comparisons mentioned above.

Rigid rules for interpreting data from factorial experiments are difficult to establish. Each experiment presents a different interpretational challenge. Usually, if interaction is negligible, the techniques described above can be used to interpret the main effect means. If interaction is sizable, these techniques are used for one factor at each level of the other factors.

Multiple comparison techniques. The required treatment comparisons should be envisioned at the planning stage of the experiment, to ensure that the treatments have a structure that permits meaningful comparisons to be made. In recent years, multiple comparison procedures such as Duncan's multiple range test have commonly been overused and misused in interpreting and reporting the results of research (16). These procedures should be reserved for cases in which no obvious treatment structure exists. Two flagrant misuses of multiple comparison procedures are 1) comparing means of treatments that consist of increasing levels of a quantitative variable and 2) comparing means of all treatment combinations in a factorial experiment (thus ignoring the factorial structure). I currently favor the Waller-Duncan Bayesian least significant difference test for comparing varieties in a variety trial. In this procedure, the size of the *F* value for the overall test of treatments is one of the arguments for looking up the tabular criterion for comparing any two means. Dunnett's procedure is useful for comparing the mean of each treatment individually with the control mean.

Reporting of Results

Most journals accept tables of means but not tables of analysis of variance. Consequently, the author of a technical article must interpret the data and then summarize the interpretations in the narrative portion of the paper. Tables of means, accompanied by standard errors or multiple comparison criteria, should also be reported. These tables should be kept simple but should be clearly annotated. One of the most common faults in the preparation of tables is an attempt to report too much information (on too many factors) in one table. In complicated tables, the reader is often bewildered about which comparisons are valid with the reported multiple comparison criterion (e.g., least significant difference).

Providing a table of means and the standard error of a treatment mean, while leaving the interpretation entirely to the reader, is not sufficient. The author is responsible for the bulk of the interpretation, using some of the techniques described above, and then for summarizing the results either graphically or in the narrative portion of the paper. Brief tables of analysis of variance, such as Table 2, are useful in many cases, especially where the treatments are factorially structured or where orthogonal comparisons are made. For quantitative factors, graphic plottings of response surfaces, such as the one in Fig. 10, are useful to give an overall impression of the relationship between the response variable and the controlled variables. If tables of means are reported, they should reflect the results of analysis of variance. For a factorial experiment, if only main effects are significant, tables of means of the main effects would be in order. Table 3 is an example of a well-designed table in which the results of a factorial experiment are clearly presented.

Table 2. Brief presentation of analysis of variance

Source of variation	df	Mean square	*F*
Blocks	3	143	4.0
Check vs. other treatments	1	379	10.5**
Fungicide *A* vs. *B*, *C*, *D*	1	100	2.8
Fungicide *B* vs. *C*, *D*	1	35	1.0
Fungicide *C* vs. *D*	1	40	1.1
Error	12	36	

***F* is significant at the 0.01 level.

Table 3. Results of a factorial experiment

	Number of plants emerged[a]	
Soil type	Fungicide f_1	Fungicide f_2
b_1	74.4	81.0
b_2	85.8	86.9
b_3	31.7	59.9

Least significant difference for comparing two fungicide means within a soil type = 9.28 plants.

[a] Each value is based on the average germination of nine observations (three species × three replications); 100 seeds were planted per plot.

SUMMARY

The logical, statistical principles that apply to the design of experiments, experimental technique, and the collection and analysis of data were reviewed, and suggestions were made for the improvement of experiments. The importance of consulting a statistician in planning an experiment was emphasized. Detailed and explicit written plans should be prepared before an experiment is executed. Experimental designs should be as simple as possible to control error variation. Plots should be long and narrow unless interplot competition is a problem; in that case, they should be square. Blocks in a randomized block design should be square or nearly square, unless another shape maximizes interblock variation and minimizes intrablock variation.

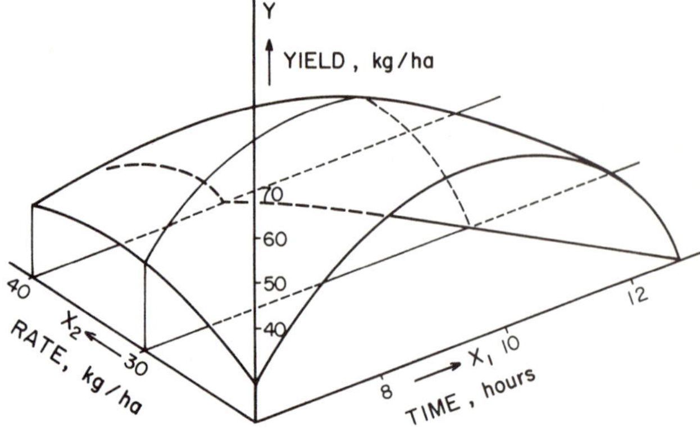

Fig. 10. Response surface fitted to yield data from plots treated with increasing levels of nematicide over time.

Data evaluation and interpretation involve careful scrutiny of data. Desk or pocket calculators permit such analysis. Electronic computers have the advantages of efficiency and accuracy in analysis. Techniques of data analysis should be appropriate for the experimental design, patterns of data variation, and knowledge of concomitant variables that affect the character being studied. An overall test of treatments is usually just the first step in comparing treatments. Orthogonal (or even nonorthogonal) comparisons are desirable if they make sense biologically.

Results can then be reported in the text of a report. A well-written research report should contain in its section on methods a detailed description of the experimental design, experimental technique, and methods used in the analysis of the data. Tables of means should be simple yet well annotated. Attempts to include too much information in a single table usually cause difficulties for readers.

LITERATURE CITED

1. Cochran, W. G., and Cox, G. M. 1962. Experimental Designs, 2nd ed. Wiley, New York.
2. Cox, D. R. 1958. Planning of Experiments. Wiley, New York.
3. Federer, W. T. 1955. Experimental Design, Theory and Application. Macmillan & Co., New York.
4. Federer, W. T., and Balaam, L. N. 1972. Bibliography on Experiment and Treatment Design Pre-1968. Oliver and Boyd, Edinburgh.
5. Federer, W. T., and Federer, A. J. 1973. A study of statistical design publications from 1968 through 1971. Am. Stat. 27:160-163.
6. Fisher, R. A. 1951. The Design of Experiments, 6th ed. Hafner, New York.
7. Gomez, K. A. 1972. Technique for Field Experiments with Rice: Layout, Sampling, Sources of Error. International Rice Research Institute, Los Baños, The Philippines.
8. Gomez, K. A., and Gomez, A. A. 1984. Statistical Procedures for Agricultural Research, 2nd ed. Wiley, New York.
9. Graybill, F. A., and Kneebone, W. R. 1959. Determining minimum populations for initial evaluation of breeding material. Agron. J. 51:4-6.
10. Hebert, T. T. 1982. The rationale for the Horsfall-Barratt plant disease assessment scale. Phytopathology 72:1269.
11. Horsfall, J. G., and Barratt, R. W. 1945. An improved grading system for measuring plant diseases. (Abstr.) Phytopathology 35:655.
12. James, W. C. 1971. An illustrated series of assessment keys for plant diseases, their preparation and usage. Can. Plant Dis. Surv. 51:39-65.
13. Laird, R. J., and Cady, F. B. 1969. Combined analysis of yield data from fertilizer experiments. Agron. J. 61:829-834.
14. LeClerg, E. L., Leonard, W. H., and Clark, A. G. 1962. Field Plot Technique, 2nd ed. Burgess Publishing, Minneapolis.
15. Marcuse, S. 1949. Optimum allocation and variance components in nested sampling with an application to chemical analysis. Biometrics 5:189-207.
16. Nelson, L. A., and Rawlings, J. O. 1983. Ten common misuses of statistics in agronomic research and reporting. J. Agron. Educ. 12:100-105.
17. Shoemaker, P. B. 1974. Fungicide testing: Some epidemiological and statistical considerations. Fungic. Nematic. Tests 29:1-3.
18. Smith, H. F. 1938. An empirical law describing heterogeneity in the yields of agricultural crops. J. Agric. Sci. 28:1-23.
19. Snedecor, G. W., and Cochran, W. G. 1980. Statistical Methods, 7th ed. Iowa State University Press, Ames.
20. Steel, R. G. D., and Torrie, J. H. 1980. Principles and Procedures of Statistics, 2nd ed. McGraw-Hill, New York.
21. Van der Plank, J. E. 1963. Plant Diseases: Epidemics and Control. Academic Press, New York.
22. Wishart, M. A., and Sanders, H. G. 1958. Principles and Practice of Field Experimentation, 2nd ed. Tech. Commun. 18. Commonwealth Bureau of Plant Breeding and Genetics, Cambridge.
23. Yates, F., and Zacopanay, I. 1935. The estimation of the efficiency of sampling, with special reference to sampling for yield in cereal experiments. J. Agric. Sci. 25:545-577.

Determining and Reporting Soil Properties in Fungicide and Nematicide Tests

WILLIAM C. NESMITH, Department of Plant Pathology, University of Kentucky, Lexington 40546, and CHARLES W. AVERRE, Department of Plant Pathology, North Carolina State University, Raleigh 27607

Many tests are performed each year to evaluate the efficacy of chemicals for controlling plant-pathogenic microorganisms in the soil. Many of these tests are reported with minimal or no reference to soil conditions. Yet the soil can have profound effects on the efficacy of a chemical in two major ways: it can affect the susceptibility of a microorganism to the chemical, and it can directly affect the toxicologic properties of the chemical (10).

For example, most microorganisms, to varying degrees, are less sensitive to toxicants when in a dormant or resting state, which is often encouraged by unfavorable soil conditions such as cold temperatures, drought, and flooding. Also, soils are very reactive chemically and physically. Many chemicals are adsorbed or absorbed on soil particles such as clay and organic matter, and thereby the activity can be altered. In other cases, the soil structure, texture, temperature, and moisture may limit the dispersal of a toxicant throughout the plow layer from the point of application; this limitation is especially critical with fumigants. The soil pH can also have a marked influence on the toxicity of the chemical.

Because soils are very complex, most of the processes that relate to the efficacy of a toxicant applied in soil are poorly understood (10). However, because soil properties can have a significant effect on the performance of a pesticide, it is very important that researchers determine and report the major soil characteristics. The failure of a product may be related to highly variable soil factors like temperature, moisture, pH, organic matter, or texture. On the other hand, a product may be highly effective only under a very narrow set of soil conditions. In either case, knowledge of soil characteristics may markedly influence the interpretation of data and decisions concerning the development of a particular product. Potentially useful products may have been prematurely dropped because they performed poorly in a particular test because of a single soil factor. For example, the highly effective nematicide 1,3-dichloropropene might not have been developed had it first been tested in a cold, wet organic soil.

This chapter has two purposes: to encourage researchers evaluating soil fungicides and nematicides to characterize the soil in which tests are conducted and to include this information in their reports, and to give specific methods for characterizing soils that can be done with a minimum amount of equipment by most plant pathologists and nematologists. The description included should be adequate for most fungicide and nematicide testing; however, we encourage readers to become acquainted with the soils literature cited, especially Black (2) or other reliable soil method text, because different soils often require different procedures.

This article first appeared as "The importance of reporting soil properties in fungicide and nematicide tests and methods of determination" in *Fungicide and Nematicide Tests* 32:262-266 (1977) and is reprinted here because of its relevance to the subject of this volume.

SITE DESCRIPTION

All reports should include a description of the site, including information on the following items: crops, weeds, hosts, and flooding history for the previous year and preferably the last three years or more; amount and condition of remaining crop residue; soil characteristics; climate; drainage characteristics; topography; presence and depth of hardpans; and any unusual or peculiar characteristics of the location. Meteorologic data, soil moisture, and soil temperature during the treatment period (time when the material is active) should also be recorded.

It is important to select uniform soils when locating a test site (2,7,11). Obviously, sites with extremely variable soils, flooding problems, old roadways or other structures, or other irregularities should be avoided. Uniform distribution of soilborne pathogens is desirable, but nonuniform distribution is common (10). The effects of site variations can be minimized by proper use of experimental design. Blocking allows one to account for variations due to these variables (2,7,11). A block should be assigned to a homogeneous soil group, and treatments should be randomized within each block.

SAMPLING SOIL

For our work, the objective of soil sampling is to obtain a unit of soil from the biologically active zone that is representative of the larger mass of soil in which the experiment is to be conducted. Because soils are often highly heterogeneous, both vertically and horizontally, this may be no small task. Because of budget and time limitations, intense sampling is usually not justified for fungicide and nematicide testing. Therefore, some type of composite sample is used for laboratory analyses (7,10).

Procedure

Samples should be taken from undisturbed soil. Equal numbers of random subsamples or borings should be collected from all blocks in the field. If the field is known to have recognizable soil patterns, each group must be characterized separately.

Small hand trowels or sampling tubes are adequate tools for sampling most soils. A minimum of 10 borings should be collected from each block and combined to obtain an acceptable composite sample. Three composite samples are then used to estimate the soil characteristics. Large rocks, sticks, etc. should be removed from the samples but must be noted in the site description. The sample should be placed in moisture-retaining containers such as plastic bags.

SOIL TEMPERATURE

Temperature affects markedly the activity and state of microorganisms in soil and their susceptibility to toxicants (6,10). It also influences the activity of many chemicals, especially toxicants with a vapor phase, like sodium azide, chloropicrin, sulfur, and other volatile fungicides and nematicides.

Procedure

Soil temperature should be taken in undisturbed soil in the biologically active zone. The bulb or sensing element should be inserted 10–20 cm deep, and the temperature should be recorded once the reading has stabilized (14). The accuracy of the thermometer should be checked, which can be done easily by verifying that ice water is at 0° C (32° F) and that mouth temperature is 37° C (98.6° F). The depth at which soil temperature was taken must be stated.

SOIL MOISTURE

Soil moisture is a critical factor in fungicide and nematicide testing because it affects the solubility of the pesticide, the vaporization and diffusion of fumigants, and the susceptibility of the target organism (6,10). Soil water, along with soil air, occupies soil pore space (6,9). Soon after heavy rain or irrigation, the soil becomes saturated and pores are filled with water. The point of field capacity is reached when free water has drained from large pores, yet medium and small pores remain full. Water is removed from these smaller pores through evaporation and plant utilization until the permanent wilting point is reached. Higher plants cannot successfully compete for water held beyond this point; they wilt, even though significant water remains in the soil. The soil texture and type and external water supplies govern these critical points of soil moisture.

Soil moisture is retained because of energy forces governed by soil factors and is expressed more accurately in energy values. Soil moisture availability is usually expressed as bars of tension, atmospheres of suction, or pF equivalents of centimeters of suction. The relationship of these terms is shown in Table 1. If these energy values are determined for various soil moisture levels, a moisture retention curve can be drawn showing this moisture-energy relationship (Fig. 1). However, because the curve depends on soil type, the percentage of soil moisture has little meaning unless the corresponding energy level is also known. The curves show that a silty clay loam at 30% soil moisture (dry weight basis) could be successfully fumigated, whereas a sand and a sandy loam soil would be nearly saturated and could not be fumigated successfully. Conversely, at 10% soil moisture the sand would be almost ideal for fumigation, but the sandy loam and silt loam would be too dry.

Procedure

Soil moisture can be determined gravimetrically, by using electrical resistance, by using a suction-pressure method, or through a rough estimation procedure (2,6,8,9). The gravimetric method is the simplest because it measures the quantity of soil water removed by drying. However, it does not measure the tension under which the water was held. The field sample is weighed to the nearest 0.001 g, dried at 105°C for 24 hr, and reweighed. The weight loss is moisture lost by evaporation, and the percentage soil moisture is calculated by dividing the weight loss by the weight of dry soil. For example,

$$\frac{20 \text{ g water loss}}{80 \text{ g dry soil}} \times 100 = 25\%$$

soil moisture expressed as a percentage of dry weight.

The resistance method employs porous blocks containing electrodes. The blocks are placed in the soil and allowed to reach equilibrium with soil moisture. Soil moisture is determined by the resistance encountered by electrons between the electrodes embedded in the blocks.

Table 1. Relationship of measures of soil moisture retention

Standard terms (critical points)	Bars of tension	Atmospheres of suction	Pf[a]
Saturation	0	0	0
Moisture equivalent (field capacity)	−1/3	1/3	2.54
Wilting point	−15	15	4.2
Hygroscopic point	−31	31	5.4

[a]pF equivalents of centimeters of suction; considered obsolete by the Soil Science Society of America.

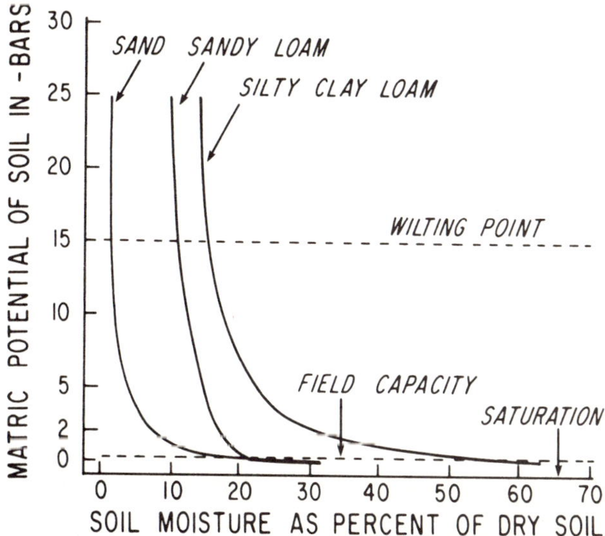

Fig. 1. Variation in soil moisture as affected by soil texture and matric potential.

The apparatus gives reasonable estimates of the soil moisture within the range of field capacity to wilting point (6). It can be used at the field site or in a laboratory. When the apparatus is properly calibrated, the resistance in ohms can be converted to percent moisture. Similarly, soil water content can be determined with tensiometers and psychrometers (9).

The suction-pressure method is most often used in soil physics laboratories to determine the matric potential of soil moisture (the energy required to remove water from soil). This method determines the tension under which soil water is held and when coupled with the gravimetric method can be used to determine the moisture content-energy relationship. The procedure employs pressure plates or membranes and a source of suction or pressure to remove water from a soil sample. Once water stops moving across the membrane, signifying that equilibrium has been reached between soil moisture and the applied suction or pressure, the sample is quickly moved to a moisture-retaining container and weighed. The moisture content is then determined by the gravimetric procedure. With proper equipment, soil moisture-energy relationships can be determined from 0 to 100 bars.

Critical points of soil moisture availability can also be roughly estimated by the following procedure. Saturate the soil sample overnight by placing it in a perforated crucible or Büchner funnel and surrounding the container with water to about half its height. After saturation, transfer a portion of the sample to a moisture-retaining container and determine its water content gravimetrically. This point is approximately 0 bars, the saturation point. To determine field capacity, place the container with saturated soil on a sidearm flask and apply one-third bar of pressure with an aspirator or vacuum pump. The funnel should be covered with a moist towel or cloth to prevent the soil surface from drying. Most soils will reach equilibrium with the suction within 30–45 min. Gravimetrically determine the moisture content, which will be approximately equal to field capacity ($-1/3$ bar). Indicator plants can be used to estimate the permanent wilting point (2). Transplant dwarf sunflower seedlings in the three-leaf stage to nonporous crucibles of soil and water frequently until the plants recover from transplanting. When the plants have two or three new leaves, stop watering and allow the plants to extract water from the soil. To prevent evaporation, cover the soil surface by tying a small plastic bag around the crucible and the base of the stem. When the plant begins to wilt, transfer the system to a dark, damp chamber. If turgor is not regained, the permanent wilting point has been reached. The moisture content is then determined gravimetrically on a sample from the root zone of these plants. Large roots should be removed.

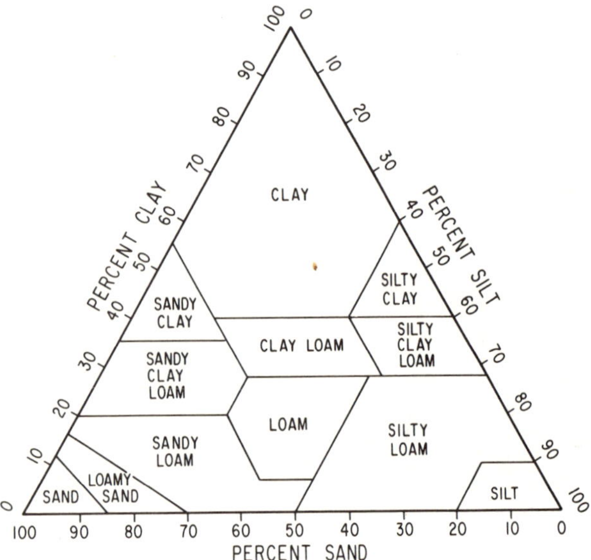

Fig. 2. Approximate relationship of sand, silt, and clay content of soil to soil texture name. To determine soil texture, first locate the percentage of silt and draw a line parallel to the percentage of clay side; next, locate the percentage of clay and draw a line parallel to the percentage of sand side; the intersection of the lines identifies the soil texture.

SOIL TYPE AND TEXTURE

Soil type is an important property to report because it precisely defines a soil, including its texture, organic matter content, pore space, and internal surface area. Clay and organic matter content affect the internal surface area, the water-holding capacity, and the activity of soil pesticides (6,9,10). The amount of space available for fumigants to occupy and the surface area for a chemical to be absorbed are directly related to soil texture (6,9,10).

The soil type is a name given a particular soil in the USDA classification system (12,13). Many soil properties are involved in determining soil type. By stating the soil type, one effectively communicates soil characteristics much as plant pathologists use Latin names to describe a species. Identifying the soil type requires the expertise of a soil scientist. However, field soil scientists have surveyed and mapped most agricultural soils of the United States to series level (12,13). These data are available from the U.S. Soil Conservation Service, which has offices in most county seats. Detailed soil type maps are available for some areas, especially experiment stations.

Soil texture is expressed by a name representing the percentage of sand, silt, and clay a particular soil possesses. Texture can be determined by simple laboratory procedures (2,5,8). Once the percentages have been determined, the class name can be read directly from the texture triangle shown in Fig. 2.

Procedure

Mechanical analysis or particle size analysis is based on the differential settling properties of sand, silt, and clay particles (8). Add 50 g of oven-dry soil, 500 ml of distilled water, and 5 ml of 1*N* sodium hexametaphosphate (in that order) to a 1,000-ml graduated cylinder. Stir the suspension vigorously to break up aggregates and allow sand, silt, and clay to fractionate freely. Stirring is best accomplished with a brass plunger. Fill the cylinder with distilled water to the full line, stopper it, and shake vigorously while inverting the cylinder to dislodge sediment. Place the cylinder on a flat surface and record the time to the nearest second. After 20 sec, gently insert a hydrometer (standard hydrometer with Bouyoucos [5] scale in grams). At exactly 40 sec, read the value on the hydrometer. For each degree above 67° F, add 0.2 to the reading; subtract

0.2 for each degree below 67°F. This corrected reading gives the grams of soil remaining in suspension. Because sand settles out within 40 sec, the percentage of sand is calculated by subtracting the 40-sec reading from the dry weight of soil (50 g). Divide that value by the total (50 g) and multiply by 100 to obtain the percentage of sand. Clay content is determined by reshaking the suspension and taking a reading at 2 hr. Because both sand and silt have now settled, the reading indicates grams of clay in suspension. The reading is converted to the percentage of clay by dividing the hydrometer reading by the total weight of dry soil (50 g) and multiplying by 100. The percentage of silt is obtained by subtracting the percentages of sand and clay from 100%. The texture name can be read directly from the texture triangle (Fig. 2). For example, if there is 56% clay, 32% silt, and 12% sand, the lines intersect in the clay section, and thus the soil texture is clay.

Although not recommended for research work, feel can be used to determine soil texture (8). It is a workable alternative for less critical studies if the following general steps are practiced and followed. Place about one teaspoon of soil in the palm of the hand and knead while adding water drop by drop until a workable putty forms. Attempt to form ribbons by pressing firmly between thumb and forefinger. Clay, sandy clays, and silty clays form good ribbons. Clay loams, sandy clay loams, and silty clay loams form medium ribbons. Loams, sandy loams, and silty loams form very poor ribbons, and sands will not form a ribbon. Interpolate using the texture triangle (Fig. 2) to get closer to the specific soil texture. For example, if the sample forms a good ribbon that is smooth, then the soil may be a silty clay loam, but if a slippery good ribbon forms, then a clay loam is likely.

SOIL pH

Soil pH is often determined by using soil and distilled water in a 1:2 ratio (2,8). The suspension is allowed to stand 5 min with occasional stirring, and then the reading is made directly from a calibrated pH meter. The sample should be swirled while the reading is taken. Soil pH dye procedures are often used in the field and can provide adequate results when carefully used but are not as accurate as a calibrated pH meter.

ORGANIC MATTER

Although it generally accounts for only a small percentage of the total soil mass in mineral soils (6), organic matter influences most physical, chemical, and biological activities in soil, because of its very high moisture-holding capacity, pore space, and exchange capacity per unit volume. Thus it is important that organic matter content be measured and reported. Most soil scientists determine organic matter by some chemical digestion procedure, especially for soils with less than 50% organic matter (2). These procedures should be used where facilities are available because of the increased accuracy obtained.

Procedure

A combustion process can provide a working estimate for soils with more than 10% organic matter. Place the soil sample in a gouch crucible and dry it overnight at 105°C to remove water. Weigh the sample, then insert it in a 400°C furnace for 8 hr; cool, reweigh, and calculate the loss in weight. Be careful not to allow the sample to rehydrate. The loss in weight divided by the air-dry weight and multiplied by 100 equals the percentage of organic matter. The error is very large for small amounts of organic matter, and the method is not suitable for soils high in clay.

Walkley-Black or Schollenberger methods are generally used to estimate organic matter in mineral soils where the organic matter is less than 5–6% (1). The organic matter is oxidized in the presence of strong oxidizing agents. Place a known volume of oven-dry soil (5–10 g) in a 500-ml open-mouth Erlenmeyer flask. (Note: the soil should be ground with mortar and pestle until it will pass a 60-mesh sieve.) Add 10 ml of $1N\ K_2Cr_2O_7$ and swirl to ensure that the soil is uniformly mixed in the solution. Rapidly direct a stream containing 20 ml of concentrated H_2SO_4 directly into the suspension and gently swirl the flask for 1 min. Because time is required for good oxidation and the reactions are exothermic, place the flask on an asbestos mat for 30 min. Add 200 ml of distilled water to the flask and swirl gently for about 1 min. Add four drops of O-phenoanthraline-ferrous indicator (available from Fredericks Smith Chemical Company, Columbus, OH). Titrate with $0.5N$ $FeSO_4$. Near the end of titration the color will begin to change from green to blue; at this point, begin to add the $FeSO_4$ drop by drop while swirling, until a sharp and sudden change from blue to red occurs. A control in which no soil is added should be run to standardize the titration scheme. The percentage of organic matter is calculated with the following formula:

$$\frac{\text{meq } K_2Cr_2O_7 - \text{meq } FeSO_4 \times 0.003 \times 100 \times 1.33}{\text{grams of oven-dry soil}},$$

where meq = milliequivalents.

TOTAL PORE SPACE

We should be very concerned about soil pore space because the dispersal and effectiveness of fumigants largely depend on it (10). Texture and structure are the main factors affecting pore space (6,9). Soil particles, water, and air occupy a given volume of soil. The total pore space is the volume not occupied by solid particles and can be determined once particle density and bulk density are known. The percentage of pore space can be calculated from the following formula:

$$\frac{\text{particle density} - \text{bulk density}}{\text{particle density}} \times 100$$

Procedure

Bulk density should be estimated from a soil sample obtained with a special sampling device that removes a core of soil with minimum disturbance of the soil. The device is driven or pressed into undisturbed soil, and the core is carefully removed. The excess soil is trimmed flush with the ends of the sample holder, and the volume is calculated from the internal dimensions of the cylinder. The sample is then weighed, heated at 105°C until a constant weight is reached, and then reweighed. Bulk density is simply weight divided by volume of oven-dry soil and is expressed in grams per cubic centimeter (4).

Particle density of soil is the density of soil particles per volume exclusive of pore space. It is determined as follows (3,15). Weigh a clean, air-dry 100-ml volumetric flask and add about 50 g of air-dry soil of known moisture. Weigh the container and soil, taking care to keep the exterior of the flask clean. Slowly add enough distilled water to fill the flask halfway, making sure that all soil is washed from the neck and inside walls of the flask. Gently boil the suspension 10 min to remove trapped air. Agitate carefully to avoid losing the liquid through foaming. Cool the flask overnight, then add boiled, cooled distilled water slowly until the water is at the 100-ml mark. Insert a fritted glass stopper, dry the flask, and determine weight and temperature. Control weight is determined by discarding the soil suspension and repeating the test in a clean flask without the soil; be sure that the temperature is the same for both control and soil suspension. Particle density is calculated from the following formula:

$$\frac{D/cm^3\,(W_1 - W_2)}{(W_1 - W_2) - (W_3 - W_4)},$$

where D = density of water, W_1 = weight of flask and soil, W_2 = weight of flask, W_3 = weight of flask, soil, and water, and W_4 = weight of flask and water. In lieu of this procedure, a standard particle density for mineral soil of 2.65 g/cm^3 is often used.

EXAMPLES OF REPORTED SOILS DATA

Example 1 (good). The test was conducted in a Norfolk sandy loam (70% sand, 20% silt, and 10% clay) having an "A-horizon" of 39 cm, pH of 5.5, organic matter of 1%, temperature of 15°C at 15-cm depth, and soil moisture of 33, 18, and 9% of dry weight of soil for saturation, field capacity, and wilting point, respectively. Total pore space was 56.7%. This site was planted to peppers with a winter cover crop of rye during the last three seasons. At the time of application, crop residue was decomposed except for a few pepper taproots. The fumigant was injected to a depth of 25 cm 14 days before planting. During the reaction period, soil temperature at 15 cm ranged from 10 to 18°C and soil moisture varied from 24 to 14% on a dry weight basis.

Example 2 (acceptable). The soil was a Norfolk sandy loam with pH of 5.5, organic matter of 1%, temperature of 15°C at 15-cm depth, and soil moisture of 23%. The site had been cropped for three years to peppers with a winter crop of rye and had a history of root-knot nematodes. The site was disked two days before treatment. The fumigant was injected 25 cm deep 14 days before planting. During this period, soil temperature ranged from 10 to 18°C and soil moisture varied from 24 to 14% on a dry weight basis.

Example 3 (minimal). Soil characteristics at treatment time were as follows: sandy loam, pH 5.5, 15°C at 10 cm, minimal crop residue, soil moisture estimated at 70% field capacity, and three-year history of pepper root-knot problem with winter rye.

ACKNOWLEDGMENTS

We acknowledge the counsel and suggestions of Dr. Keith Cassel, Mr. Paul J. Lilly, and Dr. James W. Gilliam, Department of Soil Science, North Carolina State University, Raleigh, in the preparation of this report.

LITERATURE CITED

1. Allison, L. E. 1965. Organic carbon. Pages 1367–1378 in: Methods in Soil Analysis, Part II. C. A. Black, ed. Am. Soc. Agron., Madison, WI.
2. Black, C. A. 1965. Methods in Soil Analysis. Part I: Physical and Minerological Properties, Including Statistics of Measurement and Sampling (pages 1–770); Part II: Chemical and Microbiological Properties (pages 771–1572). Am. Soc. Agron., Madison, WI.
3. Blake, G. R. 1965a. Particle density. Pages 371–373 in: Methods in Soil Analysis, Part I. C. A. Black, ed. Am. Soc. Agron., Madison, WI.
4. Blake, G. R. 1965b. Bulk density. Pages 374–390 in: Methods in Soil Analysis, Part I. C. A. Black, ed. Am. Soc. Agron., Madison, WI.
5. Bouyoucos, G. T. 1927. The hydrometer as a new method for the mechanical analysis of soil. Soil Sci. 23:343–353.
6. Brady, N. C. 1974. The Nature and Properties of Soil. 8th ed. MacMillan Publishing Co., New York. 638 pp.
7. Cline, M. D. 1944. Principles of soil sampling. Soil Sci. 58:275–288.
8. Foth, H. D., and Jacobs, H. S. 1964. Laboratory Manual for Introductory Soil Science. 2nd ed. W. C. Brown Co., Dubuque, IA. 82 pp.
9. Kohnke, H. 1968. Soil Physics. McGraw-Hill Book Co., New York.
10. Mennecke, D. E. 1972. Factors affecting the efficacy of fungicides in soil. Annu. Rev. Phytopathol. 10:375–398.
11. Petersen, R. G., and Calvin, L. D. 1965. Sampling. Pages 54–71 in: Methods in Soil Analysis, Part I. C. A. Black, ed. Am. Soc. Agron., Madison, WI.
12. Soil Survey Staff. 1951. Soil Survey Manual. U.S. Dep. Agric. Agric. Handb. 18. U.S. Gov. Printing Office, Washington, DC.
13. Soil Survey Staff. 1975. Soil Taxonomy: A Basic System of Soil Classification for Making and Interpreting Soil Surveys. U.S. Dep. Agric. Agric. Handb. 436. U.S. Gov. Printing Office, Washington, DC.
14. Taylor, S. A., and Jackson, R. D. 1965. Temperature. Pages 331–353 in: Methods in Soil Analysis, Part I. C. A. Black, ed. Am. Soc. Agron., Madison, WI.
15. Vomocil, J. A. 1965. Porosity. Pages 299–314 in: Methods in Soil Analysis, Part I. C. A. Black, ed. Am. Soc. Agron., Madison, WI.

Site Selection Procedures for Field Evaluation of Nematode Control Agents

G. W. BIRD, Department of Entomology, Michigan State University, East Lansing 48823; M. V. McKENRY, San Joaquin Valley Research and Extension Center, University of California, Parlier 93648; R. A. DAVIS, Agricultural Research Service, U.S. Department of Agriculture, Beltsville, MD 20705; R. B. MALEK, Department of Plant Pathology, University of Illinois, Urbana 61801; D. MacDONALD, Department of Plant Pathology, University of Minnesota, St. Paul 55108; J. L. TOWNSHEND, Agriculture Canada, Vineland, Ontario L0R 2E0; C. ORR, Agricultural Research Service, U.S. Department of Agriculture, Lubbock, TX 79401; and J. R. RICH, Department of Entomology and Nematology, University of Florida, Agricultural Research Center, Live Oak 32060

The process of site selection is an important part of a well-planned experiment for field evaluation of nematode control agents. Inadequate consideration of site selection is frequently responsible for the failure of field investigations. The objective of this chapter is to describe and discuss the five major components of site selection (Fig. 1): 1) research planning and associated considerations, 2) selection of a provisional test site, 3) evaluation of the nematodes associated with the provisional site, 4) evaluation of the biological, physical, and chemical parameters of the provisional test site, and 5) experimental design and final selection of the test site. The procedures for site selection and field evaluation of nematode control agents are recommended for use in conjunction with the procedures recommended in this book. Conceptual models are presented for each of the site selection components (Figs. 1–5). Examples of checklists for use in research planning and data recording are also provided (Tables 1–4).

RESEARCH PLANNING

The first two steps in research planning are to define the problem clearly and concisely and to state the experimental objectives (Fig. 2). Without fulfilling these prerequisites, a successful investigation is usually impossible. In most cases, research planning includes selecting test nematode species, test crop host, and experimental treatments. The next step involves further definition of the problem, including critical reviews of the biological, chemical, and physical properties of the control agent, and of the objectives of the experiment (4,8). Research planning information should be recorded in a data base management system (DBMS) or research project checklist (Table 1). Appropriate adjustments in the definition of the problem and experimental objectives, as well as in the desired test species, test host, or experimental treatments, must be made before the research begins.

Revision of a 1978 paper by the joint SON/ASTM E35.16 Task Force on Site Selection Procedures for Field Evaluation of Nematode Control Agents, G. W. Bird, chairman.

PROVISIONAL SITE SELECTION

Suitable test sites for field evaluation of nematode control agents may be naturally or artificially infested

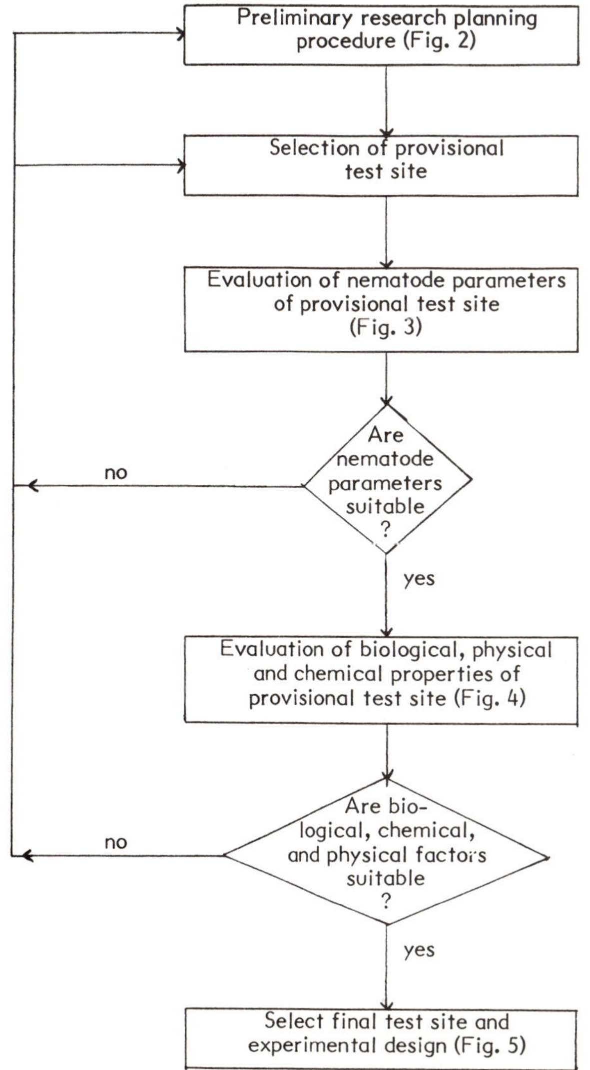

Fig. 1. Conceptual model of the selection of experimental sites and experimental designs for evaluation of nematode control agents.

with the test nematode species. The investigator may make provisional selection of a test site after considering the experimental objectives, experimental design, replication, experimental units for observation (plot size, shape, and general characteristics), and suitability of the site for experimental purposes (Fig. 3). The success of provisional site selection depends largely on the investigator's experience and desire to locate and acquire the most suitable test site.

Table 1. Prototype data base management system checklist for research planning[a]

1. Definition of problem
2. Experimental objectives
3. Proposed target nematode
4. Proposed test host (cultivar)
5. Proposed treatments
6. Proposed experimental unit
7. Proposed experimental design and replication

[a] This information should be recorded and evaluated before provisional sites are analyzed.

The initial population density of the test nematode species in the site must be stable and equal to or above the action threshold for the specific host-parasite relationship. Relatively uniform horizontal and vertical distribution of nematode populations is desirable. In most locations, however, soil texture is not uniform, and nematode species can be expected to favor zones of habitation within the soil profile, depending on the biology and physics of the soil environment. Because uniform testing conditions may not be attainable under field conditions, researchers must take steps to identify, measure, and record variables. It is helpful if the site has a history of uniform cropping sequence and productivity. No chemicals that would interfere with the experimental objectives should have been recently applied to the site. Information about the provisional research site should be recorded in an appropriate DBMS or research project checklist (Table 2).

The control that an investigator has over a test site is important. Investigator-controlled sites, such as those

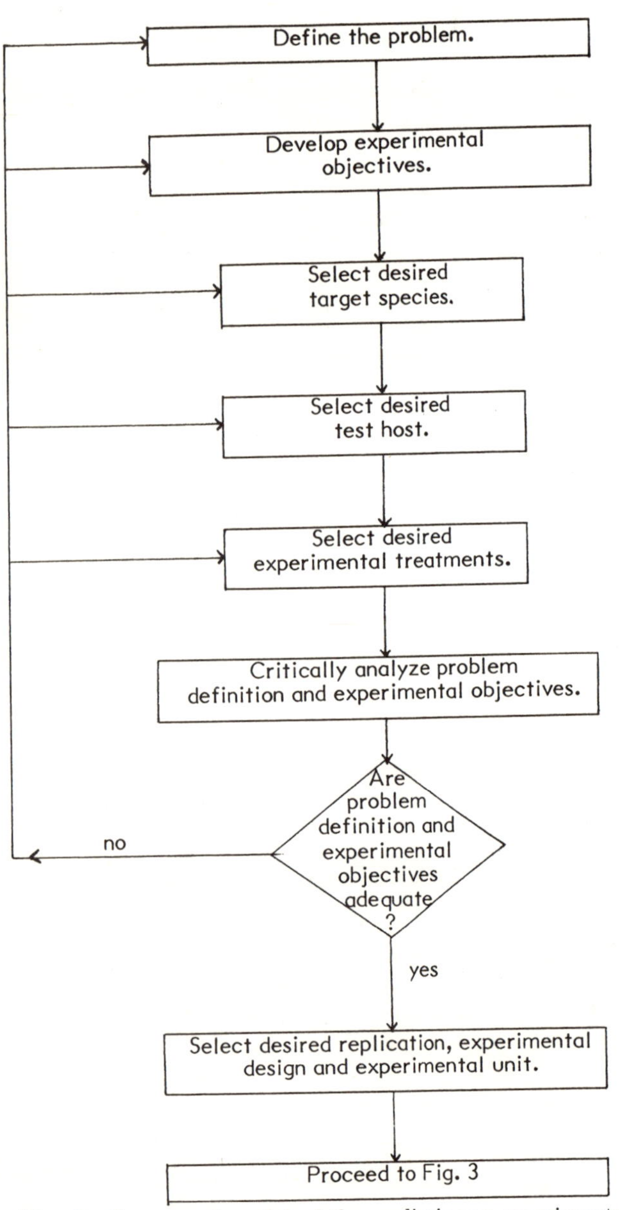

Fig. 2. Conceptual model of the preliminary experimental procedure for field evaluation of nematode control agents. The required information should be recorded in the checklist in Table 1.

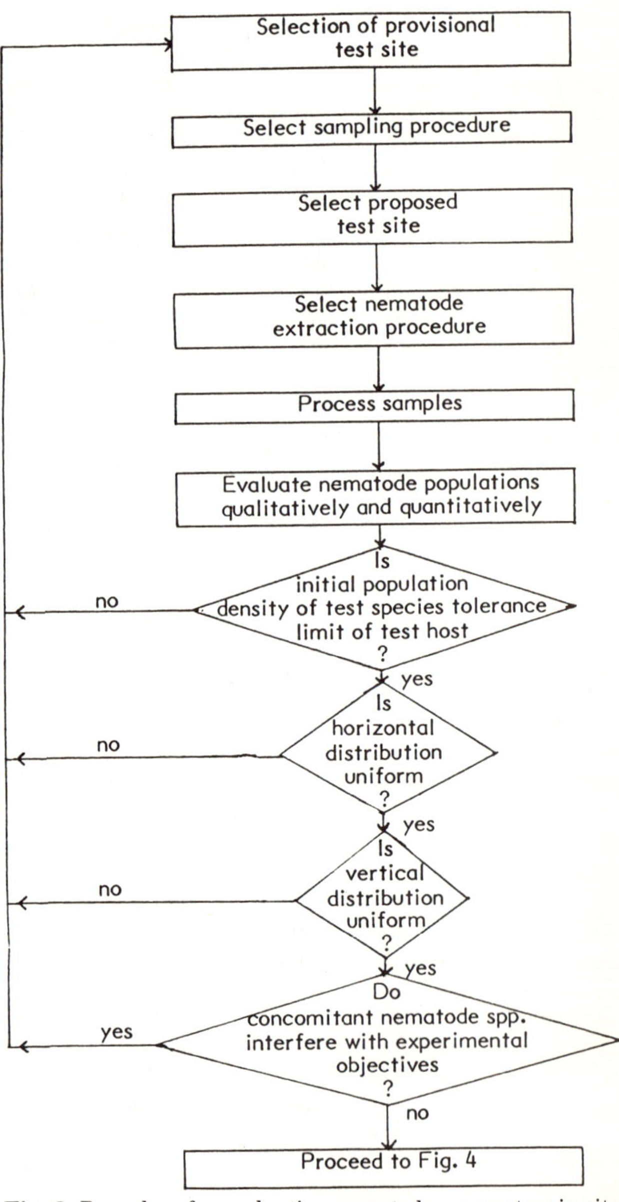

Fig. 3. Procedure for evaluating nematode parameters in site selection for evaluation of nematode control agents. The required information should be recorded in the checklist in Table 2.

at experiment stations, and grower-owned sites both have advantages and disadvantages. Good cooperators are invaluable. Although small plots and research equipment are adequate for many experiments, research sites (not demonstration plots) maintained under commercial agricultural conditions often minimize the number of years required for complete evaluation of a nematode control agent.

NEMATODE EVALUATION

The nematode parameters of a proposed test site must be evaluated in detail before final site selection (Fig. 3). The first step in this component is to select a sampling procedure suitable for the proposed test site, test host, and test species. Many experiments fail because of inadequate sampling procedures. Samples must evaluate both the horizontal and the vertical distribution of the nematode population (see Barker et al, *this volume*). Samples should be processed as soon as possible after collection or stored in a suitable environment (10–15°C) to prevent nematode mortality and population alteration. Proper assay techniques vary for different test species, sites, and hosts (7).

Sample assay is followed by identification of the nematodes and analysis of the population density. The initial population of the test species should be greater than the tolerance limit of the specific cultivar of the proposed host. Relatively uniform horizontal and vertical distribution of the test species at the site is desirable, and the population should be relatively stable. Concomitant plant-parasitic nematode species also should be distributed uniformly, and their populations should be of a type and density that will not interfere with the experimental objectives. If any of the nematode factors are not suitable for successful evaluation, a new provisional test site should be selected and evaluated with respect to its nematode parameters (Fig. 3).

Table 2. Prototype data base management system checklist for information about the provisional research site[a]

1. Location of proposed site
2. Sampling date and procedure
3. Size of unit sampled
4. Number of units sampled
5. Number of cores per sample
6. Sample processing procedure and date
7. Population density of proposed test species
8. Horizontal distribution of proposed test species
9. Vertical distribution of proposed test species
10. Concomitant nematode species
11. Population densities of concomitant nematode species

[a]These nematode data should be collected and analyzed before final selection of the test site and experimental design.

Table 3. Prototype data base management system checklist for biological, physical, and chemical factors for field evaluation of nematode control agents[a]

1. Cropping sequence and productivity history (two or more years)
2. Mechanical analysis of soil, including organic matter
3. Soil profile and uniformity
4. Soil fertility
5. Soil pH
6. History of weed problems
7. History of disease problems
8. History of insect problems
9. History of chemical use

[a]These data should be collected and analyzed before final selection of the test site and experimental design.

BIOLOGICAL, PHYSICAL, AND CHEMICAL FACTORS

The biological, physical, and chemical properties of the provisional site must be evaluated in relation to the experimental objectives (Fig. 4). The site should be capable of supporting normal growth and development of the desired cultivar of the proposed test host. Its suitability can be determined only through investigation of the cropping sequence and production records of the site for the previous two or more growing seasons. The potential for disease, insect, weed, and drought stress problems should be evaluated, and measures should be considered for their alleviation. Information about soil texture, profile, organic matter, fertility, and pH uniformity should be analyzed in relation to the

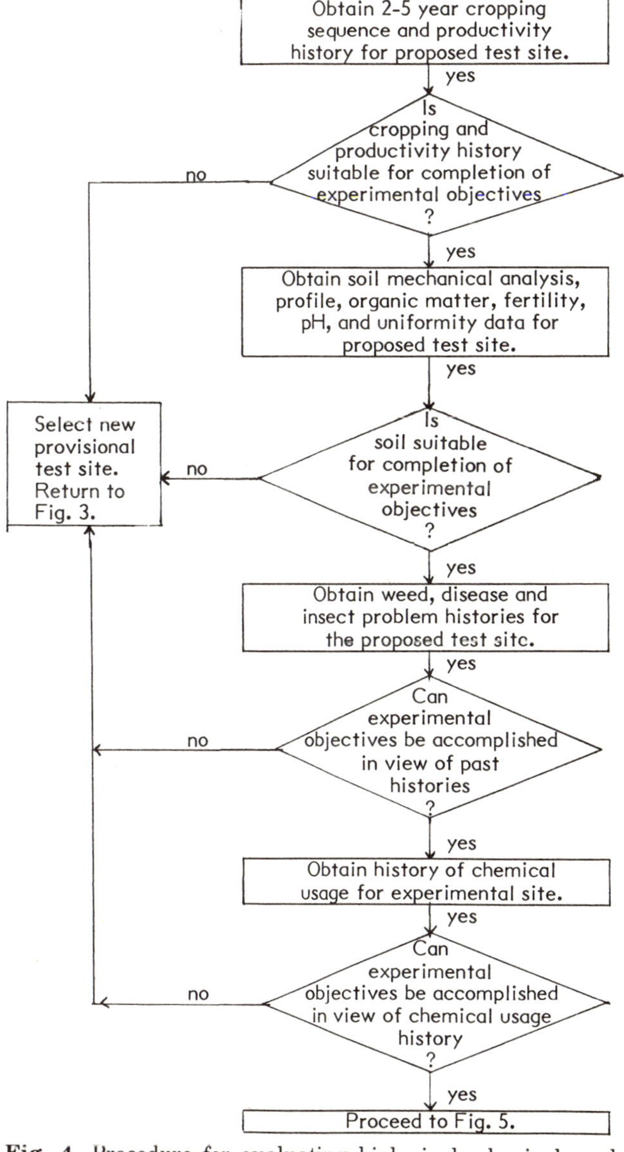

Fig. 4. Procedure for evaluating biological, physical, and chemical properties of a site before final selection of site and experimental design for evaluation of nematode control agents. This information should be recorded in the checklist in Table 3.

experimental objectives and recorded in an appropriate DBMS or research project checklist (Table 3).

A record of all edaphic chemical applications during the previous two years should be obtained. Although soil fertility and pH can be adjusted, the site should not be used unless all other edaphic chemical applications during the previous year were applied uniformly to the entire proposed experimental area. If any of these biological, physical, or chemical factors are unsuitable for fulfillment of the experimental objectives, a new provisional site should be selected and evaluated for its nematological, chemical, physical, and biological properties.

FINAL SITE SELECTION AND EXPERIMENTAL DESIGN

If a provisional site meets the requirements for evaluation of a nematode control agent, it should be selected as the test site and described in detail (Fig. 5). The description should contain the size, shape, and location of the site, including coordinates and local landmarks (Table 4). Nematologists should become familiar with the use of coordinates for geographic descriptions, because this technique most likely will be used in future computerized management of ecosystems.

The investigator should reevaluate some of the initial decisions and make the final selection of the test host species and cultivar, experimental units, number of replications, and experimental design (1-4,6-8). The initial population density of the nematode test species should be recorded for each experimental unit, and the

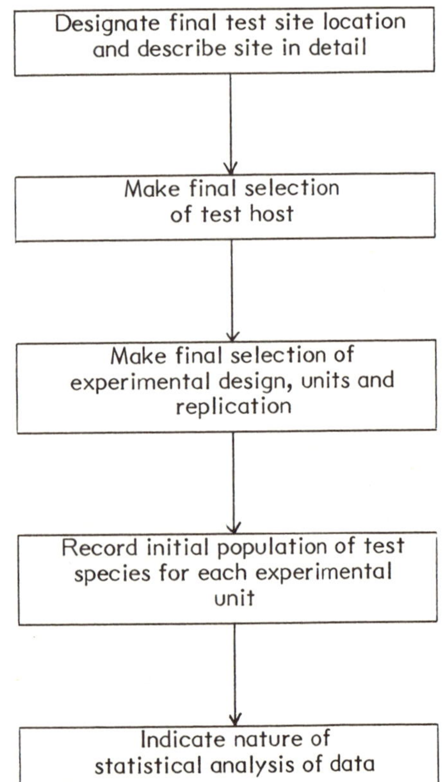

Fig. 5. Conceptual model of final selection of the test site and experimental design for evaluation of nematode control agents. The required information should be recorded in the checklist in Table 4.

nature of statistical analyses of data should be described.

Experimental unit and replication. Final selection of plot size, plot shape, and the number of replications must be related to the precision desired for the experiment. Although precision in an experiment can usually be increased through additional replication, the extent of improvement decreases as the number of replications is increased beyond a certain level. Generally, six to eight replications provide reasonable precision for evaluation of nematode control agents in a properly selected field site.

The size and shape of field plots also influence experimental precision. Increasing plot size usually minimizes variability and increases precision, but the increase in precision decreases rapidly as plot size is enlarged further because of natural variability in biological and physical factors in the field. The nature of the nematode control agents being evaluated also must be considered when plot size and shape are selected. Experience from similar tests can be helpful in selecting plot size and shape for specific crops and is also important in selecting adequate experimental units and the appropriate number of replications.

Multiple test sites. Selecting multiple test sites using the previously described criteria is often necessary to determine the utility of specific nematode control agents under the variable biological and physical conditions represented in commercial agriculture. Among the major factors that may contribute to variability in test results from one site to another are soil type, annual precipitation, irrigation, agro-ecological niches, and grower management practices. For example, heavy rainfall may render a control agent ineffective in a sandy soil, while having little effect on the agent in a heavier soil type. Statistical tests are available to determine which factors significantly affect control agent performance at differing test sites.

Experimental design. The most commonly used experimental designs and statistical analyses involve random placement of the experimental units within the experimental site. The completely randomized, randomized complete block, Latin square, and several types of factorial designs are frequently used in evaluation of nematode control agents. The suitability of a particular experimental design depends in part on the test site and the desired statistical analysis of the data. In general, the best design is the simplest one possible for the test site, experimental objectives, and desired data analyses (1–4).

The completely randomized design is the simplest of

Table 4. Prototype data base management system checklist for site selection for evaluation of nematode control agents[a]

1. Description of site location (including coordinates and local landmarks)
2. Size and shape of test area
3. Test cultivars
4. Experimental treatments
5. Experimental design
6. Replication
7. Experimental units (size and shape)
8. Initial population density of the nematode test species for each experimental unit
9. Proposed statistical analysis of data

[a]This information should be recorded during the final selection of test site and experimental design.

the designs most commonly used in tests of nematode control agents. In this design, treatments are assigned at random to a previously determined number of experimental units. Any number of treatments and experimental units can be used with this flexible design. Its main disadvantage is its low level of precision.

The randomized complete block is most commonly used in field trials. The test site is divided into blocks of experimental units, and the treatments are assigned at random to the units within each block. Each treatment is usually assigned once to each block. This arrangement minimizes variability among the experimental units within each block. The number of treatments should be as small as possible to meet the objectives of the experiment.

The Latin square is useful when two major sources of variation are present (e.g., soil variability in two directions). In this design, the randomization of treatments is grouped by both columns and rows. The design makes it possible to remove variability of experimental error associated with columns and rows. If variation is associated with the columns, this design allows greater precision than does the randomized block design. Each treatment is assigned to one experimental unit in each row and to one in each column. A Latin square design requires at least as many replications as there are treatments. Locating a suitable site is difficult if many treatments are required.

In field tests of nematode control agents, evaluation of two or more factors is sometimes necessary. This is usually accomplished through a factorial experiment. Split-plot designs and the split-plot variation are frequently used for factorial experiments. A completely randomized, randomized complete block, or Latin square design is used, and treatments of the second factor are assigned to subplots within each initial plot. These designs have numerous variations. The split-plot principle also can be used in experiments in which successive observations are made over a period of time or to compare data statistically from multiple-year tests.

Statistical data analysis. Although statistical analysis of data is not a component of site selection, it must be considered in relation to experimental design. All data should be analyzed and the results evaluated in accordance with the experimental conditions, original objectives, and previous scientific information (1,2,4–6).

Student's *t* test can be used to evaluate the probability that means of two treatments are significantly different from each other. Analysis of variance, however, is probably the statistical procedure used most frequently when more than two treatments are included in a test. If the null hypothesis is rejected at ≤ 0.05 or other desired level of probability, then the least significant difference, Tukey's honestly significant difference, and Duncan's multiple range or Student-Newman-Keuls' multiple range tests can be used to compare multiple pairs of means. The investigator should not overlook regression, correlation, probit, and numerous other types of analyses when selecting an appropriate statistical tool for assistance in interpreting experimental results.

Several other factors must be considered in regard to the overall approach of analysis. Methods of sampling for nematodes (as discussed by Barker et al, *this volume*) have strong effects on experimental results. Improper sampling may result in excessive variance, thus masking treatment effects. In fumigation trials, low population densities of nematodes frequently are associated with certain treatments. One basic assumption in analysis of variance is that variability is much the same over the whole experiment. When population densities vary markedly among treatments, however, this assumption is probably invalid. If the populations follow a Poisson or negative binomial distribution, the variance increases with the mean. If this occurs, analyzing a simple function of the data rather than the original population densities is desirable. Variance can be stabilized by transforming the data.

Another problem in data analysis arises if nematode control agents have differential effects on several target nematode species. A given control agent may reduce populations of one pathogen but fail to control a second one that continues to parasitize and damage the host plants. An analysis of growth or yield data may fail to indicate control of the test species because of the damage caused by the second species. Conversely, use of a control agent that reduces populations in general may result in growth and yield responses that cannot be attributed solely to control of the test species.

These comments tend to enforce the axiom in statistics that the manner in which the data are obtained dictates the analysis of any set of data. The analyses are exact only if the underlying assumptions are satisfied. Because this is often difficult in field trials involving nematodes, much depends on the skill of the researcher in selecting the method of analysis that best fits the circumstances of the experimental situation.

LITERATURE CITED

1. Cochran, W. G., and Cox, G. M. 1964. Experimental Designs. 2nd ed. John Wiley & Sons, New York.
2. Federer, W. T. 1955. Experimental Design: Theory and Application. The MacMillan Co., New York.
3. LeClerg, E. L., Leonard, W. H., and Clark, A. G. 1972. Field Plot Technique. 2nd ed. Burgess Publishing Co., Minneapolis.
4. Little, T. M., and Hills, F. J. 1972. Statistical Methods in Agricultural Research. Agricultural Extension Service, University of California, Davis.
5. Proctor, J., and Marks, C. F. 1974. The determination of normalizing transformations for nematode count data from soil samples and of efficient sample schemes. Nematologica 20:395–406.
6. Snedecor, G. W., and Cochran, W. G. 1967. Statistical Methods. 6th ed. Iowa State University Press, Ames.
7. Southey, J. F. 1970. Laboratory Methods for Work with Plant and Soil Nematodes. Tech. Bull. 2. Ministry of Agriculture, Fisheries and Food. Her Majesty's Stationery Office, London.
8. Unterstenhofer, G. 1963. The basic principles of crop protection field trials. Pflanzenschutz-Nachr. Bayer (Am. Ed.) 16(3):81 117.

Calibration and Use of Ground-Operated Sprayers for Applying Foliar Pesticides to Orchards and Vineyards

PAUL W. STEINER, Botany Department, University of Maryland, College Park 20742

Pesticides intended for use in orchards and vineyards to control fruit and foliar pathogens and insect pests are generally formulated to be mixed in water and applied as foliar sprays. They may be applied either as high-volume (HV) sprays requiring several thousand liters of a dilute pesticide mixture per hectare (several hundred gallons per acre) or as low-volume (LV) sprays in which only 187.0–748.0 L/ha (20–80 gal/acre) of a concentrated pesticide mixture is used.

Before any new pesticide is fully registered for general use, its efficacy is evaluated experimentally over several years in field trials conducted at many locations. The results of these trials provide the basis for establishing dosage rates, timing of the treatments, permissible residue levels, and other use restrictions. With such widespread testing, it is important that the methods used to apply these pesticides be reasonably well standardized so that the results obtained in one location by one researcher are relevant to, and can be compared with, those obtained in other locations by other investigators. Standard methodology is already well established and proven for HV spraying. For LV spraying, however, there are few guidelines for standardizing treatments, so actual dosage rates may vary. Detailed experimental protocols for evaluating pesticide efficacy on specific orchard and vineyard crops are discussed in Part III of this volume. The purpose of this chapter is to describe the theoretical bases and technical details of pesticide application methods using ground-operated sprayers.

HIGH-VOLUME SPRAYING

Method

Early in the testing period of any new or experimental material, foliar sprays are made almost exclusively as HV or dilute sprays using high-pressure, hand-held spray guns. Each tree or vine is treated individually using between 7.6 and 30.3 L (2–8 gal) of spray mixture per tree or 0.95–3.79 L (1–4 quarts) per vine, depending on the size of the plant and the amount of foliage and fruit present at the time of the application. All target surfaces are thoroughly wetted to the point where some excess spray liquid just begins to drip to the ground.

Because each tree or vine is treated individually using the drip point as a visible gauge for equivalent dosage, no part of the plant is either overdosed or underdosed. All plant targets can be treated equally, regardless of differences in size, shape, or stage of foliar canopy development. Spraying to the drip point also eliminates dosage variations that might be introduced when different types of equipment are used. As a result, assuming that the timing of the applications is correct and that the material is at least partially effective, the level of control achieved in any treatment is directly related to the concentration of the pesticide in the sprayer tank. Such tests provide the basis for the dilute concentration recommended on all pesticide product labels.

Equipment

High-pressure, hand-held spray guns are the most widely used type of spraying unit for the application of HV, dilute sprays for experimental orchard work. Spray liquid is pumped to the spray gun from a supply tank at pressures of 2,068.4–2,757.9 kPa (300–400 psi) through a flexible, high-pressure hose. The spray gun, which may or may not be equipped with an on-off trigger mechanism, contains a variable-depth swirl chamber within the barrel. This chamber opens into a nozzle fitted with an orifice plate at the end of the gun. Because the depth of the swirl chamber can be adjusted from a shallow setting (wide-angle spray pattern with fine droplets) for close work to a deep setting (narrow, HV spray stream with maximum force) for penetrating the tree canopy and reaching the tops of trees, the unit is flexible enough to be used on a variety of plant targets.

For vineyard work, where trellis structures make the use of hand-held spray guns cumbersome in close quarters, dilute sprays are sometimes applied using high-pressure, vertical-boom sprayers. These units consist of a supply tank, a high-pressure pump, and a series of hollow-cone nozzles spaced along a vertical boom mounted on each side of the sprayer. Within the body of the nozzles is a swirl plate; a short, fixed-depth swirl chamber; and an orifice plate through which the spray liquid is forced to form the spray droplets. The swirl plate is a small disk with one to four holes cut at an angle through the plate near its perimeter. Its function is to put an initial spin on the liquid stream so that the stream exits through the orifice plate swirling into a thin, hollow conical sheet that then breaks up to form spray droplets. Some swirl plates have an additional hole drilled through the center, which produces a full cone pattern. Full-cone nozzles generally produce a narrower spray pattern with greater force than hollow-cone nozzles. The volume of liquid discharged from any one nozzle depends on the operating pressure, the size of the nozzle orifice, and the number and size of holes in the swirl plate. These specifications are usually provided by the sprayer or nozzle manufacturer. Because many pesticides are formulated as wettable powders, which can be abrasive on nozzle parts, each orifice plate and swirl plate should be checked periodically for signs of wear, which can result in excessive application rates and poor droplet formation.

Vertical-boom sprayers are operated at a fixed travel speed and discharge rate, so they are much less flexible than high-pressure, hand-held spray guns in adjusting for differences among vines in the amount of spray liquid required to reach the drip point. As a result, the spray volume used per hectare (acre) with these units is frequently in excess of that needed for full coverage, simply to insure that all plant surfaces are wetted and that underdosing does not occur. Also, the only force driving the spray droplets to and depositing them on the target surfaces is hydraulic pressure on the spray liquid. For this reason, high-pressure, vertical-boom sprayers are only practical for use in vineyards where the distance between the sprayer manifold and target plants is relatively short. These units are neither practical nor adequate for testing in most orchards. Nevertheless, some material often drifts through the just-sprayed foliar canopy, so that one or more buffer rows should be included in the plot design.

Calibration

With high-pressure, hand-held spray guns, the only mechanical calibration required is to select a nozzle orifice plate and operating pressure suitable for delivering 37.9–56.8 L/min (10–15 gal/min). Dosage is controlled by simply spraying the tree or vine until the drip point is reached throughout the canopy.

For vertical-boom sprayers used in vineyards, three or four similarly sized nozzles are mounted on each side of the sprayer so that adjacent nozzle spray patterns overlap to cover all parts of the vines uniformly. Depending on the type of trellis system used and the purpose of the treatment, one or more nozzles on the lower half of the vertical boom may be closed off during the early season before foliage and fruit develop. The sprayer should be operated at a fixed speed of not more than 3.2 km/hr (2.0 mph). A mature vineyard with a fully developed canopy requires approximately 1,870.0–2,337.5 L/ha (200–250 gal/acre) to reach the drip point. Cultivars with small leaves and relatively sparse canopies can be sprayed to the drip point using lower volumes.

Because the vertical-boom sprayer is operated continuously at a fixed travel speed and discharge rate, it can be calibrated using the following formula:

$$\frac{D_V \times \text{travel speed} \times \text{swath}}{K} = \text{volume discharged per minute,}$$

where D_V is dosage volume (gal/acre or L/ha), travel speed is given in miles or kilometers per hour, swath is feet or meters between vineyard rows, and K is a constant. K is either 495 mile-feet-minutes per acre-hour or 600 kilometer-meter-minutes per hectare-hour and is used to relate area covered, time, and distance traveled to solve the equation for either gallons or liters discharged per minute with both sides of the sprayer operating. If only one side of the sprayer is to be used, then the appropriate constant should be doubled. Once the nozzles have been mounted on the sprayer according to the manufacturer's specifications, the discharge rate should be checked by operating the machine for 3–5 min and carefully measuring the amount of water required to refill the tank to the starting level. Because the speedometers on most farm tractors are generally inaccurate, especially with a fully loaded sprayer and uneven terrain, travel speed should always be checked over a measured course (1 km/hr = 16.67 m/min; 1 mph = 88 ft/min).

Application

For sprays applied with high-pressure, hand-held spray guns, the operator should approach to within 10 ft (3 m) of the tree and always work from the top of the tree down and from the inside to the outside of the foliar canopy. To avoid damage from the force of the spray stream, the fully open or deep swirl chamber setting of the spray gun should not be used on foliage and fruit nearest to the spray gun. Each tree should be visually divided into sectors that can be treated from three or four positions around the outside of the tree (very large trees may require more, very small trees fewer sectors). The job is complete only when it is apparent that some excess spray liquid is dripping from all parts of the tree.

The spray pattern and hence the coverage obtained with hydraulically operated vertical-boom sprayers are very subject to disruption caused by forward travel speed and by even a light wind. When sprays must be applied under windy conditions, operating at a slightly slower than normal travel speed sometimes improves the coverage obtained. In any case, the travel speed for this type of equipment should never exceed 3.2 km/hr (2 mph). During any application, the operator must regularly check freshly sprayed foliage and fruit for visible signs of spray drip. As the vine canopies become more dense during the course of a growing season and greater spray volumes are required to achieve the drip point, it may become necessary to make some adjustment in travel speed or operating pressure or to switch to slightly larger nozzle orifice plates to deliver sufficient spray liquid to the target.

LOW-VOLUME SPRAYING

Method

Unlike dilute spraying, which uses high volumes of water per hectare to flood or thoroughly wet all target surfaces, the basic premise of LV or concentrate spraying is that an amount of pesticide equivalent to that used in making a dilute, HV application can be applied using a relatively low volume of water per hectare. The pesticide is mixed at three to 20 times the normal dilute concentration, and the spray liquid is distributed as fine droplets over all target surfaces using a spray volume well below that required to reach the drip point. Indeed, where a fixed amount of pesticide is applied per hectare (acre), spray volumes in the range of 168.3–3,740.0 L/ha (18–400 gal/acre) give good pest control as long as coverage is adequate (4). In commercial fruit-growing operations, where LV spraying is now widely accepted as standard practice, spray volumes of 187.0–748.0 L/ha (20–80 gal/acre) are common.

By convention, the amount of pesticide applied per hectare (acre) using LV sprays is equivalent to the amount in a volume of dilute spray that, when applied to trees or vines with a fully developed foliar canopy, results in some drip from all treated surfaces. As a result,

LV spray application rates are nearly always based on an established dilute volume standard. Fifteen to 20 years ago, the most commonly used dilute standard for apples was 3,740 L/ha (400 gal/acre). This value was, and in some instances still is, widely used in state extension publications and on many pesticide product labels. Now, however, orchard management practices have shifted so that more trees are planted per acre and overall tree size has decreased markedly through the use of size-controlling rootstocks and special mechanical pruning equipment. As a result, a dilute standard of 1,870.0–2,805.0 L/ha (200–300 gal/acre) is becoming more common.

To achieve a uniform dosage per hectare (acre) with LV sprays, the spray volume and concentration are adjusted proportionately, relative to the HV or dilute spray standard. Thus, if the spray volume selected for a LV application is one-fourth or one-tenth or one-twentieth of the dilute standard per hectare (acre), then the pesticide is mixed at four or 10 or 20 times the normal dilute concentration per 100 L (gal), respectively. This method is easy to use and usually results in effective pest control. However, basing dosage on a per hectare (acre) basis lacks precision because orchards vary widely in tree size and planting density. As a result, even though two researchers may test a given product at the same rate per hectare (acre) using LV spraying methods, the volume of spray liquid discharged into the tree canopies and the concentration of pesticide in the spray droplets can vary.

An alternative approach (7) defines spray volume and concentration for LV spraying separately from each other and independently of dilute spray volume requirements. In this approach, the LV spray volume is based on the amount required to achieve adequate coverage of the target using a standard volume of spray liquid per 28.32 m^3 (1,000 ft^3) of space occupied by the target; the concentration used is based on the efficacy of control, as it is in HV spraying. An index of tree row volume (TRV) is determined by multiplying tree height times the width of the tree canopy times the row length. This is not a precise geometric measure of actual TRV, but it does offer a relevant guide that can be used to make adjustments for differences in tree size. By defining row length on the basis of linear row meters per hectare (10,000 m^2 ÷ cross row distance in meters [43,560 ft^2/acre ÷ cross row distance in feet]), adjustments for differences among orchards in planting density also can be made.

Commercial fruit growers now routinely apply the equivalent of 0.15–0.38 L per 28.32 m^3 TRV (0.04–0.10 gal per 1,000 ft^3 TRV) in making LV applications (P. W. Steiner, *unpublished*). Extension guidelines for fruit growers in Maryland (7) recommend the use of 0.34 L per 28.32 m^3 TRV (0.09 gal per 1,000 ft^3 TRV). This amount is not purported to be a minimum dosage volume requirement, but one that appears to provide coverage that is adequate for control without being excessive in many orchard situations. In effect, this method considers the orchard as a three-dimensional spray target rather than as the conventional two-dimensional area. The formula for determining a suitable dosage volume (D_V) in liters per hectare is:

$$\frac{\left[\text{tree height} \times \text{canopy diameter} \times \left(\dfrac{10{,}000\ \text{m}^2/\text{ha}}{\text{cross row distance}}\right)\right] 0.34\ \text{L}}{28.32\ \text{m}^3} = D_V$$

and in gallons per acre is:

$$\frac{\left[\text{tree height} \times \text{canopy diameter} \times \left(\dfrac{43{,}560\ \text{ft}^2/\text{acre}}{\text{cross row distance}}\right)\right] 0.09\ \text{gal}}{1{,}000\ \text{ft}^3} = D_V.$$

The principal problem with this calculation is that when the TRV for small trees in high-density plantings is less than 15,737.2 m^3/ha (225,000 ft^3/acre), obtaining good coverage using only 93.5–187.0 L/ha (10–20 gal/acre) is often difficult with the types of equipment now available. Because of this limitation, a minimum D_V of 234 L/ha (25 gal/acre) is suggested for testing purposes, regardless of calculated TRV requirements. Despite this drawback, and until some specific dosage volume per unit of target space (e.g., 4 L per 280 m^3 TRV, 1.0 gal per 10,000 ft^3 TRV) is accepted as an experimental standard, researchers should routinely report, in addition to liters applied per hectare (gal/acre), the tree dimensions and planting distances as well as the actual concentrations used in test orchards so that comparisons can be made on this basis if desired.

The significance of this change in reporting experimental parameters is illustrated by data (P. W. Steiner and L. A. Hull, *unpublished*) acquired from 17 Maryland and Pennsylvania apple growers who reported using an application rate of 467.5 L/ha (50 gal/acre). Because of differences in these 17 operations in tree size and planting density, the actual volume of spray liquid delivered per 28 m^3 (1,000 ft^3) of TRV ranged from 0.30 to 1.10 L (average of 0.42 ± 0.23 L) (0.08–0.29 gal [average of 0.11 ± 0.06 gal]). At a fixed concentration of pesticide of 1.8 kg per 380 L (4 lb per 100 gal) of finished spray, such variation would result in dosage rates of 0.15–1.12 kg per 28,300 m^3 (0.32–2.46 lb per 100,000 ft^3) of TRV.

With respect to defining LV spray concentrations more precisely, results obtained by Steiner (6) and by Michel (5) suggest that, where a fixed volume of spray liquid is applied per hectare (acre), concentrating pesticides more than three times the normal dilute rate per 100 L (gal) does not improve the level of pest control achieved. Thus, for many pesticides (both insecticides and fungicides), there appears to be an effective threshold concentration that can be defined by testing concentrations in the range of two to five times the effective dilute concentration established in earlier testing programs. Currently, depending on the spray volume selected and the HV, dilute standard used, most new products are tested using LV methods at only one concentration, which may range from three to 20 times the dilute rate.

Equipment

Except for aerial applications, LV sprays in tree fruit orchards and in vineyards are applied almost exclusively using ground-operated air-blast equipment. Despite some variations in design, all air-blast sprayers are similiar in principle and function. The spray liquid is pumped from a supply tank to a series of nozzles through which it is injected into a high-velocity airstream produced by a large fan. This airstream is responsible for carrying the spray droplets into the trees and for depositing them on the foliage and fruit. Depending on the type of nozzles used, the airstream may also assist in or be largely responsible for producing the spray droplets.

Two basic types of air-blast sprayers are used in orchards and vineyards: those with high-volume airstreams and those with low-volume airstreams. (In this instance, volume refers to the amount of air and not the amount of spray liquid involved.) Sprayers capable of developing high air volume airstreams in the range of 850–2,550 m^3 (30,000–90,000 ft^3) per minute generally use an axial flow type fan and have an initial air velocity of 160–240 km/hr (100–150 mph). Sprayers with lower air volumes in the range of 142–567 m^3 (5,000–20,000 ft^3) per minute are usually equipped with a centrifugal type fan and produce an airstream with an initial velocity of 240–290 km/hr (150–180 mph) at the discharge manifold. In practice, both sprayer types provide adequate coverage when they are adjusted properly and are used within the limits of their particular design.

Calibration and Adjustment

The purpose of sprayer calibration is to control the delivery of a fixed volume of spray liquid to a given target area (acres or hectares) or space (cubic feet or cubic meters of TRV). Once an appropriate dosage volume is selected, determine the discharge rate required for a given travel speed using the same procedure outlined earlier for use with hydraulically operated vertical-boom sprayers. With air-blast sprayers, however, several other critical adjustments must be made to achieve uniform coverage. These adjustments are necessary because of certain inherent problems with using high-velocity air to transport and deposit spray droplets.

The primary problem with all air-blast sprayers is that their airstreams, regardless of air volume or initial velocity, lose velocity rapidly with distance from the machine. With a stationary sprayer, up to 75% of the initial air velocity is lost within 8 m (25 ft) of the outlet (1). The rate of loss increases directly in proportion to the forward travel speed of the sprayer and with the amount of interference caused by the density of tree or vine foliar canopies. This loss is important in any spray application because how far a given droplet is transported and whether it can be deposited are both functions of the velocity and mass of the spray droplet. Large droplets are not transported well in airstreams because their mass causes them to settle out quickly over short distances. Small droplets may travel farther but, because of their relatively low mass, they often lack the energy or momentum at low velocities needed to impinge on target surfaces. Under conditions of low relative humidity, this latter factor assumes major importance because spray droplets of a size that might normally be transported to and deposited on foliage may be reduced by evaporation over the distance between the sprayer and the target to the point where impingement is no longer possible (3).

Because all conventional sprayers produce a range of droplet sizes, uniform coverage is possible only by discharging approximately two-thirds of the spray volume into the top one-third of the effective airstream. The effective airstream is defined as that portion of the airstream actually involved in spraying the target. For best results, the spray stream should overshoot the treetops by several feet. This disproportionate discharge profile is necessary to compensate for the droplet transport deficiencies inherent in the equipment now available. The worst possible arrangement for orchard spraying is to use the same nozzle size throughout the sprayer manifold. In vineyards, the spray stream also should be adjusted to just overshoot the top of the trellis but, because the distance between the sprayer and the vines is relatively short, overloading the top portion of the airstream is less important than in orchard spraying.

Regardless of how effective the spray stream appears in penetrating and covering target trees and vines, the thoroughness of coverage should be assessed directly using spray targets positioned throughout the tree or vine canopy. The waxy or pubescent nature of most foliage and fruit surfaces and the presence of previously applied pesticide deposits on them preclude their use in visually assessing spray coverage for LV sprays. Water-sensitive paper targets are available (Spraying Systems Co., Wheaton, IL 60187) and can be used effectively at humidities below 80% to detect droplets of 50 μm and larger. These targets are normally clipped or stapled directly onto leaves at several locations thoroughout a tree canopy. Overexposed (clear) or underexposed (black) 35-mm film transparencies can also be used for this purpose. At least two tree rows on either side of the tree row in which the targets are located should be treated before overall spray coverage is assessed.

Application and Testing

Unlike high-pressure, hand-held spray guns used to treat individual trees, air-blast sprayers are orchard sprayers and should be operated continuously as they are driven through the orchard rows, even where some open space exists between trees in a row. Up to 40% of the pesticide deposited on foliar surfaces in an apple orchard under moderate wind conditions consists of spray droplets that passed over, under, through, or around other trees in the treated orchard (2). For this reason, orchard test plots used in evaluating LV sprays need to be much larger than those where treatments are made with hand-held spray guns. The minimum replicate plot size recommended for this type of evaluation is five trees by five trees, where spray residues and disease and insect damage will be assessed on the three center trees.

Each time the sprayer is used, the operating pressure and travel speed should be monitored closely so that dosage levels are consistent. Because of the marked influence of travel speed on the terminal velocity of the spray droplets and the turbulence the sprayer can create within the tree canopy, travel speeds should never exceed 4.8 km/hr (3 mph). In most instances, a speed of 3.2–4.0 km/hr (2.0–2.5 mph) is adequate for LV spray applications. In addition, nozzle parts should be checked periodically through the season for signs of wear and replaced when necessary.

Because of the large acreage required for testing LV spray application rates, the limited amount of experimental pesticide materials available, and the restrictions that apply to their use during the early testing phase of their development, LV tests are best delayed until the product receives registration under an experimental use permit. Even then, except on a very limited scale, testing at most experiment stations is often impractical because of the acreage required. For this reason, I and my co-workers have sought cooperation with commercial growers. Ideally, unsprayed check trees should be included in such tests, but growers often will not accept this condition, so comparisons must be made to the performance of standard materials and the rates the grower normally uses. Depending on the threat posed by a target organism during the course of a season, arrangements can sometimes be made for regular, early assessments whereby particularly ineffective treatment schedules can be terminated before significant loss is incurred. As a rule, at this point in the testing program, pesticide concentration (kilograms per 100 L [pounds per 100 gal] of finished spray mixture) is the primary variable under consideration. For this reason, we select as grower-cooperators experienced individuals who normally do a good job of monitoring their orchards and timing the applications and who routinely apply spray volumes in the range of 240–470 L/ha (25–50 gal/acre).

Insect pest populations and the inoculum levels of many pathogens in well-managed commercial operations are often lower than those we encourage in test orchards at experimental stations. Nevertheless, when a problem does occur either in timing a treatment or in selecting an appropriate pesticide, detectable levels of fruit or foliar damage can occur. This situation is preferable to working with growers who routinely have problems controlling disease and insect problems because of poor management rather than incorrect concentration or spray volume. In all cases, the calibration and adjustment of the grower-cooperator's spraying equipment should be checked in detail, including the use of spray targets in the trees to ensure that spray coverage is not a limiting factor.

LITERATURE CITED

1. Brann, J. L. 1964. Factors affecting the use of airblast sprayers. Trans. Am. Soc. Agric. Eng. 7:200–203.
2. Byass, J. B., and Charleton, G. K. 1964. Spray drift in apple orchards. J. Agric. Eng. Res. 9:48–59.
3. Cunningham, R. T., Brann, J. L., Jr., and Fleming, G. A. 1962. Factors affecting the evaporation of water from droplets in airblast spraying. J. Econ. Entomol. 55:192–199.
4. Lewis, F. H., and Hickey, K. D. 1972. Fungicide usage on deciduous fruit trees. Annu. Rev. Phytopathol. 10:399–425.
5. Michel, Von H. G. 1977. Experiments for testing the effectiveness of reduced pesticide concentrations in apple plantations. Nachrichtenbl. Dtsch. Pflanzenschutzdienstes (Braunschweig) 29:82–86.
6. Steiner, P. W. 1982. Defining minimum pesticide dosage rates for low volume spraying. Proc. 112th Annu. Meeting State Hortic. Soc. Mich., pp. 24–30.
7. Steiner, P. W., Walsh, C. S., and Krestensen, E. R. 1984. 1984–1985 Maryland Commercial Tree Fruit Recommendations. Univ. Md. Coop. Ext. Serv. Bull. 134. 40 pp.

Calibration and Use of Aircraft for Applying Fungicides

B. J. JACOBSEN, Department of Plant Pathology, University of Illinois at Urbana-Champaign, Urbana 61801

Researchers unfamiliar with aerial application of fungicides often avoid research involving aerial application or reach erroneous results by using aerial application equipment incorrectly. Fundamentally, aircraft that apply pesticides are only air tractors with a spray boom. However, because of the aircraft's speed and the dynamics of the physics involved, errors can be large if mistakes are made or the equipment is used incorrectly.

In general, the most common errors involve improper calibration, excessive drift losses, improper volume, insufficient coverage, application under weather conditions adverse for deposition, application from too great an altitude, and failure to recognize application problems associated with obstructions such as trees, power lines, and buildings. Such errors can lead researchers to reach false conclusions as to fungicide efficacy, rate, and volume of carrier needed. The objective of this chapter is to assist researchers in correctly using aircraft in fungicide research.

EVALUATING AERIALLY APPLIED SPRAYS

Methods for evaluating fungicides applied aerially should take into account coverage (number of droplets deposited per unit area in various areas of the plant canopy), droplet size, effective application swath width, uniformity of application across the swath, and drift beyond the target area. Before any of these factors can be considered, proper calibration must be achieved.

Calibration of Aircraft

Calibration of aircraft is more difficult than calibration of ground sprayers, and errors caused by speed fluctuations and drift are generally greater with aircraft than with ground equipment.

Useful formulas for calibration of aircraft include the following:

$$\text{gal/acre} = \frac{\text{total boom discharge in gal/min from the boom}}{\text{acres/min}},$$

where

$$\text{total boom discharge in gal/min} = \text{flow rate per nozzle in gal/min} \times \text{number of nozzles},$$

$$\text{flow rate per nozzle} = \frac{\text{total boom discharge}}{\text{number of nozzles}}, \text{ and}$$

$$\text{acres/min} = \frac{\text{swath width (ft)} \times \text{speed (mph)}}{495}.$$

In the determination of total boom discharge in gallons per minute, the approximate flow rate for new nozzles can be obtained from charts supplied by the nozzle manufacturer. These charts give the discharge rate for each nozzle at different pressures. Because the pressure gauges in aircraft are often inaccurate, bench-tested gauges or calibrated flow meters should be used during calibration.

If nozzles are worn or flow rate charts are unavailable, the output per nozzle can be caught in a container for a given period of time (usually 15 sec) and converted to volume per minute per nozzle. Discharge should be collected from nozzles on the inboard and outboard areas of each boom in order to check flow rate across the boom.

Flow rate for the whole boom can be determined without a flow meter, although the method is rather time-consuming. Fill the spray tank, attain level flight at normal application speed, and open the spray valve fully. Continue to discharge spray until pressure drops. Land and add an amount of water calculated by the following formula:

$$\text{flow rate (L/min)} = \frac{\text{flying speed} \times \text{swath width} \times \text{volume to be applied (L/ha)}}{K},$$

where K is a conversion factor equal to 373 for flying speed in miles per hour, 600 for flying speed in kilometers per hour, or 324 for flying speed in knots.

Fly again in level flight at the same application speed. Open the spray valve and measure the time until the pressure drops. If this time is 1 min, the desired flow rate has been achieved. If the time is more than 1 min, the flow rate is too low and nozzles with greater flow rate should be used. If the time is less than 1 min, the flow rate is too high. In this case, fill the tank with the calculated amount of water and open the spray valve to less than fully open. Repeat this procedure until the calculated amount of water is discharged in 1 min. Note that this procedure can be used only for water or water-based spray mixtures. For products with a viscosity greater than that of water, a correction factor must be used.

In the determination of acres per minute, two factors are critical: the effective swath width and accurate ground speed.

EFFECTIVE SWATH WIDTH

To determine the effective swath width, the aircraft should be flown at the height, speed, power settings, nozzle position, and pressure to be used for actual application. Effective swath width can be determined by using a tracer (i.e., colored or fluorescent dyes) in the carrier and paper targets laid in a sample line across the proposed swath. Alternatively, water-sensitive papers can be used in the pattern sample line to eliminate the need for a tracer in the carrier. Fly the aircraft directly into the wind over the pattern sample line while discharging dye. After the pass, examine the sampling papers for droplet size, uniformity of deposition across the swath, and the area covered uniformly plus the area receiving 50% deposition. This area is the effective swath width.

This swath pattern type is known as trapezoidal. The gradual reduction in deposition to 50% at the swath edges allows for some margin of error in swath overlap. Although the swath width varies from its widest point when the aircraft is carrying a full load to its narrowest point when the aircraft's tank is empty, the variation is generally less than 5%. Rectangular swath patterns provide the most uniformity and accuracy but allow no margin for error in swath marking. Triangular patterns allow the greatest margin for error in swath marking but give the least uniformity over the swath.

Where deposition is not uniform over the swath, nozzles must be added, subtracted, or repositioned to achieve uniformity. Propeller vortex effects, wake turbulence from fairings, struts, and flag markers may lead to lack of uniformity across the swath. Worn nozzles, nozzle screens plugged by dirt or foreign material, and uneven pressure distribution along the boom are other potential sources of variation. In general, as the spray boom is located back and down from the trailing edge of the wing, the more uniformly nozzles can be placed on the boom. Conversely, the closer the boom to the trailing edge of the wing, the less symmetric the nozzle positioning on the boom. Booms mounted on the "toe" of the helicopter landing skids will have more symmetric nozzle positioning than those mounted at the middle or back of the skid or fuselage.

TRUE GROUND SPEED

The next step in calibration is to determine true ground speed using a calibrated electronic speed-measuring device or by measuring the time it takes the aircraft to fly over a measured distance. For example, fly the aircraft 1 mile upwind and 1 mile downwind, and average the two times. If it takes 32 sec to cover 1 mile on the upwind leg and 28 sec on the downwind leg, the average time is 30 sec. To determine the speed, divide 3,600 by the time (30 sec). In this example, the speed is 120 mph. The following formula can also be used:

$$\text{speed (mph)} = \frac{0.682 \times \text{distance traveled in feet}}{\text{seconds to travel distance}} .$$

SPRAY VOLUME, COVERAGE, DROPLET SIZE, AND DRIFT

After the effective swath width and speed have been determined, it is important to decide the desired volume of carrier (usually water) to be used and the desired droplet size, and measures must be taken to minimize drift losses. Basic considerations to these decisions are 1) the atomization of the liquid to a given droplet size range and the limitation of this size range, 2) an understanding of the role of meteorologic conditions at the time of application, 3) the needed deposition and coverage (droplet density) of the pesticide, and 4) the physical properties of the pesticide formulation.

Although drift is generally discussed in relation to the contamination of nontarget areas with a pesticide, it is important in this discussion because drifted pesticide is not deposited in the target area and therefore does not provide control in this area. Where drift is not controlled, the researcher will obtain erroneous results when determining the lowest effective application rate of a pesticide. Drift losses of 30–50% are not uncommon when proper drift control procedures are not used.

Spray Volume and Coverage

Spray volume as determined in the previous section on calibration is determined by pressure and the flow rate of the nozzles selected. Although nozzles may be selected to give any volume desired, drift losses can be minimized only by using nozzle and pressure combinations that produce only a low percentage of the spray volume in droplets smaller than 100 μm volume mean diameter (VMD).

Generally recommended fungicide application volumes should be determined by the type of fungicide (systemic or nonsystemic), the size of the target (upper leaves only or whole plants), and the need for coverage (stem, lower leaves, whole plant canopy). For example, approximately 20 droplets per square centimeter are sufficient for systemic fungicides and approximately 70 droplets per square centimeter are required for nonsystemic fungicides (2). For fungicide applications, droplet sizes should be in the range of 300–500 μm VMD. This size range is a good compromise between canopy penetration and reduced drift losses. Small droplets penetrate the canopy better in the wake turbulence than large droplets but are more subject to drift losses, as is discussed further under droplet size and drift.

Although fungicides can generally be successfully applied in 50–150 L/ha (approximately 5–16 gal/acre) for nonsystemic and 20–50 L/ha (approximately 2.1–5 gal/acre) for systemic fungicides, the actual volume needed should be calculated from the coverage needed (i.e., 70 droplets per square centimeter for nonsystemic and 20 droplets per square centimeter for systemic fungicides). For example, cereal grains have been protected with the nonsystemic fungicide mancozeb at spray volumes of 28–47 L/ha (3–5 gal/acre), whereas orchard spraying often requires 93–140 L of water carrier per hectare (10–15 gal/acre). Note that MacCollon et al (4) demonstrated that tree fruit diseases have been controlled using as little as 5–9 L of water per hectare (1.5–2.0 gal/acre).

The relationship between the volume of liquid carrier used and the density of spray deposition on the plant is important to understand because the biological efficacy of a fungicide is determined not only by rate but also by spray droplet density on the plant. Droplet size (VMD) is a critical initial consideration. For example, if 37.4 L (10 gal) of spray was uniformly sprayed over 1 ha in uniform droplets of 500, 200, and 50 μm, deposition density would be 5.7, 88.3, and 5,705.4 droplets per square centimeter, respectively. Because 20–70 droplets per square centimeter are needed for optimal fungicide efficacy, the researcher should determine the spray volume needed to achieve this deposition with the desired droplet size. If 20 droplets per square centimeter are desired, spray volume should be 50 L/ha for 400-μm VMD droplets or 370-μm VMD droplets, 30 L/ha for 330-μm VMD droplets, and 20 L/ha for 280-μm VMD droplets. The following formula can be used to determine droplet volume:

$$V = \frac{\pi d^3}{6} ,$$

where V is droplet volume, d is droplet diameter, and π is approximately 3.15. There are 10^8 cm^2 in 1 ha.

Because fungicides are generally applied to plants rather than to flat areas, it is important to understand that 1 ha of crop has more than 1 ha of plant surface for deposition of fungicide. The crop surface area to be covered, called the leaf surface index (LSI), varies with the plant and with plant density per hectare. The LSI is equal to the leaf surface area per plant times the plant population divided by the planted area. Thus, if a crop with an LSI of 4 were sprayed with 37.4 L/ha using 500-μm VMD droplets, droplet density would be 1.4 droplets per square centimeter. Using 200-μm or 100-μm VMD droplets, density would be 22 or 178.3 droplets per square centimeter, respectively. Therefore, a droplet VMD between 100 and 200 μm would give the required 20–70 droplets per square centimeter. However, to prevent excessive drift losses, droplet VMD should be between 300 and 500 μm. To achieve the desired droplet density with this droplet size, a larger volume of water carrier would be needed. The droplet density should be constant in all research trials.

The size of spray droplets can be determined reasonably accurately and most easily by examining water- or oil-sensitive papers with a dissecting microscope and optical micrometer. Alternatively, treated microscope slides can be used instead of papers. Care must be taken to avoid errors caused by evaporation when using slides. A general rule of thumb for most droplets is that VMD is equal to 0.45 times the maximum droplet diameter (note 1,000 μm).

Droplet size can be measured very precisely using laser beam equipment. This level of precision is generally not needed by fungicide researchers.

Droplet Size

Droplet size is important relative to coverage and spray volume per hectare but is a most important consideration in reducing drift losses. The size of the droplets impinging on the plant depends on the size of droplets emitted from the nozzle orifice, the wind shear angle and speed at the nozzle orifice, and the rate of evaporation of the spray droplets (for water-based sprays), which depends on temperature and relative humidity.

Nozzles do not produce uniform droplets but rather emit a spectrum of droplets. The range of the droplet size spectrum varies greatly with different types of nozzles, pressures used, and properties of the spray mixture. The range of droplet VMDs from a hollow-cone nozzle is larger than that from a rotary atomizer such as Micronair. The spectrum of droplet sizes from a hollow-cone nozzle ranges from 1 μm to 1,000 μm or more. The critical factor for the researcher is the VMD. If a nozzle produces droplets with a VMD of 350 μm, this means that half of the volume is in droplets smaller than 350 μm and half of the volume is in droplets larger than 350 μm. VMD is not the arithmetic mean but the volume mean diameter.

To minimize drift losses, it is important to minimize the percentage of the spray volume contained in droplets less than or equal to 100 μm. At 25°C and 50% relative humidity, a 100-μm droplet will fall less than 2.4 cm before evaporating; a 50-μm droplet will fall only 8 cm (3). A hollow-cone nozzle producing droplets with a VMD of 350 μm will have 2.1% of the output volume in droplets of less than 100 μm. Thus 2.1% of the spray volume is unlikely to be deposited in the test area.

The prime determinant of droplet size for aerial applications is not pressure but the angle of wind shear across the orifice of the hydraulic nozzle. Spinning speed is the prime factor for droplet size with rotary nozzles.

For hydraulic nozzles, droplet sizes are smallest where wind shear is 90° relative to the nozzle orifice (i.e., the nozzle is pointed straight down) and largest where wind shear is 0° relative to the nozzle orifice (nozzle pointed straight back). For example, a Spraying Systems D-6 nozzle with a size 46 core at 2.2 kg/cm^2 pressure in an airstream of 160 km/hr generates 4% of its output volume in droplets less than 100 μm, with a VMD of 298 μm when pointed straight down. Under the same conditions except with the nozzle pointed straight back, only 0.1% of the output volume of this nozzle will be in droplets less than 100 μm, and VMD will be 460 μm.

Droplet size is adjusted for rotary atomizers by adjusting their speed of rotation. This speed is adjusted by changing the pitch on the propeller or by using propeller blades of a different size and shape.

Spray pressure also affects droplet size, although the effects are relatively minor compared to those of wind shear at the nozzle orifice. For example, at a pressure of 7 kg/cm^2, a Spraying Systems D2-13 hollow-cone nozzle has a droplet VMD of 150 μm whereas at 14 kg/cm^2 pressure, droplet VMD is 100 μm (1). In general, researchers should operate nozzles at less than 2.2 kg/cm^2 (30 psi) to minimize drift losses. At higher pressures, more small droplets are formed.

The relationship of pressure to volume is poorly understood by many aerial applicators, who often try to change volume merely by changing pressure. The attempts fail dramatically because more than a doubling of pressure is required to double the volume of output. For example, a nozzle with a flow rate of 4 L/min at 2.1 kg/cm^2 requires 8.4 kg/cm^2 pressure to deliver 8 L/min. Because most aircraft use centrifugal pumps, which generally have maximum pressures of 5.0 kg/cm^2, large changes in volume by pressure changes are not possible. The following formula can be used to determine the pressure needed to apply a given volume per acre:

$$P = \frac{\text{gal/acre} \times \text{mph} \times \text{swath width (ft)}}{495 \times \text{no. nozzles} \times \text{gal/min at } p} \times p ,$$

where P is the necessary pressure in pounds per square inch and p is a particular pressure. For example, suppose an applicator wants to apply 5 gal/acre at 100 mph with an effective swath width of 45 ft. The applicator has 58 new Spraying Systems D-7 nozzles with an output of 1.25 gal/min at 30 psi according to the manufacturer's charts. What pressure should be used to apply the 5 gal/acre?

$$P = \frac{5 \times 100 \times 45}{495 \times 58 \times 1.25} \times 30 = 11.8 \text{ psi} .$$

This is the needed pressure at the nozzle orifice. However, the pressure necessary to open the safety check valve must be added. Most aircraft nozzles use a 5–7 psi check valve. Therefore, 16.8 psi (11.8 + 5) would be necessary if a 5-psi check valve were used.

Spray Drift

Although droplet size is important relative to deposition density and volume of spray, every researcher should recognize its importance to drift losses. Drift losses for water-based sprays depend on several interrelated factors, the most important ones being droplet size, temperature, relative humidity (RH), wind velocity, and the height from which the droplets must fall to the plant surface. For nonevaporating liquid carriers, droplet size and wind velocity are most important. Therefore, both equipment and meteorologic factors must be understood to minimize drift losses.

The interrelationship of water droplet size and RH is shown in the following data (2): a 200-μm droplet at 20°C will fall only 7.7 m at 60% RH before evaporating but will fall 25.0 m at 90% RH before evaporating. At 30°C and 50% RH, a 25-μm droplet will fall 0.025 m, a 50-μm droplet 0.075 m, and a 100-μm droplet 2.4 m before evaporating (3). Once the water has evaporated, only the pesticide particle is left. A 10-μm particle will fall at 0.18 m/min (3), so even light winds will cause deposition well outside the research area. Under conditions of 20–30°C and 50–70% RH, researchers should consider droplets of 100 μm or less for water-based sprays lost to evaporation and to drift if application heights are greater than 1.5–2.0 m.

Wind is a key factor in the drift problem. The magnitude of the effect depends on droplet size. For example, a 1,000-μm nonevaporating droplet falling only 3 m will be displaced 0.8 m by a 3.6 km/hr horizontal wind, but a 100-μm droplet will be displaced 11.5 m. At a wind velocity of 24 km/hr, the 100-μm droplet will be displaced 45.7 m.

The distance the spray droplet must fall before impact is critical to the effects of evaporation and wind on drift. The farther the droplet must fall, the greater the effect of these factors.

Fall distance also affects canopy penetration. Aircraft can apply pesticides in low volumes effectively because they are air-displacement sprayers. Spray droplets penetrate to lower levels of the plant canopy primarily because of the turbulence created by the displacement of air in the canopy by the aircraft and its violent replacement in the aircraft wake by spray-laden air. The strength of the air displacement depends on the height of the aircraft above the canopy and the weight of the aircraft. Better canopy penetration and reduced drift because of lesser effects of winds and evaporation will occur where application heights are lower. In general, the spray boom should be no more than 3 m above the crop canopy. However, application at heights less than 1.5 m causes streaking and crop damage. With the exception of orchard spraying, the turbulence created by rotor or wing-tip vortices has little to do with canopy penetration.

Another way to reduce drift losses is to avoid trapping spray droplets in wing-tip or rotor vortices. Spray droplets trapped in these vortices are lifted upward and usually evaporate before being deposited on the spray target. The vortices are most powerful near the wing tip of fixed-wing aircraft and the tip of the rotor blades of helicopters. Vortices increase in strength as aircraft weight increases and as speed decreases. They entrap spray droplets and carry them above the wing or rotor tip, and most droplets evaporate before they are carried downward. Research done cooperatively by the University of Illinois and the National Aeronautics and Space Administration has shown that even a 100-μm droplet released from the boom three-fourths of the way along the length of the wing will never reach the ground; it will evaporate because of entrapment in the wing-tip vortex. A 300-μm droplet emitted at three-fourths of the wing length will deposit from a height of 2 m. The wing-tip vortex also displaces droplets emitted from one-half the wing length, although these droplets reach the ground. It is the effect of these vortices that causes the effective swath width to be wider than the wingspan or boom length for most aircraft. No nozzles should be located beyond the length of the rotor on helicopters. The precise location of nozzles relative to wing-tip or rotor-tip vortices depends on boom location relative to these vortices and droplet size. Therefore, no rule of thumb (such as no nozzles should be located beyond three-fourths of the wing length) is justified. I have seen numerous agricultural aircraft with nozzles discharging over the full length of the wing with excellent drift control. These aircraft used booms placed down and back from the trailing edge of the wing and nozzle-pressure-wind shear combinations that produced large droplets with sufficient mass to prevent their being carried upward by the vortex.

To reduce or eliminate drift caused by entrapment in wing-tip or rotor-tip vortices, nozzles should discharge spray droplets away from the most powerful vortex area and be large enough to escape vortex forces. This is accomplished by one or more of the following techniques: 1) produce large droplets (large nozzle orifice and/or low pressure), 2) eliminate nozzles near the wing tip, and 3) locate nozzles down and in back of the trailing edge of the wing.

A final consideration in drift control is a meteorologic one. To minimize drift losses, avoid thermal drift and drift under temperature inversion conditions.

Thermal drift occurs when the air temperature decreases with altitude faster than the normal 1–1.5°C per 305 m (2–3°F per 1,000 ft). Under these conditions, spray droplets are trapped in updrafts (thermals) and drift away from the target areas. These conditions often occur in the semiarid areas of the western United States and can be avoided in these areas by making applications early in the morning or at night, when air conditions are more stable.

Drift under inversion conditions is often the most serious form of drift, because spray droplets remain concentrated where no vertical mixing of air occurs.

Droplets less than 50 μm will fall only short distances and may be deposited miles away, even where wind speeds are low. Under inversion conditions, the air layer above 152 m (500 ft) is generally warmer than at ground level and there is little wind. In general, aerial pesticide applications should be avoided when ceilings are below 152 m (500 ft) and wind velocities are less than 3.2–4.8 km/hr (2–3 mph). For example, a 50-μm nonevaporating droplet will drift 402 m before deposition when released from a height of 1.5 m. Droplets of 200 and 100 μm will drift 47.5 and 122 m, respectively, before deposition under similar conditions.

EVALUATING DRIFT POTENTIAL, DROPLET SIZE, AND COVERAGE

The following procedures should be used to determine and achieve desired droplet size, coverage, and drift potential. Pattern uniformity and swath width can be determined at the same time.

Set out a pattern of collector papers or treated slides. Place them every 0.3–0.6 m (1–2 ft) for 9–15 m (30–50 ft) right and left of the center line of the deposition area and at 1.5-m (5-ft) intervals from 17 to 23 m (55–80 ft). Place this line of collection papers or slides perpendicular to the wind so that the aircraft discharges spray while flying directly into or with the wind.

Collector papers manufactured by Ciba-Geigy (available from Spraying Systems Company, Wheaton, IL) that are sensitive to either water or oil can be used directly to determine droplet size. If dye and paper are used, the spread factor for the dye-paper combination must be determined before droplets are measured. Alternatively, slides coated with petroleum jelly, grease, wax, or silicon can be used. When using water and such slides, droplets must be protected from evaporation before measurements are made. When using slides, it is best to photograph the slides immediately. The advantage of water- or oil-sensitive paper is that pattern uniformity and swath width can be determined visually. By using a fluorescent dye, total deposition can be measured over the swath width by using a continuous collection tape, a fluorometer, and a computer printout. The operation SAFE equipment uses a combination of water- or oil-sensitive collector papers and the fluorometric deposition determination methods (5). This equipment provides excellent data on deposition and droplet size and can be obtained for use from chapters of the National Aerial Applicators Association. Dyes that can be used may be visible only (food coloring, Ag Mark P2 agricultural dye) or fluorescent (Rhodamine FB by BASF Wyandotte).

Load the aircraft to the desired level with water, water plus dye, or water plus pesticide to be tested plus dye (if needed). Where possible, the pesticide formulation plus adjuvants to be tested should be used since these materials can significantly change nozzle flow rates and droplet size spectra.

Special precautions should be taken when using oil-based sprays. The viscosity and therefore the flow rate of oils vary with temperature. Therefore, calibration and application tests should be made at the same temperature. This is a critical factor since flow rates can double with a 10°C temperature rise with some oils and nozzle combinations.

The addition of drift control additives such as Target, Driftgon, Driftless, Nalcotrol, or Orthotrol generally does not require changes in calibration since these materials only reduce the production of small droplets and do not increase the production of large droplets. They may, however, cause slight changes in flow rates.

Choose nozzle size and number to deliver the desired volume in an even pattern at pressures between 20 and 30 psi. All nozzles should be set in the straight-back position. The aircraft operator should be sure that all nozzles are charged with spray material. The operator should begin the spray run in level flight, at operational height and speed, at least 100 m before the target and should continue in level flight discharging spray until well past the target. The operator should be sure that the center of the spray swath is located in the center of the pattern of collector papers.

Measure droplet size and density from cards in all areas of the spray swath. Droplet sizes should be determined with an optical micrometer and a dissecting microscope, or a hand lens and a finely graduated rule can be used. On each card randomly select a square centimeter and determine droplet density and size. Excessively large droplets generally indicate a system leak.

Rotate the nozzles downward if droplet density is too low or droplet size too large, so as to increase the wind shear on the nozzle orifice. Remember to maintain droplet VMD in the 300–500 μm range. In general, nozzles need to be turned down no more than 45° from the straight-back position. For Micronair nozzles, larger droplets are formed by slowing the screen rotation (i.e., increasing the pitch on the fan blades).

Determine effective swath width as the width over which the pattern is uniform plus the distance where at least 50% deposition occurs. If spray droplets are found more than 3–4.6 m (10–15 ft) beyond the 50% deposition point, drift caused by entrapment in wing-tip or rotor-tip vortices is occurring. If this is happening, remove nozzles from the ends of the boom one at a time until effective swath width remains the same and no visible deposition occurs more than 3–4.6 m (10–15 ft) beyond the effective swath width.

Recalibrate the aircraft to be sure discharge rates are compatible with previous calibration. Minor changes can be made by altering pressure. Do not exceed 30 psi pressure in the boom. Major changes can be made by changing nozzle orifice size; refer to the nozzle manufacturer's flow rate charts or use an accurate flow meter.

Remember that research plots must be sprayed at the same height and pressure as used for calibration.

EVALUATING COVERAGE AND DEPOSITION WITHIN THE PLANT CANOPY

Several methods have been used to evaluate deposition within the plant canopy. Water- or oil-sensitive collection papers may be attached to leaves, stems, or support frames within the canopy. Or plastic collectors or glass slides may be distributed to sample within the plant canopy; these are collected, the pesticide is washed off, and the residue is analyzed. Occasionally copper sulfate or copper hydroxide is used, and a direct colorimetric analysis is done on the material removed from the collector. Or fluorescent dyes may be used in the spray solution; collector slides, plastic films, or leaves themselves are collected after spraying and analyzed for total fluorescence and for deposit size. They may be photographed to provide a permanent record.

Each of these three methods has its benefits and shortcomings. The use of water- or oil-sensitive papers is not directly quantitative but does give a quick visual estimate of droplet size and density from various points in the plant canopy. This method is often used in conjunction with a quantitative method. At present the fluorescent dye method is most widely used. This method is quantitative and visual (with the use of ultraviolet light), the analysis is relatively easy, and the procedure can be automated (7).

CONCLUSION

Whenever aerially applied pesticides are evaluated it is important that calibration be done correctly and that drift losses be minimized. Accurate records must be kept on initial deposition in various strata of the plant canopy, spray droplet size and density, volume applied, and weather conditions. Poor application techniques such as poor swath marking, inaccurate calibration, and drift problems have often resulted in poor control or excessive pesticide rate or volume recommendations.

Researchers doing aerial application studies are using very expensive equipment that belongs to a pesticide applicator and should remember that the applicator uses this equipment to make a living (6). In general, aerial applicators are skilled workers who are supportive of university research. Researchers who develop a rapport with aerial applicators based on mutual respect for each other's goals and limitations usually achieve the best results.

LITERATURE CITED

1. Akesson, N. B., and Yates, W. E. 1974. The uses of aircraft in agriculture. FAO Dev. Pap. 94. Food and Agriculture Organization of the United Nations, Rome. 217 pp.
2. Ciba-Geigy. Undated. Correct and safe aerial application of pesticides. Ciba-Geigy Ltd., Basel, Switzerland. 43 pp.
3. Kuhlman, D. K., ed. 1981. Aerial Application Handbook for Applicators. MF-622. Cooperative Extension Service, Kansas State University, Manhattan. 73 pp.
4. MacCollon, G. B., Gottlieb, A. R., and Calahan, C. L. 1980. Air spraying for orchard pest control in Vermont. Agric. Exp. Stn. Univ. Vt. Bull. 685. 15 pp.
5. National Agricultural Aviation Association. Undated. Operation SAFE. Information Kit. National Agricultural Aviation Association, Washington, DC.
6. Rester, D. 1981. Extension service working with aerial applicators. Paper AA-81-001 presented at the National Agricultural Aviation Association annual meeting, Las Vegas, NV.
7. Sistler, F. E. 1981. An image analyzer for aerial application rates and patterns. Paper 81-1001 presented at the American Society of Agricultural Engineers summer meeting, Orlando, Florida.

Safe Use and Calibration of Irrigation Systems for Chemigation

E. DALE THREADGILL, Department of Agricultural Engineering, Coastal Plain Experiment Station, University of Georgia, Tifton 31793

Chemigation is defined as the application of a chemical (nutrient, herbicide, insecticide, fungicide, nematicide, plant growth regulator, etc.) to a crop via an irrigation system. The chemical is injected into the water flowing through the system. The rapidly increasing acceptance and adoption of chemigation as a crop production management tool have stimulated considerable research among scientists to determine the criteria for successful chemigation and to evaluate the efficacy of pesticides applied by chemigation.

The management advantages of chemigation have been described by Threadgill (14). The effectiveness, limitations, and problems associated with the application of pesticides by chemigation have been described in the proceedings of two national symposia on this topic (17,18) and in other publications (2,3,5,8–13,16,19). In general, this research has demonstrated that chemigation is feasible and can be very effective when the irrigation system and the chemical injection system are properly installed and operated and when the proper chemical, formulation of the chemical, and water application rates are used.

Although much has become known about chemigation in the last few years, much more research is needed on chemical formulations, chemical application rates, water application rates, and chemigation system design. Two of the most important factors to be considered in conducting research on chemigation and in applying this technology are safety and calibration.

CHEMIGATION SAFETY

The absolute minimum equipment required for chemigation is an irrigation system, an irrigation pumping plant, a chemical injection pump, and a chemical reservoir. However, chemigation should not be attempted with only this minimum equipment. Properly installed and maintained safety devices to prevent backflow of chemicals into the water source or backflow of water into the chemical reservoir are considered essential. Some states have adopted legislation mandating certain chemigation safety measures (6,15), and other states are considering such legislation. In addition, the Environmental Protection Agency (EPA) is considering requiring a statement concerning safety devices on the label of pesticides that are registered for chemigation (P. Gray, *unpublished*).

Several descriptions of recommended safety device systems have been published (4,7), and the American Society of Agricultural Engineers (ASAE) has approved an engineering practice defining appropriate safety devices (1). Proper safety devices are categorized into two groups, each addressing a specific hazard. Backflow-prevention devices prevent flow of chemical or a mixture of water and chemical back into the water supply. Interlocking injection devices insure that if the irrigation pumping plant stops, the chemical injection pump will also stop, to prevent pumping the entire chemical mixture from the supply tank into the irrigation pipeline. They also prevent water from flowing back through the injection pump and overflowing the chemical supply tank if the injection pump stops and the irrigation pump continues to operate. Figure 1 illustrates the arrangement of a complete system of both backflow-prevention and interlocking injection devices and identifies appropriate components of the system.

Backflow-Prevention System

Five types of backflow-prevention devices can be used, as described by the ASAE (1). The device illustrated in Fig. 1 is the only commonly used system, primarily because it is most economical. The backflow-prevention system illustrated in Fig. 1 consists of a check valve in the irrigation mainline, a vacuum breaker, and a drain point. The check valve and vacuum relief valve keep chemical or a mixture of water and chemical from draining or siphoning back into the irrigation water supply. Both of these valves must be located between the irrigation pump discharge and the point where the chemical is injected into the irrigation mainline.

Check valves for the irrigation mainline vary greatly in reliability and durability. In general, the less reliable and durable valves also cost less and, thus, are frequently installed in chemigation systems to minimize capital costs. These less expensive check valves frequently develop minor leaks with use. Thus, a low pressure-activated drain point installed between the antisiphon device and the irrigation pump serves as an additional safety feature. This low-pressure drain will drain any water-chemical mixture that does leak slowly past the check valve into a storage reservoir or onto the ground rather than allowing it to contaminate the water supply.

Interlocking Injection System

The interlocking injection system consists of a check valve and a solenoid valve located in the chemical injection line and the interlocking of the power supplies for the irrigation system and the chemical injection pump. In systems where the irrigation pump and the chemical injection pump are powered from an internal combustion engine (Fig. 1A), the chemical injection pump can be belt-driven from the drive shaft or an accessory pulley on the engine to provide interlocking. In systems where an internal combustion engine provides power for the irrigation pump and also generates electricity for powering an electric motor-operated chemical injection pump, interlocking is inherent in the design. In systems using an electric motor-powered irrigation pump and an electric motor-powered chemical injection pump, the electric controls on the two motors should be interlocked so that the injection pump motor will stop when the irrigation pump motor stops (Fig. 1B).

A check valve in the chemical injection line is necessary to prevent the flow of water from the irrigation system into the chemical supply tank. Without this check valve, if the injection pump stops, irrigation water could flow back through the chemical line into the chemical supply tank, overflow the tank, and cause a chemical spill. The spilled chemical may be concentrated or diluted but in any case is a potentially serious contaminant.

A small, normally closed solenoid valve is recommended for all chemigation systems and is essential for chemigation in certain types of irrigation systems, particularly those systems where the elevation of any discharge point in the downstream irrigation system is lower than the elevation of the chemical supply tank. This automatic solenoid valve is electrically interlocked with the engine or motor that powers the injection pump. This interlocking provides a positive shutoff in the chemical injection line, thereby preventing the flow of chemical or water in either direction if the chemical injection pump stops. This solenoid valve should be located on the suction side of the chemical injection pump rather than on the pressure side as shown by ASAE (1) to prevent the chemical injection line from rupturing should the solenoid valve inadvertently close while the chemical injection pump is operating.

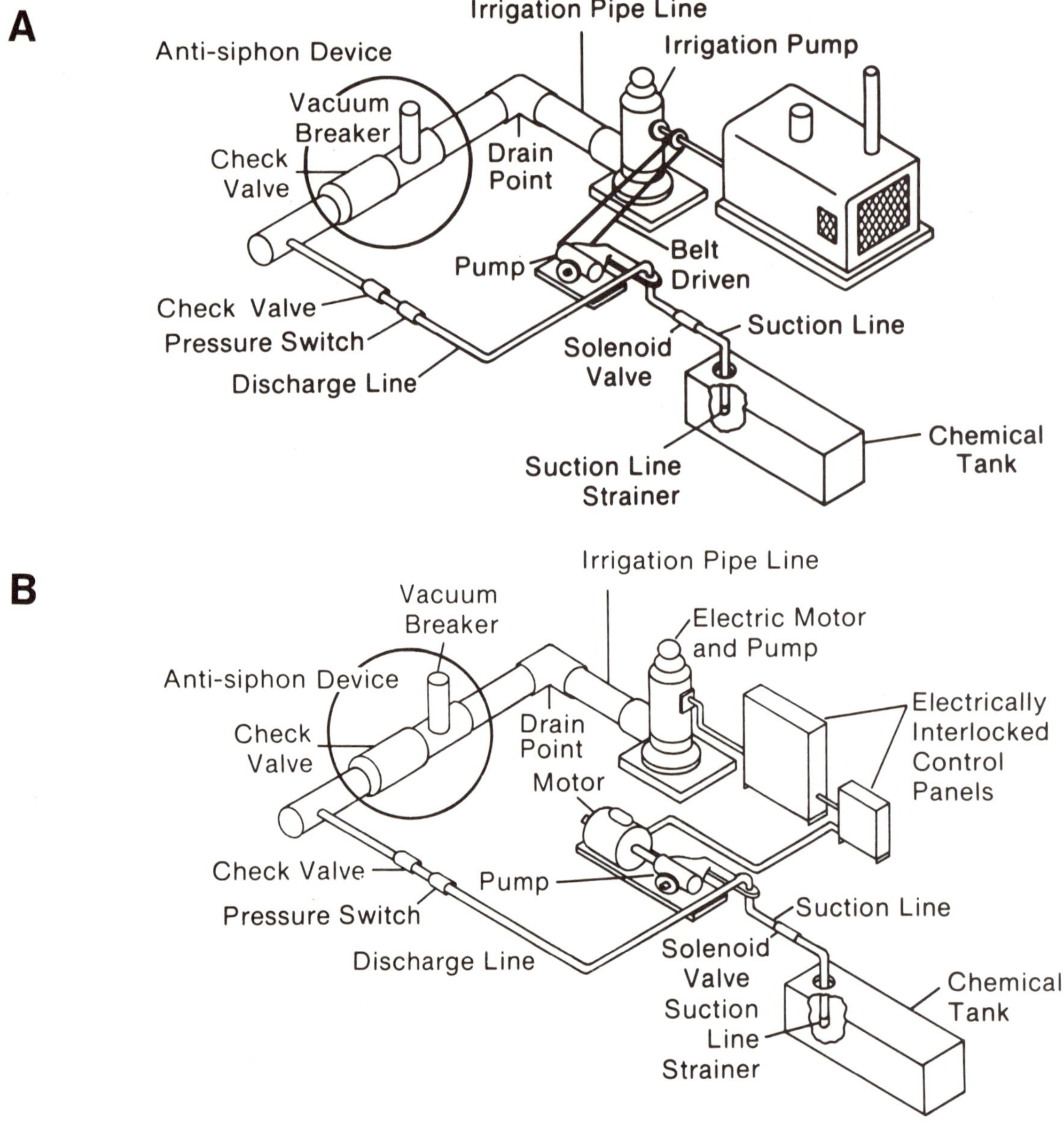

Fig. 1. Arrangement of safety devices for chemigation through **(A)** an engine-powered irrigation system and **(B)** an electric motor-powered irrigation system. (Reprinted, by permission of the American Society of Agricultural Engineers, from "Chemigation via sprinkler irrigation: Current status and future development" by E. D. Threadgill, Appl. Eng. Agric. 1:16-23.)

Injection Pump

The chemical injection pump should be accurate (to within 0.5–1%), easily adjustable to different injection rates during operation, constructed of noncorrosive materials (internally and externally), and mechanically rugged. A major disadvantage of some pump designs is corrosion and wear, which lead to leakage of the chemical onto the pump and soil. Also, pumps that cannot be easily adjusted during operation are very difficult to calibrate, greatly increasing the probability of worker contact with the chemical during calibration.

Many chemicals have components that make them highly combustible. For electric motor-powered injection pumps, a totally enclosed and sealed motor is essential to minimize the possibility of combustion when combustible chemicals or chemical vapors are present.

Equipment Construction

All equipment, hoses, and accessories that come in direct contact with chemicals must be resistant to all formulations of agricultural chemicals being applied, including emulsifiers, solvents, and other carriers as well as the active material. Hoses, seals, gaskets, and the like should be constructed of polypropylene, polyethylene, EPDM (ethylene-propylene-diene elastomer), EVA, Teflon, Hypalon, or Viton. In general, products that contain polyvinyl chloride, neoprene, butadiene, or styrene-butadiene rubber are not satisfactory for many agricultural chemicals.

Supply Tanks

Chemical supply tanks must be constructed of materials that withstand the corrosive action of agricultural chemicals. Stainless steel, fiberglass, nylon, and polyethylene are common construction materials. Iron, steel, copper, and brass should be avoided. The supply tank should be large enough so that a single mix will treat the irrigated acreage, to reduce the risk of accidental spills and repeated mixing errors.

Allow at least 1 gal (3.78 L) of water for each pound (454 g) of wettable powder added. Liquid or flowable formulations require less water.

Agitation in the chemical supply tank is necessary when wettable powders, dry flowables, flowables, or any other suspended formulation is used. Mechanical and hydraulic agitation are the two most common types. Experience has shown that mechanical agitation is highly desirable. Both types of agitation require a separate power source.

Other Components

A pressure switch in the chemical injection line may be desired for automated control. Electrically interlocking this switch with the safety shutdown system for the irrigation system will assure automatic shutdown of both the irrigation system and the injection pump if pressure is lost in the injection discharge line. Depletion of chemical in the chemical supply tank is a primary cause of pressure loss.

Additional Safety Considerations

Safety precautions should be taken to protect workers against danger from accidental spillage or discharge of chemicals. These precautions include a water source located near the chemical supply tank and injection pump for washing off chemicals that may contact the skin, and the use of protective goggles, face shields, and clothing by workers handling chemicals. Field workers should be cautioned against being sprayed with water from the irrigation system during chemigation and against reentering the field until enough time has gone by for safe reentry.

In the preparation of the chemical dilutions, concentrated chemicals should always be added to the water rather than adding water to concentrated chemicals. The chemical injection pump and associated components should be thoroughly flushed with clear water or other appropriate solvents after chemigation to prevent clogging of the injection system.

Because many chemicals are classified as flammable, workers should be cautioned against the hazards of flame, heat, and sparks, and irrigation-chemigation system components should be appropriately sealed or shielded (engine or motor) and made of proper materials to minimize the potential for combustion.

The irrigation system should be flushed with water for an appropriate period of time after the chemigation system has been shut down. Signs should be posted against drinking or bathing in water from an irrigation system used for chemigation, and other appropriate measures should be taken to warn workers and visitors.

CALIBRATION OF CHEMIGATION SYSTEMS

Calibration is an extremely important step in chemigation and involves both the irrigation system and the chemigation system. To uniformly distribute a chemical, the irrigation system sprinklers must distribute water uniformly. Check the uniformity of the irrigation system at least once each cropping season and correct any deficiencies. Also, the water application rate of the irrigation system must be accurately known, and devices (such as precision timers) for setting the application rate must be reliable and capable of a repeatable setting. To inject the proper amount of chemical into moving irrigation systems (center pivot, linear move, and traveling gun), both the rate of system travel and the rate of injection must be accurately known. The time required to irrigate an area with a continuously moving irrigation system is the precise time available for chemical injection, because injection will occur continuously over the entire irrigation period. Speed control devices on moving irrigation systems must be calibrated periodically. Some precision timers on center pivots have been found to be very unreliable with respect to repeatability of a given speed setting.

Calibration for chemigation with moving irrigation systems involves the following six basic steps: 1) determine the area (in hectares; acres) to be chemigated; 2) determine the amount of chemical (in liters; gallons) to be applied per hectare (acre); 3) determine the total amount of chemical (in liters; gallons) required to chemigate the entire area (step 1 × step 2); 4) determine the time (in hours) for the irrigation system to cover the area determined in step 1; 5) determine the amount of water to be applied (centimeters; inches, etc.), and be sure this amount does not exceed the label instructions for the chemical; and 6) determine the chemical injection rate (in liters per hour; gallons per hour) needed to apply the chemical (step 3 ÷ step 4). This procedure gives the amount of chemical that must be injected per unit of time. If the chemical is diluted in water or other diluents, the total volume of the mixture must be used in step 3.

Calibration for chemigation with stationary irrigation systems (solid-set, trickle, and periodic lateral move) may differ slightly from that for chemigation with moving irrigation systems. With stationary irrigation systems, chemical injection may occur either continuously during the entire period of irrigation system operation or periodically during one (or perhaps more) period of time while the irrigation system continues to operate. For continuous injection over the entire period of operation of a stationary irrigation system, the calibration procedure outlined above for moving irrigation systems is appropriate. For periodic chemical injection in a stationary irrigation system, step 4 of the calibration procedure should be modified as follows: 4) determine the total time (in hours) that the chemical injection pump will operate during the irrigation period.

Calibrate the equipment (including the irrigation system) yourself rather than relying on the manufacturer's or dealer's recommendations. The setting for injection pumps suggested by manufacturers eliminates much of the need for trial and error in calibration, but you still should determine the exact injection pump setting required, because conditions at the work site are not the same as at the factory.

An accurate timing device, such as a stopwatch or wristwatch with second hand, and a collection container with easily read units, such as a graduated cylinder, can be used to calibrate the injection pump. The calibration procedure should be done with the irrigation system running and the chemical supply tank placed as it would be during the actual chemigation. If operation of the irrigation system is not practical or desired during calibration, the injection pump can also be calibrated by attaching a back-pressure regulating system between the check valve in the chemical injection line and the injection port in the irrigation line, adjusting the back pressure to the pressure at which the irrigation system will operate, and then collecting the chemical output per unit of time. Calibration of the chemigation system without either the irrigation system operating or the use of a back-pressure regulating system will not be accurate.

Monitoring the complete irrigation-chemigation system at regular intervals is the most effective way to ensure accurate and uniform application. Calibration tubes have been designed to facilitate checking injection output during operation and may be useful. However, the most effective and accurate way to check injection pump calibration is close monitoring of the chemical supply tank.

The injection pump should be operated long enough before calibration to ensure that all air has been expelled from the injection system. Calibration measurement periods should begin and end with the injection pump running. Calibration with the actual chemical or chemical-diluent mixture to be used is recommended; however, calibration with water may be desired when highly toxic chemicals are used, recognizing that some differences may occur between the calibrated injection rate for the water and the actual injection rate for the chemical.

Frequently, the desired injection rate will be the same as that used in the most recent operation of the system and, thus, recalibration may not be necessary. However, checking the calibration of the chemigation system is strongly recommended before each use to ensure that the injection rate is indeed accurate.

SUMMARY

Chemigation can be an effective and economical method for applying agricultural chemicals. Knowledge about chemigation has rapidly expanded in the past few years, and much additional research will be conducted in the future. To assure the long-term benefits and value of this chemical application technology, appropriate consideration must be given to both human and environmental safety. Safety considerations for chemigation are well defined, and appropriate, practical safety components for chemigation systems are commercially available. Certain safety devices are now required by law in some states, and the EPA is considering chemical label requirements that may result in universal use of safety devices for the legal application of chemicals via chemigation.

Proper calibration of the chemigation system is essential for effective use of this technique. The calibration procedure is relatively simple and quite analogous to that used with any conventional application system, such as a boom sprayer.

LITERATURE CITED

1. ASAE. 1983. Safety devices for applying liquid chemicals through irrigation systems. Engineering Practice ASAE EP409. Pages 522–523 in: Agricultural Engineers Yearbook of Standards. American Society of Agricultural Engineers, St. Joseph, MI.
2. Chalfant, R. B., and Young, J. R. 1982. Chemigation, or application of insecticide through overhead sprinkler irrigation systems, to manage insect pests infesting vegetable and agronomic crops. J. Econ. Entomol. 75:237–241.
3. Dowler, C. C., Rohde, W. A., Fetzer, L. E., Scott, D. E., Jr., Sklany, T. E., and Swann, C. W. 1982. The effect of sprinkler irrigation on herbicide efficacy in selected crops and distribution and penetration in some Coastal Plain soils. Univ. Ga. Coll. Agric. Res. Bull. 281. 27 pp.
4. Fischbach, P. E. 1982. Applying chemical through irrigation systems—Safety and environmental considerations. Pages 80–87 in: Proc. 2nd Natl. Symp. Chemigation, University of Georgia, Tifton.
5. Gascho, G. J., Hook, J. E., and Mitchell, G. A. 1984. Sprinkler-applied and side-dressed nitrogen for irrigated corn grown on sand. Agron. J. 76:77–81.
6. Georgia General Assembly. 1981. Agriculture—Rules, etc. regulating irrigation systems. Georgia General Assembly, House Bill 521.
7. Harrison, K. A., and Skinner, R. E. 1981. Why use chemigation. Pages 109–112 in: Proc. Natl. Symp. Chemigation, University of Georgia, Tifton.
8. Heikes, E. 1979. Herbigation: Applying herbicides through sprinkler systems. Weeds Today 10:7–9.
9. Johnson, A. W., Rohde, W. A., and Wright, W. C. 1982. Soil distribution of fenamiphos applied by overhead sprinkler irrigation to control *Meloidogyne incognita* on vegetables. Plant Dis. 66:489–491.
10. McMaster, G. M., and Douglas, D. R. 1976. Fungicide application through sprinkler irrigation systems. Trans. ASAE 19:1041–1044.
11. Ogg, A. G., Jr., Dowler, C. C., Martin, A. R., Lange, A. H., and Heikes, P. E. 1983. Application of herbicides through

irrigation systems. Ext. Serv. U.S. Dep. Agric. Item AD-FO-2280. 8 pp.
12. Smith, D. T., Berner, R. C., and Walter, J. P. 1973. Nitralin and trifluralin incorporation by rainfall and irrigation. Weed Res. 13:359–366.
13. Sumner, D. R., and Phatak, S. C. 1982. Control of root diseases of snapbean with soil fungicides and metham sodium applied through overhead irrigaton. Pages 61–62 in: Report of Bean Imp. Coop. and Natl. Dry Bean Council Res. Conf., Gainesville, FL, Jan. 5–7.
14. Threadgill, E. D. 1981. Why chemigate? Pages 1–6 in: Proc. Natl. Symp. Chemigation, University of Georgia, Tifton.
15. Wisconsin. 1979. Agriculture Admin. Code NR 112.15.
16. Young, J. R. 1980. Suppression of fall armyworm populations by incorporating insecticides into irrigation water. Fla. Entomol. 73:447–450.
17. Young, J. R., ed. 1981. Proceedings of National Symposium on Chemigation. University of Georgia, Tifton. 126 pp.
18. Young, J. R., and Sumner, D. R., eds. 1982. Proceedings of 2nd National Symposium on Chemigation. University of Georgia, Tifton. 98 pp.
19. Young, J. R., Kiesling, T. C., and Stansell, J. R. 1981. Insecticide application with sprinkler irrigation systems. Trans. ASAE 24:121–123.

Isolating, Identifying, and Producing Inoculum of *Rhizoctonia solani*

G. C. PAPAVIZAS and J. A. LEWIS, Agricultural Research Service, U.S. Department of Agriculture, Soilborne Diseases Laboratory, Plant Protection Institute, Beltsville, MD 20705

Rhizoctonia solani Kühn is a ubiquitous soilborne plant pathogen that causes many types of diseases on numerous plants under diverse environmental conditions (2). The characteristics of the anamorphic state of this pathogen were established by Parmeter and Whitney (23) at the 1965 Miami Symposium entitled "*Rhizoctonia solani*, Biology and Pathology." Talbot (28) established *Thanatephorus cucumeris* (Frank) Donk as the teleomorphic state and treated it as a collective species with "no fundamental systematic importance," pointing out that its nomenclature was unlikely to remain stable. A further breakdown of the anamorph was suggested by a series of investigators (22,24,25,29). At present, *R. solani* is composed of seven known anastomosis groups (AG-1, AG-2, AG-3, AG-4, AG-5, AG-6, AG-B1) (13), but most research has been done on the first four groups (1).

R. solani is known to exist in many, if not most soils (20,21). Some strains of the species are virulent pathogens whereas others exist simply as soil saprophytes. The presence of *R. solani* is usually detected by symptoms on infected plants. The pathogen is seldom recovered on soil dilution plates, but it has been detected with the hyphal isolation method, improved immersion plate, plate-profile method, glass-cloth and nylon-mesh traps, and immersion tubes. These techniques were described by Sinclair (26). In the late 1960s, a technique was described for microscopic examination of debris that passed through a 0.045-mm sieve (U.S. Standard Sieve Series, No. 325) (3). Most of these techniques have not been used extensively in subsequent years, during which so much work has been done to understand the ecology and survival of *R. solani* in soil.

The saprophytic ability and survival of *R. solani* in soil has been well documented (16). Most of the methods of isolation from soil used at present depend for their success on the saprophytic ability of the pathogen and its ability to grow out rapidly from soil or from a colonized substrate. These characteristics of *R. solani* were first exploited by Papavizas and Davey (19–21), who developed a colonization method using buckwheat stem segments to isolate *R. solani* directly from soil. This technique was later modified by the use of table beet seed instead of buckwheat (17). Three other methods—the debris particle method (4), the multiple pellet soil-sampler method (10), and the paper disk method (11)—also depend on the ability of *R. solani* to grow out competitively from soil particles, plant debris particles, or other kinds of traps. These techniques are described in detail.

METHODS OF ISOLATION

Colonization of table beet seed. The colonization method of isolating *R. solani* directly from soil as originally described by Papavizas and Davey (19–21) involved the use of internodal stem segments (5 mm long) of mature buckwheat (*Fagopyrum esculentum* L.). Since buckwheat or other plant materials may not be available readily, table beet seed may be used with the same degree of success (17,18). One gram of autoclaved table beet (*Beta vulgaris* L.) seed, suggested for use by D. J. DeZeeuw of Michigan State University, is mixed with 100 g of soil in 9-cm-diameter petri plates. The soil is brought to 50% of its water-holding capacity before the addition of the seed. After 2 two days of incubation in soil at about 23–26° C, the seeds are recovered on a 1.4-cm sieve (No. 14), washed for 20 min in running tap water, and transferred to petri plates (8–10 seeds per plate) containing 15 ml of 2% water agar (WA). Chlorotetracycline hydrochloride and streptomycin sulfate (50 mg each) are added to a liter of agar after autoclaving and cooling to 50–55° C. One hundred seeds are placed on the medium per replication. After 20–24 hr of incubation at 23–25° C, plates are examined first with the naked eye by holding each plate against a fluorescent light and observing the characteristic branched growth of *R. solani* originating from the colonized seed. The plates are also placed on the stage of a compound microscope, and the areas around the seeds are examined to verify the presence of *R. solani*. The results are expressed as percentage of the seed colonized by *R. solani*. Hyphal-tip isolation of the fungus growing from colonized seeds on the antibiotic medium (AWA) are made directly to potato-dextrose agar (PDA) slants.

After the original report on the colonization method of isolation (19), several reports appeared in the literature describing similar methods. El Zarka (8) used dry, mature stems of Jew's mallow plant (*Corchorus olitorius* L.) to isolate *R. solani* from soil. Sneh et al (27) also used bean and cotton stem segments to isolate *R. solani* from soil and to predict the degree of parasitism from the degree of saprophytism.

In 1962, the colonization method was compared simultaneously with the immersion tube method, the debris particle method, and the infected host method (7). The colonization method gave the most complete and consistent information and the best estimate of activity of *R. solani* in soil. The colonization method in conjunction with the debris particle method gave the most complete information on inoculum density.

Paper disk method. A relatively new and interesting method of isolation based on the saprophytic ability of *R. solani* was described and used by Herr (11). Paper disks (S & S, No. 740E, 6.35 mm in diameter) are immersed in Richard's solution (grams per liter of water: KNO_3, 10; KH_2PO_4, 5; $MgSO_4 \cdot 7H_2O$, 0.25; $FeCl_3$, 0.02; sucrose, 50) containing (micrograms per milliliter): rose bengal, 100; streptomycin sulfate, 100; Cu from cupric sulfate, 5; benomyl, 5 (active ingredient). The disks are dried and stored in a desiccator. Aluminum plates (7.6 × 10.2 cm, 26 gauge) are drilled with three rows of holes (4 mm in diameter), six holes per row, and the disks are placed over the holes and covered on one side with vinyl tape. The plates are autoclaved, inserted into the soil, and withdrawn after 5–7 days. After the excess soil is removed, the disks are placed on AWA in petri plates. The percentage of colonization of the disks is determined after 24–36 hr.

Debris particle method. The method first described by Boosalis and Scharen (4) was later modified slightly by Davey and Papavizas (7). A 100-g sample of soil is suspended in 2.5 L of tap water with a large stirring bar. The suspension is allowed to settle for 30 sec and decanted onto a 0.25-mm sieve (No. 60). The settled soil is resuspended in 1.0 L of tap water with a stirring bar, allowed to settle for 30 sec, and again decanted onto the 0.25-mm sieve. The soil is repeatedly resuspended in 1.0 L of tap water until the supernatant that is decanted onto the screen is clear. This requires five to eight resuspensions. Small forceps are used to remove pieces of wood, straw, roots, lumps of soil, and other foreign objects from the screen. A strong jet of tap water is then run through the screen to remove soil adhering to the organic debris particles. The washed particles are spread between sterile paper towels, blotted dry, and 100 of them are then transferred to 25 petri plates each containing 15 ml of AWA. The plates with the debris particles are incubated at 23–25° C. *Rhizoctonia* spp. are observed growing from the debris particles on the agar, and hyphal tip isolations are made and transferred to PDA slants usually about 48 hr after the debris particles are placed on the agar.

Modifications of the debris particle method. To determine the distribution of *R. solani* in various soil fractions and to isolate the pathogen from such fractions, Papavizas (15) used a dry sieving and sedimentation method. Twelve kilograms of soil in aliquots are passed through a 0.045-mm sieve (No. 325), and the screenings and residue (Res I) are kept separately. Res I is further separated into three fractions (debris particles, Res II, and Res III) by a modification of the debris particle method of Boosalis and Scharen (4). Several 100-g samples of Res I are suspended in 2 L of tap water in a glass jar (28 cm deep and 12 cm in diameter) by means of a large stirring bar. As in the debris particle method, the settled soil is resuspended in water until the supernatant that is decanted onto the sieve is free of debris. After the last suspension, the settled soil—referred to as Res II— is air-dried to about 50% of its water-holding capacity. All of the tap water used for the soil suspensions is collected in large plastic pans and allowed to settle for 5–6 hr. The water is carefully decanted from the pans and centrifuged for 20 min at 16,000 *g*. The silt that settles in the bottom of the pans and the sediment from the centrifuge tubes are combined and air-dried to about 50% of the water-holding capacity. This fraction is referred to as Res III. The organic debris particles collected on the 0.25-mm sieve (No. 60) are washed and plated out on the AWA medium (400 debris particles on 40 petri plates). The pathogen is isolated from the screenings that pass the 0.045-mm sieve and from Res II and Res III by the table beet colonization method. Most of the *R. solani* is found in Res I. In the fractions of Res I, most *R. solani* is found in the debris particles. The fungus is found only rarely in Res II.

Weinhold (30) modified the debris particle method further in order to develop a quantitative assay of *R. solani* in field soil. Soil samples are mixed thoroughly, subsamples removed for moisture determinations, and three 50-g samples are wet-sieved separately through a 0.355-mm mesh sieve (No. 45). The residue on the sieve is washed into a beaker of water and the screenings are kept separately. The residue particles are evenly spread on sterile filter paper by means of a Büchner filter system. Six to eight filter papers are used for 9-cm-diameter plates. Liquid WA (1.0% agar) is cooled to 52° C, acidified with lactic acid to prevent bacterial growth, and poured into petri plates to form a layer 2.5 mm deep. Immediately after dispensing the agar, each filter paper is inverted in the agar and agitated to dislodge and disperse the particles. The plates are incubated at room temperature for 18–24 hr, and suspect colonies are transferred to PDA for identification. A quantitative estimation of *R. solani* can be made by screening known amounts of field soil. The portion that passes through the sieve is tested with the colonization method using cotton stem segments. According to Weinhold (30), the pathogen is rarely detected in soil that passes through the 0.355-mm sieve.

Clark et al (6) described another modification of the soil debris particle method for isolating *R. solani* from soil. The modification involves the use of a semiautomatic elutriator whereby debris particles are collected on a 0.425-mm sieve, cultured on WA, and examined after 24–36 hr. The assay can be conducted simultaneously for nematode detection and isolation.

The multiple-pellet soil-sampler (MPSS) method. Henis et al (10) described the MPSS method for isolating and counting propagules of *R. solani* in soils. The MPSS is a circular plate containing 15 stainless steel tubes 5 mm in diameter and 28 mm in length. Stainless steel pistons are fitted into the tubes. The pistons are held in an upright position by a spring.

Soils to be assayed with the MPSS are adjusted to 12–15% moisture content, and about 30 g is placed in an empty 9-cm petri plate. The surface of the soil is pressed lightly and smoothed with a bent spatula. Soil is pressed into the tubes by tapping the MPSS lightly in the prepared soil sample. The soil surface at the tips of the tubes is leveled off with a sterile spatula, and the loaded sampler is placed on a support that is fitted over a 9-cm petri plate bottom containing an agar medium. By pressing the knob two to four times, uniform soil pellets are delivered on the agar surface (four plates per replication, 15 pellets each). Additional pellet samples are dried at 30–32° C for 24 hr to determine the average dry weight of the pellet samples. The sampler tubes are dipped in ethyl alcohol and flamed between samplings. For a culture medium, Henis et al (10) used a medium developed by Ko and Hora (12). We routinely use the AWA medium with the MPSS as described in the table beet colonization method. The AWA medium is simple to prepare and does not contain fungicides.

The plates with the pellets are incubated at room temperature for 18–24 hr, and the pellets are examined for the presence of typical *R. solani* mycelium. Isolations can be transferred into PDA slants. The number of *R. solani* propagules per gram of soil is estimated from the percentage of pellets colonized by applying the first order of the Poisson distribution (10):

$$\log_e \frac{1}{1-X},$$

where X is the proportion of the pellets colonized. This transformation depends on the assumption that the propagules are distributed randomly in the soil to be assayed. Propagules of *R. solani*, however, may not be distributed randomly in naturally infested soils. Random distribution may be assumed only for artificially infested soils.

METHODS OF IDENTIFICATION

Species identification. To accurately identify *R. solani*, one should develop a complete picture and understanding of the species concept of the pathogen. The species concept with reference to the anamorphic and teleomorphic states was described by Parmeter and Whitney (23) and Talbot (28), respectively. Parmeter and Whitney said that "it is difficult if not impossible to describe a mycelium with absolute assurance that another worker can identify that mycelium from the written description. It is possible only to provide a description with sufficient detail as to exclude most other mycelia." For identification purposes, the unknown fungus is cultured on PDA, WA, or other media in petri plates or coated glass slides. Slides are placed on the stage of a compound microscope and examined.

According to Parmeter and Whitney (23), *R. solani* is identified by its pale to dark brown, rapidly growing mycelium of relatively large diameter (usually greater than 5 μm) with branching near the distal septum of cells in young vegetative hyphae, constriction of the branch hyphae at the point of origin, and formation of a prominent septal pore apparatus in the branch near the point of origin. Young vegetative hyphae are multinucleate, and the pathogen possesses a basidiomycetous teleomorphic state (not always easy to obtain in culture) referred to as *Thanatephorus cucumeris* by Talbot (28). *R. solani* never develops conidia, rhizomorphs, or clamp connections, and it never has pigments other than brown. Several isolates may be pathogenic to a wide variety of hosts; isolates may develop sclerotia (without differentiated rind and medulla) or monilioid cells.

Subspecific divisions and identification. First, reference strains of confirmed anastomosis groups are obtained and grown on PDA in petri plates. The unknown isolates of *R. solani* are tested for anastomosis with the tester strains by a method described by Parmeter et al (22). The unknown isolates also are grown on PDA and opposed with the tester strains on WA (2% agar) or on cellophane resting on 2% WA in 9-cm petri plates by transferring mycelia from the margins of actively expanding, young cultures on PDA and placing them on the agar or cellophane 2–4 cm apart. Usually one pair of isolates is tested per plate. The plates are incubated at 22–26°C until advancing hyphae from both members of the pair make contact and slightly overlap. Depending on the anastomosis group, this may take 24–48 hr. Hyphal growth may be promoted by dipping the cellophane strips in melted PDA (13 g of PDA per liter of water) before transferring to the WA (5). A 2- to 3-cm portion of the agar or the cellophane where contact is made is removed, placed on a microscope slide, stained with 0.001% (w/v) cotton blue in dilute lactophenol (lactophenol:water, 1:9, v/v) or phloxine B in water, and examined at 500× for hyphal fusions. If an unknown isolate of *R. solani* fuses cytoplasmically with anastomosis group 4 (AG-4), then it belongs to AG-4, and so on. Most of the subspecies identifications and genetic research have been done with AG-1, AG-2, AG-3, and AG-4. Castro et al (5) recently developed a new method for identifying anastomosis group 3 (AG-3) by colony color on Stewart's medium without the use of hyphal anastomosis as the criterion of identification.

METHODS FOR PRODUCTION OF INOCULUM

Although many methods for production of inoculum of *R. solani* have been described, quantitative methods have yet to be developed and comparisons of existing methods have not been made. Development of quantitative methods is extremely difficult with a fungus lacking asexual spores. We selected three methods for inoculum production based on simplicity of preparation and repeated use and description in published articles.

Sand-cornmeal. This method, used by Papavizas and Davey (21) and by many other scientists, is simple but does not allow for an accurate estimation of the mycelial and sclerotial biomass to be added to soil. Washed quartz sand is passed through a 2-mm sieve, mixed with ground cornmeal and water (96% sand + 4% cornmeal; water to 18–20%, v/w), and added to 300-ml, wide-mouth storage dishes. The preparations are autoclaved for 60 min, seeded with mycelial agar chunks of *R. solani*, and incubated at 23–26°C for 18 days. The inoculum is mixed thoroughly and added to soil at rates from 0.25 to 2% (w/w). Similar preparations may include sand-oatmeal, soil-oatmeal, or sand-barley meal. The method can be adapted easily to obtain small-scale field infestations. Seven-kilogram batches of the sand-cornmeal mixture are autoclaved for 70 min in polypropylene pans (12.5 × 23 × 43 cm), infested with a ground culture of *R. solani* (grown in 250 ml of potato-dextrose broth), and incubated at room temperature for 4–5 weeks. The inoculum is mixed and spread in the field at a rate of 0.5 kg/m^2 and incorporated by disking it to a depth of 7.5 cm (14).

Whole grain. Gaskill (9) developed a whole-grain method. Approximately 520 ml of dry, whole barley grain and 300 ml of water are placed in a 1-L flask, stirred, and allowed to stand 24 hr. The mixture is stirred again and autoclaved for 2 hr. The grain is infested in two places using mycelial agar chunks, and the cultures are incubated for 3 weeks on a bench without special temperature control. The cultures are air-dried and ground in a mill to pass through a 3-mm sieve. The ground inoculum is mixed, placed in paper bags, and stored at 2–4°C. Inoculum should be used within 2 weeks, but it remains highly pathogenic for more than a year. In addition to barley, other grains such as oats, sorghum, corn, wheat, and millet can be used with various amounts of water, length of incubation, and incorporation rates. Barley inoculum is added in the field at 14–56 kg/ha.

Soil-potato. Ko and Hora (12) used potato pieces to produce inoculum. A fine sandy loam is mixed with chopped potato pieces (10:1, w/w), moistened to about 50% of its moisture-holding capacity, and autoclaved. The soil-potato mixture is infested with mycelial agar chunks and incubated for 7 days at 26°C. Infested soils are prepared by mixing different ratios of inoculum and natural soil as desired. The ratio of sand to chopped potato pieces may be decreased (4:1, w/w), and incubation may extend to 20 days. The inoculum may be dried, passed through a 1-mm sieve, and stored in a gauze-covered flask at room temperature.

LITERATURE CITED

1. Anderson, N. A. 1982. The genetics and pathology of *Rhizoctonia solani*. Annu. Rev. Phytopathol. 20:329-347.
2. Baker, K. F. 1970. Types of Rhizoctonia diseases and their occurrence. Pages 125-148 in: *Rhizoctonia solani*: Biology and Pathology. J. R. Parmeter, Jr., ed. University of California Press, Berkeley.
3. Baker, K. F., Flentje, N. T., Olsen, C. M., and Stretton, H. M. 1967. Effect of antagonists on growth and survival of *Rhizoctonia solani* in soil. Phytopathology 57:591-597.
4. Boosalis, M. G., and Scharen, A. L. 1959. Methods for microscopic detection of *Aphanomyces euteiches* and *Rhizoctonia solani* and for isolation of *Rhizoctonia solani* associated with plant debris. Phytopathology 49:192-198.
5. Castro, C., Davis, J. R., and Wiese, M. V. 1983. Differential medium for identification of *Rhizoctonia solani* AG-3. Plant Dis. 67:1069-1071.
6. Clark, C. A., Sasser, J. N., and Barker, K. R. 1978. Elutriation procedures for quantitative assay of soils for *Rhizoctonia solani*. Phytopathology 68:1234-1236.
7. Davey, C. B., and Papavizas, G. C. 1962. Comparison of methods for isolating *Rhizoctonia* from soil. Can. J. Microbiol. 8:847-853.
8. El Zarka, A. M. 1963. A rapid method for the isolation and detection of *Rhizoctonia solani* Kühn from naturally infested and artificially inoculated soils. Meded. Landbouwhogesch. Opzoekingsstn. Staat Gent 28:877-885.
9. Gaskill, J. O. 1968. Breeding for Rhizoctonia resistance in sugarbeet. J. Am. Soc. Sugar Beet Technol. 15:107-119.
10. Henis, Y., Ghaffar, A., Baker, R., and Gillespie, S. L. 1978. A new pellet soil-sampler and its use for the study of population dynamics of *Rhizoctonia solani* in soil. Phytopathology 68:371-376.
11. Herr, L. J. 1973. Disk-plate method for selective isolation of *Rhizoctonia solani* from soil. Can. J. Microbiol. 19:1269-1273.
12. Ko, W., and Hora, F. K. 1971. A selective medium for the quantitative determination of *Rhizoctonia solani* in soil. Phytopathology 61:707-710.
13. Kuninaga, S., and Yokosawa, R. 1980. A comparison of DNA base composition among anastomosis groups in *Rhizoctonia solani* Kühn. Ann. Phytopathol. Soc. Jpn. 46:150-158.
14. Lewis, J. A., and Papavizas, G. C. 1980. Integrated control of Rhizoctonia fruit rot of cucumber. Phytopathology 70:85-89.
15. Papavizas, G. C. 1968. Survival of root-infecting fungi in soil. VIII. Distribution of *Rhizoctonia solani* in various physical fractions of naturally and artificially infested soils. Phytopathology 58:746-751.
16. Papavizas, G. C. 1970. Colonization and growth of *Rhizoctonia solani* in soil. Pages 7-19 in: *Rhizoctonia solani*: Biology and Pathology. J. R. Parmeter, Jr., ed. University of California Press, Berkeley.
17. Papavizas, G. C. 1973. Saprophytic activity of *Rhizoctonia solani* and *Sclerotium rolfsii*. Pages 13-17 in: The Relation of Soil Microorganisms to Soilborne Plant Pathogens. G. C. Papavizas, ed. Southern Cooperative Ser. Bull. 183. Virginia Polytechnic Institute and State University, Blacksburg.
18. Papavizas, G. C., Adams, P. B., Lumsden, R. D., Lewis, J. A., Dow, R. L., Ayers, W. A., and Kantzes, J. G. 1975. Ecology and epidemiology of *Rhizoctonia solani* in field soil. Phytopathology 65:871-877.
19. Papavizas, G. C., and Davey, C. B. 1959. Isolation of *Rhizoctonia solani* Kuehn from naturally infested and artificially inoculated soils. Plant Dis. Rep. 43:404-410.
20. Papavizas, G. C., and Davey, C. B. 1961. Saprophytic behavior of *Rhizoctonia* in soil. Phytopathology 51:693-699.
21. Papavizas, G. C., and Davey, C. B. 1962. Isolation and pathogenicity of *Rhizoctonia* saprophytically existing in soil. Phytopathology 52:834-840.
22. Parmeter, J. R., Jr., Sherwood, R. T., and Platt, W. D. 1969. Anastomosis grouping among isolates of *Thanatephorus cucumeris*. Phytopathology 59:1270-1278.
23. Parmeter, J. R., Jr., and Whitney, H. S. 1970. Taxonomy and nomenclature of the imperfect state. Pages 7-19 in: *Rhizoctonia solani*: Biology and Pathology. J. R. Parmeter, Jr., ed. University of California Press, Berkeley.
24. Richter, H., and Schneider, R. 1953. Untersuchungen zur morphologischen und biologischen Differenzierung von *Rhizoctonia solani* K. Phytopathol. Z. 20:167-226.
25. Schultz, H. 1937. Vergleichende Untersuchungen zur Ökologie, Morphologie, und Systematik des "Vermehrungpilzes." Arb. Biol. Reichsanst. Land Forstwirtsch. Berlin-Dahlem 22:1-41.
26. Sinclair, J. B. 1970. *Rhizoctonia solani*: Special methods of study. Pages 199-217 in: *Rhizoctonia solani*: Biology and Pathology. J. R. Parmeter, Jr., ed. University of California Press, Berkeley.
27. Sneh, B., Katan, J., Henis, Y., and Wahl, I. 1966. Methods for evaluating inoculum density of *Rhizoctonia* in naturally infested soil. Phytopathology 56:74-78.
28. Talbot, P. H. B. 1970. Taxonomy and nomenclature of the perfect state. Pages 20-31 in: *Rhizoctonia solani*: Biology and Pathology. J. R. Parmeter, Jr., ed. University of California Press, Berkeley.
29. Watanabe, B., and Matsuda, A. 1966. Studies on the grouping of *Rhizoctonia solani* Kühn pathogenic to upland crops. Designated Exp. (Plant Dis. Insect Pests) 7, Agric. For. Fish. Res. Counc. and Ibaraki Agric. Exp. Stn. (In Japanese with English summary)
30. Weinhold, A. R. 1977. Population of *Rhizoctonia solani* in agricultural soils determined by a screening procedure. Phytopathology 67:566-569.

Isolating, Identifying, and Producing Inoculum of Pathogenic Species of *Fusarium*

PAUL E. NELSON and T. A. TOUSSOUN, Fusarium Research Center, Department of Plant Pathology, Pennsylvania State University, University Park, PA 16802; L. W. BURGESS, Department of Plant Pathology and Agricultural Entomology, University of Sydney, Sydney, N.S.W., Australia 2006; W. F. O. MARASAS, National Research Institute for Nutritional Diseases, Tygerberg 7505, Republic of South Africa; and C. M. LIDDELL, Department of Plant Pathology and Agricultural Entomology, University of Sydney, Sydney, N.S.W., Australia 2006

Fusarium spp. are a widespread group of fungi that consist of both saprophytes and plant pathogens. They are common in soil and on aerial and subterranean plant parts, in which they may act either as primary or secondary colonizers. It is a rare occurrence when some species of *Fusarium* is not found in a soil used for crop production. These fungi also occur in soil in many noncultivated areas, and they are found in such contrasting areas and climates as deserts and arctic regions. In addition to being plant pathogens, some are pathogens of animals and humans, and many are important in the production of mycotoxins in food and feedstuffs.

The techniques described below are taken primarily from two recent publications (2,18) that provide the reader with general information on techniques for handling all *Fusarium* species. Many of these techniques were developed at the Fusarium Research Center, Pennsylvania State University (18), and the Fusarium Research Laboratory, University of Sydney, Australia (2). Individual papers are cited where appropriate for special techniques.

ISOLATION OF *FUSARIUM* SPECIES FROM DIFFERENT SUBSTRATES

Culture media. *Natural media.* Isolation of *Fusarium* spp. from plant tissue should be done on a medium low in nutrients. Natural media such as those recommended by Hansen and Snyder (9) and Snyder and Hansen (26) are excellent for this purpose. They used plant materials such as wheat straw, barley straw, pea straw, and seeds embedded in water agar. Carnation-leaf agar may also be used (5).

Selective media. Nash and Snyder (16) devised an excellent medium for the direct isolation of *Fusarium* spp. from plant tissue and freshly collected field soils. The formula for modified Nash-Snyder medium is 15.0 g of Difco peptone, 1.0 g of KH_2PO_4, 0.5 g of $MgSO_4 \cdot 7H_2O$, 20.0 g of agar, 1.0 g of pentachloronitrobenzene (PCNB; Terraclor), and 1.0 L of water. The medium is adjusted to pH 5.5–6.5 and autoclaved, and 20 ml of a streptomycin sulfate stock solution and 12 ml of a neomycin sulfate stock solution are added to each liter after the medium has cooled and just before being poured into petri dishes. The stock solutions are prepared by adding 5 g of streptomycin sulfate to 100 ml of water and 1 g of neomycin sulfate to 100 ml of water. It is best to allow the medium to dry in the petri dishes for 5 days before use.

This medium is useful in isolating *Fusarium* spp. from plant material, especially material that is badly rotted or infested with fast-growing contaminants. As the fungus colonies develop on this medium, large amounts of ammonia are given off, which may kill the fungi if they are left on it for more than 20–30 days.

Another useful medium, Selective Fusarium Agar, is a modified Czapek-Dox medium (2,28). The formula for this medium is 20.0 g of dextrose, 0.5 g of KH_2PO_4, 2.0 g of $NaNO_3$, 0.5 g of $MgSO_4 \cdot 7H_2O$, 1.0 g of yeast extract (Vegemite or Marmite may be substituted), 1.0 ml of 1% ferrous sulfate solution, 20.0 g of agar, and 1.0 L of water. The following components are added from sterile stock solutions after the medium has been autoclaved and cooled to 48°C: 5.0 ml of 1% suspension (0.05 g/L) of 50% (w/w) 2,6-dichloro-4-nitroaniline (Botran), 0.1 g of streptomycin sulfate, and 0.01 g of aureomycin sulfate. Aureomycin sulfate is dissolved in 95% ethyl alcohol and further diluted in water before addition to the medium. This antibiotic is difficult to dissolve if added directly to the medium. The antibiotics are added from stock solutions to each liter after the medium has cooled and just before pouring. The medium should be stored in a cool, dark place.

This medium is used for the isolation of *Fusarium* spp. from plant roots and soil debris. It has also been used for isolating *F. graminearum* from soil using the dilution plate technique. It is less inhibitory to other microorganisms than the Nash-Snyder medium described above, but it is a very effective selective medium that allows the development of distinctive colonies by some *Fusarium* spp. Colonies of *Fusarium* spp. show more color on this medium than on the Nash-Snyder medium.

A relatively nonselective, low-nutrient medium, called

Reprinted, with modifications for the purposes of this publication, from *Fusarium Species: An Illustrated Manual for Identification*, by Paul E. Nelson, T. A. Toussoun, and W. F. O. Marasas, The Pennsylvania State University Press, University Park, Pennsylvania, 1983, by permission from The Pennsylvania State University Press, and from *Laboratory Manual for Fusarium Research*, by L. W. Burgess and C. M. Liddell, Department of Plant Pathology and Agricultural Entomology, The University of Sydney, New South Wales, Australia, 1983, by permission from the Department of Plant Pathology and Agricultural Entomology, The University of Sydney.

Wayd medium, has been used successfully for the isolation of *Fusarium* spp. and many other soil fungi from sorghum roots and soil debris at the University of Sydney (S. Jeffery, F. Hanson, and L. W. Burgess, *personal communication*). The formula is 10.0 g of dextrose, 1.0 g of yeast extract, 0.01 g of aureomycin sulfate, 0.1 g of streptomycin sulfate, 20.0 g of agar, and 1.0 L of water. The aureomycin sulfate is dissolved in 95% ethyl alcohol and further diluted in water before addition to the medium. If bacteria are a problem as contaminants, the amount of streptomycin sulfate used should be increased to 1.0 g/L. This medium has also been used successfully without the dextrose. However, fungus colonies are more difficult to differentiate because of slow growth and poor development of color in the colonies.

This medium shows considerable promise for use in isolation studies of *Fusarium* species and other fungi from roots and soil debris. The low level of nutrients permits more accurate differentiation of colonies earlier than is possible on water agar. The excessive growth by fast-growing fungi that occurs on carbohydrate-rich media such as PDA—which can conceal slow-growing species—does not occur.

Isolation from plant material. The most common method for isolating *Fusarium* spp. from plant material is by plating colonized material on agar media. The tissue selected should be typical of the diseased material under study. Portions of tissue that show extensive advanced necrosis should be avoided because this tissue is more likely to be colonized by secondary organisms. Tissue samples should be treated with a surface disinfestant to reduce the numbers of secondary organisms that may grow out on the agar medium and interfere with the isolation of the primary organism. Sodium hypochlorite (Clorox) is a suitable chemical disinfestant to use on most tissue. Concentrations of sodium hypochlorite used range from 1 to 5%, and the time of treatment varies with the nature of the tissue. Treated tissue should be damp-dried on absorbent sterile paper toweling before being placed on agar to reduce the amount of bacterial contamination. Fine roots or feeder rootlets that are too small to be surface disinfested may be washed in running tap water for several hours and then rinsed in several changes of sterile water, damp-dried, and placed on the agar medium.

Several other techniques are available for use in isolating *Fusarium* spp. from plant material. If the fungus is producing sporodochia on the surface of the plant tissue, the sporodochia can be removed and used to prepare a spore suspension in sterile water. The *Fusarium* spp. in question may be isolated using the single-spore technique for obtaining pure cultures.

If fertile perithecia are being produced on the plant tissue, a small piece of tissue with perithecia is well washed, the excess water removed, and the piece of tissue placed on the inner side of an inverted petri dish containing water agar or carnation-leaf agar; the inverted petri dish is then incubated under conditions of high humidity. The tissue piece may be held in place with Vaseline or similar material. After incubation for about 24 hr, ascospores are released and land on the agar surface where they germinate and produce colonies. The fungus can be transferred to other media by a single-spore transfer or by means of a hyphal tip.

Fusarium spp., especially slow-growing ones, may be difficult to isolate directly from necrotic root tissue that is extensively colonized by other fungi and bacteria. However, these species can sometimes be isolated using a combination baiting and plating technique. After the necrotic tissue is washed thoroughly and cut into small pieces, the pieces are mixed with steam/air-treated soil. Seeds of the plant involved are planted in this mixture and incubated under conditions favorable for disease development. The plants are sampled as soon as root lesions develop, and the root lesion is plated on suitable media after thorough washing or surface disinfestation. This procedure has been used successfully to isolate *F. avenaceum* from old necrotic roots of subterranean clover (3).

Isolation from soil. *Fusarium* spp. may be isolated directly from soil either by the dilution plate method or by plating organic debris on selective media; they may be isolated indirectly from soil by using a root-baiting technique or sterile baits such as pieces of straw (2,3,15,28).

Soil dilution plate technique. The soil sample is ground in a mortar and diluted 1:1,000 to 1:10,000 in 0.1% water agar. This dilution range is satisfactory for most soils, but in some cases it may be necessary to run a preliminary test to find the most suitable dilution rate. In some soils from tropical and semitropical areas, dilutions of 1:50 to 1:500 are satisfactory. A 1-ml sample of the soil suspension is dispersed uniformly over the surface of a selective medium in a petri dish. Propagules in the soil suspension usually germinate in a few days and produce small colonies in a week. The dilution plates should be incubated in the light to ensure sporulation. It is important that the soil be air-dried before the suspension is prepared to reduce bacterial contamination. The agar medium should also be allowed to dry for 5–7 days before use for the same reason.

Both selective media described earlier can be used for direct isolation from soil suspension. The Nash-Snyder medium (16) suppresses most contaminants, but the colonies of *Fusarium* spp. do not show vivid color on it. Colonies of *Fusarium* spp. on Selective Fusarium Agar (2) exhibit more vivid colors, but this medium does not suppress contaminants as well as the Nash-Snyder medium. Komada's (12) medium is useful for isolating *F. oxysporum* but may suppress the growth of some other *Fusarium* spp. The formula for the basal medium is 1.0 g of K_2HPO_4, 0.5 g of KCl, 0.5 g of $MgSO \cdot 7H_2O$, 0.01 g of Fe-Na-EDTA, 2.0 g of L-asparagine, 20.0 g of D-galactose, and 1.0 L of water. When the basal medium is melted and cooled, the following antimicrobial supplement is added and mixed thoroughly: 1.0 g of PCNB (75% W.P.), 0.5 g of oxgall, 1.0 g of $Na_2B_4O_7 \cdot 10H_2O$, and 0.3 g of streptomycin sulfate. The pH is adjusted to 3.8 ± 0.2 with an approximately 10% solution of phosphoric acid.

Debris isolation technique. A technique has been developed at the Fusarium Research Laboratory, University of Sydney, for the isolation of *Fusarium* spp. from small pieces of debris from soils (2,28). This technique consists of washing the soil sample through a nest of three sieves, respectively 4.0 mm, 2.0 mm, and 0.5 mm in aperture. The first sieve retains identifiable plant remains, such as pieces of roots and crowns that can be surface disinfested and plated on a selective medium. The smaller pieces of debris retained on the other two sieves are also plated on a selective medium. Since the small pieces of debris are porous and cannot be surface sterilized easily, the sieves with retained debris are placed under a fine spray of filtered tap water for 2 hr or until the soil adhering to the debris has been removed. The debris is allowed to dry on sterile paper toweling before plating. It can also be air-dried before plating to further reduce bacterial contaminants or dried over silica gel before plating. Colonies of *Fusarium* spp. develop in 5–7 days; a wide variety of *Fusarium* spp. can be isolated with this technique, which has been evaluated by McMullen and Stack (15). The Nash-Snyder medium (16), Selective Fusarium Agar (2,28), and the Wayd medium can be used.

Isolation from the atmosphere. *Fusarium* spp. can be isolated from the atmosphere by any method that enables the collection of living propagules. A number of these methods have been used by aerobiologists for many years, but they have rarely been applied to the study of *Fusarium* aerobiology. None of the methods described below impair the viability of the trapped propagule, and they thus permit the recovery and identification of the living fungus. All of these methods have been used successfully to isolate *Fusarium* spp., but they vary in efficiency.

Exposed petri dish technique. This is the simplest and most widely used technique. Petri dishes containing a selective agar medium are exposed for varying amounts of time (5–60 min) depending on the concentration of conidia and wind conditions. Nelson et al (17) used Nash-Snyder medium to recover *Fusarium* spp. from the greenhouse atmosphere and adjacent areas. The petri dish should be located in an open-sided shelter if the exposure is made in direct sunlight or rain, and care should be taken not to overexpose the petri dish. This technique is qualitative, of variable efficiency, and selective with respect to spore size. It does allow recovery of the fungi for accurate identification and is simple and inexpensive. One of the authors (LWB) has used this technique successfully in semiarid and desert areas but found that flies and other insects are attracted to the medium and may contaminate it during the exposure period. Ants may also be a problem if the petri dish is exposed close to the ground.

Spore traps. Several spore traps have been developed to compensate for the shortcomings of the exposed petri dish technique, and two of these have been used to trap *Fusarium* spp. The Andersen spore sampler (1) is nonselective with respect to spore size, and it is designed to sort the propagules according to size. It is volumetric and permits collection directly onto agar plates. Ooka and Kommedahl (20) used an Andersen spore sampler fitted with Nash-Snyder medium in petri dishes to trap propagules of *F. moniliforme*.

The Hirst spore trap (10) is designed to capture airborne particles on a sticky microscope slide. Isolations can be made from the trapped spores if the slide is covered with acetate tape or agar that is subsequently placed on a selective medium. This trap also allows the trapped spores to be examined under the microscope before the propagules are cultured. It is volumetric, nonselective, and very efficient under most conditions. Lukezic and Kaiser (14) used a Hirst trap to detect conidia of *F. semitectum* (*F. roseum* 'Gibbosum').

Various other spore traps have been designed over the years, some of which may be suitable for collecting *Fusarium* spp., but a description of these devices is beyond the scope of this article. Interested readers should consult publications by Davies (4) and Gregory (8) for additional information. Two other spore-trapping devices are worthy of mention because of their special features. Ogawa and English (19) described a cyclone collector that samples large volumes of air into a liquid trap and is suited for use when spore concentrations are low. The most important feature of this collector is the principle of operation, which can be readily adapted to suit many experimental situations. Collection may be made into a single volume of liquid or, more usefully, into graduated aliquots that are changed according to a time base. The spore suspension can then be cultured using the dilution plate method and an appropriate medium. Schwarzbach (23) described a collector that is very efficient and captures spores on media in petri dishes or living plant tissue. It is nonselective, small, and simple to operate.

Any attempt to isolate *Fusarium* spp. from the atmosphere involves selecting a method of collecting propagules that suits the objectives of the experiment. In many cases, the exposed petri dish technique is satisfactory; but in quantitative studies one must use a well-designed collector like one of those described above, realizing the limitations and advantages of each design.

GROWING *FUSARIUM* SPECIES FOR IDENTIFICATION

Members of the genus *Fusarium* are variable in culture because changes in the environment in which they grow cause morphological changes in both the culture and conidia. Since the morphology of the macroconidia is the primary basis for identification, all possible steps should be taken to standardize procedures to make the task of identification easier. In the following paragraphs we deal with the techniques, culture media, and environmental conditions necessary to accomplish this task.

Culturing methods. *Single-spore method.* This technique was devised by H. N. Hansen and modified by others, and these publications should be consulted for additional information about the methods (2,18). Basically the method consists of pouring 3 ml of 2% water agar into unscratched petri dishes and allowing the agar to solidify. A suspension of conidia is prepared in a 10-ml sterile water blank so that it contains 1–10 conidia per low-power (10×) microscope field when a drop from a 3-mm-diameter loop is examined on a slide. This suspension of conidia is poured over the solidified agar so that the entire surface is covered, and the excess is drained off. The petri dishes are then incubated in an inclined position at room temperature (18–24°C) for 16–24 hr, after which they are opened, shaken to remove any accumulated moisture, and examined under a dissecting microscope. Small squares of agar containing single germinating conidia are cut out with a suitable needle and transferred to the desired growth medium. If the original culture is contaminated with bacteria, a drop of 25% lactic acid may be added to the 10-ml sterile water blank to inhibit bacterial growth. The acid spore suspension should be allowed to stand for 10 min before being poured on agar in a petri dish. This technique may delay germination of conidia of some *Fusarium* spp. for 24 hr or more.

Hyphal-tip method. Mutant colonies may develop from single conidia taken from sporodochial cultures. New sporodochial cultures often can be obtained from such cultures from hyphal tips from the original colony (2,18). This does not mean a mass transfer from the growing edge of the colony, but rather the transfer of a single hyphal tip, under the dissecting microscope, in the same manner as one would transfer a single germinating conidium. Mycelium from the sporodochial parent culture is used to initiate a colony on a medium low in nutrients, such as water agar. Only enough water agar is poured into the petri dish to form a very thin layer so that a sparse thallus of the fungus develops. This allows hyphal tips to be removed without difficulty and used to initiate a new colony.

Culture media. *Potato-dextrose agar (PDA).* This is a valuable medium principally for gross morphological appearances and the coloration of colonies. *Fusarium* spp. show their full diversity and color on this medium. Because of its high available carbohydrate content, PDA generally emphasizes growth over sporulation. Cultures grown on PDA may sporulate poorly, frequently taking more than a month to do so, and the conidia produced are often misshapen and atypical. Consequently, except for microconidia produced by species in the section Sporotrichiella, cultures grown on PDA are not used for microscopic observations. Cultures grown on PDA are used only in a secondary role and are never used for long-term storage of cultures or for preparation of cultures for lyophilization.

Mutations also increase when *Fusarium* spp. are grown on PDA and similar media. Mass transfers of single conidia from such cultures may gradually sector and mutate, or the changes may occur very suddenly. These changes are the principal reason for the degeneration of laboratory cultures over time. They make identification of *Fusarium* spp. difficult because culture morphology, types of conidia produced, and, in extreme cases, even the production and morphology of the macroconidia are altered. The problem may be reduced by keeping subculturing to a minimum, by using the single-spore or hyphal-tip techniques, and by not subculturing or storing *Fusarium* spp. on media high in carbohydrates, such as PDA.

Potato-dextrose agar (18) should always be prepared from raw ingredients rather than by using any of the commercial preparations available. Red-skinned potatoes should not be used. Baking grade, white-skinned potatoes are washed and sliced, unpeeled, and 250 g of potatoes is added to 500 ml of water and placed in an autoclave along with a flask containing 20 g of agar in 500 ml of water. The potatoes are cooked and the agar melted by operating the autoclave on steam bypass for 45 min at 3.6 kg of pressure. The potato broth is strained through several layers of cheesecloth into the flask containing melted agar. The remaining potato pulp is squeezed through several layers of cheesecloth until 114 ml of potato pulp is obtained; this pulp is added to the melted agar and potato broth along with 20 g of dextrose. If necessary, the total amount is brought up to 1 L by adding water, and all of the ingredients are mixed thoroughly. The medium can then be dispensed into test tubes, autoclaved, and slanted. Properly made, each tube should have a small button of sediment at its base.

Carnation-leaf agar (CLA). We base our identification procedures almost entirely on CLA for the reasons given in the previous paragraphs. The great advantage of CLA is that it promotes sporulation rather than mycelial growth. Conidia and conidiophores are produced in abundance, their morphology closely resembles that seen under natural conditions, and phenotypic variation is reduced. All of the macroconidia in a given microscope field will be quite similar, and this uniformity makes the work of identification simpler and avoids many difficulties. CLA is a valuable growth medium for *Fusarium* spp. because it is low in available carbohydrates and contains complex, naturally occurring substances of the type encountered by *Fusarium* spp. in nature. Consequently, the fungi grow and sporulate in a manner similar to that found on a host plant or natural substrate. Direct microscopic observation of a *Fusarium* spp. growing on CLA may also indicate the manner in which conidia are borne on conidiophores. This medium also favors the development of perithecia of some homothallic *Fusarium* spp. (30).

Carnation leaves are eminently suitable for growth and sporulation of *Fusarium* spp., and they should be used whenever possible. Other plant materials may be used provided they are sterilized and handled in the same manner. We have found that leaves and stems of corn, wheat, alfalfa, and various other grasses can be used. Seeds are not recommended because they are generally too high in available carbohydrates and are more difficult to sterilize. It is best to be consistent in the use of a given substrate, as different plant materials may modify the shape of the conidia.

The preparation of carnation leaves and CLA are described by Fisher et al (5). Young carnation (*Dianthus caryophyllus* L.) leaves are harvested from actively growing, disbudded plants free from pesticide residues. The leaves are cut into pieces measuring approximately 5 mm^2 and dried in an oven at 45–55°C for 2 hr. When dry, the leaves are green and crisp. Loss of green pigmentation indicates that the drying temperature was too high. The leaf pieces are placed in aluminum canisters 5 cm deep and 9 cm in diameter and sterilized with 2.5 megarads of γ-irradiation from a cobalt 60 source. Propylene oxide fumigation may be used as another method of sterilization (9), but sterilization of the green leaves is not as thorough as with γ-irradiation, and repeated fumigations may be required.

Carnation-leaf agar is prepared by placing several sterile leaf pieces in a petri dish or culture tube and floating them on 1.5–2% water agar cooled to 45°C (5). The medium is left at room temperature for 3–4 days before use to check for growth of possible contaminants from the leaf pieces.

KCl medium. This medium is used to facilitate the observation of the formation of microconidia in chains by *Fusarium* spp. in the section Liseola (6). It is prepared by adding 4–8 g of KCl to 1 L of 1.5% water agar. Those species that form chains of microconidia form more abundant, longer chains on this medium.

Formation of conidia. *Fusarium* spp. produce macroconidia in sporodochia and in the aerial mycelium. Macroconidia in sporodochia are produced on monophialides. Depending on the species, macroconidia and microconidia in the aerial mycelium are produced on monophialides or on polyphialides. Although macroconidia produced in sporodochia are highly diagnostic, the observation of the structures borne in the aerial mycelium is also vital to species determination in some cases. A complete discussion of the types of conidia and conidiogenous cells in *Fusarium* is given by Nelson et al (18).

Conditions for growth and sporulation. Sporulation and pigmentation in cultures of *Fusarium* spp. are favored by light, including ultraviolet wavelengths, and by fluctuating temperature conditions (2,26,31). If possible, all cultures should be incubated in an alternating temperature of 25°C by day and 20°C by night (2). Although fluctuating temperatures are best, *Fusarium* spp. grow well at constant temperatures of 21–22°C (18). Cultures are incubated in diffuse daylight from a north window or in light from fluorescent tubes. We have found that F4055/SX fluorescent tubes (Consumer Lighting Products, Inc., P.O. Box 5760, Baltimore, MD 21208) are excellent substitutes for natural light. Two of these tubes in an ordinary 40-W fluorescent fixture are suspended 40–45 cm above the laboratory bench or shelf supporting the cultures. A day length of 12 hr is sufficient. Cultures that do not sporulate readily may be placed under a fixture containing one 40-W black light tube and one of the 40-W fluorescent tubes described above to enhance sporulation. This light combination also enhances the formation of the perfect state in culture. A useful mobile light bank for growing cultures has been described by Burgess and Liddell (2).

Cultural mutation. The majority of *Fusarium* spp. isolated from nature produce their macroconidia on sporodochia. The sporodochial type often mutates in culture and occasionally in nature. These mutants may give rise to others, creating a mutational sequence. In pathogenic isolates the mutants frequently exhibit a loss in virulence, and loss of toxin production may also occur. Variability and its effect on virulence and taxonomy have been discussed in detail by several workers (18,21,24,27). The mutational sequence has never been shown experimentally to reverse itself. Starting with the sporodochial type, mutation in general proceeds toward forms producing abundant aerial mycelium but few macroconidia, termed *mycelial types,* and toward forms producing little or no aerial mycelium but abundant macroconidia, termed *pionnotal types*. Mycelial types often lack sporodochia and pigmentation and thus have a white, featureless look. The colonies of pionnotal types are often more highly colored than the parent type and have a wet, shiny appearance because the macroconidia are borne in sheets over the surface of the colony. The macroconidia of pionnotal types may be longer and thinner or shorter than those of the parental types.

Mutation can be avoided by observing the following rules: initiate cultures by the single-spore or hyphal-tip method, avoid media rich in carbohydrates, keep subculturing to a minimum, and store for long periods by lyophilization or in liquid nitrogen.

Control of culture mites. Mites can be excluded from culture tubes by means of cigarette paper barriers fixed to the mouth of the tube with a simple gelatin glue that contains $CuSO_4$ as a microbial growth inhibitor (25).

MASS PRODUCTION OF INOCULUM OF *FUSARIUM* SPECIES

Chaff-grain medium. This medium is used for preparing inoculum that is suitable for addition to soil in pathogenicity tests. It is particularly appropriate for species that do not form chlamydospores and that normally persist in soil as hyphae in plant residues.

Cereal chaff or ground corn husks are used together with ground cereal grain, usually 50 g of chaff and 10 g of ground grain. The chaff-grain mixture is soaked in water overnight at 5°C, drained, autoclaved, and added to polyethylene bags that have been sterilized by γ-irradiation; the bag opening is then closed around a large, sterile cotton plug. It is inoculated with a conidial or mycelial suspension that can be mixed through the bag by shaking. The bag should be incubated under standard conditions described earlier and shaken regularly to encourage colonization. When the medium has been thoroughly colonized, it is air-dried and crushed to the required size (2,28,29).

Ground cornstalk medium. Mature cornstalks may be collected from field-grown plants, the leaves removed, the stalks dried in an oven at 45°C for 24 hr, and the dried stalks ground in a hammer mill fitted with a 6.4-mm screen and stored at 2°C in a plastic bag until needed. Ground cornstalks (GCS) are prepared for use as inoculum by being soaked in water overnight, drained, and autoclaved for 15 min at 121°C and 103 kPa (15 psi). Conidia of *F. moniliforme* are washed from a 2-week-old culture grown on a PDA slant with 10 ml of sterile distilled water, and 0.2 ml of this conidial suspension is added to 1 g of GCS. Within 5–7 days, mycelial growth should cover the medium. The GCS colonized by *F. moniliforme* may be incorporated into the soil at the rate of 200 g/35 L prior to planting. All GCS inoculum should be used within 3 weeks of inoculation with the fungus. Autoclaved, noninoculated GCS may be added at the same rate to the potting medium used for plants grown in noninfested soil (11).

Grass clipping medium. This medium was developed by Fulton et al (7). It is prepared from Kentucky bluegrass clippings that are collected and placed in 0.5-L jars. The clippings are treated with propylene oxide (1.0 ml/jar per 24 hr) (9). The clippings are then inoculated with the appropriate *Fusarium* sp., and the medium is ready to use after 5 days of incubation. This medium was used to infest sod for studies of Fusarium blight of turfgrasses.

Carnation-leaf agar technique. This medium can be used to produce large quantities of conidia for use in spore suspensions as inoculum. The appropriate *Fusarium* sp. can be grown on a large number of CLA slants to produce the required number of conidia. Another method involves growing the *Fusarium* sp. on a layer of CLA in 500-L Erlenmeyer flasks. The spore suspension is made by adding sterile water to the tubes or the flasks. Sterile water containing glass beads may be more useful in detaching the conidia when the fungus is grown in Erlenmeyer flasks.

Cornmeal-sand technique. The cornmeal-sand medium as described by Riker and Riker (22) can be used to produce large amounts of inoculum to add to soil. In some cases it is desirable to reduce the amount of cornmeal used in this medium.

Chlamydospore inoculum. Locke and Colhoun (13) described a method for preparing inoculum of *F. oxysporum* f. sp. *elaeidis* consisting almost entirely of chlamydospores. A brief description of this technique is given below. Cultures of the fungus are grown on a solid medium containing 9% malt extract and 1.2% agar at

27°C for 14 days under black light illumination for 18 hr continuously in each 24-hr cycle. The intensity of light should be about 116 $\mu W/cm^2$ at culture level. Spore suspensions are prepared by flooding each culture with water and scraping the conidia from the surface of the agar with a scalpel. The agar is then cut into small pieces and the entire suspension filtered through lens tissue in a Buchner funnel. The filtrate is collected and immediately mixed with talc at the rate of 1 ml of spore suspension per 2 g of talc. The thick paste that results is placed on a tray and air-dried for 3 days. At this time the mixture is set hard and should be ground with a mortar and pestle to produce a powder fine enough to pass through a 500-μm sieve, which ensures complete mixing of talc and spores. After 40 days, most of the macroconidia will be converted into chlamydospores and the microconidia will be dead. The chlamydospore concentration of the talc-chlamydospore inoculum can be determined using a standard dilution series on a selective medium, such as the Nash-Snyder medium.

LITERATURE CITED

1. Andersen, A. A. 1958. New sampler for the collection, sizing, and enumeration of viable airborne particles. J. Bacteriol. 76:471-484.
2. Burgess, L. W., and Liddell, C. M. 1983. Laboratory Manual for *Fusarium* Research. University of Sydney, Australia. 162 pp.
3. Burgess, L. W., Ogle, H. J., Edgerton, J. P., Stubbs, L. L., and Nelson, P. E. 1973. The biology of fungi associated with root rot of subterranean clover in Victoria. Proc. R. Soc. Victoria 86 (Part I):19-28.
4. Davies, R. R. 1971. Air sampling for fungi, pollens and bacteria. Pages 367-404 in: Methods in Microbiology. Vol. 4. C. Booth, ed. Academic Press, New York. 795 pp.
5. Fisher, N. L., Burgess, L. W., Toussoun, T. A., and Nelson, P. E. 1982. Carnation leaves as a substrate and for preserving cultures of *Fusarium* species. Phytopathology 72:151-153.
6. Fisher, N. L., Marasas, W. F. O., and Toussoun, T. A. 1983. Taxonomic importance of microconidial chains in *Fusarium* section *Liseola* and effects of water potential on their formation. Mycologia 75:693-698.
7. Fulton, D. E., Cole, H., Jr., and Nelson, P. E. 1974. Fusarium blight symptoms on seedling and mature Merion Kentucky bluegrass plants inoculated with *Fusarium roseum* and *Fusarium tricinctum*. Phytopathology 64:354-357.
8. Gregory, P. H. 1973. The Microbiology of the Atmosphere. 2d ed. Leonard Hill, Aylesbury. 377 pp.
9. Hansen, H. N., and Snyder, W. C. 1947. Gaseous sterilization of biological materials for use as culture media. Phytopathology 37:369-371.
10. Hirst, J. M. 1952. An automatic volumetric spore trap. Ann. Appl. Biol. 39:257-265.
11. Kaiser, R. P. 1980. Initial stages of infection of *Zea mays* L. by *Fusarium moniliforme* Shelden emend. Snyd. & Hans. causing stalk rot. Ph.D. thesis, Pennsylvania State University, University Park. 100 pp.
12. Komada, H. 1975. Development of a selective medium for quantitative isolation of *Fusarium oxysporum* from natural soil. Rev. Plant Prot. Res. 8:114-125.
13. Locke, T., and Colhoun, J. 1974. Contributions to a method of testing oil palm seedlings for resistance to *Fusarium oxysporum* Schl. f. sp. *elaeidis* Toovey. Phytopathol. Z. 79:77-92.
14. Lukezic, F. L., and Kaiser, W. J. 1966. Aerobiology of *Fusarium roseum* 'Gibbosum' associated with crown rot of boxed bananas. Phytopathology 56:545-548.
15. McMullen, M. P., and Stack, R. W. 1983. Effects of isolation techniques and media on the differential isolation of *Fusarium* species. Phytopathology 73:458-462.
16. Nash, S. M., and Snyder, W. C. 1962. Quantitative estimations by plate counts of propagules of the bean root rot *Fusarium* in field soils. Phytopathology 52:567-572.
17. Nelson, P. E., Pennypacker, B. W., Toussoun, T. A., and Horst, R. K. 1975. Fusarium stub dieback of carnation. Phytopathology 65:575-581.
18. Nelson, P. E., Toussoun, T. A., and Marasas, W. F. O. 1983. *Fusarium* Species: An Illustrated Manual for Identification. Pennsylvania State University Press, University Park. 193 pp.
19. Ogawa, J. M., and English, H. 1955. The efficiency of a quantitative spore collector using the cyclone method. Phytopathology 45:239-240.
20. Ooka, J. J., and Kommedahl, T. 1977. Wind and rain dispersal of *Fusarium moniliforme* in corn fields. Phytopathology 67:1023-1026.
21. Oswald, J. W. 1949. Cultural variation, taxonomy and pathogenicity of *Fusarium* species associated with cereal root rots. Phytopathology 39:359-376.
22. Riker, A. J., and Riker, R. S. 1936. Introduction to Research on Plant Diseases. John S. Swift Co., St. Louis. 117 pp.
23. Schwarzbach, E. 1979. A high throughput jet trap for collecting mildew spores on living leaves. Phytopathol. Z. 94:165-171.
24. Snyder, W. C., and Hansen, H. N. 1939. The importance of variation in the taxonomy of fungi. Proc. 6th Pacific Sci. Congr. 4:749-752.
25. Snyder, W. C., and Hansen, H. N. 1946. Control of culture mites by cigarette paper barriers. Mycologia 38:455-462.
26. Snyder, W. C., and Hansen, H. N. 1947. Advantages of natural media and environments in the culture of fungi. Phytopathology 37:420-421.
27. Snyder, W. C., and Hansen, H. N. 1954. Variation and speciation in the genus *Fusarium*. Ann. N.Y. Acad. Sci. 60:16-23.
28. Tio, M., Burgess, L. W., Nelson, P. E., and Toussoun, T. A. 1977. Techniques for the isolation, culture and preservation of the Fusaria. Aust. Plant Pathol. Soc. Newsletter 6:11-13.
29. Trimboli, D. S., and Burgess, L. W. 1983. Reproduction of *Fusarium moniliforme* basal stalk rot and root rot of grain sorghum in the greenhouse. Plant Dis. 67:891-894.
30. Tschanz, A. T., Horst, R. K., and Nelson, P. E. 1975. A substrate for uniform production of perithecia in *Gibberella zeae*. Mycologia 67:1101-1108.
31. Zachariah, A. T., Hansen, H. N., and Snyder, W. C. 1956. The influence of environmental factors on cultural characters of *Fusarium* species. Mycologia 48:459-467.

Isolating, Identifying, and Producing Inoculum of Pathogenic Species of *Cylindrocladium*

G. J. GRIFFIN and G. S. TOMIMATSU, Department of Plant Pathology, Physiology and Weed Science, Virginia Polytechnic Institute and State University, Blacksburg, VA 24061

Species of *Cylindrocladium* incite diseases of many field and vegetable crops, ornamentals, and forest nursery or large trees in temperate, subtropical, and tropical areas of the world. Many species cause damping-off, root rot, crown rot or stem canker, and leaf spot or needle blight. A wide host range is typical for the most important species in the genus. Twenty large-spored species and four small-spored ones (now considered *Cylindrocladiella* species) are known (5). Most of the present methodology has been developed for the large-spored species—*C. crotalariae* (Loos) Bell & Sobers, *C. floridanum* Sobers & Seymour, and *C. scoparium* Morgan—that are most commonly encountered as plant pathogens in the United States.

ISOLATION

Cylindrocladium species can be isolated quite readily from some diseased plant tissue, whereas it is difficult to recover the species from others. For *C. crotalariae*, and in many instances for *C. scoparium* and *C. floridanum*, acidified potato-dextrose agar (PDA), semiselective agar, or *Cylindrocladium*-selective agar media (2,8,13,16,18,20,24) can be used with high success. Marginal necrotic tissues, usually surface-sterilized in 0.1–1.0% NaClO for 1 or 2 min, are plated on the medium; distinctive cottony, tan or yellow brown to brown or red brown colonies, with a white to light-colored margin, develop within 3–7 days at room temperature (25–28° C) depending on the medium used for isolation. Pigmented colonies typically produce adequate numbers of conidia for identification. Colonies of *C. floridanum* often have less distinctive coloration and texture than *C. crotalariae*, and acidified Czapek-Dox agar has proved more successful for their routine isolation than acidified PDA. Colonies may be white or with little pigment and have a loose, cottony texture, and wet mounts of all individual colonies of this type generally must be examined initially to confirm identification.

Special techniques are often required to induce sporulation of isolates of *C. floridanum* for identification (see below). Where isolation is difficult, such as from badly decayed plant tissue or with root lesions on black walnut seedlings, baits, inserted into the lesion (azalea leaf trap of Linderman [15]), can be used to recover *C. scoparium* or *C. floridanum* and small-spored species. The baits are then plated on semiselective or selective media. As an alternative, selective media (8,13) may be used directly, and subculturing to induce sporulation of poorly sporulating isolates may be required. Diagnostic colony characteristics for some species are also often more apparent upon subculturing.

Use of a selective medium is recommended for quantitative studies of pathogen populations in plant tissue or soil. Several selective or semiselective media are available (2,8,13,16,18,20,24). Interference from undesired fusaria is the most important problem when assaying the low soil populations (0.1–10 propagules or microsclerotia per gram of soil) that are commonly encountered, or when assaying badly decayed plant material. The sucrose-QT medium (8), which is selective for *Cylindrocladium*, utilizes thiabendazole and three quaternary ammonium compounds to control undesired fusaria. Other media, some semiselective, use thiabendazole (2,13,20) alone or in combination with a pretreatment of soil with 0.25% NaClO (w/v) for 0.5–1 min prior to wet-sieving. The latter treatment, however, reduces recovery of the fungus from soil (13). For quantitative root infection studies, washed, asymptomatic roots have been plated on sucrose-QT medium, and good statistical relationships between microsclerotial populations in soil and infection density, and between disease incidence and infection density, have been obtained in field and greenhouse studies (10,25).

In quantitative soil assays, microsclerotial populations are most often assayed because these propagules (or chlamydospores in some species) are the primary survival structures in soil. Wet-sieving of soil (25- to 45-μm pore-size sieve) for 5–8 min is used to reduce the numbers of undesired fungi and other propagules (conidia or ascospores) of the *Cylindrocladium* species before standard dilution plating on a selective or semiselective medium (2,8,13,20,24). The smaller sieve (25-μm pore-size) retains more microsclerotia but has a greater resistance to water flow. For *C. crotalariae*, total soil population (ascospores plus microsclerotia) assays (standard dilution plating) at planting time often give similar or identical population estimates to those obtained with a wet-sieving pretreatment (microsclerotial assay), suggesting that ascospores and conidia are of little importance. A large pore-size sieve (420- to 600-μm) can be used on top of the small pore-size sieve during washing to separate microsclerotia in plant debris from "free" microsclerotia.

Although manual washing of the soil is effective and economical for moderate numbers of soil samples, a semiautomated elutriation procedure facilitates the wet-sieving of large numbers of soil samples (20). Generally, 0.1–1.0 g of wet-sieved soil (dry weight equivalent) suspension is plated on each of 10–20 dilution plates per soil sample. Because this is a concentrated soil suspension, use of a magnetic stirrer is recommended to keep the washed soil in suspension during pipetting. Also, the greater the amount of soil per plate, the more effective must be the selective medium, as large numbers of undesired fungi (10^3–10^4 propagules per gram of soil) must be controlled. Very rapid and economical, qualitative or semiquantitative isolation of *Cylindrocladium* species from soil can be done by placing about 10 azalea leaves (15) in each of several moist soil samples in 200-ml plastic cups or beakers. This is followed by plating colonized leaves on semiselective or selective media.

Samples of soil should be stored moist, but not saturated, at room temperature and assayed as soon as possible after collection. Soil and plant samples should not be frozen, refrigerated, or dried extensively, because these treatments are detrimental to *Cylindrocladium* recovery (9).

IDENTIFICATION

As indicated above, subculturing from isolation media to Czapek-Dox agar, PDA, glucose-lima bean agar (12), or glucose-asparagine agar (12) to induce sporulation and diagnostic cultural characteristics is often required initially to identify *Cylindrocladium* species. Incubation in room light (cool white, fluorescent) and at room temperature (25–28°C) is adequate. Multipoint inoculation (flooding plates with 1 ml of a dilute, blended mycelial and/or conidial suspension) on Czapek-Dox agar or other medium has proven successful in inducing sporulation of some freshly isolated, wild-type *C. floridanum* isolates. Glycerol has also been used at 25 g/L to induce sporulation (12). Pigment production in agar, observed from the reverse side, is often helpful in the preliminary identification of *C. floridanum* (reddish brown to reddish purple) and *C. scoparium* (reddish brown) when sporulation is poor and aerial mycelium is lightly or not pigmented. The distinctive vesicles, borne terminally on hyphae, and cylindrical conidia are usually sufficient to identify isolates to the genus. Pigmented microsclerotia in the agar also aid in identification at the genus level. Conidial length, conidial septation, and vesicle morphology for five large-spored species are given in Table 1, but different media typically were used by the different workers. In comparative studies using the same media for all species (12), conidia were typically one-septate and 35–63 μm long for *C. scoparium*, one-septate and 40–58 μm long for *C. floridanum*, three-septate and 34–85 μm long for *C. crotalariae*, three-septate and 51–73 μm long for *C. ilicicola*, and three-septate and length not measured for *C. theae*. These length measurements, however, are not completely consistent (Table 1) with those given in previous reports (1,3,4,7,17,19,23).

Distinctive vesicles produced terminally on hyphae in association with conidiophores are regarded as relatively constant and an important taxonomic or morphological criterion by some workers (6,17,23) but not by others (12,22). Vesicles (Table 1) of *C. floridanum* are described as globose to obovate (17) and of *C. scoparium* as oval to ellipsoid (4,17). Globose vesicles have been indicated in some instances for the latter species in other work (12). Bell and Sobers (3) indicated that *C. crotalariae* has globose vesicles, and this has been consistently observed for several hundred isolates (P. M. Phipps, *personal communication*). *C. theae* was reported to have narrowly clavate vesicles (19). *C. ilicicola* was described as having clavate vesicles (19) or, in other work, as having globose (4,12) to ellipsoid (12) vesicles. Peerally (19) indicated, however, that he observed no globose vesicles on the type material of *C. ilicicola*. In one study, *C. crotalariae* and *C. theae* have been considered to be equivalent to *C. ilicicola*, based in part on the assumption that vesicle morphology is not a reliable characteristic (22).

INOCULUM PRODUCTION

Microsclerotia or chlamydospores are most commonly used as inoculum because they are the principal structures present in soil at planting time for most economically important *Cylindrocladium* species. Microsclerotia of eight isolates, representing five species, were produced earlier and more abundantly on glucose-lima bean agar than on four other media, including PDA (12). Microsclerotia may be produced as early as 4 days, but maximum numbers are produced after longer incubation periods (1–5 months). A high C:N ratio and addition of tyrosine to glucose-yeast extract agar favored microsclerotial production (11). Other workers have successfully harvested microsclerotia for inoculum purposes from 14-day-old PDA

Table 1. Conidial length, number of septa per conidium, and vesicle morphology for five *Cylindrocladium* species

Species	Conidial length (μm)[a]	Number of septa[a]	Vesicle morphology[a]
C. crotalariae	60.4–105.2 (19) 58–107 (3)	Mostly 3 (19) 1–3 (3,19)	Subglobose to globose (19) Globose (3)
C. theae	63.0–102.8 (19) 57.1–95.2 (1)	Mostly 3 (1,19) 1–3 (19)	Narrowly clavate (19) Clavate (1)
C. ilicicola	37.4–68.0 (19) 49–69 (4)	Mostly 3; 1–3 (4,19)	Clavate (19) Globose (4)
C. scoparium	50–60 (7) 33–66 (17)	1 (7,17) Occasionally 3 (7)	Obovoid (7) Oval to ellipsoid (17)
C. floridanum	36.4–57.2 (19) 34–55 (23) 33–59 (17)	1 (17,19) Mostly 1 (23)	Globose (19,23) Oval or subglobose to globose on secondary conidiophores (23) Globose to obovate (17)

[a]Numbers in parentheses indicate literature citations.

cultures (21). Incubation in the dark at room temperature is preferred.

Microsclerotia that are relatively free of hyphae and other structures can be prepared by homogenizing mycelial mats in a Waring Blendor or similar device intermittently (to prevent heating up) at high speed for 20 min and washing on a sieve of 50-μm pore size. For ecological studies, microsclerotia can be grown on sterilized stem tissue or stem tissue extract for about 4 months, then blended and washed on a sieve (14). A sieve with large pores (105-μm or larger pore size) placed over the smaller pore-size sieve aids in excluding debris and larger propagules. Washed microsclerotia are suspended in water or a mineral salt solution, counted under the microscope in a 1-ml aliquot, and used to infest *Cylindrocladium*-free, nonsterile soil (for ecological studies) or to infest pasteurized soil or potting mix (for pathogenicity trials) at the desired population densities (usually 0.01–100 microsclerotia per gram of soil). As an alternative, roots may be dipped in a microsclerotial suspension or microsclerotia on water agar disks may be applied to root surfaces, with or without wounding, for pathogenicity trials. Root dips appear to be most effective, as microsclerotial germination is stimulated by exudates produced by the root tip (14).

Conidial or ascospore suspensions may be used for root-dip inoculation, for spraying on plant surfaces, or for propagule survival studies by using the techniques described above to induce sporulation. Conidia may be scraped off agar plates with a rubber policeman, separated from mycelium by filtration through cheesecloth, and washed free from culture media constituents by centrifugation or Millipore filtration with water or dilute mineral salt solution. A hemacytometer can be used to determine conidial density, and appropriate dilutions can be prepared for inoculations (usually 10^2–10^5 conidia per milliliter). For *Cylindrocladium* species producing the *Calonectria* teleomorph stage, ascospores may be obtained from squashed perithecia grown for 2–3 weeks in room light on glucose-lima bean agar (12), V-8 Juice agar, sterilized leaf or stem tissue of the host, or from perithecia removed from diseased plant tissues collected in the field. Ascospore suspensions can be washed and spore densities adjusted as indicated for conidia.

LITERATURE CITED

1. Alfieri, S. A., Jr., Linderman, R. G., Morrison, R. H., and Sobers, E. K. 1972. Comparative pathogenicity of *Calonectria theae* and *Cylindrocladium scoparium* to leaves and roots of azalea. Phytopathology 62:647-650.
2. Almeida, O. C. de, and Bolkan, H. A. 1982. Selective medium for quantitative determination of microsclerotia of *Cylindrocladium* species in soil. Phytopathology 72:300-301.
3. Bell, D. K., and Sobers, E. K. 1966. A peg, pod and root necrosis of peanuts caused by a species of *Calonectria*. Phytopathology 56:1361-1364.
4. Boedijn, K. B., and Reitsma, J. 1950. Notes on the genus *Cylindrocladium* (Fungi: Mucedinaceae). Reinwardtia 1:51-60.
5. Boesewinkel, H. J. 1982. *Cylindrocladiella*, a new genus to accommodate *Cylindrocladium parvum* and other small-spored species of *Cylindrocladium*. Can. J. Bot. 60:2288-2294.
6. Boesewinkel, H. J. 1982. Heterogeneity within *Cylindrocladium* and its teleomorphs. Trans. Br. Mycol. Soc. 78:553-555.
7. Booth, C., and Gibson, I. A. S. 1973. CMI Descriptions of Pathogenic Fungi and Bacteria. No. 362. Commonw. Mycol. Inst., Kew, Surrey, England.
8. Griffin, G. J. 1977. Improved selective medium for isolating *Cylindrocladium crotalariae* microsclerotia from naturally infested soils. Can. J. Microbiol. 23:680-683.
9. Griffin, G. J., Roth, D. A., and Powell, N. L. 1978. Physical factors that influence the recovery of microsclerotium populations of *Cylindrocladium crotalariae* from naturally infested soils. Phytopathology 68:887-891.
10. Griffin, G. J., and Tomimatsu, G. S. 1983. Root infection pattern, infection efficiency, and infection density-disease relationships of*Cylindrocladium crotalariae* on peanut in field soil. Can. J. Plant Pathol. 5:81-88.
11. Hunter, B. B., and Barnett, H. L. 1976. Production of microsclerotia by species of *Cylindrocladium*. Phytopathology 66:777-780.
12. Hunter, B. B., and Barnett, H. L. 1978. Growth and sporulation of species and isolates of *Cylindrocladium* in culture. Mycologia 70:614-635.
13. Krigsvold, D. T., and Griffin, G. J. 1975. Quantitative isolation of *Cylindrocladium crotalariae* microsclerotia from naturally infested peanut and soybean field soils. Plant Dis. Rep. 59:543-546.
14. Krigsvold, D. T., Griffin, G. J., and Hale, M. G. 1982. Microsclerotial germination of *Cylindrocladium crotalariae* in the rhizospheres of susceptible and resistant peanut plants. Phytopathology 72:859-864.
15. Linderman, R. G. 1972. Isolation of *Cylindrocladium* from soil or infected azalea stems with azalea leaf traps. Phytopathology 62:736-739.
16. Morrison, R. H., and French, D. W. 1969. Direct isolation of *Cylindrocladium floridanum* from soil. Plant Dis. Rep. 53:367-369.
17. Morrison, R. H., and French, D. W. 1969. Taxonomy of *Cylindrocladium floridanum* and *C. scoparium*. Mycologia 61:957-966.
18. Newhouse, J. R., and Hunter, B. B. 1983. Selective media for the recovery of *Cylindrocladium* and *Fusarium* species from roots and stems of tree seedlings. Mycologia 75:228-233.
19. Peerally, A. 1974. CMI Descriptions of Pathogenic Fungi and Bacteria. Nos. 421, 424, 425, and 429. Commonw. Mycol. Inst., Kew, Surrey, England.
20. Phipps, P. M., Beute, M. K., and Barker, K. R. 1976. An elutriation method for quantitative isolation of *Cylindrocladium crotalariae* microsclerotia from peanut field soil. Phytopathology 66:1255-1259.
21. Phipps, P. M., Beute, M. K., and Hadley, B. A. 1977. A microsclerotia-infested soil technique for evaluating pathogenicity of *Cylindrocladium crotalariae* isolates and black rot resistance in peanut. (Abstr.) Proc. Am. Phytopathol. Soc. 4:146.
22. Rossman, A. Y. 1983. The phragmosporous species of *Nectria* and related genera (*Calonectria, Ophionectria, Paranectria, Scoleconectria,* and *Trichonectria*). Mycological Papers. No. 150. 147 pp.
23. Sobers, E. K., and Seymour, C. P. 1967. *Cylindrocladium floridanum* sp. n. associated with decline of peach trees in Florida. Phytopathology 57:389-393.
24. Thies, W. G., and Patton, R. F. 1970. An evaluation of propagules of *Cylindrocladium scoparium* in soil by direct isolation. Phytopathology 60:599-601.
25. Tomimatsu, G. S., and Griffin, G. J. 1982. Inoculum potential of *Cylindrocladium crotalariae*: Infection rates and microsclerotial density-root infection relationships on peanut. Phytopathology 72:511-517.

Isolating, Identifying, and Producing Inoculum of *Phytophthora* spp.

D. J. MITCHELL and M. E. KANNWISCHER-MITCHELL, Department of Plant Pathology, University of Florida, Gainesville 32611, and G. A. ZENTMYER, Department of Plant Pathology, University of California, Riverside 92521

Species of *Phytophthora* cause many devastating plant diseases, and it has been estimated that the costs for the chemical control of these and other plant pathogens in the order Peronosporales represent about one-third of the total fungicide market (8). Methods of evaluating new and sophisticated chemical controls can be based on reliable methods for isolating these important pathogens and on techniques for producing inoculum with qualitative and quantitative refinements (3,4). The objective of this chapter is to recommend a few of the many techniques available for isolation, identification, and production of inoculum of *Phytophthora* spp. Other techniques obviously may prove more useful with different pathosystems or technical objectives. Several papers present expanded discussions or background information on these topics (5,6,9,11,16).

ISOLATION

The development of media selective for the isolation of *Phytophthora* spp. has greatly enhanced the ability of researchers to gain insight into the ecology, epidemiology, and control of diseases caused by these fungi. Although techniques that do not employ selective media (e.g., nonselective media or baits) may in some circumstances be better for isolation of *Phytophthora* spp. from soil or plant tissues, methods incorporating selective media generally provide more efficient use of time and facilities after experience is gained by the worker. The medium developed by Tsao and Ocaña (12) is widely used for the isolation of *Phytophthora* spp. from infected plant tissue or infested soil. The medium (PVP) contains 10 mg of pimaricin (Delvocid, 50% a.i., Gist-Brocades N. V., Delft, Holland), 200 mg of vancomycin (Vancocin HCl, Eli Lilly & Co.,Indianapolis, IN 46285), and 100 mg of pentachloronitrobenzene (PCNB; Terraclor, 75% a.i., Olin Mathieson Chemical Corp., Little Rock, AR 72203) added to 1 L of cornmeal agar (17 g of Difco cornmeal agar per liter of deionized water) previously autoclaved and cooled to 43–45°C.

Although this medium provides inhibition of most fungi other than pythiaceous ones by pimaricin supplemented with PCNB and inhibition of bacteria by vancomycin, its value in selective isolation of *Phytophthora* spp. from soil is compromised by the lack of inhibition of *Pythium* spp., which are commonly found in soils in high numbers and grow faster on the medium than do *Phytophthora* spp. This deficiency may be overcome by the addition of 50 ppm of hymexazol (99.4% a.i., Sankyo Co. Ltd., Tokyo, Japan) to PVP or to other selective media; both *Pythium* spp. and *Mortierella* spp., which typically are the only fungi that compete with *Phytophthora* spp. on PVP, are inhibited, and *Phytophthora* spp. generally develop free of interference of other fungi (11). Hymexazol, however, may not strongly inhibit all *Pythium* spp. (e.g., *P. vexans*) and may inhibit some *Phytophthora* spp. (e.g., *P. infestans*) (11).

Other media developed for the selective isolation of *Phytophthora* spp. are listed and discussed by Tsao (11). Vancomycin is very expensive and may be replaced by other antibiotics, such as a combination of 10 ppm rifampicin (Rifamycin SV, 100% a.i., Sigma Chemical Co., St. Louis, MO 63178) and 250 ppm ampicillin (100% a.i., Sigma Chemical Co.). If pimaricin is difficult to obtain, 25–50 ppm nystatin plus 10 ppm benomyl may be used as a less effective substitute.

Selective media are useful for the isolation of *Phytophthora* spp. from infected plant tissue and infested soil by the general methods presented in this book for the selective isolation of *Pythium* spp. (Mitchell and Rayside, *this volume*). However, *Phytophthora* spp. typically grow more slowly than *Pythium* spp., and isolation plates should be incubated longer for growth of the former from plant material or soil; 48–72 hr are usually suitable for initial observations of selective media containing infected tissue or soil dilutions. *Phytophthora* spp. generally are found at low inoculum densities in soil, especially in soils not in the immediate vicinity of diseased plants, and thus very low dilutions of soil in water or dilute agar (initial dilutions of 1:2–1:10, w:v, of soil to dilute agar) typically are required to detect these fungi. Frequently inoculum densities are below a level in soil reliably detected quantitatively by dilution-plating methods (less than approximately 0.1 propagule per gram of soil); but selective media may be used qualitatively by placing small clumps or blocks of soil directly on the agar surfaces.

When selective media are not effective or available, baits may be used to isolate *Phytophthora* spp. Many different baiting materials and techniques have been used to isolate these pathogens from soil, and long lists of materials and incubation conditions are provided by Ribeiro (6) and Tsao (11). In general, the ideal bait for the selective isolation of *Phytophthora* spp. should be susceptible to a broad range of soilborne species of *Phytophthora,* highly sensitive at low inoculum levels, reasonably sized, inexpensive, readily available geographically and seasonally, and convenient to set up for assay and isolation (2).

Although no single bait fulfills all of these criteria, several have been used commonly. Avocados or apples may be partially embedded in soil samples in a container with a water level maintained at several centimeters over the soil surface but not covering the fruit. Similarly, small pieces of leaves (e.g., pine needles or citrus leaves) or seedlings (e.g., alfalfa or soybean seedlings) may be floated in water several centimeters over the surface of a flooded soil sample (5–100 g of soil) for several days. High water-to-soil ratios in baiting procedures enhance recovery of the fungus, especially when inoculum densities are very low; ratios of at least 4:1 water to soil have been recommended (11). Only water low in ions toxic to *Phytophthora* spp. should be used (rainwater, glass-distilled water, or high-quality deionized water). Thoroughly moistened soil can be placed in holes bored in fruit (e.g., apples or watermelons) with a small cork borer and covered with tape. Any of the above baits should be incubated for several days (typically at least 3 and up to 10 days) under fluctuating conditions of light (natural day and night) and temperature (20 and 28°C) to induce spore formation and germination. Small pieces of tissue from the margins of firm, brown rots of large baits or intact small baits may be transferred to a selective medium for isolation. If a selective medium is not available, cornmeal agar or other culture media may be used; but interference with the isolation of *Phytophthora* spp., especially by *Pythium* spp. and bacteria, often becomes a problem.

In contrast to the difficulties encountered in isolating *Phytophthora* spp. from soil because of low inoculum densities, these pathogens may often be easily isolated from advancing margins of active lesions on plant tissues with normal techniques for fungal isolation on water agar or common agar media. Moribund or dead plant tissue, from which the fungi are difficult to isolate, may be washed in running water for several hours and then plated on nonselective media. Selective media, however, typically will greatly enhance the probability of recovery of *Phytophthora* spp. from such tissue.

IDENTIFICATION

Pure cultures of *Phytophthora* spp. may be obtained from isolation media by transferring hyphal tips from actively growing colonies to plates of fresh selective medium. It is helpful to place the hyphal tips plus as little adhering agar as possible under the selective medium to encourage growth of the fungus up through the medium rather than over the surface, which sometimes allows bacterial contaminants to be carried on the hyphae. Cultures free of contaminants can be transferred to various rich substrates, such as commercially prepared cornmeal agar or potato-dextrose agar, for storage. Freshly prepared media such as oatmeal or cornmeal agar should be used, however, for the production of spores for identification of isolates (6,16). Cornmeal agar may be prepared by boiling 60 g of freshly ground corn kernels in 1 L of distilled water for 1 hr; the suspension is drained through cheesecloth, 17 g of agar is added, and the suspension is adjusted to 1 L in volume and autoclaved (6).

Production of sporangia and zoospores may be induced by transferring small pieces from the agar culture or from infected plant tissue to an autoclaved mixture of pond water and distilled water (1:2, v:v) or to salt solutions (14) such as that developed by Chen and Zentmyer (1). Hempseed (*Cannabis sativus*) may be boiled for 30 min and added to autoclaved salt solutions or distilled water and inoculated with pieces of pure cultures to induce formation of sporangia, oogonia, oospores, chlamydospores, and hyphal swellings. Zoospore formation and emission, which can be induced by chilling cultures with sporangia at 10°C for 15–20 min, must be observed to ensure that zoospores are formed in the sporangium and thus verify that a species of *Phytophthora*, and not of *Pythium*, is being observed. If oospores are not formed in single culture on rich media (e.g., cornmeal, oatmeal, or V-8 Juice agars) or in the infected tissues observed, it may be necessary to grow the isolate with an oospore-inducing strain (opposite compatibility type) of the same or related species.

After sporangia and oospores have been formed in culture, the key and descriptions of *Phytophthora* spp. developed by Waterhouse (14,15) may be referred to for attempted identifications. Six main groups of species were presented by Waterhouse (14) and retained by Newhook et al (5) in a tabular key. The characters used to separate these groups include the extent of apical thickening of sporangia and the width of the exit pore, the caducity of sporangia and the pedicel length, and the attachment of antheridia to oogonia (amphigynous, paragynous, or both). Other characters that are recognized by Waterhouse (14,15) as important aids for the differentiation of species include sporangial size, sporangial length-to-breadth ratio, oogonial size, oospore wall characteristics, presence of hyphal swellings in water, presence of chlamydospores, pathogenicity (for host-restricted species), and growth at various temperatures. Additional characters are discussed by Waterhouse et al (16), Tucker (13), and Savage et al (7). Unfortunately, identification of isolates to species of *Phytophthora* is often difficult because of extreme ranges of variability in isolates and complexities within the methods of classification presently used, which are based on morphological characteristics. Although many isolates can be identified, at least tentatively, by the observation of an aggregate of as many characters as possible, there is an urgent need for an updated key to the species of *Phytophthora*.

INOCULUM PRODUCTION

Inoculum for various types of fungicide evaluations should be selected with several considerations in mind. Preparation of inoculum should allow efficient use of available facilities, materials, and time. The type and quality of inoculum should not be critically atypical of the inocula that function in disease situations under normal conditions, or the efficacy of the fungicide may ultimately be inappropriately rated. The function of the fungicide under field conditions (e.g., eradicants, chemotherapeutants, or long-lasting protectorants) may dictate need for evaluation with one or more types of propagules associated with survival, infection, and disease development. For example, when host plants will be introduced into a field after treatment with an eradicant for *Phytophthora* spp., inoculum selected for the evaluation of the eradicant might be oospores, chlamydospores, or infested debris, which are generally the forms of primary inoculum encountered under natural conditions. Conversely, tests with protectorant fungicides might use zoospores, sporangia, or propagules in soil (e.g., chlamydospores, oospores, zoospores, sporangia, infested debris) that may be splashed onto the host and function as transitory or secondary inoculum.

Very low levels of inoculum of *Phytophthora* spp. are required for infection and subsequent disease development under natural conditions or in greenhouse or laboratory tests carefully manipulated to simulate natural conditions (4). Less than one propagule of several different *Phytophthora* spp. per gram of soil may result in severe disease of various hosts, such as pepper or tobacco, under field conditions. Under optimum conditions in the greenhouse or laboratory, for example, only 0.1–0.9 chlamydospore per gram of soil was required for infection of 50% (ID_{50}) of various hosts exposed to *P. cinnamomi, P. citrophthora, P. palmivora*, or *P. parasitica* var. *nicotianae*. Similarly, low levels of oospores of homothallic species result in disease. Low concentrations of zoospores are required for infection if suspensions are treated carefully; for example, 50% infection of tobacco seedlings occurred at 42 zoospores per plant in flooded surface water. In contrast, 10^4–10^5 zoospores per plant may be required for ID_{50} levels with several pathosystems when suspensions are drenched into soil.

Chlamydospores of some *Phytophthora* spp. (e.g., *P. cinnamomi* and *P. parasitica* var. *nicotianae*) are formed after 3–4 weeks on mycelial mats grown on natural media such as V-8 Juice or lima bean broths. To prepare V-8 Juice broth, a mixture of 350 ml of V-8 Juice plus 5 g of $CaCO_3$ is centrifuged for 20 min at 4,000 rpm, and the supernatant is then diluted 1:4 with distilled water and autoclaved. Lima bean broth may be prepared by steaming 250 g of frozen lima beans in 1 L of distilled water for 30 min, filtering the suspension through cheesecloth, adjusting the volume to 2 L, and autoclaving the broth. Other species of *Phytophthora*, such as *P. parasitica* and *P. palmivora*, require nutrient dilution or replacement methods for abundant production of chlamydospores. Tsao (10), for example, incubated each culture of *P. parasitica* in 25 ml of V-8 Juice broth in a 325-ml prescription bottle at 25°C in the dark for 24 hr; shook the bottle to fragment the hyphae; incubated the horizontally positioned bottle for an additional 6 days to induce a large mycelial mat; added 100 ml of sterile, deionized water to submerge the mat; and incubated the vertically positioned bottles for an additional 2–3 weeks in the dark at 18°C. Chlamydospores may also be produced in defined media or other natural media, as well as in in vitro plant systems such as roots of lupine infected with *P. drechsleri* (6).

Various blenders or tissue grinders may be used to separate the chlamydospores from mycelial mats after they have been washed to remove residual nutrients (4,6). Care should be taken when using high-speed mixers (e.g., Sorvall Omni-Mixer) or blenders to prevent heat injury to chlamydospores by protecting suspensions with ice or cold-water baths. Excess mycelial fragments, which interfere with counting and may function as unnatural inoculum propagules in inoculations, may be removed by centrifugation or filtration (6). For refined experiments in which only chlamydospores are used as inoculum, any viable mycelial fragments, zoospores, or sporangia surviving filtration or centrifugation generally can be eliminated by sonication (Bronsonic 1510 sonicator made by B. Braun Instruments, San Francisco, CA 94080). Chlamydospores of *P. cinnamomi, P. citrophthora, P. palmivora, P. parasitica*, and *P. parasitica* var. *nicotianae* have germinated in excess of 90% after elimination of other viable propagules by sonication of cooled suspensions for two 30-sec exposures at 60% of the maximum wattage of a Bronsonic 1510 sonicator (D. J. Mitchell, *unpublished*).

Chlamydospores can be counted easily on a hemacytometer or by the microdrop method (Mitchell and Rayside, *this volume*). Dilutions can then be prepared for addition to soil or application directly to plant parts; it is important, however, to verify that accurate inoculum levels have been established. The viability of chlamydospores can be ascertained by evaluation of a 0.05-ml sample of the spore suspension mixed with 0.3 ml of an aqueous solution of rose bengal at 50 ppm (6); injured and nonviable spores are stained selectively after 6–12 hr at 24°C in darkness. As an alternative, either samples of the dilutions placed on selective media or chlamydospores placed on agar for microdrop counts may be retained and observed for spore germination after 18 hr at 25°C in the dark. Samples of infested soil should be diluted and plated on a selective medium to validate the actual inoculum levels established. Use of suppressive soils often results in levels of inoculum much lower than those calculated, and chlamydospores may germinate and result in increased populations in wet, conducive soils.

Until experimental methods that result in plant infection from oospores of heterothallic *Phytophthora* spp. are available, oospores of these species, even though they may be produced in paired cultures, are not recommended for inoculation in evaluations of fungicides. Oospore inoculum of homothallic species may be produced and quantitated by methods similar to those used for chlamydospores. Various natural and synthetic media have been used for oospore production. Clarified V-8 Juice broth supplemented with 30 ppm b-sitosterol, 20 ppm tryptophan, 1 ppm thiamine HCl, and 100 ppm of $CaCl_2 \cdot 2H_2O$ allows production of oospores by isolates that normally do not form oospores in standard media (6). Oospores may be separated from mycelial mats by sonication as indicated above or by treatment with enzymes such as snail-gut enzyme (Helicase) (6). It is difficult as yet to validate prepared inoculum levels because oospores apparently do not germinate synchronously, and recovery on selective media is often low.

Sporangia of various species of *Phytophthora* may be produced on seeds (e.g., hempseed) in water or salt solutions or on natural (lima bean or V-8 Juice agar or broth) or artificial media with or without flooding of cultures subsequent to growth. Many of the media and methods reported for sporangium production are reviewed by Ribeiro (6). Abundant sporangia are produced by several species with the technique using rapid growth in V-8 Juice broth for 24–48 hr followed by several washes and a 24-hr incubation in distilled water or salt solution (1) at 25°C under light. Sporangia can be induced to produce and release zoospores by washing of the cultures, exposure to 5–10°C for 15–20 min, and incubation for 1 hr at 25°C. The use of a buffered solution instead of water for washes and incubation, procedures for quantitation of zoospores, and precautions recommended for handling zoospore inoculum have been discussed by Mitchell and Rayside (*this volume*).

Naturally infested soils or nonquantitated, artificially infested soils may be used to evaluate fungicides on hosts planted in the greenhouse or field. *Phytophthora* spp. grown in axenic broth cultures, on seeds (e.g., pearl millet), or in infected roots of susceptible hosts may be comminuted, mixed with sterile coarse sand, and added in measured amounts by weight or volume to soil. Populations in naturally infested soil can be increased by planting susceptible crops in soil maintained under conditions favorable for disease. By this method, populations of *P. parasitica* var. *nicotianae*, for example, increase from less than one to several hundred propagules per gram of soil during the growth of one generation of tobacco plants in the field or the greenhouse (3,4). This method may be complicated by the stimulation of other microorganisms in the soil that might induce suppressiveness to disease development; but, in general, favorable test conditions are created with a careful selection of soil conducive to disease development.

Estimations of inoculum density in infested soil with a selective medium allow better comparisons among tests or with related studies and allow manipulation of fungal populations to levels typical for natural conditions. Populations in infested soil for greenhouse tests may be adjusted to desired levels by assaying the infested soil with a selective medium, diluting the soil with noninfested soil, maintaining the soil for two or more weeks at room temperature to allow populations to stabilize, and assaying the soil on the selective medium to obtain a final determination of inoculum density.

In conclusion, many techniques for isolation and production of inoculum of *Phytophthora* spp., despite being time-consuming and sometimes expensive, are available and easily mastered. Attempts at the identification of isolates, especially isolates of heterothallic species when compatibility strains are not available, may be difficult and frustrating.

LITERATURE CITED

1. Chen, D.-W., and Zentmyer, G. A. 1970. Production of sporangia by *Phytophthora cinnamomi* in axenic culture. Mycologia 62:397-402.
2. Dance, D. H., Newhook, F. J., and Cole, J. S. 1975. Bioassay of *Phytophthora* spp. in soil. Plant Dis. Rep. 59:523-527.
3. Kannwischer, M. E., and Mitchell, D. J. 1978. The influence of a fungicide on the epidemiology of black shank of tobacco. Phytopathology 68:1760-1765.
4. Mitchell, D. J., and Kannwischer-Mitchell, M. E. 1983. Relationship of inoculum density of *Phytophthora* species to disease incidence in various hosts. Pages 259-269 in: *Phytophthora*: Its Biology, Taxonomy, Ecology, and Pathology. D. C. Erwin, S. Bartnicki-Garcia, and P. H. Tsao, eds. American Phytopathological Society, St. Paul, MN.
5. Newhook, F. J., Waterhouse, G. M., and Stamps, D. J. 1978. Tabular key to the species of *Phytophthora* de Bary. Mycol. Pap. 143. Commonw. Mycol. Inst., Kew, Surrey, England. 20 pp.
6. Ribeiro, O. K. 1978. A Source Book of the Genus *Phytophthora*. Cramer, Vaduz, Liechtenstein. 417 pp.
7. Savage, E. J., Clayton, C. W., Hunter, J. H., Brenneman, J. A., Laviola, C., and Gallegly, M. E. 1968. Homothallism, heterothallism and interspecific hybridization in the genus *Phytophthora*. Phytopathology 58:1004-1021.
8. Schwinn, F. J. 1983. New developments in chemical control of *Phytophthora*. Pages 327-334 in: *Phytophthora*: Its Biology, Taxonomy, Ecology, and Pathology. D. C. Erwin, S. Bartnicki-Garcia, and P. H. Tsao, eds. American Phytopathological Society, St. Paul, MN.
9. Tsao, P. H. 1970. Selective media for isolation of pathogenic fungi. Annu. Rev. Phytopathol. 8:157-186.
10. Tsao, P. H. 1971. Chlamydospore formation in sporangium-free liquid cultures of *Phytophthora parasitica*. Phytopathology 61:1412-1413.
11. Tsao, P. H. 1983. Factors affecting isolation and quantitation of *Phytophthora* from soil. Pages 219-236 in: *Phytophthora*: Its Biology, Taxonomy, Ecology, and Pathology. D. C. Erwin, S. Bartnicki-Garcia, and P. H. Tsao, eds. American Phytopathological Society, St. Paul, MN.
12. Tsao, P. H., and Ocaña, G. 1969. Selective isolation of species of *Phytophthora* from natural soils on an improved antibiotic medium. Nature (London) 223: 636-638.
13. Tucker, C. M. 1931. Taxonomy of the genus *Phytophthora* de Bary. Mo. Agric. Exp. Stn. Bull. 153. 208 pp.
14. Waterhouse, G. M. 1963. Key to the species of *Phytophthora* de Bary. Mycol. Pap. 92. Commonw. Mycol. Inst., Kew, Surrey, England. 22 pp.
15. Waterhouse, G. M. 1970. The genus *Phytophthora* de Bary. Mycol. Pap. 122. Commonw. Mycol. Inst., Kew, Surrey, England. 104 pp.
16. Waterhouse, G. M., Newhook, F. J., and Stamps, D. J. 1983. Present criteria for classification of *Phytophthora*. Pages 139-147 in: *Phytophthora*: Its Biology, Taxonomy, Ecology, and Pathology. D. C. Erwin, S. Bartnicki-Garcia, and P. H. Tsao, eds. American Phytopathological Society, St. Paul, MN.

Isolating, Identifying, and Producing Inoculum of *Pythium* spp.

D. J. MITCHELL and P. A. RAYSIDE, Department of Plant Pathology, University of Florida, Gainesville 32611

Many methods and materials have been evaluated for routine isolation of *Pythium* spp. from soil and infected plants, for identification of many of the isolates to species or at least to groups of species, and for production of inocula of phytopathogenic species. The objective of this chapter is to introduce efficient and reliable techniques for the preparation of inocula of identified pathogens isolated for use in studies on the evaluation of fungicides. Only a few methods and alternatives are recommended for initial approaches. Many other alternatives of various refinement are available and obviously may prove more useful for specific studies. Several reviews are particularly helpful for background information and additional approaches for isolating and producing inocula of *Pythium* spp. (3,6,11–13,15,16).

ISOLATION

Most *Pythium* spp. are easily isolated from partially diseased plant tissues by surface disinfesting pieces of infected parts showing symptoms of disease (necrosis, water soaking, etc.) and placing them on 2% water agar or weak nutrient media such as cornmeal agar (11,15,16). These fungi are often not recovered from moribund or dead plant tissue. Although submerging pieces of such tissue in running water for several hours sometimes allows recovery when the pieces are subsequently placed on agar media, the process is time-consuming and not always successful. Isolation of *Pythium* spp. from soil or plant debris with general laboratory media is also difficult because these fungi often do not compete well with other microflora. Consequently, selective isolation techniques are commonly used today to promote growth of *Pythium* spp. while inhibiting growth of bacteria, actinomycetes, or fungi that interfere with germination of spores and growth of the desired fungi (12). The selectivity of such techniques for isolation of pythiaceous fungi may be based on many factors, but the most commonly used approaches involve manipulations of components of selective media and of the conditions under which such media are incubated (7,12,13).

The medium developed by Tsao and Ocaña (14) for selective isolation of *Phytophthora* spp. from soils has been modified in various ways, but still it is used widely and efficiently for the isolation of species of *Pythium* and *Phytophthora* from infected plants and infested soil. The medium (PVP) is prepared by adding 10 mg of pimaricin (Delvocid, 50% a.i., Gist-Brocades N. V., Delft, Holland), 200 mg of vancomycin (Vancocin HCl, Eli Lilly & Co., Indianapolis, IN 46285), and 100 mg of pentachloronitrobenzene (PCNB; Terraclor, 75% a.i., Olin Mathieson Chemical Corp., Little Rock, AR 72203) to 1 L of cornmeal agar (17 g of Difco cornmeal agar per liter of deionized water) previously autoclaved and cooled to 43–45°C. Pimaricin and several other polyene antibiotics are effective antifungal agents that generally are selectively nontoxic to fungi in the Pythiaceae and a few other groups (12,13). Nystatin (another polyene antibiotic), benomyl plus PCNB, and gallic acid have been used in various combinations to replace or supplement pimaricin for the inhibition of nonpythiaceous fungi (13). Various antibacterial agents that are effective in selective media, but do not inhibit *Pythium* spp. greatly, may be used to replace vancomycin, which is very expensive. Based on our experience, ampicillin (100% a.i., Sigma Chemical Co., St. Louis, MO 63178) at 250 mg plus rifampicin (Rifamycin SV, 100% a.i., Sigma Chemical Co.) at 10 mg per liter of medium provide an effective and less expensive replacement for vancomycin. A selective medium with a defined, artificial basal medium that may be a useful alternative to PVP was reported by Schmitthenner (11).

After thorough mixing, the molten medium may be dispensed into petri plates (approximately 15 ml/plate), which, after cooling, may be stored for short periods of time at room temperature or preferably under refrigeration (plates should be stored upside down when cooled to prevent water condensation on the agar surface). However, because antibiotics such as pimaricin are unstable, the selective medium is most effective when used as soon after preparation as possible. It is important to maintain the medium during storage and incubation periods in the dark to avoid photoinactivation of polyene antibiotics and possibly of other ingredients.

Pieces of roots or other organs suspected of containing *Pythium* spp. may be washed, blotted dry on paper towels, and placed on or in the selective medium. The plant tissue may be surface disinfested before plating, but this is generally not recommended because it rarely improves recovery and may reduce or eliminate the pathogen in fine roots or other tissues.

For qualitative determinations of *Pythium* spp. in soil, small clumps of soil suspected to be infested may be placed directly on the selective medium. Quantitative estimations of populations of *Pythium* spp. may be based on the detection of colony-forming units with the soil dilution plate method. Dilutions for efficient estimates will vary with soil populations, but initial attempts may employ dilutions of 1:20 and 1:300 (w/v, based on equivalent air-dried soil weight) of soil in dilute agar (2.5 g of Difco Bacto agar per liter of deionized water, autoclaved and cooled to room temperature before use). Because populations of *Pythium* spp. are typically not uniformly distributed in soil, soil sample sizes as large as can be economically and effectively processed should be used for dilutions (e.g., a minimum of 10 g of soil per dilution). The dilutions should be stirred continuously on a magnetic stirrer or some other device to maintain a uniform distribution of propagules in the dilute agar.

Samples of 1 ml of the soil-agar dilution should be spread over each of a minimum of 10 plates per sample with a glass rod or similar instrument. Dilute agar is usually superior to water for dilutions because propagules are suspended better in the dilute agar, and the agar remains spread over the selective medium better (water often tends to flow and carry propagules to low sides of plates); thus, crowding of colonies at edges of some plates, which interferes with differentiation and enumeration, can be avoided with the use of dilute agar. After one to a few days of incubation in the dark, the soil-agar suspension can be washed off the surface of the selective medium with a gentle stream of water and the colonies observed and counted.

Small pieces of agar from the margins of colonies growing from propagules in the soil suspension or from plant pieces in other isolations may be placed under the agar of fresh PVP plates. Hyphal tips of colonies growing up through the agar are generally free of contaminants and may be transferred aseptically to cornmeal agar or other common culture media to initiate pure cultures.

The efficiency of any selective medium depends on many factors, such as differences in the nature and population of *Pythium* spp., soil characteristics, and the personnel preparing and using the medium. Selective media can be highly refined and very effective when one or a few known species of *Pythium* are targeted for isolation. Burr and Stanghellini (1), for example, used a refined selective medium and a high incubation temperature to isolate *P. aphanidermatum* from soil without interference from other *Pythium* spp.

Although baiting is generally less efficient and more time-consuming than techniques using selective media for isolation of pythiaceous fungi, various plant parts, such as apple fruits or potato tubers, may be used to recover *Pythium* spp. from infected plants or soil (3,16). Hendrix and Campbell (3) listed many baiting techniques and recommended that any bait should be selected for efficiency of isolation of the *Pythium* spp. of interest to the individual researcher. Their approach to total qualitative estimations of *Pythium* spp. from many diverse soils incorporated the use of two selective media and three baiting techniques for each sample.

IDENTIFICATION

Pure cultures of *Pythium* spp. may be readily prepared for identification by standard hyphal-tipping techniques. Single-spore isolates may also be obtained easily for many species, but consideration should be given to restriction of genetic variation within such isolates before they are considered for use in identification procedures. Isolates should be transferred to substrates recommended in the several common taxonomic keys (6,16,17) to allow comparison of characteristics with those established for the keys and to avoid variation induced in some media (4). Small pieces of cornmeal agar cultures, for example, may be transferred to small petri plates containing a sterile solution of one part pond water and one part autoclaved, distilled water plus several segments of grass-blades (1–2 cm long) previously boiled for 10 min in distilled water (16). Although many types of grass-blades may be used effectively, bent grass (*Agrostis alba*) and bluegrass (*Poa pratensis*) have been recommended (11). Hempseeds boiled for 30 min may be used instead of grass-blades, and a synthetic salt solution may be used as an alternative to the diluted pond water (11). Cultures should be incubated in the light at approximately 25°C unless species suspected to have low or high cardinal temperatures are being observed. Structures such as antheridia and sporangia may be formed rapidly and then degenerate, so the cultures should be observed about 24 hr after inoculation and then periodically until all possible characteristics have been observed.

Van der Plaats-Niterink (16) published the most recent monograph on the genus *Pythium*, which contains a taxonomic key and descriptions of species. The key is usable and current, especially in reference to known heterothallic species for which compatibility strains are required for oospore production. Many unacceptable species were omitted from the key and relegated to a section in the monograph on doubtful and excluded species. Keys and descriptions by Waterhouse (17,18) and Middleton (6) should also be consulted for comparison or confirmation of identifications. Middleton's (6) monograph is out-of-date in reference to species described since 1940, but it includes host records and excellent illustrations of many important species; the key to species recognizes important morphological characters and emphasizes relationships based on broad groupings. The key provided by Waterhouse (17) is difficult to use because it utilizes small differences in size and emphasizes some nonmorphological characters. As Hendrix and Papa (4) noted, the key may lead to misidentifications in some cases because variation in size of various structures within an isolate or single species can be greater than the size differences used to separate species.

Another approach to identification of *Pythium* spp. was presented by Hendrix and Papa (4), who questioned the value of attempting to separate certain described species in groups that are extremely difficult to differentiate. Based on many years of experience, they presented 15 groups of species; in some of the groups, species could not be separated or were differentiated only with difficulty. Although this classification has not gained acceptance for publication of identified organisms, the tabular presentation of the characteristics of the species complexes provides a very useful comparison to the above keys.

PRODUCTION OF INOCULUM

The differing needs for qualitative or quantitative information will dictate the selection of form of inoculum best suited for tests on chemical control of diseases caused by *Pythium* spp. Although simple pathogenicity tests may be conducted by adding cultures or nonquantitated suspensions of mycelium or spores to host plants or soil in which they are planted, this type of inoculation may best be reserved for preliminary fungicide screening. Since hosts are exposed by these methods to atypical levels and often atypical types of inoculum compared with natural conditions, disease may take place at abnormally high rates or levels. Consequently, the efficacy of fungicides under conditions approximating those in nature may not be accurately determined.

Natural exposure of plants to *Pythium* spp. may be simulated by growing the host plants in soil artificially infested with the fungi plus organic substrate. The pathogen may be added to soil conducive to disease development containing seeds or seedlings of the susceptible host, which can then be replanted in successive crops until high inoculum levels are attained. As an alternative, the fungus can be grown on an organic substrate that is then added to soil and allowed to age before planting the host. Soil, for example, has been infested routinely with these fungi by inoculating a sterilized cornmeal-sand mixture (3% cornmeal plus 97% sand, w/w, adjusted to 20% water content, w/v), incubating the mixture for about 2 weeks at room temperature, incorporating the mixture into moist soil, and air-drying the soil for 1–2 weeks before planting. Pearl millet, wheat, and other seeds or fruits have also been used commonly, but care should be taken to ensure that other organisms are excluded from the seed (e.g., autoclave moist seeds twice for 1 hr at 24-hr intervals).

This method may be refined by assaying the infested soil with a selective medium and diluting the soil with noninfested soil to produce a defined level of inoculum. The diluted soil lots should be incubated at room temperature (air-dried) for 2 weeks or more to allow populations to stabilize and then preferably assayed on selective media again before introducing test plants into the infested soil. This method generally provides quantitative reproducibility of pathosystems for fungicide tests, but it may require fairly long periods of time for stabilization of populations in the infested soil.

Untreated field soil, naturally infested with one or more *Pythium* spp. known to be pathogenic to a designated crop, may serve as an adequate inoculum source for routine screening of fungicides in the greenhouse or field. Tests conducted with naturally infested soil are better defined quantitatively and can be compared with other evaluations when the populations of the fungi are assayed with a selective medium before and during the progress of the exposure of plants to the pathogen. After the initial estimation of populations of *Pythium* spp., soil may be diluted with noninfested soil or other planting media to establish desired inoculum levels, which allow more accurate reproducibility of test conditions. The use of naturally infested soil is simplistic, but it provides several advantages in addition to efficient use of time, materials, and space. Ecological implications, such as natural interactions of *Pythium* spp. with other organisms that may greatly affect pathogenesis and consequently control measures (2,3), are automatically included in the tests. The inoculum is also in a nearly natural state in relation to specific types of propagules and stage of activity or survival.

Production and quantitation of specific forms and qualities of inoculum are required for a critical evaluation of the impact of chemicals or other control measures on the host-pathogen interaction. When specific forms of inoculum are to be used in studies on the effects of fungicides on quantitative associations of *Pythium* spp. and hosts, the type of inoculum used should be based on natural inoculation. Primary inoculum typically includes oospores or pieces of plant debris containing mycelia or spores of the fungus, but zoospores under flooded conditions and sporangia or chlamydospores of some species may also be important. When selected forms of inoculum, concentrations of fungicides, and incidences or severities of disease are all quantitated carefully under defined environmental conditions, finite information on the impact of fungicides on the ecological and epidemiological balance of the pathosystem may be ascertained.

Oospores of many *Pythium* spp. may be produced in natural media, such as V-8 Juice broth (20% Campbell's V-8 Juice plus 4.5 g of $CaCO_3$ per liter of deionized water clarified by centrifugation at 1,275 *g*) or synthetic media (2,8,10,11,16). Culture flasks containing a shallow volume of medium (15 ml of medium per 250-ml Erlenmeyer flask) may be inoculated with a plug cut from a fresh agar culture of the fungus and should be maintained in the dark for at least 2–3 weeks to generate abundant oospores with a minimum of viable mycelia. Oospores of many *Pythium* spp. are produced at 25–30°C, but lower or higher temperatures may be required for others. Pairings of compatibility types will be required for abundant oospore production by heterothallic *Pythium* spp., but quantitative information on oospore production by heterothallic pathogens is not yet available. Similarly, little is known about the differences in oospores produced by artificial methods and those produced naturally in diseased plants.

Oospores may be freed of mycelia to prevent altering the ultimate defined inoculum levels by comminuting washed cultures in a blender for 30 sec, passing them through a glass tissue grinder (2,8,10), and subjecting the resulting suspension to sonication (40 sec at 20–30°C with maximum intensity of a Bronsonic 1510 ultrasonic system) or enzymes (cellulase plus hemicellulase or Glusulase) (10). Suspensions of oospores may also be frozen to eliminate mycelia, but some oospores, such as those of *P. aphanidermatum* and *P. myriotylum*, may be killed by freezing. Regardless of the techniques used, the oospores should be evaluated for viability by chemical tests or observations of spores on solid culture media (8,10).

Oospores can be counted directly on a hemacytometer, eelworm counting cell, or in microdrops under a microscope (5,8,9). Once the inoculum is quantitated, soil can be infested at defined rates to result in desired levels of disease incidence or severity, which is ascertained by testing a range of inoculum densities (8). Specific inoculum densities that result in defined disease levels (ID_{50} = the inoculum density at which 50% of exposed seedlings are infected or diseased) can be used for fairly uniform reproducibility of tests such as fungicide evaluations. With *P. aphanidermatum*, *P. myriotylum*, and *P. polymastum*, for example, ID_{50} levels of 15–45 oospores per gram of soil were observed with several crops over many tests (8).

Large numbers of zoospores of many *Pythium* spp. may be produced easily by a number of methods. A method that has been effective in our studies is initiated by inoculating petri plates containing 15 ml of V-8 Juice broth with plugs (e.g., eight 6-mm plugs) from the margin of a 1-day-old culture on V-8 Juice agar (8). The medium is drained from the plates after incubation for 24 hr in the dark and washed three times over a 1-hr period with 25-ml aliquots of autoclaved, distilled or deionized water. High numbers of zoospores have been produced more consistently in our studies when the wash water was replaced with an autoclaved, buffered solution of 10^{-4} M 2-(morpholino)-ethanesulfonic acid (MES) in deionized water that is adjusted to pH 6.2 with 1 *N* KOH before use (8,9). After 18–24 hr under light (4,000 lux at the level of the cultures), the cultures are washed three times with water or MES and suspended in 15 ml of the same solution. Zoospores are released from sporangia after additional incubation of 2–5 hr. Some *Pythium* spp. may require specialized salt solutions for the production of sporangia or zoospores, and solutions such as that recommended by Schmitthenner (11) may be used in place of water.

If motile zoospores are to be used, the suspensions of zoospores must be handled carefully and rapidly to avoid encystment (9). Small samples of zoospore suspensions in test tubes can be treated on a vortex mixer for 30 sec to induce encystment for counting on a hemacytometer. The microdrop method of quantifying zoospores may be used either with samples of encysted zoospores or samples with motile zoospores in suspensions diluted to give a few zoospores per microdrop (e.g., two to five zoospores per 10-μL drop) (5,8,9). Prepared concentrations and viabilities of zoospores may be confirmed by placing small samples of zoospore dilutions on water agar or selective media and observing the plates after 12–18 hr for germination and growth.

Defined numbers of zoospores may be added directly to plants or parts of plants treated with fungicides, plant parts may be floated in suspensions of zoospores containing fungicides, or zoospores may be added to water or soil containing plants. Although zoospore suspensions have been poured over soil in many studies, this method results in unnaturally high concentrations of zoospores being required for infection, probably because of high attrition rates, and poor reproducibility of disease incidence. Much lower concentrations of zoospores are required for infection, and ecological implications in disease are more accurately simulated, when zoospore suspensions are added directly to flooded soil (8,9). The soil can be maintained in the flooded stage for several hours to expose stems, crowns, or flagging foliage, and then the infested water may be drained through the soil to expose roots to the inoculum. Zoospore suspensions may be sprayed onto plant foliage or poured into crowns of plants to approximate dispersal of inoculum in wind-driven or splashing rain.

We have presented only a few examples of the many techniques and approaches for isolating, identifying, and producing inocula of *Pythium* spp. We hope, however, that, within the constraints of time and space involved in conducting this type of background work for tests on fungicide evaluation, researchers will be more aware than in the past of ecological and epidemiological considerations in studying *Pythium* spp.

LITERATURE CITED

1. Burr, T. J., and Stanghellini, M. E. 1973. Propagule nature and density of *Pythium aphanidermatum* in field soil. Phytopathology 63:1499-1501.
2. Garcia, R., and Mitchell, D. J. 1975. Interactions of *Pythium myriotylum* with *Fusarium solani, Rhizoctonia solani*, and *Meloidogyne arenaria* in pre-emergence damping-off of peanut. Plant Dis. Rep. 59:665-669.
3. Hendrix, F. F., Jr., and Campbell, W. A. 1973. Pythiums as plant pathogens. Annu. Rev. Phytopathol. 11:77-98.
4. Hendrix, F. F., Jr., and Papa, K. E. 1974. Taxonomy and genetics of *Pythium*. Proc. Am. Phytopathol. Soc. 1:200-207.
5. Ko, W. H., Chase, L. L., and Kunimoto, R. K. 1973. A microsyringe method for determining concentration of fungal propagules. Phytopathology 63:1206-1207.
6. Middleton, J. T. 1943. The taxonomy, host range and geographic distribution of the genus *Pythium*. Mem. Torrey Bot. Club 20:1-71.
7. Mircetich, S. M., and Kraft, J. M. 1973. Efficiency of various selective media in determining *Pythium* populations in soil. Mycopathol. Mycol. Appl. 50:151-161.
8. Mitchell, D. J. 1978. Relationships of inoculum levels of several soilborne species of *Phytophthora* and *Pythium* to infection of several hosts. Phytopathology 68:1754-1759.
9. Mitchell, D. J., Kannwischer, M. E., and Moore, E. S. 1978. Relationship of numbers of zoospores of *Phytophthora cryptogea* to infection and mortality of watercress. Phytopathology 68:1446-1448.
10. Sauve, R. J., and Mitchell, D. J. 1977. An evaluation of methods for obtaining mycelium-free oospores of *Pythium aphanidermatum* and *P. myriotylum*. Can. J. Microbiol. 23:643-648.
11. Schmitthenner, A. F. 1973. Isolation and identification methods for *Phytophthora* and *Pythium*. Proc. Woody Ornamental Dis. Workshop, 1st, 24–25 January 1973. University of Missouri, Columbia. 128 pp.
12. Tsao, P. H. 1970. Selective media for isolation of pathogenic fungi. Annu. Rev. Phytopathol. 8:157-179.
13. Tsao, P. H. 1983. Factors affecting isolation and quantitation of *Phytophthora* from soil. Pages 219-236 in: *Phytophthora:* Its Biology, Taxonomy, Ecology, and Pathology. D. C. Erwin, S. Bartnicki-Garcia, and P. H. Tsao, eds. American Phytopathological Society, St. Paul, MN. 392 pp.
14. Tsao, P. H., and Ocaña, G. 1969. Selective isolation of species of *Phytophthora* from natural soils on an improved antibiotic medium. Nature (London) 223:636-638.
15. Tuite, J. 1969. Plant Pathological Methods—Fungi and Bacteria. Burgess Publishing Co., Minneapolis, MN. 239 pp.
16. Van der Plaats-Niterink, A. J. 1981. Monograph of the genus *Pythium*. Stud. Mycol. 21. 242 pp.
17. Waterhouse, G. M. 1967. Key to *Pythium* Pringsheim. Mycol. Pap. 109. Commonw. Mycol. Inst., Kew, Surrey, England.
18. Waterhouse, G. M. 1968. The genus *Pythium*: Diagnosis (or descriptions) and figures from the original papers. Mycol. Pap. 110. Commonw. Mycol. Inst., Kew, Surrey, England.

Preparation of Inoculum for Brown Rot of Stone Fruit and Apple Scab

MICHAEL SZKOLNIK, Department of Plant Pathology, New York State Agricultural Experiment Station, Geneva 14456

Successful orchard and greenhouse research on the biology and control of plant diseases is inseparably linked with suitability of inoculum. Whenever researchers can exercise reasonable control of inoculum as to type, quality, concentration, and timeliness of use, they can minimize unusual variability and maximize reproducibility and interpretations of experimental results.

The procedures described here for the procurement and preparation of inoculum in plant disease studies are simple and effective, and they have been proven through their wide usage for brown rot of stone fruits caused by *Monilinia fructicola* (Wint.) Honey (2,7,8) and *M. laxa* (Aderh. & Ruhl.) Honey and apple scab caused by *Venturia inaequalis* (Cke.) Wint. (anamorph: *Spilocaea pomi* Fr.) (1,5,6).

MONILINIA FRUCTICOLA AND *M. LAXA*

Production of inoculum. Conidia of *M. fructicola* and *M. laxa* can be produced in abundance on canned fruit as needed for inoculation of stone fruit (*Prunus* spp.) in studies of either the blossom blight or fruit rot phases of brown rot. The conidia can be produced in transparent plastic boxes with lids, each containing a wire mesh tray with crimped or rolled edges to support it about 1 cm (0.5 in.) above the bottom. Good cleaning of this equipment with detergent is usually adequate, but contamination can be minimized by an added rinse in 1:10 chlorine bleach (5.25% sodium hypochlorite) in water.

Chances for contamination by *Rhizopus* spp. can be reduced appreciably by painting the inside walls and lid of the containers with 2,6-dichloro-4-nitroaniline (Botran 75W) at 6 g/L. This precaution usually discourages extensive sporulating aerial mycelium from colonizing the container and rendering the *M. fructicola* inoculum undesirable. Approximately 0.5 cm of water in the bottom of the container, placed beneath but not touching the wire rack, maintains a satisfactory level of humidity.

Canned peaches or apricots processed in heavy sugar syrup, which are readily available in grocery stores, provide an excellent medium for spore production. Drained fruit halves or slices are placed on the wire mesh rack. Absolutely sterile transfer conditions are not necessary, but to avoid contamination one should work in a clean room without much air movement; treat the top of the fruit can, can opener, and forceps or tongs with 10% chlorine bleach; and wash down immediate work area and hands.

The canned fruit is inoculated by the transfer of small segments of mycelial culture or loopfuls of spores to several sites from some uncontaminated culture, fruit, or other source. Upon incubation at about 22°C (72°F), the fungus will spread over the fruit and produce 4×10^6 or more spores per square centimeter of fruit surface in about 4 or 5 days. The spores can be used at this time for preparation of inoculum. Otherwise, the containers can be stored at about 6°C (43°F) for 3 or 4 weeks without loss of spore viability. Considering the simplicity of the procedure, the production of a new source of supply of conidia is preferable to long-term storage. Such storage may encourage contaminants, especially *Penicillium* spp.

Preparation of inoculum suspension. The mass-produced conidia from canned fruit are of comparable age and are very satisfactory for biological or fungicide evaluation studies on either the blossom blight or fruit rot phase of brown rot. The spores are easily washed off the fruit with water into a beaker by use of a hand-held atomizer fitted to an air line, first at 69 kPa (10 psi) to wet the spore mass and avoid blowing away spores and then at 172 kPa (25 psi) to remove the spores more effectively from the fruit. The heavy suspension is then poured into a blender for high-speed agitation for up to a minute to break up spore chains and to aid in final spore inoculum standardization. The addition of one drop of a 1:10 dilution of Triton X-100 surfactant per 100 ml of spore concentrate during high-speed agitation helps to break up the spore chains.

The spore concentrate is diluted in distilled water without additives and standardized by use of a hemacytometer to the concentration of spores needed for blossom or fruit inoculation. A suspension of 60,000 spores per milliliter is usually adequate for infection without excessive inoculum pressure.

Some strains of *M. fructicola* and *M. laxa* may produce predominantly mycelial growth and rather few spores on artificial media and on canned fruit. A greater number of fruit may need to be inoculated to produce enough spores for large-scale tests. Some increase in spore productivity has been noted on fruit in transparent containers with fluorescent lights directly above them.

Dry conidia can be removed from the peaches or apricots with the use of a cyclone spore trap described by Ogawa and English (3). Handling of these spores for inoculum preparation would be the same as described above.

M. fructicola and *M. laxa* growing and sporulating on canned fruit can be stored in tight plastic containers in the freezer at about −20.5°C (−5°F) for over a year.

VENTURIA INAEQUALIS

The apple scab pathogen grows too slowly and sporulates too sparsely in artificial culture media to provide the abundance of conidial inoculum often needed for biological or fungicide evaluation tests on apple trees in the field or greenhouse. An adequate and continuing supply of comparable-aged conidia with unquestioned pathogenicity can be made from inoculated apple seedlings or nursery stock. This procedure is the most desirable for long-term projects on apple scab. Once the preparatory steps described here are put into effect, an ongoing source of conidial inoculum is assured.

Preparation of plant material for inoculum source. *Apple seeds.* Apple seeds are procured from McIntosh or Red Delicious apples (seeds from polyploid cultivars are much less desirable), dried, and stored in closed glass containers at 1°C (34°F). At apple harvest time each fall, enough seeds should be obtained for research needs during the next year or two. Seeds are usable after 6–8 weeks of apple or seed storage at the indicated low temperature.

Seed pregermination. A continuing supply of pregerminated apple seeds can be maintained for future needs. A determined number of good, viable seeds are first mixed in a water slurry of captan (50WP, 2.5 g/100 ml) for 3–12 hr. The fungicide minimizes fungal overgrowth of seeds during germination. A water-saturated pad of about 10 thicknesses of paper toweling is placed into a petri dish. About 200 captan-coated seeds are placed on the toweling in each dish, which is then covered and kept in a refrigerator at about 7°C (44°F for germination to take place. In 5 or 6 weeks most seeds will have germinated, with the primary root up to about 16 mm (0.7 in.) long.

Planting seedlings. The pregerminated seedlings are transplanted into flats or pots as desired. The substrate should be a well-draining greenhouse medium (e.g., a 1:1:2 mixture of good field soil, sand, and shredded peat moss). Seedlings are planted in dowel holes to a depth just barely covering the seeds. In normal maintenance in the greenhouse, care must be taken to keep the substrate moist and to avoid excessive watering. Once the seedlings break through the surface, a weekly watering of the substrate with a complete soluble fertilizer (including minor elements) helps in establishing vigorous plants that will serve, after inoculation, as valuable sources of inoculum for research.

Inoculation and infection. Once the seedlings have reached the 8- to 10-leaf stage (even earlier, if needed), they are satisfactory for inoculation with spores of *V. inaequalis*. The spores may be primary ones (ascospores) discharged onto the seedlings from pseudothecia-bearing, overwintered apple leaves (at the mature ascospore stage), which are first wetted and placed on a wire mesh above the apple seedlings; or they may be conidia from infected leaves from the field or greenhouse, artificial culture, or stored frozen spores (4). When possible, the conidial suspension should be standardized to at least 50,000 spores per milliliter or 100 spores per low-power (×100) microscope field. The number of seedlings inoculated should meet inoculum needs as determined by experience.

Inoculated seedlings should be infected in a mist chamber at 20°C (68°F) where leaves are kept wet for 24–30 hr. The seedlings are then returned to the greenhouse for normal maintenance. In 8–10 days (sometimes less), sporulating scab symptoms on the leaves indicate the readiness of this inoculum source. Usually the conidia are produced abundantly on both upper and lower leaf surfaces of seedlings.

Preparation of inoculum for experiments. The conidia are easily washed off the leaves into a beaker by use of a hand-held atomizer fitted to an air line at 172 kPa (25 psi). The spore concentrate is further diluted with distilled water fortified with V-8 juice at 10 ml/L. Inoculum should be standardized by use of a hemacytometer. Spore suspensions should be prepared shortly before they are needed for inoculation. Spores that settle from the suspension on standing can readily be resuspended by vigorous shaking of the suspension. Allowing the suspension to stand too long may result in unwanted spore germination, intertwined germ tubes, and change in the concentration and quality of effective inoculum.

In many research investigations on scab, especially the evaluation of fungicides for control, the optimal conidial concentration has been found to be 70,000 per milliliter. Concentrations substantially below 50,000 often result in a very low level of scab infection; those substantially above 100,000 can provide too much inoculum and reduce fungicidal efficacy.

LITERATURE CITED

1. Hamilton, J. M., Szkolnik, M., and Nevill, J. R. 1964. Greenhouse evaluation of fruit fungicides in 1963. Plant Dis. Rep. 48:295-299.
2. Nevill, J. R., Szkolnik, M., Gilpatrick, J. D., and Ogawa, J. M. 1978. Mass production of conidia of brown rot fungi on canned fruit pieces. Plant Dis. Rep. 62:966-969.
3. Ogawa, J. M., and English, H. 1955. The efficiency of a quantitative spore collector using the cyclone method. Phytopathology 45:239-240.
4. Rich, A. E. 1971. A simple method for maintaining *Venturia inaequalis* inoculum. Plant Dis. Rep. 55:976.
5. Szkolnik, M. 1978. Techniques involved in greenhouse evaluation of deciduous tree fruit fungicides. Annu. Rev. Phytopathol. 16:103-129.
6. Szkolnik, M. 1978. Evaluation of physical modes of action of fungicides against the apple scab fungus. Proc. Apple and Pear Workshop, Kansas City, MO, 1976; N.Y. State Agric. Exp. Stn. Spec. Rep. 28, pp. 22-27.
7. Szkolnik, M. 1981. Protective and eradicative fungicidal action in the control of cherry brown rot blossom blight. Fungic. Nematic. Tests 36:30.
8. Szkolnik, M., and Gilpatrick, J. D. 1977. Tolerance of *Monilinia fructicola* to benomyl in western New York State orchards. Plant Dis. Rep. 61:654-657.

Monitoring Resistance to Benomyl in *Venturia inaequalis*, *Monilinia* spp., *Cercospora* spp., and Selected Powdery Mildew Fungi

K. S. YODER, Department of Plant Pathology, Physiology and Weed Science, Winchester Fruit Research Laboratory, Virginia Polytechnic Institute and State University, Virginia Agricultural Experiment Station, Winchester 22601; and A. E. DAVIS and B. A. HADLEY, Agricultural Chemicals Department, Stine-Haskell, E. I. du Pont de Nemours & Company, Newark, DE 19711

The development of resistance in fungi to specific fungicides has been well recognized (1,3) and continues to be a matter of practical concern in plant disease management. Monitoring the fungal population for resistance to preferred fungicides can reduce severe disease losses by enabling appropriate changes in the spray program to control the resistant strain (7).

This chapter describes procedures for monitoring resistance to benomyl (methyl 1-[butylcarbamoyl]-2-benzimidazolecarbamate) of *Venturia inaequalis* (Cke.) Wint. and *Podosphaera leucotricha* (Ell. & Ev.) Salm. on apples (*Malus sylvestris* Mill.), *Monilinia* spp. on stone fruits (*Prunus* spp.), *Cercospora apii* Fres. on celery (*Apium graveolens* L.), *C. arachidicola* Hori on peanut (*Arachis hypogaea* L.), *C. beticola* Sacc. on beet (*Beta vulgaris* L.), and *Erysiphe cichoracearum* DC. on cucurbits (*Cucumis* and *Cucurbita* spp.).

METHODS AND PROCEDURES

General sampling procedures. When fungicide resistance is suspected as a cause of inadequate disease control, diseased material should be selected from well-sprayed areas that show the highest infection. When the effects of altered spray programs on the proportion of resistant and sensitive strains in the population are being investigated, however, samples should be randomly collected from large, replicated plots that are spaced to avoid interplot interference.

Collection of fresh, sporulating lesions facilitates isolation and testing. Cross-contamination, overheating, or delaying the processing of samples should be avoided. The isolated organism must be identified as the pathogen, perhaps through host inoculation studies.

In vitro concentration for determining resistance. Although earlier experience from monitoring benomyl resistance problems had indicated that the single in vitro benomyl rate of 5 mg/L was usually adequate for most monitoring purposes, occasional isolates of some fungi (e.g., *Monilinia* sp.) have been found to be sensitive to 5 mg/L but resistant to lower rates (1–2.5 mg/L). Therefore, an in vitro rate of 2.5 mg/L is more commonly used to screen for resistance when only one rate is used; but a range of rates from 1 to 5 mg/L may be useful to determine the resistance levels of individual isolates in a population.

Apple scab (*Venturia inaequalis*). *Sampling procedures.* Sporulating lesions on the youngest leaves or fruits are best for testing. For fruit in storage, small lesions that have not broken the cuticle are ideal. Leaf samples should be collected when the leaf surface is dry and placed into plastic bags without the addition of extra moisture. Refrigeration immediately after collection helps to prolong conidia viability and reduce contamination. The sample should include 20 or more leaves or fruits with lesions from different trees.

Determination of resistance. In vitro resistance to benomyl correlates closely with in vivo resistance. *V. inaequalis* can be assayed on Difco potato-dextrose agar (PDA, 39 g/L + 5 g of agar per liter) acidified to pH 3.7 with eight drops (1.05 ml) of 85% lactic acid per liter and containing benomyl at 2.5 mg/L. This dosage provides approximately five times the amount of toxicant required to inhibit sensitive strains under identical conditions. The lactic acid should be added after autoclaving when the medium has cooled to 60° C. For monitoring purposes, commercial formulations of benomyl may be added to PDA as a water stock suspension before autoclaving or as an ethanol stock suspension after autoclaving.

Divided plastic petri dishes containing benomyl-amended and nonamended PDA in the opposite compartments can be used to simplify culture identification, record keeping, and fungal growth comparison. After the medium has solidified, the plates are returned to the sterile plastic sleeve in which they were purchased, where they may be stored upside down for several months at ambient room temperature.

Because *V. inaequalis* is comparatively slow growing, care is required in the isolation technique. Sporulating lesions from unsterilized leaves or fruits are bisected with a sterile scalpel and excised with sterile forceps. Conidia are dislodged by rubbing one-half of the excised lesion across the fungicide-amended agar and the other half across the control agar. By aligning the conidial streaks on opposite sides of the plate perpendicular to the center divider, treated and control conidia from six to eight lesions may be placed on a single plate. A small, sterile spatula may also be used to scrape conidia from the lesion and deposit them on the agar surface. The concentration of conidia is greatest where the lesion or spatula first touches the agar surface. This initial contact should be made to accommodate microscopic examination.

Germination of conidia is observed after 24 hr at 16–19°C. Benomyl does not greatly reduce germination. Sensitive germ tubes grow only one or two cells in length, however, and appear distorted in the presence of the fungicide. Germ tubes from resistant strains are uninhibited; growth on medium containing benomyl at 2.5 mg/L is the same as on nonamended medium. Since some conidia may have germinated on moist leaves before contact with the agar, a second transfer is needed to confirm resistance. Spores that appear to be uninhibited are transferred as single-spore isolates onto sterile, benomyl-amended PDA plates for confirmation. A resistant isolate will continue to grow and sporulate on medium containing benomyl at 2.5 mg/L, but a sensitive isolate will not grow. Germinated conidia transferred with the agar block may grow from the top of the block without contacting the fungicide-amended PDA. Thus one must ascertain that the mycelium is in direct contact with the fungicide-amended PDA before confirming resistance. Sporulation of the fungus in pure culture is desirable to confirm identity of the fungus.

In storage, scab lesions that have not broken the fruit cuticle may be surface sterilized by being wiped lightly with a paper tissue soaked with 95% ethanol, bisected, and placed on divided plates as described, with the cut edge of the lesion in contact with the agar.

Conidia from a single, uncontaminated lesion are usually either all resistant or all sensitive.

Brown rot of stone fruits (*Monilinia* spp.). *Sampling procedures.* Twenty or more infected fruits, blossoms, or twig cankers or a combination of these from different trees are collected in plastic bags and refrigerated immediately. Dry fruit mummies collected in paper bags are kept dry until processing.

Determination of resistance. Sporulation of nonsporulating blossom infections and mummies may be induced by alternating daylight and dark conditions on 1.5% water agar. Cultures apparently will sporulate in variable day-length periods. Twig cankers are cut to 3 cm in length to include a canker margin (1 cm for the dead portion, 2 cm for the live portion), dipped for 2 sec in 95% ethanol, rinsed in sterile distilled water, dissected longitudinally, and placed on 1.5% water agar. A selective medium that Phillips and Harvey (8) described has also been useful for isolation from heavily contaminated samples.

Conidia from sporulating cultures or lesions are transferred with a sterile needle to fungicide-amended PDA prepared as described for *V. inaequalis*. Uninhibited growth on medium with benomyl at 2.5 mg/L confirms resistance, but sensitive germ tubes of *Monilinia* spp. may grow several cells in length before complete inhibition of sensitive strains occurs. Sporulation with benomyl at 2.5 mg/L,therefore, is a more reliable indication of resistance than is the percentage of germination or germ tube length. *M. fructicola* will sporulate within several days of the first transfer of conidia to PDA. *M. laxa* may not sporulate readily on the first transfer, but if an agar block from the first colony is transferred again to PDA, sporulation on the agar block will occur.

Leaf spots of celery, peanut, and sugar beet (*Cercospora* spp.). *Sampling procedures.* Fifty or more leaves with young, sporulating lesions from near the shoot apex are selected. Contaminated, senescent leaves near the soil surface should be avoided. Samples may be collected in plastic or paper bags, but crushing the sample may flatten the spores against the leaf surface and make isolation with a transfer needle more difficult. *Cercospora* spores are resistant to drying, so leaf samples may be either dried or refrigerated.

Determination of resistance. Lesions without spores may be induced to sporulate by placing them in a humid atmosphere. The long conidia stand well above the lesion surface and can be readily transferred with a sterile needle to divided plates containing fungicide-amended PDA; the amended and the control PDA are prepared as described for *V. inaequalis*. Because *C. arachidicola* is slow growing, special care should be taken to touch only the tips of the conidia to avoid contamination in the transfer process.

The sensitive germ tube grows only a few cells in length on the benomyl-amended PDA, but growth of resistant isolates is uninhibited.

C. beticola and *C. apii* do not sporulate well on PDA but will do so on transfer to a sugar beet leaf extract medium. The medium is prepared by steaming 16 g of dry-weight or 130 g of fresh-weight sugar beet leaves in 500 ml of distilled water for 20 min. The liquid is strained through four layers of cheesecloth, its total volume is brought to 500 ml, 10 g of agar is added, and it is autoclaved for 20 min. Cultures incubated 17 cm below 20-W General Electric cool white fluorescent tubes under 12-hr alternating light and dark periods at 25°C will begin to sporulate in about 2 days.

Powdery mildews of cucurbit and apple (*Erysiphe cichoracearum* and *Podosphaera leucotricha*). *Sampling procedures.* Twenty or more cucurbit leaves or apple shoots with lesions showing freshly sporulating areas are selected. The samples are placed in plastic bags and refrigerated until processing.

Determination of resistance. To monitor for powdery mildew resistance, susceptible plants must be isolated to maintain mildew-free conditions without the use of fungicides. Cucurbit or apple powdery mildews can be tested on young potted seedlings of susceptible varieties such as Straight Eight cucumber or Rome Beauty apple. The youngest fully expanded apple leaves are excised from seedlings or grafted stock and maintained throughout the incubation period by immersing the cut end of the petiole in a small vial containing a 3% (w/v) sucrose solution. Covering the opening of the vial with Parafilm M (American Can Company, Neenah, WI) retards evaporation from the vial and supports the excised plant parts. The sucrose solution is conveniently replenished with a narrow-spout laboratory wash bottle through a small slit in the Parafilm M.

Cucumber foliage is sprayed to runoff with active benomyl (Benlate 50% WP) at 50–100 mg/L. Because no resistant apple powdery mildew has been confirmed, lower rates (20 mg of active ingredient per liter) are used to assure adequate sensitivity of the test method.

Inoculum may be transferred from the disease sample by gently blowing across the sporulating surface toward the healthy plant or by preparing a spore suspension containing from 5×10^4 to 10×10^4 conidia per milliliter of water and one drop of Tween 20 per 750 ml of water. Because the viability of conidia in the suspension decreases rapidly with time, such suspension should be prepared immediately before spray inoculation with a DeVilbiss atomizer. After inoculation, the plants are incubated in translucent plastic containers covered with four layers of cheesecloth. Percentage of area infected and lesions per leaf are compared on treated plants and controls. Sensitive strains cause little or no infection on plants treated at the suggested rates. If a few lesions appear on the treated plants, reinoculation of treated plants from these lesions confirms resistance or sensitivity of the organism.

DISCUSSION

The methods described have been proven effective in detecting resistance to benomyl with numerous samples of each organism except apple powdery mildew. No resistant strain of the apple powdery mildew fungus has been detected. Related methods have been reported for *Cercospora* spp. (2,6), *M. fructicola* (4), and *V. inaequalis* (5). The methods reported here for *V. inaequalis* have also been used to test resistance of peach and pecan scab fungi (*Cladosporium carpophilum* Thuem. and *Fusicladium effusum* Wint.) and are also adaptable to the pear scab fungus (*Venturia pirina* Aderh.).

The in vitro rate of 2.5 mg of benomyl per liter is about five times the amount required to inhibit sensitive strains of most fungi and can adequately differentiate resistant and sensitive strains. Higher rates in vitro sometimes inhibit organisms resistant to 2.5 mg/L. Germination and initial germ tube development in sensitive strains are not greatly affected by benomyl at 2.5 mg/L and therefore are poor indicators of resistance. Extended germ tube development is a quick, fairly reliable criterion of resistance if the possibility of pregermination is eliminated and if the spore is still visible for identification. Since numerous fungi are not inhibited, confirmation of the pathogenic organism is necessary.

In our experience, a relatively small sample size (20 lesions) may be used to determine whether ineffective control is the result of high frequencies of resistant propagules. Larger samples are needed to detect low frequencies of resistant strains in population dynamics studies. If all the spores on one lesion arise from one parent spore, each lesion represents a sample size of one. Thus, one million viable spores from 20 lesions would represent a sample size of 20, not one million.

To have valuable predictive qualities, monitoring methods must detect resistant pathogenic propagules at a frequency lower than that capable of causing visible disease losses beyond those expected in the absence of resistant strains. This information must be available soon enough to prevent the organism from reaching economic significance before the control program can be altered to accommodate the resistant strain.

Convenient mass screening techniques that are rapid, more predictive, and capable of detecting 1 resistant spore in 1,000 need to be developed. Such techniques could involve mass trapping of airborne spores or composite spore suspensions plated onto highly selective media. Detection methods less sensitive than this may not forecast the threat of resistance soon enough to avert serious economic loss. Because spontaneous mutations to benomyl resistance may occur in some fungi at a frequency of 1 in 5×10^7 conidia (9), highly sensitive methods would likely detect spontaneous mutants if 10^7 conidia were screened. Such information could be misleading, however, unless resistant spores are tested for pathogenicity.

LITERATURE CITED

1. Dekker, J. 1976. Acquired resistance to fungicides. Annu. Rev. Phytopathol. 14:405-428.
2. Georgopoulos, S. G., and Dovas, C. 1973. A serious outbreak of strains of *Cercospora beticola* resistant to benzimidazole fungicides in Northern Greece. Plant Dis. Rep. 57:321-324.
3. Georgopoulos, S. G., and Zaracovitis, C. 1967. Tolerance of fungi to organic fungicides. Annu. Rev. Phytopathol. 5:109-130.
4. Jones, A. L., and Ehret, G. R. 1976. Isolation and characterization of benomyl-tolerant strains of *Monilinia fructicola*. Plant Dis. Rep. 60:765-769.
5. Jones, A. L., and Walker, R. J. 1976. Tolerance of *Venturia inaequalis* to dodine and benzimidazole fungicides in Michigan. Plant Dis. Rep. 60:40-44.
6. Littrell, R. H. 1974. Tolerance in *Cercospora arachidicola* to benomyl and related fungicides. Phytopathology 64:1377-1378.
7. Littrell, R. H. 1976. Techniques of monitoring for resistance in plant pathogens. Proc. Am. Phytopathol. Soc. 3:90-96.
8. Phillips, D. J., and Harvey, J. M. 1975. Selective medium for detection of inoculum of *Monilinia* spp. on stone fruits. Phytopathology 65:1233-1236.
9. Polach, F. J., and Molin, W. T. 1975. Benzimidazole-resistant mutant derived from a single ascospore culture of *Botryotinia fuckeliana*. Phytopathology 65:902-904.

Testing for Streptomycin Resistance in *Erwinia amylovora*

SHERMAN V. THOMSON, Department of Biology, Utah State University, Logan 84322

Streptomycin, which was found to control fire blight in the early 1950s, came into commercial use in the late 1950s and early 1960s (13). However, the widespread use of streptomycin was critically examined and consequently reduced in some areas when resistance to streptomycin was discovered in several different geographic areas in the early 1970s. Moller et al. (10) and Miller and Schroth (9) reported in 1972 that streptomycin-resistant strains (str^r) of *Erwinia amylovora* were isolated from pear orchards in California where fire blight was serious despite the repeated use of streptomycin. Subsequent reports of resistance of *E. amylovora* to streptomycin were made from Oregon and Washington (4). The development of str^r in *E. amylovora* has not occurred in all areas where streptomycin has been used. For example, resistant strains were not found in Michigan (14) or New York (2).

The widespread use of streptomycin is not necessary for the development of resistance, as shown by the isolation of resistant strains from areas where streptomycin has not been used (13), and is not correlated with the previous history of streptomycin usage (10). However, it is more likely that resistant strains will become a significant problem in areas having the selective pressure of repeated streptomycin usage.

There is considerable variability in the characteristics of the resistant strains isolated from orchards. The level of resistance ranged from 10 to 800 μg/ml in direct isolations from Oregon and Washington, but in California orchards the isolates were all resistant to approximately 800–1,000 μg/ml (4). The earlier reports suggested that the str^r strains were more virulent than streptomycin-sensitive strains (str^s) (4,10), but subsequent tests with larger numbers of isolates showed a considerable range of virulence in both str^r and str^s strains and no consistent difference between the two groups of strains (13). Generation times among str^r strains also showed considerable variability and were not significantly different from str^s strains (13). The diversity of str^r strains with respect to colony morphology, virulence, and generation times indicates that they arose by separate mutations from a heterogeneous assortment of strains rather than from one resistant strain that then spread from a single location.

The str^r strains are capable of surviving in nature without the constant selective pressure of streptomycin (4,13). Coyier and Covey (4) reported that the number of str^r strains in pear orchards in the Yakima Valley declined to about 50% of the total 2 years after discontinuance of streptomycin. Schroth et al. (13) found that the population of str^r strains declined from 68% of the infections 2 years after streptomycin applications were discontinued to 16% after 6 years. This is in contrast to previous reports that clinical mutants resistant to the aminoglycoside antibiotics are less capable of surviving than wild types (13). There does not appear to be strong competition between str^r and str^s strains because isolations from epiphytically colonized flowers and insects as well as those from infected pear flowers revealed a mixture of both str^r and str^s isolates (13).

Streptomycin resistance in *E. amylovora* does not appear to be plasmidborne but most likely is a chromosomal mutation (12). The str^r property was not transmissible by conjugation even though the strains tested had conjugative fertility. The absence of aminoglycoside phosphorylase and adenylylase suggests that the resistance is not the result of the common R plasmid-encoded resistance mechanisms (12). Streptomycin resistance has been transferred on R factors from strains of *Escherichia coli* to *E. amylovora* in the laboratory (12). However, none of the isolates from the field appears to be carrying these R factors.

Laboratory-derived isolates resistant to streptomycin are easily obtained (3,13), but mutants resistant to 1–5 μg/ml are frequently unstable and may have a reduced virulence or survivability (3,13). They are usually genetically unfit and are considered to be metabolic cripples.

There is some question whether streptomycin resistance in *E. amylovora* would have become a problem in fruit production if the early antibiotic formulations had been retained. A widely used formulation in the early 1950s contained a combination of streptomycin sulfate (100 μg/ml) and oxytetracycline (10 μg/ml) (16). This combination was used initially but was changed to streptomycin alone in the 1960s, apparently because of production costs and the lack of evidence of increased efficacy in the field. English and Van Halsema (5) demonstrated in laboratory studies that resistant strains of *E. amylovora* and *Xanthomonas vesicatoria* developed much more frequently when they were exposed to progressively increasing concentrations of streptomycin alone than when a combination of streptomycin plus oxytetracycline was used. The occurrence of resistant strains of *E. amylovora* developed more rapidly when streptomycin plus 1% oxytetracycline was used when compared with streptomycin plus 10% oxytetracycline. The mutation rates of *E. amylovora* were reported to be 1.0×10^{-8} and 4.1×10^{-9} c.f.u. respectively for streptomycin at 10 μg/ml and 500 μg/ml (13) and were 6×10^{-8} c.f.u. for oxytetracycline at 1.6 μg/ml (7). However, field applications of streptomycin are recommended at 25–100 μg/ml. Therefore, small amounts of oxytetracycline incorporated in the mixture would result in very low amounts actually applied. Thus at lower rates the level of oxytetracycline was not likely to inhibit the growth of *E. amylovora*, and resistance could have developed to streptomycin just as if streptomycin were applied alone.

Standardized techniques to determine the presence and level of resistance of *E. amylovora* to streptomycin are needed so that comparisons can be made with different geographic areas. Growers should be aware of the potential or presence of str^r strains and be prepared to modify control applications as necessary.

METHODS

Sampling procedures. Infected tissue should be sampled from orchards where streptomycin has been used and resistance is suspected because of poor control or repeated usage. Samples should be taken from representative areas of the orchard. The number of samples required will vary depending on the incidence of resistance and the level of confidence required. A sample of 10–20 infections per area is adequate. Infected tissue is surface sterilized with 10% sodium hypochlorite for 5 min, plated on Miller-Schroth (MS) medium (15) or other suitable selective medium, and incubated at 28° C. Resistance can also be determined in epiphytic populations by making washes of healthy flowers, insects, or terminals (15). Isolates should be verified to be *E. amylovora* by inoculating green pear fruit or etiolated pear seedlings.

Replicate plating. Single colonies from the selective media are spotted on master plates of King's medium B (6). The number of colonies per plate can range from a few to 25 or more. Known str^r and str^s isolates of *E. amylovora* should be included for comparison. These colonies are transferred via the replica-plating technique (1) to plates of MS medium (surface-dry) with streptomycin sulfate (78.1% streptomycin; Sigma Chemical Co., St. Louis, MO 67178) incorporated at 0, 10, 100, or 500 μg a.i./ml or at whatever level of resistance is of interest. Inverted plates are incubated at 28°C for 36–48 hr.

Large numbers of *E. amylovora* isolates can be screened by making various dilutions of blossom washes or washes from other healthy tissues and plating the washes on MS medium. Plates with well-dispersed colonies can be used as master plates for the replica-plating technique.

Mass screening. The presence or absence of streptomycin resistance can also be tested directly from infections by macerating infected tissue in a small quantity of sterile 0.05 M potassium phosphate, pH 6.5, and streaking loopfuls on plates of MS medium amended with streptomycin (as sulfate) at 0, 10, and 100 μg/ml.

Levels of resistance. Recommended testing standards must be used in comparing levels of streptomycin resistance from various geographic areas and from other laboratories. Previous comparative tests have shown a marked influence of media on the size of inhibition zones produced around disks impregnated with various levels of streptomycin (13). The antibiotic disk susceptibility test established by the National Committee for Clinical Laboratory Standards (NCCLS) (11) recommends the use of Mueller-Hinton medium (Difco Laboratories, Detroit, MI 48201). Mueller-Hinton plates are prepared ahead of time with a depth of approximately 4 mm of media (25 ml in a 100-mm plate). Just before use, the plates are placed in an incubator (35° C) with lids ajar until excess surface moisture is lost by evaporation (10–20 min). There should be no droplets of moisture on the surface of the medium or on the cover.

Isolates used in these tests should be from single-colony transfers and should be previously or concurrently tested for pathogenicity. Colonies are suspended in 4 or 5 ml of sterile saline and the turbidity is adjusted to visually match a $BaSO_4$ standard (0.5 ml of 0.048 M $BaCl_2$ [1.175% $BaCl_2 \cdot 2H_2O$] added to 99.5 ml of 0.36 *N* H_2SO_4 [1%, v/v]). The inoculum should be used within 15–20 min of suspension.

The agar medium is inoculated by dipping a sterile cotton swab on a wooden applicator stick into the suspension. Excess broth is expressed by pressing and rotating the swab firmly against the inside of the tube above the fluid level. The swab is streaked three directions over the entire surface of the agar plate, and excess inoculum is removed by making a circular sweep around the rim of the agar. Inoculation plates should be allowed to dry for 3–5 min but no longer than 15 min before the disks are applied.

Filter paper disks specifically certified for susceptibility testing are impregnated with streptomycin at 1, 10, or 50 μg (BBL, Cockeysville, MD 21030; Difco Laboratories, Detroit, MI 48201). Disks should be stored under refrigeration (2–8° C) or in the freezer at −14° C or lower. They should not be used past the manufacturer's expiration date. Disks are gently pressed onto the agar within 15 min of inoculation. No more than four or five disks should be placed on a single plate. Disks should be no closer than 15 mm from the edge of the plate and arranged so that zones of inhibition do not overlap. Plates are incubated inverted at 28° C for 24 hr.

Results are interpreted by measuring the diameter of the inhibition zone on the back of the plate. Measurement is taken from the line where there is complete inhibition of growth, and small, isolated colonies within the zone are disregarded; they may require reidentification and retesting. Zones should be compared with the standards established by the NCCLS (11). For 10 μg of streptomycin, an inhibition zone of $\leq$ 11 mm indicates a resistant isolate, 12–14 mm indicates intermediate, and $\geq$ 15 mm indicates susceptible.

The testing of levels of streptomycin resistance other than those available with the sensitivity disks can be accomplished but should be reserved primarily for in-house laboratory use. Blank disks can be obtained from supply houses and dipped in freshly prepared solutions of streptomycin. Impregnated disks should be consistently blotted on filter paper to remove excess antibiotic solution by briefly touching the edge of the disk to the filter paper. Disks are then gently pressed onto the agar with forceps to ensure complete contact with the agar surface. The diameter of the complete inhibition zone is measured.

INTERPRETATION OF RESULTS AND DISCUSSION

The comparison among laboratories of levels of str^r in isolates of *E. amylovora* is difficult because standardized procedures are generally not used. The use of the approved performance standards for antimicrobial disk susceptibility published by the NCCLS should alleviate some of the problems. However, the levels of resistance of importance in agricultural crops require the testing of streptomycin at levels higher than the suggested standards. It therefore becomes necessary to use disks prepared with higher levels of streptomycin or to incorporate streptomycin into the medium. These other techniques provide valuable information and should continue to be used, but the accepted standard procedure should always be performed for purposes of comparison.

Known str^r and str^s isolates of *E. amylovora* should always be used in the comparisons. Pathogenicity tests are also necessary when testing field isolates to be sure they are virulent *E. amylovora*.

The rapid screening of numerous *E. amylovora* isolates directly from the field is an important procedure that can be used to aid fruit growers in decisions regarding the type of bactericide to use. In areas where streptomycin is used on a regular basis, a resistance survey should be made annually, especially in orchards where control is not as good as expected. In areas where streptomycin resistance has not yet been detected, growers should be alerted to the potential problem and recommendations made to limit the repeated use of streptomycin. In areas where streptomycin resistance is a spotty problem, orchards should be closely monitored to detect resistance. In orchards where resistance is currently present, growers should be informed of alternative bactericides and advised not to use oxytetracycline exclusively. Fixed copper bactericides should be used when varieties or environmental conditions are not conducive to russetting.

LITERATURE CITED

1. Barry, A. L., and Thornsberry, C. 1980. Susceptibility testing: Diffusion test procedures. In: Manual of Clinical Microbiology. 3d ed. E. H. Lennette, ed. American Society for Microbiology, Washington, DC. 1,044 pp.
2. Beer, S. V., and Norelli, J. L. 1976. Streptomycin-resistant *Erwinia amylovora* not found in western New York pear and apple orchards. Plant Dis. Rep. 60:624-626.
3. Bennett, R. A., and Billing, E. 1975. Development and properties of streptomycin resistant cultures of *Erwinia amylovora* derived from English isolates. J. Appl. Bacteriol. 39:307-315.
4. Coyier, P. L., and Covey, R. P. 1975. Tolerance of *Erwinia amylovora* to streptomycin sulfate in Oregon and Washington. Plant Dis. Rep. 59:849-852.
5. English, A. R., and Van Halsema, G. 1954. A note on the delay in the emergence of resistant *Xanthomonas* and *Erwinia* strains by the use of streptomycin plus terramycin combinations. Plant Dis. Rep. 38:429-431.
6. King, E. O., Ward, M. K., and Raney, D. E. 1954. Two simple media for the demonstration of pyocyanin and fluorescin. J. Lab. Clin. Med. 44:301-307.
7. Lacy, G. H., Cannon, N. P., and Stromberg, V. K. 1983. *Erwinia amylovora* mutants and transconjugants resistant to oxytetracycline. (Abstr.) Phytopathology 73:967.
8. Miller, J. H. 1972. Experiments in molecular genetics. Cold Spring Harbor Laboratory, Cold Spring Harbor, NY. 466 pp.
9. Miller, T. D., and Schroth, M. N. 1972. Monitoring the epiphytic population of *Erwinia amylovora* on pear with a selective medium. Phytopathology 62:1175-1182.
10. Moller, W. J., Beutel, J. A., Reil, W. O., and Zoller, B. G. 1972. Fireblight resistance to streptomycin in California. (Abstr.) Phytopathology 62:779.
11. National Committee for Clinical Laboratory Standards. 1979. Performance standards for antimicrobic disk susceptibility tests. Approved standard, ASM-2. 2d ed. National Committee for Clinical Laboratory Standards, Villanova, PA. 11 pp.
12. Panopoulos, N. J. 1978. Genetic nature of streptomycin resistance in *Erwinia amylovora*. Proc. Int. Conf. Plant Pathol. Bacteriol., 4th, 2:467-469.
13. Schroth, M. N., Thomson, S. V., and Moller, W. J. 1978. Streptomycin resistance in *Erwinia amylovora*. Phytopathology 69:565-568.
14. Sutton, T. B., and Jones, A. L. 1975. Monitoring *Erwinia amylovora* populations on apple in relation to disease incidence. Phytopathology 65:1009-1012.
15. Thomson, S. V., Schroth, M. N., Moller, W. J., and Reil, W. O. 1975. Occurrence of fire blight of pears in relation to weather and epiphytic populations of *Erwinia amylovora*. Phytopathology 65:353-358.
16. Van der Zwet, T., and Keil, H. L. 1979. Fire blight: A bacterial disease of rosaceous plants. Agric. Handbook 510. U.S. Government Printing Office, Washington, DC. 200 pp.

A Yeast Manometric Bioassay Useful for Detecting Fungicide Residue

W. E. MacHARDY, University of New Hampshire, Department of Botany and Plant Pathology, Durham, 03824

Fungicide residue is assayed most commonly by standard laboratory spectrophotometric or chromatographic techniques (1). Bioassay procedures that relate the percentage of spore germination or a zone of inhibition of a test organism to fungicide concentration have also been developed (5,7,11,13,14,16,17), but they have been used primarily as initial laboratory screening procedures for selecting potential fungicides. The standard bioassay techniques have been adapted to detect fungicide residue (2,3,5,6,8,12), but they have features that limit their usefulness as a routine procedure in commercial agriculture: a test organism is required that demands special culturing techniques and equipment, a sterile medium and sterile test conditions are needed, and one to several days must elapse to complete the bioassay. The spore germination procedure also requires an expensive microscope, some training and expertise in microscopy, and training in counting germinated spores and interpreting the data. The standard analytical procedures are even less suited for routine use in commercial agriculture, since they require a trained technician, a well-equipped laboratory with expensive and delicate instrumentation, and chemical extractants and reagents that are potentially harmful to the analyst.

The yeast bioassay (designated BIO-Y) described in this chapter is a manometric technique that relates fungicide concentration in a broth medium to CO_2 production by yeast, *Saccharomyces cerevisiae*. The procedure and apparatus were developed and designed for routine use by scientists or by extension personnel, agricultural consultants, crop management specialists, and even growers. BIO-Y offers several advantages over the standard bioassay and analytical assay tests that make it practical for widespread use in commercial agriculture: it is fast (residue determination in approximately 1 hr), safe (no harmful chemicals are used), and easily performed (no culturing, sterilization, specialized laboratory facilities, or laboratory training required).

BIO-Y has two important limitations: only a few fungicides are applicable, and the test organism is a nonpathogen of plants. Thus, it is not suitable for laboratory screening of potential fungicides. BIO-Y can, however, detect the residue of selected test fungicides on either leaves or nonplant targets, and this has several potential uses in commercial agriculture: calibrating sprayers (9); evaluating the performance of sprayers; detecting the distribution pattern of pesticides in a crop canopy; monitoring pesticide deposition in the canopy of a wide range of orchard, row, and field crops; detecting pesticide contamination of nontarget objects; monitoring exposure of applicators and field workers to pesticides; and providing residue data contributing to the decision-making process for selecting the rate and timing of pesticides in crop management programs (10).

METHODS AND PROCEDURES

Bioassay conditions. The yeast concentration, buffer pH, broth ingredients, and incubation temperature described below were selected because they provided conditions for a rapid production of CO_2 and the development of a suitable standard dosage/response curve for the fungicide captan (Fig. 1). With other fungicides, adjustments in test conditions may be necessary to achieve the desired response, particularly adjustments in yeast concentration, buffer pH, and wetting agent if one is required to remove the fungicide residue from a sprayed target.

Buffer. The test medium is double-distilled water buffered at pH 5.0 with potassium biphthalate, prepared

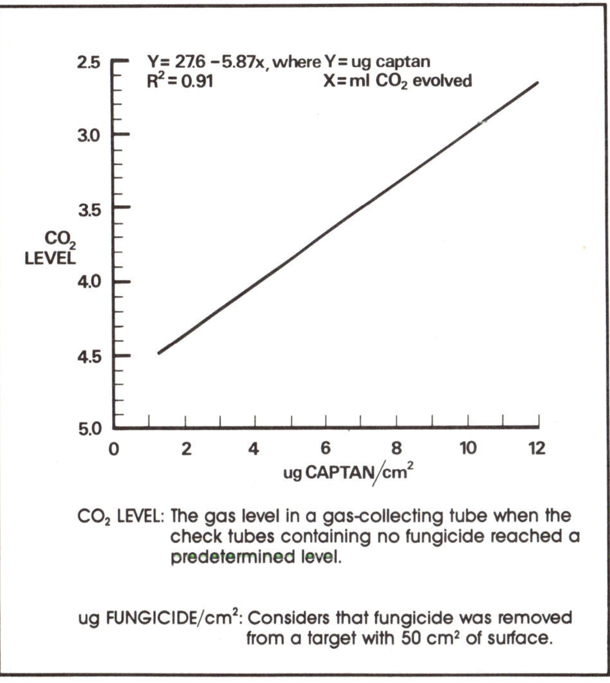

Fig. 1. Standard dosage/response curve for the fungicide captan.

by placing 10.21 g of potassium biphthalate in a 1-L volumetric flask, adding double-distilled water to volume, and adjusting to pH 5.0 with a 10-molal solution of KOH. Approximately 3.5 ml of KOH is required per liter of buffer.

The 10-molal stock solution of KOH is prepared by placing a beaker containing 100 ml of distilled water in ice and slowly adding 56.11 g of KOH, swirling the beaker between additions. The solution should be prepared under a fume hood with protective goggles and gloves, because the solution will burn if it contacts the skin. The stock solution is stored in an airtight container until needed.

The buffer is usually prepared in 20- to 25-L quantities and then measured into 175-ml aliquots that are placed in 200-ml plastic bottles and stored in a refrigerator until needed.

Wetting agent. Two drops (0.05 ml) of X-77 D(Chevron Chemical Company, Ortho Division), a nonionic spreader and activator, is added to each bottle containing buffer. The X-77 enhances the removal of captan from sprayed targets, but it also affects the yeast, as indicated by a reduction in CO production. Thus, the concentration of X-77 selected is the least amount that will facilitate the removal of 90–95% of the captan residue from a target. A different concentration of X-77 or a different wetting agent may be necessary with other fungicides. The wetting agent does not interfere with the accuracy of bioassay measurement of residue on sprayed targets because the standard curve was developed with the wetting agent included in the broth. The captan residue remaining on targets washed with the broth and wetting agent was determined by a spectrophotometric assay technique for captan (15). A comparison of washed and unwashed targets sprayed with the same concentration of captan indicated the percentage of removal of captan.

Broth ingredients. Broth ingredients are preweighed in aliquots sufficient for preparing 175 ml of broth (the amount needed for one run of the bioassay procedure), placed in airtight containers (used film canisters), and stored in a freezer until needed. Each canister contains dextrose (6.6 g), peptone (1.65 g), yeast extract (0.83 g), and yeast (1.1 g).

Yeast. The bioassay organism is baker's yeast, *S. cerevisiae*, purchased as a granular preparation in 907-g (2-lb) packages. Fleishmann's active dry yeast (Standard Brands) is used, but other commercial granular preparations of baker's yeast should be suitable. The expiration date should be several months beyond the date purchased and should extend beyond the time the yeast is used for bioassay. Yeast aliquots of 1.1 g are weighed and added to the canisters containing the dried broth ingredients and stored in a freezer. Yeast remaining in the package should also be stored in a freezer.

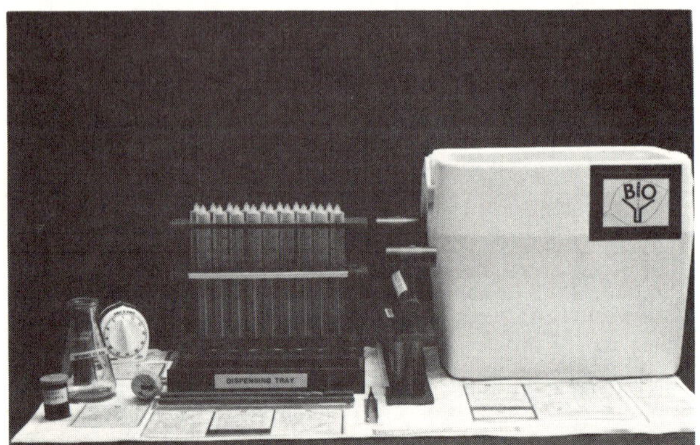

Fig. 2. BIO-Y apparatus positioned on the procedural layout sheet.

The general procedure is to obtain a package just before the growing season, develop a standard curve for the yeast in that package, and use the yeast only during the next growing season. One package provides enough yeast for more than 800 standard runs. Since each run can bioassay 15 separate targets, one package is sufficient for 12,000 residue samples.

Incubation temperature. The water bath temperature is 45 ± 1°C, regulated by a thermostatic heater located in the rear center of the bath. A thermometer placed in the rear corner of the bath is used to adjust the thermostat knob to the proper setting and to monitor the bath temperature.

Apparatus. All apparatus required to conduct a bioassay run is pictured (Fig. 2). Construction blueprints are not presented as they are too detailed and vary depending upon the water bath dimensions and construction materials. The apparatus shown is a working prototype constructed at the University of New Hampshire that has been useful in preliminary studies designed to develop and field-test the technique and to evaluate its potential in commercial agriculture. All construction materials are inexpensive and are available at most hardware stores or scientific supply houses. The framework for all constructed items is polyvinyl chloride (PVC), which is well suited for developing a working prototype; but precision in dispensing broth into the incubation tubes and measuring CO_2 levels in the gas-collecting tubes would be increased considerably if a material with less flexibility were used or if the incubation and gas-collecting racks and the plunger racks were each constructed of a rigid molded plastic. Also, a more uniform and less variable bath temperature could be obtained if the single aquarium heater were replaced with an improved heating system such as a thermostatically controlled heating strip encircling the bath. The apparatus comprises the following items:

Purchased items

Water bath, 15 L (Model 5274 cooler, Coleman Co., Wichita, KS)
Screw cap plastic bottle, 200 ml (for storing buffer aliquots)
Airtight container (film canister for storing buffer ingredients and yeast)
Thermostatic heater, 100 W (Supreme Heatmeter, E. G. Danner Mfg., Central Islip, NY; available at many aquarium supply houses)
Silicone lubricant spray (Cat. No. 316, Dow Corning Corp., Midland, MI)
Polyurethane paintbrush (for applying lubricant to plunger tips)
Plastic syringes, 5 cc (1/5-cc gradation; B-D Plastipak)
Plastic syringes, 20 cc (1-cc gradation; B-D Plastipak)
Forceps (for removing targets from incubation tubes)
Flask brush (for cleaning mixing flask and incubation tubes)

Thermometer (C)
Kitchen timer

Constructed items

Incubation tube rack and plunger rack housing 18 plastic syringes (20 cc) and plungers
Gas-collecting tube rack and plunger rack housing 18 plastic syringes (5 cc) and plungers
Dispensing tray
Mixing flask holder (metal hook screwed into a cork stopper)
Syringe rack cover (PVC sheet with strips of sticky-backed polyurethane weather stripping positioned to align with the incubation tubes)
Water bath inserts (bottom plate and two side rails with connecting strip attached to the bath with anchored polypropylene screws)
Vinyl targets (Brite-poly file folders, Augler's Co., Flushing, NY, cut into appropriate size and shape)
Target holder (metal hook, PVC rod, alligator clips)
Procedural layout sheet (vinyl tablecloth with laminated instructions attached using spray adhesive and plastic for tape; covered with clear plastic for additional protection)

Standard bioassay procedure. The procedure described below is the standard one for developing a standard curve. It has been used in preliminary field studies to evaluate the potential of the bioassay procedure for detecting residue on targets placed in a crop canopy and sprayed with an orchard sprayer (see Step 3). Modifications in this standard procedure and the apparatus will probably be necessary to accommodate specific applications of BIO-Y (see discussion of applications below).

Step 1. Set up apparatus on the procedural layout sheet (Fig. 2). The sheet is not necessary but is recommended. Positioning the apparatus at specific locations on the sheet and providing step-by-step instructions automatically organizes the procedure and practically ensures that it will proceed smoothly without delays, confusion, or mistakes that might otherwise occur.

Step 2. Warm the buffer (Fig. 3). Pour a 1,715-ml aliquot of buffer (containing the wetting agent) into the mixing flask, attach the flask to the flask holder, and hook the holder to the center of the water bath carrying bar. Allow at least 5 min for the buffer to warm to bath temperature.

Step 3. Place the targets in the incubation tubes (Figs. 4 and 5). This step may be completed at any time after the targets have been sprayed and allowed to dry.

A. Remove the plunger rack from the incubation tube rack (Fig. 4).
   1. Insert the handles into the four guideposts; then grasp the handles and pull upward until the plungers are free of the tubes.
   2. Remove the handles; then remove the plunger rack and set aside.

B. Insert the targets into the incubation tubes (Fig. 5).
   1. Grasp a target by the edges and bend the two halves together; then squeeze the edges so that the two halves bow outward.
   2. Insert the target into a tube, with the "bridge" end nearest you, and push to the bottom of the tube.
   3. Use the following sequence of target placement for targets placed in apple trees (given by tube number, target color, and target location in tree canopy): 1–3, green, unsprayed checks; 4–8, red, upper canopy; 9–13, yellow, middle canopy; 14–18, blue, lower canopy.

C. Replace the plunger rack (Figs. 6 and 7).
   1. Insert the plunger rack into the four guideposts; then grasp one end of the plunger and push downward gently to force the rubber plunger tips into the incubation tubes. Continue pushing along the plunger support until the lower lip of each rubber tip is exactly on the 15-cc mark.
   2. Position the incubation tube rack on the procedural layout sheet.

Step 4. Prepare the yeast/nutrient broth (Fig. 8).

A. Empty all the yeast and dried nutrients in the storage canister into the mixing flask containing the buffer warmed to bath temperature.
B. Set the timer for 30 min.
C. Swirl the flask gently for 5–10 sec to wet the dried nutrients thoroughly, then attach the flask to the flask holder and set the flask in the water bath again for 5 min.
D. After 5 min, remove the mixing flask and swirl gently to make sure there are no clumps of yeast remaining (all yeast granules must be completely dissolved); then proceed immediately to Step 5.

Step 5. Add the broth to the dispensing tray (Fig. 9). Swirl the mixing flask gently for 5 sec to mix the contents thoroughly (no settlement of yeast on the bottom of the flask); then quickly but carefully pour the contents into the dispensing tray.

Step 6. Draw broth into the incubation tubes (Fig. 10). Immediately place the incubation tube rack on the dispensing tray; then grasp the handles and plunger rack and pull upward until the rubber plunger tips contact the lip of the incubation tubes. There should now be 8 ml of broth in each tube.

Step 7. Wash the residue from the targets (Fig. 11).

A. Lift the incubation tube rack from the dispensing tray and place it on the syringe rack cover so that the syringe tips are resting on the centers of the two strips.
B. Remove the handles; then grasp the incubation tube rack and syringe rack cover, turn them over, and shake them vigorously for 4 min to remove the residue from the targets.

Step 8. Incubate the yeast in the water bath (Figs. 12 and 13).

A. Set the incubation tube rack on the back of the rack holder located on the bottom of the bath (with syringe tips facing the bath surface and the green targets in the front left corner of the bath); then

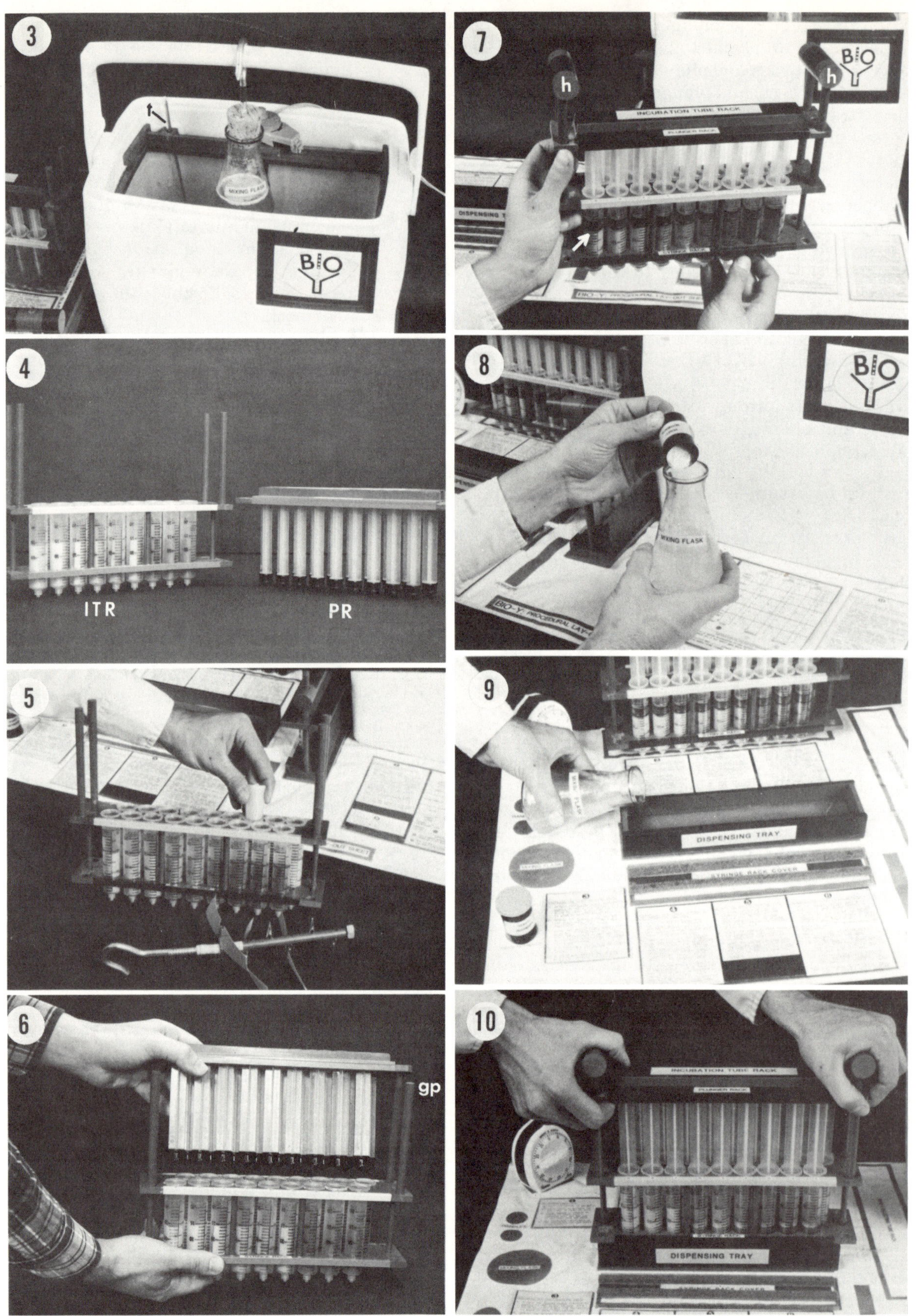

Figs. 3-10. Steps in bioassay procedure: **(3)** Mixing flask set in water bath to warm buffer; t = thermometer. **(4)** Incubation tube rack (ITR) and plunger rack (PR) separated. **(5)** Inserting a vinyl target into an incubation tube. **(6)** Inserting the plunger rack into the guideposts (gp) of the incubation tube rack. **(7)** Plunger rack inserted into the incubation tube rack with plunger tips aligned at the 15-cc marks (arrow); h = removable handles. **(8)** Adding the yeast and nutrients to warmed buffer in the mixing flask. **(9)** Pouring warmed broth and yeast into the dispensing tray. **(10)** Drawing broth and yeast into the incubation tubes.

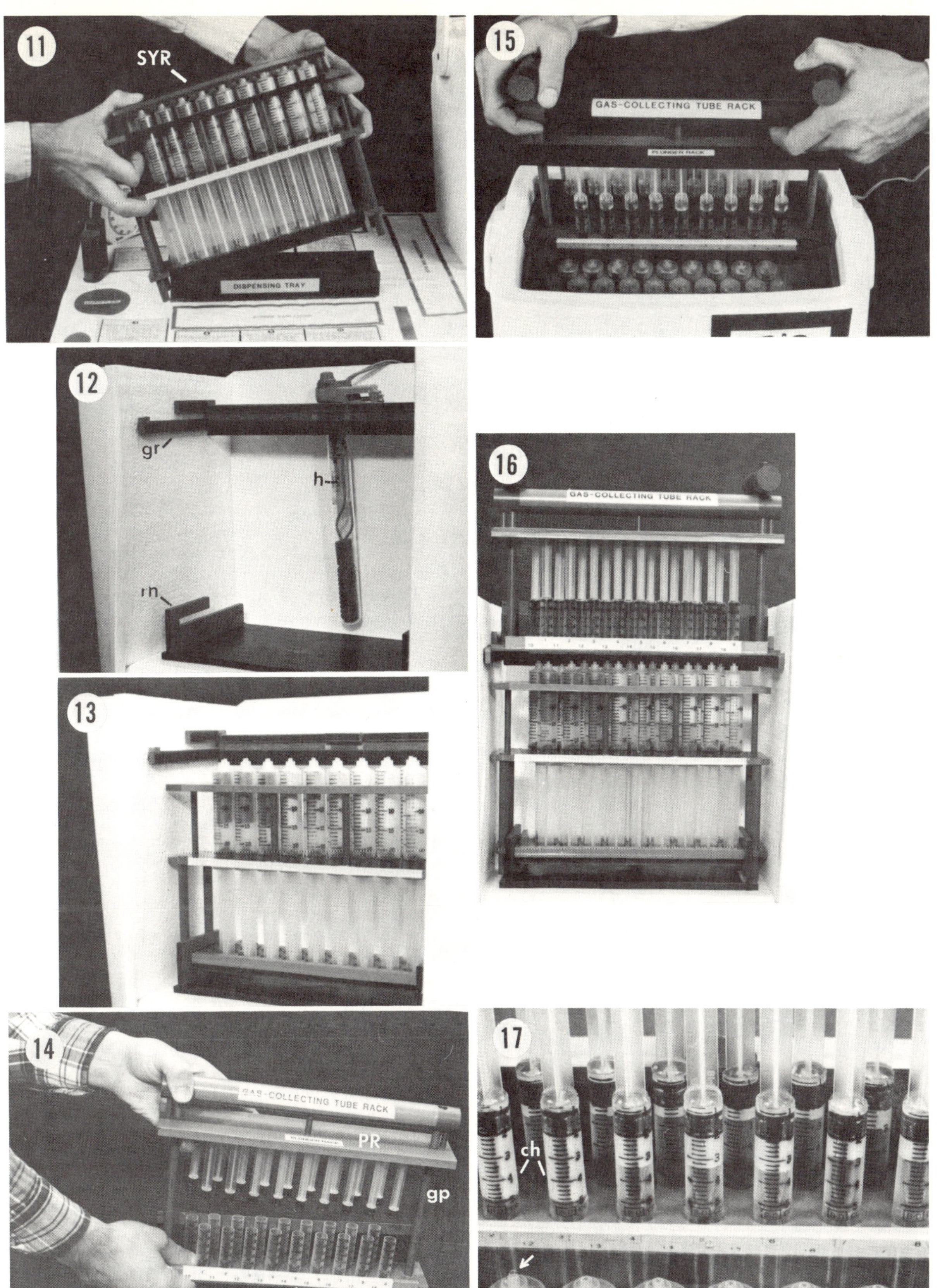

Figs. 11-17. Steps in bioassay procedure (continued). **(11)** Shaking the incubation tube rack to remove residue from the targets; SYR = syringe rack cover. **(12)** Mock-up of water bath with front removed to show the rack holder (rh) and a guide rail (gr) with connecting strip that supports the heater (h). **(13)** Cutaway view showing the incubation tube rack in forward position and the gas-collecting tube rack in rear position. **(14)** Inserting the plunger rack (PR) into the two guideposts (gp) of the gas-collecting tube rack (GCTR). **(15)** Drawing water into the gas-collecting tubes. **(16)** Cutaway view showing incubation tube rack in forward position and gas-collecting tube rack in rear position. **(17)** Gas levels in the gas-collecting tubes at completion of bioassay. White portion of each tube contains gas, dark portion contains water. The collection of gas ends when the gas level in the two check (ch) tubes reaches the 5-cc mark. Note gas bubble at tip of a check incubation tube (arrow).

slide the rack forward until it stops.

B. Allow the yeast to incubate until the timer rings.

Step 9. Draw water into the gas-collecting tubes (Figs. 14–16).

A. Set up the gas-collecting tube rack on the guide rails located along the upper sides of the bath; then slide the rack to the rear until it stops.

B. Push the plunger rack downward until the rubber plunger tips are completely through the tubes; then pull upward until the plunger tips are aligned with the 2-cc mark on the tubes.

Step 10. Collect and record the gas released by the yeast (Fig. 17).

A. When the timer rings, slide the gas-collecting tube rack forward until it stops. The gas-collecting tubes should be centered over the incubation tubes, and gas bubbles should be displacing water in the gas-collecting tubes.

B. Observe the gas level in tubes 1–3 (containing unsprayed targets). When any two of these tubes reach the 5-cc mark (as pictured), slide the rack to the rear. This completes the bioassay.

C. Record the gas level in each tube. An example of a chart that could be prepared for a quick analysis of the residue is pictured in Figure 18.

Step 11. Cleanup and maintenance.

A. Allow 5–10 min for cleanup immediately after recording the gas levels.

B. Do not wash the gas-collecting tube rack. Just push down on the plunger rack until the plunger tips expel all the water; then remove the rack from the bath and place it on the procedural layout sheet.

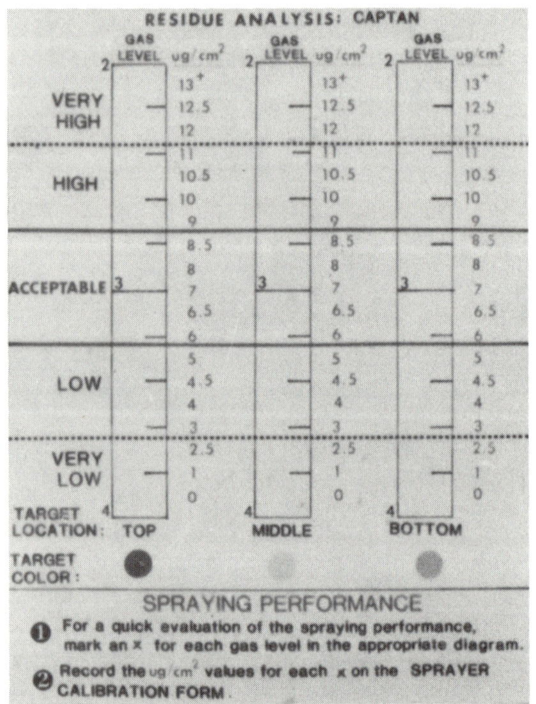

Fig. 18. Residue analysis chart for captan.

C. Remove the plunger rack from the incubation tube rack; then remove the targets using the forceps. This step should be done in a sink, flushing the tubes thoroughly with running water.

D. Rinse the rubber tips in running water; then return the plunger rack to the procedural layout sheet.

E. Rinse the targets, incubation tube rack, mixing flask, dispensing tray, and syringe rack cover in running water; then wash these items in hot, soapy water. Use the flask brush to clean the inside of the mixing flask and the incubation tubes.

F. Rinse all items thoroughly in running water. Make sure there is no soap remaining in the incubation tubes or the mixing flask. Allow everything to air-dry; then return the items to the procedural layout sheet.

G. Lubricate the plunger tips if the plungers do not slide smoothly within the incubation tubes or gas-collecting tubes. To lubricate, spray the polyurethane paintbrush with the spray lubricant and then apply in an even, light coating of lubricant on the rubber tips. Insert the plunger rack into the incubation tube rack and slide the tips up and down until the plungers slide easily.

Developing a standard curve. The procedure developed for screening pesticides to determine which are applicable and establishing standard curves is described below, using captan as an example. Each pesticide is screened initially at twice the mean density of active ingredient that is expected to be deposited on a 50-cm² vinyl target placed in a crop canopy and sprayed with the pesticide at the recommended rate. If the deposition density is not known, it can be estimated by comparing the formulation and recommended rate of application with Captan 50 WP applied at 2.4 g/L (2 lb/100 gal). Twice the expected mean density of captan when Captan 50 WP is applied dilute to the apple trees at 2.4 g/L is 12 μg/cm² (15) or 600 μg/50 cm² of target. Since 8 ml of broth is drawn into each incubation tube, captan should have a measurable effect on yeast CO_2 production at 75 ppm (600 μg/8 ml). Captan was screened by pipetting 600 μg a.i. (captan) of an appropriate concentration of Captan 50 WP onto vinyl targets (Fig. 19). The sprayed targets were allowed to dry and were then stored in the freezer if not used immediately.

If the gas levels associated with the treated targets are between the 1- and 3-cc marks at the completion of the test, the pesticide is tested further to develop a standard curve. The 3-cc mark is an arbitrary cutoff selected to identify chemicals with a high probability of success for developing a standard curve. Chemicals associated with CO_2 levels between 3 and 4 cc may be useful with modifications in target sampling and test conditions, but when the gas levels are between 4 and 5 cc the chemical is considered unsuitable. The mean gas level for captan tested at 75 ppm was 2.8 cc.

Field studies have shown that mean deposition densities of captan below 1–2 μg/cm² or above 13 μg/cm² are unlikely on apple leaves after spraying Captan 50 WP at 2.4 g/L (15), so vinyl targets were treated with captan solutions that provide six selected broth concentrations between 10 and 100 ppm. Each set of targets receiving a specified amount of captan was subdivided and assayed by a standard spectrophotometric procedure (15) or bioassayed as described above. The mean density of captan residue on the targets assayed by the analytical technique was plotted against the levels of CO_2 for the development of a standard curve (Fig. 1). Three concentrations of pesticide, each replicated five times, were bioassayed during each run, and each concentration was included in five runs.

Standard curves have also been developed for dodine (Cyprex 65 WP) and thiram (Thylate 65 WP). However, analytical procedures were not available, so the actual amount of active ingredient of fungicide pipetted onto the targets could not be verified. Several fungicides did not pass the initial screening test. The initial screening is preliminary, but it does not appear that the benzimidazoles, dithiocarbamates, or the new steroid-inhibiting compounds are applicable to this bioassay. Additional screening is under way to determine what fungicides or insecticides, acaricides, and herbicides are applicable.

Applications. Preliminary field tests were conducted to determine whether BIO-Y could be used to detect fungicide residue in a crop canopy following spraying. This was an important initial test because if BIO-Y could detect residue with acceptable accuracy on sprayed targets, field applications such as calibrating sprayers or evaluating the distribution of pesticide within a canopy would be primarily a matter of modifying the target, the sampling technique, or the apparatus.

Residue can be determined either by examining leaves or by examining nonplant targets placed in the canopy before spraying. Leaves are desirable since they are the actual targets to be protected, but they have several disadvantages with respect to the bioassay: pesticides applied previously may affect the yeast and alter CO_2 production; surface characteristics may cause too much variability in the removal of the test chemical; whole leaves may be too flexible or variable in size and shape for easy, consistent removal of residue or to accommodate the extraction apparatus; and the leaf surface area must be known, thus requiring either an accurate but inexpensive means to estimate the area of whole leaf quickly and easily or a device such as a cork borer for cutting sections of known surface area.

An artificial (nonplant) target would eliminate the major disadvantages of a real leaf discussed above because it would have a defined size and shape and would not be contaminated with chemicals that could interfere with yeast metabolism. The artificial target must have surface characteristics that allow it to receive the same density of test chemical as a real leaf and allow the easy removal of 90–100% of the chemical. It must also be flexible for insertion into the incubation tubes and durable for repeated use. The vinyl targets pictured (Fig. 20) meet these criteria. Artificial targets may not be suitable, however, for monitoring residue density over time. Pesticide retention, redistribution, and decay within a crop canopy following spraying are altered by the interaction of weather variables, phylloplane microorganisms, leaf surface characteristics, and the pesticide formulation. Thus, before artificial targets replace leaves for monitoring residue levels over time within a crop canopy, extensive field tests conducted under a wide range of weather conditions would be necessary to establish a reliable relationship between the residue on artificial targets and real leaves.

The vinyl targets were field-tested in an apple orchard to determine their usefulness as a substitute for leaves in determining residue in a crop canopy following spraying. Targets were placed in the upper, middle, and lower portions of the canopy (Fig. 20) and sprayed with Captan 50 WP at the recommended rate with an air-blast orchard sprayer. Targets from each target holder

Fig. 19. Pipetting fungicide onto vinyl targets for development of a standard curve.

Fig. 20. Target holder with vinyl targets placed in lower limb of an apple tree.

were then assayed by the spectrophotometric technique or bioassayed, and leaves from the terminal with the target holder were assayed by the spectrophotometric assay. The three analyses were compared.

The three analyses for each location were comparable, but the tests were preliminary and must be repeated with samples that include an increased number of targets. The analytical assay is a more accurate and sensitive assay, with residue assessed in increments of 0.01 $\mu g/cm^2$ compared with bioassay increments of 1.0 $\mu g/cm^2$. The analytical assay also accurately analyzes the residue on leaves or leaf disks. However, for all the reasons stated at the beginning of the chapter, the analytical procedure does not offer itself as a practical, routine procedure to be conducted by nonscientists in commercial agriculture. The preliminary studies indicate that although BIO-Y is less precise and is best used with nonplant targets, it nevertheless does have a sensitivity to captan (and other fungicides) that allows it to be used as a routine procedure for calibrating sprayers (9) and evaluating the performance of sprayers and the spraying procedure.

With any crop, field tests designed similarly to Chiba's (4) must be conducted to determine the number of leaves or artificial targets that constitute an acceptable sample. It is likely that a large number of targets will be required (4), necessitating modifications in sampling and extracting procedures and in the apparatus. For example, assume that a field study showed that 25 targets from each region of a crop canopy were required for an acceptable statistical sample for estimating the mean density ($\mu g/cm^2$) of captan and that we wish to sample five regions. The new sampling apparatus needed would be a residue extraction chamber that would accommodate 25 targets (Fig. 21) and a dispensing tray subdivided into six compartments (Fig. 22). One compartment would receive broth only, whereas each of the remaining five compartments would receive broth plus residue from an extraction chamber. Each compartment would accommodate three incubation tubes when the incubation tube rack is placed on the dispensing tray.

The standard bioassay procedure would be modified as follows: (i) place 25 targets from each crop region into a residue extraction chamber, add 200 ml of warmed broth, seal with an airtight cover, and stack and shake the chambers to remove the residue from the targets; (ii) pour warmed broth from a mixing flask into the front left compartment of the dispensing tray until the compartment is filled to a marked level; (iii) pour the broth from each extraction chamber into a designated compartment; (iv) place the incubation tube rack on the dispensing tray and draw 4 ml of broth into each tube; and (v) pour a warmed, double-strength yeast solution into a regular dispensing tray, place the incubation tube rack on the tray, and draw another 4 ml into the incubation tubes.

The remaining steps for incubating the yeast and collecting the gas are similar to the standard bioassay procedure described above. It would not be necessary to record the gas level for each of the three tubes that had received broth from a compartment and then calculate the mean level. Instead, simply record the middle level. Thus, each of the following combinations of gas levels would be recorded as 4.1 cc: 4.0, 4.1, 4.2 cc; 4.0, 4.1, 4.1 cc; 4.0, 4.1, 4.3 cc; and 3.9, 4.1, 4.2 cc. Differences within three tubes greater than those listed would suggest poor technique.

If it is necessary to extract the residue from 50 targets to obtain a reliable residue assessment, it would only be necessary either to add another rack of 25 targets to the chamber or simply to remove the rack of targets after shaking and replace with a new rack and shake again. Thus, extracting the residue from a large number of targets is a simple matter of apparatus design if the targets are batched to constitute one sample as described.

RESULTS

A procedure for detecting fungicide residue within the crop canopy has been developed. It is designed to reduce errors and to be used by nonscientists as well as scientists. It does not have the accuracy of analytical methods, but it has features that make it practical for routine use in commercial agriculture. However, with each crop, considerable field testing is needed to determine sampling size and residue patterns and to identify and define the limits and specific applications of the bioassay. At this time, only a few fungicides have been found applicable to BIO-Y. Despite this limitation, it is important that a formulated, recommended fungicide is available as a test chemical for residue detection and evaluation in those applications listed earlier. With each crop, field studies must be undertaken

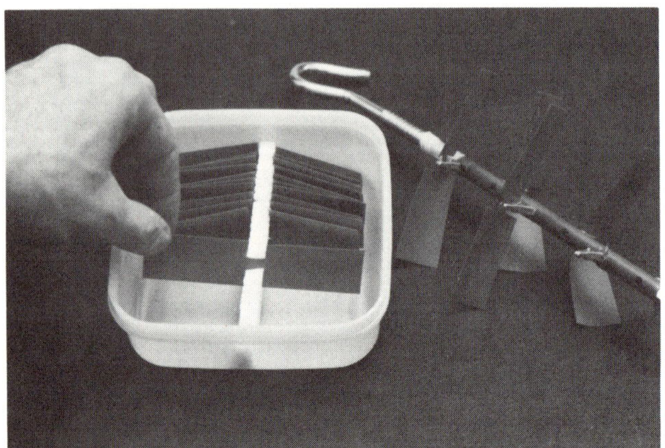

Fig. 21. Residue extraction chamber with holder for 25 targets.

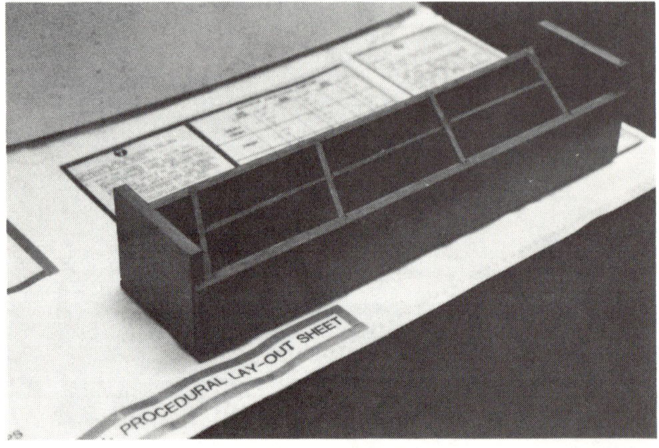

Fig. 22. Compartmented dispensing tray.

to determine whether residue analyses of the test chemical can be interpolated to estimate the deposition densities and distribution patterns of other pesticides.

LITERATURE CITED

1. Association of Official Agricultural Chemists. 1975. Official Methods of Analysis of the Association of Official Agricultural Chemists. W. Hortwits, ed. George Banta Co., Menasha, WI. 1,094 pp.
2. Bardinelli, T. R., McCarthy, C. W., and Manning, W. J. 1980. Biodetermination of fungicidal residues on apple leaves. (Abstr.) Phytopathology 70:459.
3. Blazquez, C. H. 1971. Residue monitoring, a method for timing spray schedules. (Abstr.) Phytopathology 61:885.
4. Chiba, M. 1973. Distribution of spray deposits in peach trees and number of leaves that constitute an adequate sample. J. Econ. Entomol. 67:529-534.
5. Committee on Standardization of Fungicidal Tests of the American Phytopathological Society. 1943. The slide-germination method of evaluating protectant fungicides. Phytopathology 33:627-632.
6. Hagan, A., and Larsen, P. O. 1981. Bioassay of fungicide residue persistence on leaves of Kentucky Bluegrass. (Abstr.) Phytopathology 71:222.
7. Heuberger, J. W. 1940. A laboratory biological assay of tenacity of fungicides. Phytopathology 30:840-847.
8. Ko, W. H., Lin, H., and Kunimoto, R. K. 1975. A simple method for determining efficacy and weatherability of fungicides on foliage. Phytopathology 65:1023-1025.
9. MacHardy, W. E. 1983. BIO-Y: A new, simplified one-hour fungicide bioassay useful for calibrating sprayers. (Abstr.) Phytopathology 73:800.
10. MacHardy, W. E. 1983. Residue analysis spraying: A new approach for determining the concentration of pesticides applied. (Abstr.) Phytopathology 73:1344.
11. McCallan, S. E. A. 1930. Studies of fungicides. II. Testing protective fungicides in the laboratory. N.Y. Agric. Exp. Stn. (Ithaca) Mem. 128:8-24.
12. Neely, D. 1978. A cellophane transfer bioassay to detect fungicides. Pages 15-18 in: Methods for Evaluating Plant Fungicides, Nematicides, and Bactericides. E. I. Zehr, ed. American Phytopathological Society, St. Paul, MN.
13. Peterson, P. D. 1941. The spore-germination method of evaluating fungicides. Phytopathology 31:1108-1116.
14. Sharvelle, E. G., and Pelletier, E. N. 1956. Modified paper-disk method for laboratory fungicidal bioassay. (Abstr.) Phytopathology 46:26.
15. Smith, F. D. 1982. The retention and redistribution of captan on apple. M.S. thesis, University of New Hampshire, Durham. 65 pp.
16. Thornberry, H. H. 1950. A paper-disk plate method for the quantitative evaluation of fungicides and bactericides. Phytopathology 40:419-429.
17. Torgeson, D. C. 1967. Determination and measurement of fungitoxicity. Pages 93-123 in: Fungicides, an Advanced Treatise. Vol. 1. D. C. Torgeson, ed. Academic Press, New York.

Evaluating Fungicides for Control of Postharvest Decay of Pome Fruits

D. A. ROSENBERGER, Department of Plant Pathology, New York State Agricultural Experiment Station, Hudson Valley Laboratory, Highland 12528; R. A. SPOTTS, Department of Botany and Plant Pathology, Oregon State University, Mid-Columbia Agricultural Experiment Station, Hood River 97031; W. S. CONWAY, Horticultural Crops Quality Laboratory, Horticultural Science Institute, USDA, ARS, Beltsville, MD 20705; and K. S. YODER, Virginia Polytechnic Institute and State University Fruit Research Laboratory, Winchester 22601

As early as the 1930s, more than 90 species of fungi were recognized as postharvest pathogens of apple (14), and 60 species of fungi were identified as pathogens of pear fruit in 1940 (13). Numerous additional decay-causing fungi have been isolated from pome fruits since these early reports were published. With modern storage conditions, however, only a relatively few fungi cause commercially important postharvest losses in pome fruits. The relative importance of the major postharvest pathogens varies with geographic location. Thus, blue mold caused by *Penicillium expansum* Link and gray mold caused by *Botrytis cinerea* Pers. ex Fr. are the major postharvest apple decay pathogens in North America (16), whereas *Gloeosporium* spp. and *Monilinia fructigena* (Adehr. & Ruhl.) Honey are more important in England and Europe (12).

Both blue mold and gray mold cause problems in pears. In England, losses from these molds may be less severe in pears than in apples (12), but in the U.S. Pacific Northwest, decay in pears is more serious than in apples. Other postharvest pathogens causing significant losses in pears in the United States include *Phialophora malorum* (Kidd & Beaum.) McColloch (2), *Mucor piriformis* Fischer (3), and a *Coprinus* sp. (24).

Numerous methods have been described for applying fungicides for control of postharvest diseases (10). Control of *P. expansum* on apple has been achieved by fumigation with 2-aminobutane (11), although the usefulness of this treatment is limited because 2-aminobutane is effective only at room temperature. Spread of Botrytis rot in packed pears has been controlled by impregnating fruit wrappers with a 2.5% solution of $CuSO_4$ (9). Some postharvest decays, such as those caused by *Gloeosporium* spp. and *M. fructigena*, develop primarily from infections initiated in the field, and chemical control is therefore largely dependent on use of preharvest sprays. In North America, however, postharvest fungicides are usually needed to control blue and gray molds on pome fruits and are generally applied to harvested fruit either as dips or drenches or in line sprays before storage or shipment.

In this chapter we will describe commonly used methods for evaluating postharvest fungicides on pome fruits. The methods outlined have been used primarily for testing fungicides for control of *P. expansum*, but they could easily be adopted or modified slightly for tests with other pathogens.

PLANNING THE EXPERIMENT

Defining objectives. Small variations in procedures used for postharvest fungicide tests can significantly affect fungicide performance. No single procedure is appropriate for all tests, but the procedures used should be selected and evaluated carefully to ensure that results will provide information relevant to the experimental objectives. If objectives are clearly defined before the experiment is initiated, selecting appropriate procedures should be reasonably simple. For example, if the objective is to determine the efficacy of fungicides applied immediately after harvest for protecting fruit from inoculum introduced during storage or packing, then introduction of inoculum in the experiment should also occur after fungicides are applied, and fungicides known or suspected to have good residual activity would be the most useful. However, if the objective is to use postharvest treatments to control infections originating in the field during or before harvest, then experimental fruit should be inoculated before fungicide treatment, and fungicides with good eradicant or disinfectant capabilities would be the most useful.

Experimental results are generally most applicable to commercial situations if experimental procedures closely simulate current or proposed commercial practices. In some cases, however, commercial problems initially should be separated into components, and each component should be addressed in separate experiments. Thus, in the example used above, inoculum in commercial situations may originate from both field and postharvest sources. Combining both preharvest and postharvest inoculations in a single experiment could cause difficulty in interpreting data and might not provide useful information on fungicides effective for only one of the two inoculation timings. Separate experiments that evaluate fungicides for residual and eradicant activity would provide more detailed information on individual fungicide activity, and the results might suggest possible fungicide combinations that would prove effective against both preharvest and postharvest inoculations.

Selecting appropriate experimental designs. Useful fungicide efficacy data can usually be obtained with 10–50 fruit per replicate and four replicates per treatment. If fruit have not been graded for uniformity, a randomized block experimental design should be used with fruit of different quality (e.g., size, maturity, source tree) assigned to different blocks. Harvesting fruit from three or four commercial orchards may also be a useful basis for dividing it into blocks because fruit from different orchards may differ significantly in susceptibility to decay. Use of factorial experimental design may allow investigation of several related factors and their interactions in a single experiment and provide more information with fewer experimental units than are required for separate experiments. Factorial experiments are ideally suited for investigations designed to evaluate effects of increasing inoculum on the efficacy of several fungicides or the effects of increasing fungicide concentrations on different isolates of the same fungus (17).

All tests should include inoculated controls treated only with water and a common standard fungicide for comparison with the test fungicides. Because disease pressure in postharvest treatments is dependent on a complex group of interrelated variables, disease pressure is sometimes difficult to predict and control; use of standard fungicides with known disease-control capabilities is therefore essential for interpreting experimental results. Benomyl is an effective standard for comparison in trials with benomyl-susceptible isolates of *B. cinerea* and *P. expansum*. Etaconazole (Vangard or CGA 64251) has been used as a standard in experiments with benomyl-resistant *P. expansum* (17), and iprodione (Rovral) might be an effective standard in trials with benomyl-resistant *B. cinerea*.

Collecting needed materials. Fruit used for postharvest treatments should be harvested at a commercially acceptable stage of maturity. Immature fruit is resistant to many decay fungi. Fruit should not have fungicide residues that might interfere with postharvest treatments. Fungicides applied in the orchard within a month of harvest sometimes reduce the incidence of postharvest decays of pears in the Pacific Northwest. In areas with higher humidity and more precipitation, contact fungicides applied several weeks before harvest probably have little effect on development of postharvest decay, but fruit treated before harvest with systemic or locally systemic fungicides should certainly be avoided. Attempts to remove fungicide residues or surface microorganisms from experimental fruit after harvest may also affect the fruit cuticle and the fruit's natural susceptibility to decay. However, surface sterilization with sodium hypochlorite followed by a water rinse has proven useful in reducing contamination from other decay organisms in storage trials with pears (23).

Conidial inoculum of most postharvest pathogens can be collected either from sporulating lesions on decayed fruit or from pure cultures grown on artificial media. Conidia should usually be collected from actively growing or recently matured colonies because spores from old cultures or fruit lesions may differ from younger spores in germinability and degree of resistance to fungicides. If isolates from culture collections are used, pathogenicity of the isolates should be checked by inoculating several fruit before initiation of the trial. In tests using fungicide-resistant isolates, isolates should also be retested on fungicide-amended agar. Retesting stored isolates for pathogenicity and fungicide resistance must be initiated 7–10 days before inoculum will be used. Pretesting of inoculum can prevent possible later confusion resulting from mutation or contamination of cultures during transfer and storage.

Using small quantities of fruit and small containers for testing fruit minimizes pesticide disposal problems after treatments are completed. Wooden, flat-bottomed, half-bushel baskets conveniently hold about 50 apples. They can be submersed in 30 L of solution held in 75-L plastic garbage cans, and they allow solutions to drain quickly after treatment. Lightweight plastic flats with good drainage are also available. All wooden and plastic containers can be reused for many experiments if they are rinsed in water and soaked in ammonia solution to remove pesticide residues between treatments. A wooden-framed, plastic mesh cover held across the top of baskets or flats prevents fruit from floating out during treatment.

All chemicals should be preweighed in the precise quantities required for the volume of solution to be prepared. If wettable powders are weighed into screw-capped, wide-mouthed bottles, water can be added to the bottles to make a slurry before chemicals are added to the larger containers. This process minimizes exposure to dusts otherwise encountered in mixing treatment solutions and allows thorough rinsing of containers.

CONDUCTING THE TEST

Wounding the fruit. Many postharvest fungi, including *P. expansum* and *B. cinerea*, are primarily wound pathogens. The most common wounds in commercially harvested apples and pears are punctures caused by stems of adjacent fruit in the picking buckets, boxes, or bulk bins. To ensure adequate disease pressure in small-scale trials, fruits usually are wounded before inoculation. Uniform wounds similar to stem punctures may be made using nails or other objects inserted through a cork or wood block to control depth of penetration. Uniformity is essential because depth of wounding may affect the ability of fungicides to control infection (15). If wounds are of narrow diameter, air bubbles may prevent fungicides from penetrating the wound during treatment. Wounds may more nearly simulate stem punctures if blunted nails or small bolts are used instead of sharply pointed objects. Using three to five wounds per fruit increases chances for infection and allows more observations with fewer experimental units. The stem end of pears is also a type of wound and provides an entry point for some decay fungi.

Applying inoculum. As suggested earlier, inoculum may be applied before or after fungicide treatments depending on the experimental objectives. Inoculum may be applied by dipping fruit in conidial suspensions (17,19), by introducing inoculum with the wounding tool (15,25), by placing measured droplets of conidial suspensions directly into wounds (23), by misting the wounded face on fruit with a conidial aerosol (20), or by dipping the stem end of pears into petri dishes containing inoculum. Each method has its own advantages and disadvantages. The choice of inoculation method is not critical provided inoculum is applied uniformly and the time from inoculation to fungicide treatment (or treatment to inoculation) is similar for all experimental units.

Spore concentrations should be determined with a

hemacytometer. Optimum concentrations for postharvest tests vary with the pathogen and the inoculation method, but inoculum concentrations should generally approach the highest levels normally found in commercial packinghouses. Excessively heavy inoculum loads may cause failure of fungicides that are effective under less severe commercial conditions (6).

Applying treatments. Dipping fruit into treatment solutions is probably the easiest way to uniformly treat small quantities of fruit. The fruit should be submersed 20–30 sec with some agitation to dislodge air bubbles and to allow penetration of wounds by the fungicide solution. Fungicides should be thoroughly mixed before starting treatments, and additional agitation between replicates may be needed to keep fungicides in suspension. Fruit temperature, treatment solution temperatures, and depth of submersion should be uniform throughout the experiment to minimize variation resulting from hydrostatic pressure differences between treatments (1). In some commercial storages, fungicides or disinfectants used in dump tanks are rinsed off as fruits leave the tank or flume. If rinsing is incorporated into postharvest tests, its duration and turbulence should closely simulate commercial conditions. The type and extent of rinsing significantly affect results because rinsing removes inoculum as well as some fungicide residue.

Applying treatments as drenches rather than dips may more closely simulate commercial application methods (6), but an adequate flow rate and volume of solution are required to thoroughly wet fruit. In many commercial storages, a postharvest, prestorage bin drench is applied to fruit, and spore levels build up in the recycled drench water. This situation may be simulated by applying inoculum and fungicide at the same time in a small recirculating drencher. Inoculum concentrations must be carefully measured to ensure that all treatments are exposed to the same inoculum concentration.

Storing fruit after treatment. Postharvest fungicide treatments can be completed more quickly if fruit is held at 16–21°C for 7–14 days after treatment instead of being stored at 0°C for the 3–6 months required for decay to develop at cold storage temperatures. However, fungicides may perform differently at cold storage temperatures than at warmer temperatures (17). Captan has generally provided better control of *P. expansum* at cold storage temperatures than at warmer temperatures (20), and some benomyl-resistant isolates of *P. expansum* are controlled in cold storage but not at 20°C by the combination of benomyl and diphenylamine (18). Relative humidity is also important: some pathogens cause decay only under conditions of high relative humidity. Storage conditions should therefore be as similar as possible to commercial conditions.

Fruit may be stored in bulk or on spring cushion trays. Bulk storage in bags or crates more nearly simulates commercial bulk storage conditions than does storage on spring cushion trays, but bulk storage may allow decay to spread from infected fruit to adjacent healthy fruit. If secondary spread during the course of the experiment is not desired, progression of decay should be regularly monitored and evaluations made before secondary spread occurs. Storing fruit with the wounded face up on spring cushion trays minimizes chances for secondary spread and allows for rapid evaluation, but this method requires more storage space than bulk storage.

If high relative humidity cannot be maintained in the cooling unit used for storage, treatments can be placed in tightly closed polyethylene bags. However, fruit stored in polyethylene bags will be exposed to gradually lowered oxygen and increased carbon dioxide levels as a result of respiration during storage. Low-oxygen storage alone inhibits development of stem end decay in pears (7) and may affect other pathogens as well.

Collecting data. The time for termination of the experiment and evaluation of fruit may be predetermined by the experimental design or may be determined by observing the progression of decay in control and standard treatments. Because fruits become increasingly susceptible to decay as they senesce in storage, the incidence of decay is likely to increase as the storage period is prolonged. If possible, an experiment should not be terminated until a significant proportion (> 25%) of control fruits are decayed. Prolonging storage until some fruits in the standard treatment also develop decay allows separation of the standard material from experimental materials that may be even more effective. However, if fruits are held longer than the maximum commercial storage life, results may not be applicable to commercial situations. Extensive decay in fruit treated with standard fungicides suggests that inoculum levels were unrealistically high, that fruit was overmature when harvested or treated, or that storage was prolonged unnecessarily. Where fruit is ripened after storage, as with winter pears, or where shelf life after storage is a concern, fruit should be evaluated immediately after removal from cold storage and again at the end of the ripening period or shelf life test.

In addition to the usual fungicide test data listing fungicides, application methods, and evaluations, the following items should be noted: (a) Fruit—cultivar; size, condition, grade, etc.; maturity at harvest, at time of treatment, and at termination of the experiment. Fruit maturity may be described using results from pressure testing (4) and measurements of soluble solids with a refractometer. (b) Inoculum—age of cultures, source, media used for culturing; application method; concentration; results of pretrial pathogenicity and fungicide resistance testing. (c) Posttreatment storage—duration; temperature; storage container (i.e., bulk, polyethylene bag, spring cushion tray). (d) Observations on phytotoxic effects of treatments on fruit.

Including as much data as possible on all of the above will facilitate explanations of similarities and differences in results obtained from different experiments and researchers.

EVALUATING INTERACTION EFFECTS

Chemical interactions. When two or more compounds are combined in the same treatment, as they often are in commercial situations, unexpected interactions may occur. Diphenylamine (DPA), which is commonly used to control superficial scald (22) on apples, interacts with benomyl to control some benomyl-resistant isolates of *P. expansum* (18). Used with captan, however, DPA may decrease fungicide efficacy (19). Some fungitoxic chemicals such as chlorine are pH-dependent and are affected by the sodium salts used to float pears because the salts change solution pH. Compounds to be combined in commercial treatments should therefore be tested together as well as independently to determine whether any adverse or beneficial interactions occur. Researchers should exercise caution in combining experimental compounds, however: some—such as sodium hypochlorite and DPA—may interact to produce undesirable by-products.

Postharvest calcium treatments have been shown to increase resistance of apple fruit to decay (8). Although calcium does not interact directly with other postharvest chemicals (6), fungicide efficacy readings may be altered because of changes in host susceptibility to decay. Thus, benomyl with $CaCl_2$ was more effective than benomyl alone for controlling decay by *P. expansum* (5).

In some cases, efficacy of treatments can be intentionally improved by adding surfactants or using alternative formulations (19,24). Fungicides with marginal activity but other positive attributes, such as low cost or favorable toxicology, may warrant extensive testing to determine whether efficacy can be improved by using surfactants or other formulations.

Interactions resulting from experimental variables. Effects of storage temperature, duration of storage, hydrostatic factors during dip treatments, wound size, inoculum concentration, and order of inoculation and treatment have all been mentioned or discussed earlier in this chapter. Such other factors as pH of treatment solutions, duration of exposure to treatment solutions, effect of fruit size and maturity, and effects of controlled atmosphere storage are also important and have been discussed elsewhere (10,12,21).

ACKNOWLEDGMENT

We thank M. Benedict for reviewing the manuscript and providing helpful suggestions and editorial advice.

LITERATURE CITED

1. Bartz, J. A. 1982. Infiltration of tomatoes immersed at different temperatures to different depths in suspensions of *Erwinia carotovora* subsp. *carotovora*. Plant Dis. 66:302-306.
2. Bertrand, P. R., MacSwan, I. C., Rockhan, R. L., and Moore, B. J. 1977. An outbreak of side rot in Bosc pears in Oregon. Plant Dis. Rep. 61:890-893.
3. Bertrand, P. R., and Saulie-Carter, J. 1980. Mucor rot of pears and apples. Oreg. Agric. Exp. Stn. Spec. Rep. 568. 21 pp.
4. Blanpied, G. D., Bramlage, W. J., Dewey, D. H., LaBelle, R. L., Massey, L. M., Jr., Mattus, G. E., Stiles, W. C., and Watada, A. E. 1978. A standardized method for collecting apple pressure test data. N.Y. Agric. Exp. Stn. (Ithaca) N.Y. Food Life Sci. Bull. 74. 8 pp.
5. Burton, C. L. 1979. Evaluation of fungicides for controlling postharvest benomyl-tolerant and -sensitive blue mold rots. Fungic. Nematic. Tests 34:2.
6. Burton, C. L., and Dewey, D. H. 1981. New fungicides to control benomyl-resistant *Penicillium expansum* in apples. Plant Dis. 65:881-883.
7. Chen, P. M., Spotts, R. A., and Mellenthin, W. M. 1981. Stem end decay and quality of low oxygen stored d'Anjou pears. J. Am. Soc. Hortic. Sci. 106:695-698.
8. Conway, W. S. 1982. Effect of postharvest calcium treatment on decay of Delicious apples. Plant Dis. 66:402-403.
9. Cooley, J. S., and Crenshaw, J. H. 1931. Control of Botrytis rot of pears with chemically treated wrappers. U.S. Dep. Agric. Circ. 177. 10 pp.
10. Eckert, J. W. 1977. Control of postharvest diseases. Pages 269-352 in: Antifungal Compounds. Vol. 1: Discovery, Development, and Uses. M. R. Siegel and H. D. Sisler, eds. Marcel Dekker, New York. 600 pp.
11. Eckert, J. W., and Kolbezen, M. J. 1970. Fumigation of fruits with 2-aminobutane to control certain postharvest diseases. Phytopathology 60:545-550.
12. Edney, K. L. 1973. Fungal disorders. Pages 135-172 in: The Biology of Apple and Pear Storage. Research Review No. 3. Commonw. Agric. Bur., East Malling. 235 pp.
13. English, W. H. 1940. Taxonomic and pathogenicity studies of the fungi which cause decay of pears in Washington. Ph.D. thesis, Washington State University, Pullman. 270 pp.
14. Heald, F. D., and Ruehle, G. D. 1931. The rots of Washington apples in cold storage. Wash. Agric. Exp. Stn. Bull. 353. 48 pp.
15. Hickey, K. D., Garretson, M., and May, J. 1980. Fruit decay control with postharvest fungicide dips, 1978-79. Fungic. Nematic. Tests 35:8-9.
16. Pierson, C. F., Ceponis, M. H., and McColloch, L. P. 1971. Market diseases of apples, pears, and quinces. U.S. Dep. Agric. Handb. 376. 112 pp.
17. Rosenberger, D. A., and Meyer, F. W. 1981. Postharvest fungicides for apples: Development of resistance to benomyl, vinclozolin, and iprodione. Plant Dis. 65:1010-1013.
18. Rosenberger, D. A., and Meyer, F. W. 1983. Use of diphenylamine with benzimidazole fungicides improves control of benomyl-resistant *Penicillium expansum* in stored apples. (Abstr.) Phytopathology 73:1346.
19. Rosenberger, D. A., and Meyer, F. W. 1984. Effect of captan formulation and addition of diphenylamine on control of blue mold in stored apples, 1982-83. Fungic. Nematic. Tests 39:20-21.
20. Rosenberger, D. A., Meyer, F. W., and Cecilia, C. V. 1979. Fungicide strategies for control of benomyl-tolerant *Penicillium expansum* in apple storages. Plant Dis. Rep. 63:1033-1037.
21. Segall, R. H. 1968. Fungicidal effectiveness of chlorine as influenced by concentration, temperature, pH, and spore exposure time. Phytopathology 58:1412-1414.
22. Smock, R. M. 1961. Methods of scald control in apples. N.Y. Agric. Exp. Stn. (Ithaca) Bull. 970. 56 pp.
23. Spotts, R. A., and Peters, B. B. 1982. Use of surfactants with chlorine to improve pear decay control. Plant Dis. 66:725-727.
24. Spotts, R. A., Traquair, J. A., and Peters, B. B. 1981. d'Anjou pear decay caused by a low temperature basidiomycete. Plant Dis. 65:151-153.
25. Tepper, B. L., and Yoder, K. S. 1982. Postharvest chemical control of Penicillium blue mold of apple. Plant Dis. 66:829-831.

Evaluation of Postharvest Fungicide Treatments for Citrus Fruits

JOSEPH W. ECKERT, Department of Plant Pathology, University of California, Riverside 92521, and
G. ELDON BROWN, Scientific Research Department, Florida Department of Citrus, Lake Alfred 33850

Certain postharvest diseases of citrus fruits are initiated by infection of the fruit during the growing season, whereas other diseases arise in superficial injuries that are inflicted during harvest and subsequent handling of the fruit. The stem-end rots of citrus fruits, incited by *Diplodia natalensis* Pole-Evans (*Physalospora rhodina* Berk & Curt.), *Phomopsis citri* Fawcett (*Diaporthe citri* Wolf), and *Alternaria citri* (Ell. & Pierce), arise from incipient infections in the fruit button (calyx + disk) that remain quiescent until the fruits are harvested and the buttons gradually lose vitality and their resistance to these pathogens. Infections of *Diplodia* and *Phomopsis* are localized on the surface of the disk or the ventral side of the sepals (8). *Alternaria* is also present at these sites and in the exocarp, beneath the calyx (32). Environmental conditions and treatments that accelerate button senescence usually increase the incidence of the stem-end rots.

Brown rot of citrus fruits is caused by several species of *Phytophthora* (*P. citrophthora* (R. E. Sm. & E. H. Sm.) Leonian, *P. parasitica* Dast., *P. syringae* (Kleb.) Kleb., and *P. hibernalis* Carne) that infect the fruit in the grove before harvest. Zoospores of the pathogens develop in sporangia on the surface of moist soil and are transported by rain splash onto the fruit hanging on the lower skirt of the tree. *P. citrophthora* is the most pathogenic species in the temperature range of 20 to 30° C, whereas *P. hibernalis* is most active at 10–15° C (10,30). Protective fungicides (e.g., fixed coppers) are effective only when they are applied before or soon after inoculation of the fruit. Immersion of lemons in a copper sulfate solution (1 g of $CuSO_4 \cdot 5H_2O$ per liter) 3 hr (at 12–30° C) after inoculation with *P. citrophthora* did not reduce the incidence of infection (24).

Anthracnose is caused by *Colletotrichum gloeosporioides* (Penz.) Sacc., which infects immature fruit in the grove before harvest. Spores of the pathogen carried in water from dead twigs in the tree canopy germinate on the fruit surfaces. Short germ tubes terminate with the production of darkly colored appressoria. The appressoria may produce infection hyphae that become latent after penetrating a few surface cells of the exocarp, or the appressoria may remain ungerminated until fruits are mature or are treated with ethylene. The decay may occur in mature fruit of all citrus cultivars in association with injuries or bruises inflicted during handling, or it may develop in healthy peels of certain mandarin cultivars (*Citrus reticulata* Blanco cv. Robinson) after ethylene degreening.

Green and blue mold caused by *Penicillium digitatum* Sacc. and *P. italicum* Wehmer, respectively, and sour rot caused by *Geotrichum candidum* Lk. usually are initiated at wounds inflicted to the peel during harvesting and handling of the fruit. The incidence of infection of both diseases increases with the number of spores in the inoculum and the depth of the injury (1,34,39). Shallow wounds that are limited to the flavedo (outer, orange-colored portion of the pericarp) give low and erratic infection rates because such wounds develop resistance to infection resulting from lignification or desiccation. Wounds that extend to a depth of 2–3 mm, penetrating into the albedo (white, spongy portion of the pericarp), give high infection rates provided the inoculum concentration is sufficiently large. Effective postharvest fungicides can eradicate incipient infections of *Penicillium* and *Geotrichum* when they are applied to the fruit within about 30 hr (at 20–25° C) after inoculation. Conidia of *P. digitatum* and *P. italicum* begin to form on diseased lesions after 5–7 days of incubation at this same temperature. Sporulation of *Penicillium* spp. is usually quite heavy on diseased fruit 10–14 days (at 20–25° C) after inoculation; these spores are shed onto adjacent fruit in the container, giving rise to an objectionable condition known in the fruit trade as soilage.

Postharvest fungicides for citrus fruits may be evaluated for several attributes: eradication of latent or incipient infections; inactivation of germinating spores of pathogens developing in wounds; protection of the fruit pericarp from infection at injuries inflicted after application of the fungicide treatment; inhibition of sporulation of *Penicillium* on the surface of decaying fruit; prevention of the contact spread of *P. italicum, G. candidum,* and *Phytophthora* from diseased to adjacent healthy fruits; possession of volatile or fumigant characteristics; and compatibility with wax formulations.

SELECTION AND PREPARATION OF FRUIT

Fruits of intermediate sizes are preferred so that the storage containers will accommodate 50–100 per replication. Fruits of three sizes are selected by eye and combined. When the crop consists mainly of larger and smaller fruits than those preferred, the predominant sizes will have to be used. The fruit should be horticulturally mature, but not overripe. Fruit should be relatively free of insect infestation and damage, and blemishes such as melanose, scab, wind scar, and frost damage should be minimal. Usually, the fruit should be cleaned in a detergent solution, rinsed in tap water, and surface-sterilized by being dipped for 1 min in 70% (v/v) ethanol:water or sodium hypochlorite (1 g/L), pH 8 (1,000 ppm active chlorine). When many fruits are surface-sterilized in the same solution, periodic checks should be made of the concentration of alcohol (by hydrometer) or active chlorine (test paper). The fruits must be surface-sterilized when they are contaminated with one or more pathogens that will not be controlled by any of the fungicide treatments being tested (e.g., *Geotrichum* or fungicide-resistant *Penicillium*). The test will be meaningless if a significant number of the fruits are infected by these fungi. After disinfestation, the fruits are rinsed, partially air-dried, and placed in large polyethylene garbage bags in citrus cartons or in cartons in storage rooms at 88–92% relative humidity (RH).

DISEASE PRODUCTION

Diplodia stem-end rot. Most investigators prefer to use fruit that has been naturally inoculated in the grove for fungicide tests. Oranges grown in Florida develop 30–40% or more Diplodia stem-end rot following degreening with 5–10 ppm ethylene for 3 days at 30°C (94–97% RH) and storage for 3 weeks at 21°C (3–5,33). The fruits are distributed (several at a time) to the experiment replications, usually three or more replications of 100 fruits each, either before or after the degreening treatment. Postharvest fungicide treatments are invariably more effective against Diplodia stem-end rot when they are applied before the degreening treatment, although benomyl, thiabendazole, and carbendazim gave a substantial reduction in Diplodia stem-end rot even when they were applied to the fruit 72 hr after the onset of degreening (5).

Phomopsis stem-end rot. Since the incidence of this disease under natural conditions is variable, fruit for fungicide tests should be inoculated in the laboratory with *P. citri* (3,4). The pathogen is grown in PDA broth in shake culture at 25°C for 5 days. The mycelium is chopped in a blender, and the mycelial suspension is thickened by addition of cooled, molten agar to a final concentration of 0.2% (w/v). Drops of the thickened mycelial suspension are placed on and around the calyx of each fruit. The fruits are treated with ethylene (10 μl/L) in a ventilated cabinet at 24–26°C and 94–97% RH for 48 hr before application of the fungicide treatment. After the fungicide treatment, the fruits are stored for 3 weeks at 21°C and 88–92% RH for disease development.

Alternaria stem-end rot. Natural infestations of this pathogen that arise in the grove in the button of the fruit can produce a relatively high incidence of Alternaria stem-end rot in mature fruit stored 8–16 weeks at 10–20°C and 90–95% RH (6,23,32). The fruits are coated with a storage wax containing benomyl at 1 g/L, but not 2,4-D. The wax retards water loss and appears to stimulate the incidence of *Alternaria* (32). Benomyl suppresses the development of Diplodia and Phomopsis stem-end rots and Penicillium molds on the fruit during storage, but it has no effect upon the development of Alternaria stem-end rot. Since the incidence of Alternaria stem-end rot is correlated with the frequency of dead and deteriorating buttons (black buttons) (23), strong fruit may be exposed to ethylene for 1–2 days at 25°C before storage to stimulate the development of Alternaria stem-end rot. The fruits are not treated with 2,4-D, which is applied commercially to lemons to maintain the vitality of the button and thereby reduce the incidence of Alternaria stem-end rot.

If experience has shown that natural inoculations will not provide the desired incidence of disease, the fruit may be artificially inoculated with *A. citri*. The fruit stem near the calyx (peduncle) is punctured to a depth of 3 mm. A suspension of mycelial fragments, prepared in the same manner as for *P. citri* except suspended in a suspension of benomyl at 100 μg/ml, is placed around the calyx and over the wounded site (3).

After application of the fungicide treatment, the fruits are placed in an environment that simulates long-term commercial storage. As an alternative, the test may be accelerated by holding the fruit at a somewhat higher temperature. Appropriate temperatures for the development of Alternaria stem-end rot are, for lemons, 10–20°C; grapefruit, 10–15°C; and oranges, 5–15°C. When the fruits are stored at the lower temperatures for 8–16 weeks, they may be moved to a higher temperature (e.g., 20°C) for several days to simulate a market environment and to increase the level of decay.

Anthracnose. Chemical treatments for the control of anthracnose may be evaluated with naturally infected fruit from the grove or with fruit artificially inoculated with the pathogen.

Spores of *C. gloeosporioides* are obtained from 7-day-old cultures grown on Difco oatmeal agar at 26°C under 6,000–7,000 lux of continuous fluorescent light. Spores removed in sterile water are concentrated by centrifugation at 2,500 *g* for 5 min and resuspended to a concentration of approximately 250,000 spores per milliliter of sterile water. Drops of the spore suspension are placed on the fruit equator, and the fruits are held overnight near 100% RH at 26°C for formation of the appressoria. Artificially inoculated or naturally infected fruit are treated (degreened) with ethylene (10–50 μl/L) for 24–36 hr before being washed and treated with experimental fungicides.

Phytophthora brown rot. Since zoospores of *P. citrophthora* are inactivated by Cu^{++}, it is advisable to dip clean fruit (yellow lemons preferred) for 3 min in 0.1 *N* HNO_3 containing 0.05% Triton X-100 (6.35-ml conc. HNO_3 and 0.5 g of Triton X-100 per liter) to remove residues of copper fungicides that may be on the fruit surface. The fruits are rinsed in tap water, surface-sterilized as described above, and finally rinsed in distilled and deionized water.

Inoculum production. Mother cultures of *P. citrophthora* are produced by adding 5 ml of sterile salt solution comprised of 10 mM $Ca(NO_3)_2$, 5 mM KNO_3, and 4 mM $MgSO_4$, pH 6.4 (9), onto a 7- to 10-day-old culture of *P. citrophthora* growing on V-8 juice agar. Zoospores are produced to seed cultures for sporangial production by following steps 1–6 below. Other investigators have used similar methods for the production of zoospore inoculum of *P. citrophthora* (31), *P. parasitica* (36), and *P. syringae* (10).

Encysted zoospores for inoculation of fruit are produced as follows:

1. Add 5 ml of zoospore suspension (produced above and diluted to 500 zoospores per milliliter) to each plate containing V-8 juice agar and incubate the cultures at 27°C for 72 hr.
2. Cover the mycelium for 30–60 min with 5 ml of salt solution (described above). Decant. Cover the culture for 30–60 min with 5 ml of deionized water. Decant.
3. Incubate the cultures at 27°C for 24 hr for production of sporangia. Check the cultures for bacterial

contamination by spotting liquid from each culture onto a premarked nutrient agar plate and incubate at the same temperature.

4. Inspect the nutrient agar plates for bacterial growth. Discard contaminated *Phytophthora* cultures. Briefly wash the *Phytophthora* cultures with sterile deionized water. Decant water. Incubate cultures at 27°C for 42 hr for maximum production of sporangia. Wash cultures briefly with deionized water.

5. Chill cultures at 5°C for 30 min.

6. Add 5 ml of deionized water (27°C) to each culture and incubate at 27°C for 30 min. Zoospores (ca. 2×10^6/ml) should be released.

Inoculation of fruit. Zoospore suspensions from several cultures (depending upon the concentration and volume of inoculum required) are combined. The zoospore concentration is determined by diluting a 1-ml portion of the pooled suspensions to 10 ml with deionized water. Several drops of FAA fixative (8 ml of glacial acetic acid, 13 ml of formalin, 100 ml of 95% ethyl alcohol, and 90 ml of water) are added to encyst the zoospores. The concentration is determined by means of a hemacytometer, and the suspension is diluted with water and cool sterile water agar to give a final concentration of 2×10^4 encysted zoospores per milliliter of 0.25% water agar. The zoospores are not very motile in the agar solution and encyst within a few minutes. A nontoxic dye marker, such as Schilling Red Food Color, is added to the inoculum. One hundred microliters (2,000 encysted zoospores) is transferred with a suitable pipette (e.g., Pipetteman) to a site on the fruit equator. The inoculated fruit is stacked in a stainless steel tray in a manner that does not disturb the inoculum. The fruit is incubated in a water-saturated atmosphere at 20°C for 24 hr to produce an incipient infection of *P. citrophthora* in the peel of the fruit. The inoculum is then removed from the surface of the fruit with a soft brush dipped in 50% alcohol, taking care not to distribute viable inoculum over the fruit. The inoculated fruit are randomly distributed to the fungicide treatment replication lots.

Green and blue mold. The same test procedures are used for both of these diseases. Primary infections by *Penicillium digitatum* and *P. italicum* usually occur through fresh wounds in the peel of the fruit. Fruit tends to become more susceptible to infection with physiological age (34), but this fact should not lead to excessive variability in the incidence of disease from one experiment to another if the recommended procedures are followed.

Fungicides are often evaluated for effectiveness against three aspects of Penicillium decay: eradication of 18- to 24-hr-old inoculations; protection of the peel against infection through injuries inflicted one to several days after the fungicide treatment is applied; and inhibition of *Penicillium* sporulation on decaying fruits. The first aspect is usually the primary evaluation of a new fungicide; effective compounds are subsequently evaluated for activity against the other two aspects of the disease.

The major factors that influence the success of inoculations and the difficulty of inactivating them by fungicide treatment are depth and width (diameter) of the wound in the peel; the number of spores placed in that wound; and incubation conditions that are more favorable to the growth of *P. digitatum* or *P. italicum* than to the healing process in the wounded tissue.

Stock cultures of *P. digitatum* and *P. italicum* are conveniently maintained on silica gel (27,37) or PDA slants. Spores from 7- to 14-day-old cultures on PDA slants are suspended in 0.01% (w/v) Triton X-100 and filtered through two layers of cheesecloth to remove most of the hyphal fragments. The concentration of spores in the suspension is measured by its optical density (O.D.) compared with that of a series of standard spore suspensions validated by direct spore count with a hemacytometer (25). The spore suspension is diluted to give a final concentration of 1×10^6 spores per milliliter. This spore concentration in a 13-mm-diameter tube has an O.D. of 0.1 (at 425 nm) in a Bausch and Lomb Spectronic 20. The spore concentration may also be adjusted after a direct count of the initial suspension with a hemacytometer. Dry spores are sometimes used for inoculum, but standardization of the inoculum level is less accurate.

The eradicant action of fungicides in water or wax formulations is evaluated by wounding the fruit with a tool that punctures or scratches the peel of the fruit and simultaneously introduces the spore suspension (3,14,15,20,22). The wound should be approximately 2–3 mm deep and about 1 mm in breadth (34) and penetrate into the albedo tissue, but not into the juice sacs. The depth of wounding should be adjusted to match the average peel thickness of the experimental fruit. Brown (5) reported that postharvest fungicides in a wax-formulation gave better control of decay when fruits were inoculated through wounds 2 mm long $\times$ 1 mm in diameter than through wounds 4 mm $\times$ 0.3 mm. Roistacher and Klotz (28) described a device, consisting of a circular saw blade rotating slowly in a spore suspension, for rapid inoculation of citrus fruits with *Penicillium* spores. The wound inflicted by the device was about 1–2 mm deep, 1 mm wide, and 10 mm long. The inoculated fruit is incubated for 18 hr (overnight) at 20°C and high humidity (e.g., in polyethylene bags in shipping cartons) before application of the fungicide treatment.

In some tests, wounding the fruit and introducing the spores are separate operations. This method, which is intended to simulate natural inoculation of the fruit in a packinghouse where the fruit may be injured on the processing line (29,38), has been used to evaluate fumigation treatments, especially those that are applied in the container after the fruit are packed (13,15,18). Typically, the fruits are scratched by the saw technique and, within a few hours, showered with dry spores of *P. digitatum* (28,29). This procedure is more time-consuming than the simultaneous wound-inoculation technique, and a relatively large quantity of dry spores is required. Roistacher et al (29) used 0.5–1.0 cm^3 of *P. digitatum* spores in their inoculation chamber to inoculate 150 wounded fruits. A dry spore load of 1 cm^3 deposited approximately 4,000 spores per square centimeter on the fruit. The dry spore inoculation technique has also been used to evaluate the effectiveness of a fungicide in preventing infection through injuries inflicted after application of the fungicide treatment (protective action). The fruits are wounded and inoculated with dry spores 1 day or more after application of the fungicide treatment (7,19,20).

Fruits for evaluation of sporulation inhibition by a test fungicide are inoculated by hypodermic injection

(0.2 ml) of a spore suspension (1.0×10^6 spores per milliliter) into the central cavity or inner albedo of the fruit (7,16,22). It is usually most convenient to deep-inoculate the fruit after fungicide application. After inoculation, the fruits are placed on a steel rack in a tray and covered with a sheet of polyethylene.

Fruit inoculated with *P. digitatum* or *P. italicum* are incubated at 20°C and high humidity for 14 days before evaluation as described below. For extensive testing of effective fungicides, the environmental conditions of a market exposure may be used.

Sour rot. *Geotrichum candidum* is cultured on PDA slants or in medicine bottles for 7–14 days at 25°C. The spores are suspended in 0.05 M KH_2PO_4 in 0.01% (w/v) Triton X-100, pH 6, and filtered through two layers of cheesecloth to remove hyphal fragments. The concentration of the spore suspension is standardized (O.D. 0.1 at 425 nm $\cong 1 \times 10^6$ spores per milliliter) in the same manner as for *Penicillium* (25). The procedure for inoculation with *G. candidum* is similar to that for *Penicillium* spp., but some modifications are necessary because *G. candidum* is a less aggressive pathogen and the incidence of infection can be relatively low and variable (12). The uncertainty in the incidence of infection is mainly caused by variability in the susceptibility of different lots of fruit (1). Fruit susceptibility increases with physiological age. Fruits that are just mature should be exposed to 10–20 ppm ethylene at 25°C for several days before inoculation with the pathogen. Fruits of advanced maturity usually show a high level of infection following puncture-inoculation to a depth of 2–3 mm with a 1-mm-diameter tool that is dipped in a suspension of 10^7–10^8 spores per milliliter before each inoculation. In view of the high degree of variability in susceptibility of fruit lots, it is advisable to test inoculate samples of 25 representative fruits each with the following inocula (spores per milliliter): 10^6, 10^7, 10^8, 10^7 + cycloheximide, and 10^8 + cycloheximide. Cycloheximide (10 μg/ml) inhibits wound healing but has no significant effect upon the growth of *Geotrichum* (1). For large-scale tests, decayed fruits produced under aseptic conditions and homogenized in a blender with water to give a thick puree of diseased tissue can be used as inoculum (2,21). The fruit is punctured to a depth of 2–3 mm with a 1-mm-diameter pin and either dipped in the inoculum or a drop of thick inoculum is placed on the wound. Benomyl (100 μg/ml) may be added to the inoculum to suppress potential contaminants such as *Penicillium* spp. The main disadvantage of this procedure is the difficulty of standardizing the concentration of pathogen propagules in the inoculum.

Fruit inoculated with *G. candidum* should be placed in a plastic bag, misted lightly with water, and incubated at 20°C for 18 hr (overnight) before application of the fungicide treatments.

EXPERIMENTAL DESIGN

The major sources of variability in postharvest fungicide tests are fruit susceptibility and uniformity of inoculation. In experiments with the stem-end rots and anthracnose (*C. gloeosporioides*), it is advisable to apply each treatment to a minimum of three replications of 100 fruits each. Inoculation of fruit with *Phytophthora, Penicillium,* and *Geotrichum* is fairly reproducible, and a minimum of three replications of 50 fruits each should be sufficient. A reliable measure of sporulation inhibition can be derived from 10 fruits per treatment. The fruits with incipient infections should be systematically distributed (in groups of four) to all of the experimental sublots. The sublots (treatment replications) are randomly selected for each fungicide treatment.

The usual fungicide test includes a no-fungicide control and an effective-fungicide standard, as well as the fungicide treatments to be evaluated. The active infections (especially wound inoculations) on the fruit must be treated with a fungicide within a finite time (maximum 24–30 hr, depending upon the pathogen and the incubation environment) to prevent disease development. Since several hours might be required to apply all of the fungicide treatments, it is possible that the last treatment could suffer a handicap. To overcome this uncertainty, it is advisable to apply the no-fungicide control treatment first and the effective-fungicide standard treatment last. The fungicides under evaluation can then be fairly compared with the standard treatment.

FUNGICIDE TREATMENTS

Liquid formulations. The usual methods (11) for commercial application of postharvest fungicides are dipping or drenching the fruit in a relatively large volume of the recirculated aqueous formulation, spraying the fruit on brushes with a nonrecoverable water spray or wax formulation of the fungicide, or mist application of a low-volume spray of the fungicide in a hydrocarbon-solvent wax formulation onto the fruit on a roller conveyor.

Water dispersions. Invariably, nonvolatile fungicides are most effective when they are applied in a volume of water that thoroughly wets the fruit. Therefore, most new fungicides are tested first as solutions, suspensions, or emulsions in water, and the inoculated fruits are immersed in the treatment solution. Fungicides that are effective when applied in this manner are tested later in the form of a water spray or in a wax formulation to determine the most economic and convenient method of application in a commercial packinghouse.

Most new fungicides are received in a form (wettable powder or emulsifiable concentrate) that is readily dispersible in water. Treatment solutions should be checked to be sure that they thoroughly wet the waxy cuticle of citrus fruits. Wetting and capillary action are very important in controlling the stem-end rots and wound pathogens that are located in narrow crevices in the fruit peel. Routinely, it is advisable to add an inert nonionic surfactant such as Triton X-100 or Tergitol XD to all of the treatment solutions at the rate of 0.5 g/L. Fungicides that are available only as technical material can often be dispersed in water by dissolving them in a water-miscible organic solvent (methanol, ethanol, acetone, dimethylsulfoxide, etc.) containing a nonionic surfactant and pouring this solution slowly into an aqueous solution of the same surfactant while stirring. The organic fungicide often forms fine particles or droplets that can be maintained in a dispersed state for the period of fruit treatment. Comparatively inert, nonionic surfactants are the best choice when preparing a dispersion of a compound whose chemistry is not fully known. Fungicides such as guazatine and quaternary ammonium compounds are precipitated by anionic surfactants, and the decomposition of benomyl appears to be accelerated in the presence of certain surfactants.

Wax formulations. Some fungicides appear to be less effective when applied in a water-wax formulation than in water only. Coverage of the fruit and penetration to the site of infection seem to be the main reasons for the reduced activity of fungicides in wax formulations. The smaller volume and higher viscosity of the wax formulation applied, compared with a drench or dip treatment (aqueous), reduce the amount of fungicide that actually reaches the site of infection (5).

Typically, water-wax formulations vary from about pH 8 to pH 10.5. The cyclization of benomyl to *s*-triazinobenzimidazole (STB), a compound that is inactive as a postharvest fungicide, proceeds at a faster rate at pH 10 than at pH 8. In addition, benomyl decomposes at pH 8 to carbendazim, which is considerably less active than benomyl in suppressing *Penicillium* sporulation on decaying fruit (16).

Some investigators have reported that postharvest fungicides formulated in organic solvent waxes are less active than the same fungicide in a water spray or drench (26,35). Solvent waxes are applied at a low volume and dry rapidly on the fruit. Therefore, the deposit of fungicide on the fruit tends to be minimal and the coverage and penetration is less efficient than in the case of water-based formulations that are applied in a comparatively large volume. Furthermore, the water-based formulations dry more slowly, allowing more time for the fungicide to diffuse to the infection site.

Vapor treatments. *Space fumigation.* The fruits are inoculated with *P. digitatum* by methods described above (usually the dry spore shower); after a suitable incubation period, the fruits are exposed for various time periods to several concentrations of the gaseous fungicide. Techniques for preparing and administering the different concentrations of the fungicide gas and other methods associated with fumigation tests are available in the literature (15,29).

Microfumigation in shipping containers. Treated fruit wraps and paper inserts have been developed that release fungicide vapors into the atmosphere of the shipping container. The main objective of these treatments is to prevent infection by *Penicillium* spp. in injuries sustained by the fruit during handling in the packinghouse and to inhibit fungus sporulation on decaying fruit. The dry spore shower method of inoculation has been most frequently used in tests of this type (13,17,18,29).

EVALUATION OF RESULTS

Decay control. Following fungicide treatment, the number of diseased fruits in each treatment will reach a near-maximal value after a certain period of time, depending mainly on the pathogen and incubation temperature or temperature regimen. Disease is evaluated at that time as a quantal response to the fungicide treatment—each fruit is recorded as diseased or healthy. The level of disease in each fungicide treatment is recorded as percentage of decay (100 × number of diseased fruits ÷ number of fruits in that replication). Variance analysis is usually carried out on the transformed variate [y = arc sin (% decay)$^{1/2}$] and mean differences are evaluated using Duncan's multiple range test or a similar test of mean separation. When the results of several independent tests are compiled, it is convenient to express the data as percentage of decay control (% decay in control − % decay in treated fruit ÷ % decay in control).

***Penicillium* sporulation.** After a suitable incubation period (14 days at 20°C) or a simulated shipment and market environment, the degree of *Penicillium* sporulation on the surface of decayed fruits can be evaluated on a 0–5 scale (sporulation index) in which 0 = negligible sporulation and 5 = dense fungal sporulation over the entire surface of the fruit. The index value for each fruit is treated as a replication, and the treatment means are subjected to an analysis of variance (7,12,16,22).

LITERATURE CITED

1. Baudoin, A. B. A. M., and Eckert, J. W. 1985. Development of resistance against *Geotrichum candidum* in lemon peel injuries. Phytopathology 75:174-179.
2. Brown, G. E. 1979. Biology and control of *Geotrichum candidum*, the cause of citrus sour rot. Proc. Fla. State Hortic. Soc. 92:186-189.
3. Brown, G. E. 1981. Investigations with experimental citrus postharvest fungicides in Florida. Proc. Int. Soc. Citriculture 2:815-818.
4. Brown, G. E. 1983. Control of Florida citrus decays with guazatine. Proc. Fla. State Hortic. Soc. 96:335-337.
5. Brown, G. E. 1984. Efficacy of citrus postharvest fungicides applied in water or resin solution water wax. Plant Dis. 68:415-418.
6. Brown, G. E., and McCornack, A. A. 1972. Decay caused by *Alternaria citri* in Florida citrus fruit. Plant Dis. Rep. 56:909-912.
7. Brown, G. E., Nagy, S., and Maraulja, M. 1983. Residues from postharvest nonrecovery spray applications of imazalil to oranges and effects on green mold caused by *Penicillium digitatum*. Plant Dis. 67:954-957.
8. Brown, G. E., and Wilson, W. C. 1968. Mode of entry of *Diplodia natalensis* and *Phomopsis citri* into Florida oranges. Phytopathology 58:736-739.
9. Chen, D., and Zentmyer, G. A. 1970. Production of sporangia by *Phytophthora cinnamomi* in axenic culture. Mycologia 62:397-402.
10. Chitzanidis, A., and Lascaris, D. 1981. Research on Phytophthora rots of oranges in Greece. Proc. Int. Soc. Citriculture 2:796-799.
11. Eckert, J. W. 1977. Control of postharvest diseases. Pages 269-352 in: Antifungal Compounds. Vol. 1. M. R. Siegel and H. D. Sisler, eds. Marcel Dekker, New York.
12. Eckert, J. W., Bretschneider, B. F., and Ratnayake, M. 1981. Investigations on new postharvest fungicides for citrus fruits in California. Proc. Int. Soc. Citriculture 2:804-810.
13. Eckert, J. W., and Kolbezen, M. J. 1963. Control of Penicillium decay of oranges with certain volatile aliphatic amines. Phytopathology 53:1053-1059.
14. Eckert, J. W., and Kolbezen, M. J. 1964. 2-Aminobutane salts for control of post-harvest decay of citrus, apple, pear, peach, and banana fruits. Phytopathology 54:978-986.
15. Eckert, J. W., and Kolbezen, M. J. 1970. Fumigation of fruits with 2-aminobutane to control certain postharvest diseases. Phytopathology 60:545-550.
16. Eckert, J. W., and Kolbezen, M. J. 1977. Influence of formulation and application method on the effectiveness of benzimidazole fungicides for controlling postharvest diseases of citrus fruits. Neth. J. Plant Pathol. 83 (Suppl. 1):343-352.
17. Eckert, J. W., Kolbezen, M. J., and Garber, M. J. 1963. Evaluation of ammonia-generating formulations for

control of citrus fruit decay. Phytopathology 53:140-148.

18. Eckert, J. W., and Rahm, M. L. 1983. Azomethines of sec-butylamine for the control of Penicillium decay of oranges. Pestic. Sci. 14:220-228.
19. Gutter, Y. 1969. Effectiveness of preinoculation and postinoculation treatments with sodium orthophenylphenate, thiabendazole, and benomyl for green mold control in artificially inoculated Eureka lemons. Plant Dis. Rep. 53:479-482.
20. Gutter, Y. 1970. Influence of application time on effectiveness of fungicides for green mold control in artificially inoculated oranges. Plant Dis. Rep. 54:325-327.
21. Gutter, Y. 1978. A new method for inoculating citrus fruits with *Geotrichum candidum*. Phytopathol. Z. 91:359-361.
22. Gutter, Y., Shachnai, A., Schiffmann-Nadel, M., and Dinoor, A. 1981. Chemical control in citrus of green and blue molds resistant to benzimidazoles. Phytopathol. Z. 102:127-138.
23. Harvey, E. M. 1946. Changes in lemons during storage as affected by air circulation and ventilation. U.S. Dep. Agric. Tech. Bull. 908:1-32.
24. Klotz, L. J., and DeWolfe, T. A. 1961. Limitations of the hot water immersion treatment for the control of Phytophthora brown rot of lemons. Plant Dis. Rep. 45:264-267.
25. Morris, S. C., and Nicholls, P. J. 1978. An evaluation of optical density to estimate fungal spore concentrations in water suspensions. Phytopathology 68:1240-1242.
26. Pelser, P. du T. 1972. Decay control in Washington Navel and Valencia oranges by application of TBZ or benomyl suspended in water or incorporated into waxes. Citrus Grower Subtrop. Fruit J. 463:12-13 and 464:23.
27. Perkins, D. D. 1962. Preservation of *Neurospora* stock cultures with anhydrous silica gel. Can. J. Microbiol. 8:591-594.
28. Roistacher, C. N., and Klotz, L. J. 1955. A device for rapid inoculation of citrus fruits. Phytopathology 45:517-518.
29. Roistacher, C. N., Klotz, L. J., Kolbezen, M. J., and Staggs, E. A. 1958. Some factors in the control of blue-green mold decay of citrus fruit with ammonia. Plant Dis. Rep. 42:1112-1122.
30. Schiffmann-Nadel, M. 1969. Research on Phytophthora diseases of citrus fruits in Israel. Proc. Int. Citrus Symp. (Riverside, CA), 1st, 3:1201-1205.
31. Schiffmann-Nadel, M., and Cohen, E. 1968. Method for production of sporangia of *Phytophthora citrophthora*. Phytopathology 58:550.
32. Schiffmann-Nadel, M., Waks, J., Gutter, Y., and Chalutz, E. 1981. Alternaria rot of citrus fruit. Proc. Int. Soc. Citriculture 2:791-793.
33. Smoot, J. J. 1977. Factors affecting market diseases of Florida citrus fruits. Proc. Int. Soc. Citriculture 1:250-254.
34. Smoot, J. J., and Melvin, C. F. 1961. Effect of injury and fruit maturity on susceptibility of Florida citrus fruit to green mold. Proc. Fla. State Hortic. Soc. 74:285-287.
35. Smoot, J. J., and Melvin, C. F. 1974. Decay control of oranges with benomyl by three methods of postharvest application. Proc. Fla. State Hortic. Soc. 87:234-236.
36. Timmer, L. W. 1979. Preventative and systemic activity of experimental fungicides against *Phytophthora parasitica* on citrus. Plant Dis. Rep. 63:324-327.
37. Trollope, D. R. 1975. The preservation of bacteria and fungi on anhydrous silica gel: An assessment of survival over four years. J. Appl. Bacteriol. 38:115-120.
38. Tuset, J. J., Carreres, R., and Hinarejos, C. 1977. Control de las podredumbres causadas por hongos en los frutos de satsuma despues de la recoleccion. Proc. Int. Soc. Citriculture 1:320-322.
39. Wild, B. L., and Eckert, J. W. 1982. Synergism between a benzimidazole-sensitive isolate and benzimidazole-resistant isolates of *Penicillium digitatum*. Phytopathology 72:1329-1332.

Investigating Physical Modes of Action of Tree Fruit Fungicides

MICHAEL SZKOLNIK, Department of Plant Pathology, New York State Agricultural Experiment Station, Geneva 14456

The performance of fungicides for control of an orchard fungal disease depends upon the choice of fungicide and formulation, rate of application, timing, and completeness of plant coverage. Apart from its biochemical mode of action against a fungal pathogen, each fungicide is governed by other properties bearing on chemical persistence; autodegradation by light, temperature, moisture, and other factors; and an undetermined number of properties associated with what is loosely identified as *physical modes of action*. The most effective planning of orchard sprays, whether for research or commercial disease control, lies in the availability of adequate knowledge on the strengths and weaknesses of each formulated fungicide as determined through responsible research on modes of action.

Modes of action studies on fruit fungicides are laboratory and greenhouse oriented. Control of such diseases as apple scab in the orchard requires a series of sprays under wide ranges of fungal inoculum, fungicide residue levels, foliar and fruit development, and environmental conditions; therefore, fair and accurate evaluations of fungicides for individual modes of action are almost impossible in orchard settings. Fungicide mixtures or changes to unrelated fungicidal products during the growing season add to the complexity of such studies. Laboratory and greenhouse studies are feasible because the researcher can exercise controls over plant materials; fungicide formulation and rates; precision spraying; timing of sprays; artificial rainfall; source, choice, concentration, and timing of inoculum; and temperature and moisture control for infection.

Table 1. Modes of action of apple scab fungicides

Mode of action	Comment
Protection	Fungicide effectiveness during infection period may come from original residues or those remaining after erosive action of rains.
After-infection	Fungicide activity is limited in time (hours, days) after infection has been established; this action prevents symptom manifestation. Synonyms: eradication, kickback, curative, intervention.
Presymptom	Fungicide effectiveness beyond after-infection limits but before appearance of symptoms; symptoms are abnormal, such as chlorotic apple scab lesions devoid of conidia.
Postsymptom	Fungicide effective after appearance of symptoms; examples may be "burning out" of scab lesion or prevention of new spore production on existing scab lesions.
Movement	Relocation of fungicide from original application site: (i) redistribution of particulate or of dissolved fungicide on plant surface; (ii) ingress and systemic movement (general or local) of fungicide within plant tissue; (iii) vapor-phase distribution of fungicide in plant environment.

Laboratory and greenhouse studies enable one, in effect, to dissect the "total fungicide" of field performance into its component modes of activity. Among the known physical modes of action for which study procedures have been reported are those listed in Table 1; they are essentially regarded as chemical-pathogen relationships. It is conceivable that other modes of action may apply, such as the predominant chemical effect on host surface and on host physiology and disease resistance factors. Although emphasis here centers on apple scab (*Venturia inaequalis* (Cke.) Wint.; imperf. *Spilocaea pomi*), the definitions apply to other diseases and the techniques are adaptable to other fruit and nonfruit disease control studies.

The significance of timing of fungicidal modes of action in relationship to the progress of fungal pathogen activity is depicted in Fig. 1. The time span for the infection period depends upon temperature during one or more periods of wetness, and the duration of the incubation period is influenced primarily by temperature.

Fungicidal strengths in one or more modes of action become significant factors in the choice of fungicides and in scheduling their time of application for most effective disease control in orchard spray programs. It may seem confusing that, for any infection period, the modes of action *after-infection, presymptom*, and *postsymptom* are all actually postinfection treatments. The after-infection definition (Table 1), which dates back to about 1950, refers to prevention of symptom expression. The presymptom and postsymptom modes of action were determined and defined years later. Perhaps the greatest overlap occurs with fungicides in the after-infection and presymptom modes, in which case the prevention of lesions or the occurrence of abnormal chlorotic lesions is determined by the level of fungicide at the infection site and the duration of time following infection.

Laboratory-greenhouse investigative techniques for fruit fungicides are not designed for immediate implementation unless suitable plant material and facilities for spraying, inoculation, and infection are readily available. They serve well in planned, long-term projects. Much of the information about laboratory-greenhouse techniques on modes of action of tree fruit fungicides has been assembled into composite publications that are cited here for the researcher to consult.

SUMMARY OF LITERATURE

Reference (1). "Techniques involved in greenhouse evaluation of deciduous tree fruit fungicides." This reference deals mostly with apple scab and covers the following subject matter: (i) preparing and handling potted apple trees; (ii) technique for foliar spraying; (iii) preparation and use of conidial inoculum; (iv) infection of plants in mist chambers; (v) definition and techniques on modes of action, including fungicide retentiveness in protection, after-infection activity, presymptom activity, postsymptom activity, fungicide redistribution (combined orchard-greenhouse method), and systemic activity.

Reference (2). "Evaluation of physical modes of action of fungicides against the apple scab fungus." This reference covers much of the information provided in reference (1) above, but with somewhat less detail. Photographs on the turntable and rainfall system may prove helpful.

Reference (3). "Physical modes of action of sterol-inhibiting fungicides against apple diseases." Definitions and procedures in this publication cover protection, after-infection, presymptom, and postsymptom activities in the comparative evaluation of seven sterol-inhibiting fungicides against apple scab.

The paper also introduces procedures for the evaluation of fungicides for their vapor-phase action in confined chambers using apple powdery mildew (*Podosphaera leucotricha* (Ell. & Ev.) Salm.) as the test pathogen.

Reference (4). "Unique vapor activity by CGA-64251 (Vangard) in the control of powdery mildews roomwide in the greenhouse." Comprehensive details are given here on techniques used in the evaluation of a sterol-inhibiting fungicide in roomwide protection of some 22 fruit, vegetable, and ornamental hosts against their respective powdery mildew pathogens.

Reference (5). "Redistribution of fungicide residues in the protection of unsprayed apple foliage against scab." Modern high-concentrate, low-volume spraying leaves substantial areas of susceptible fruit and foliage with little or no coverage. Fortunately, the wet conditions favoring apple scab infections also favor redistribution of fungicidal residues to unprotected areas and to new growth occurring after spraying. Thus, redistribution is a vital key to successful orchard performance by fungicides.

EVALUATING REDISTRIBUTION OF RESIDUES

Reference (5) provides only a brief description of procedures for a newly developed laboratory-greenhouse method for evaluating fungicides for their redistributive impact in apple scab control. A much more detailed description of facilities and techniques is offered here.

Equipment design. The basic structural apparatus designed for fungicide redistribution studies is an aluminum carrousel, the hub of which slides up and down on a vertical shaft from the center of a steel turntable 2.1 m (7 ft) in diameter (Fig. 2). It can be locked in place on the shaft according to the height desired in each experiment. Six radiating angle-aluminum T-arms extend from the carrousel hub, each equipped distally with an adjustable leg. These legs provide vertical support for the carrousel and maintain it in a level position. Two adjoining T-arms form a wedge designed to hold an aluminum redistribution frame measuring 38 × 80 cm (15 × 31.5 in.)

Each redistribution frame is made from 0.6 × 1.9-cm (0.25 × 0.75-in.) aluminum screen framing with a linear groove. Lumite No. 5182100 Saran shading fabric (natural) is placed over the frame and pressed into the groove with a special tool that also presses into the groove a vinyl screen spline that holds the screen taut and in place on the frame. The Saran fabric was selected over aluminum fly screen, polypropylene shading fabric, or cheesecloth because it allows uniformly falling droplets under rainfall rather than the water puddling and irregular droplet development experienced with the other materials. The complete screen panels are washed

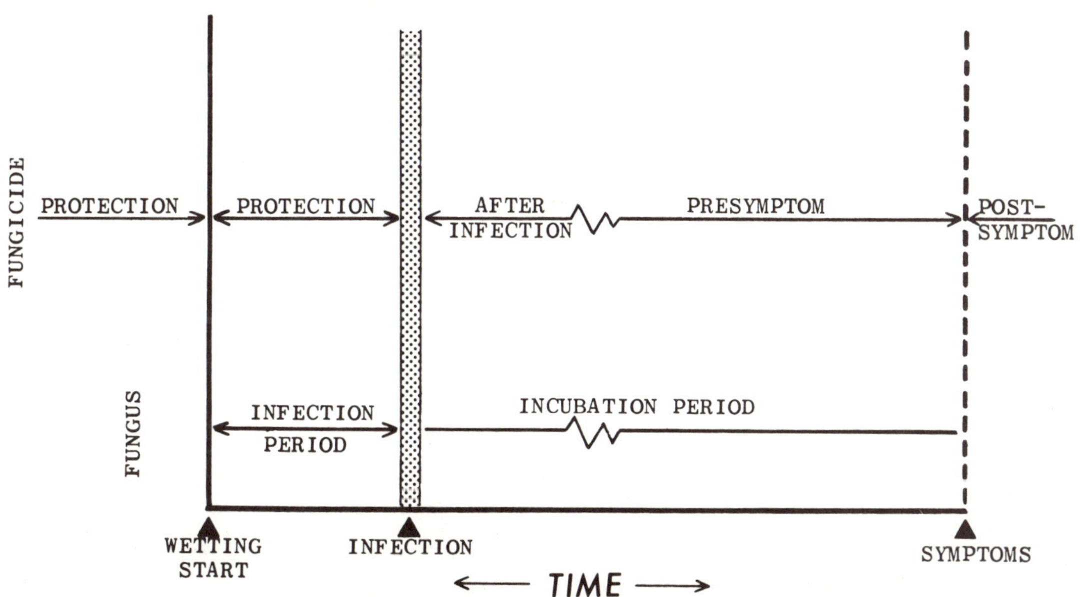

Fig. 1. Fungicidal activity in apple scab control in relationship to the scab pathogen during the infection, incubation, and postsymptom periods.

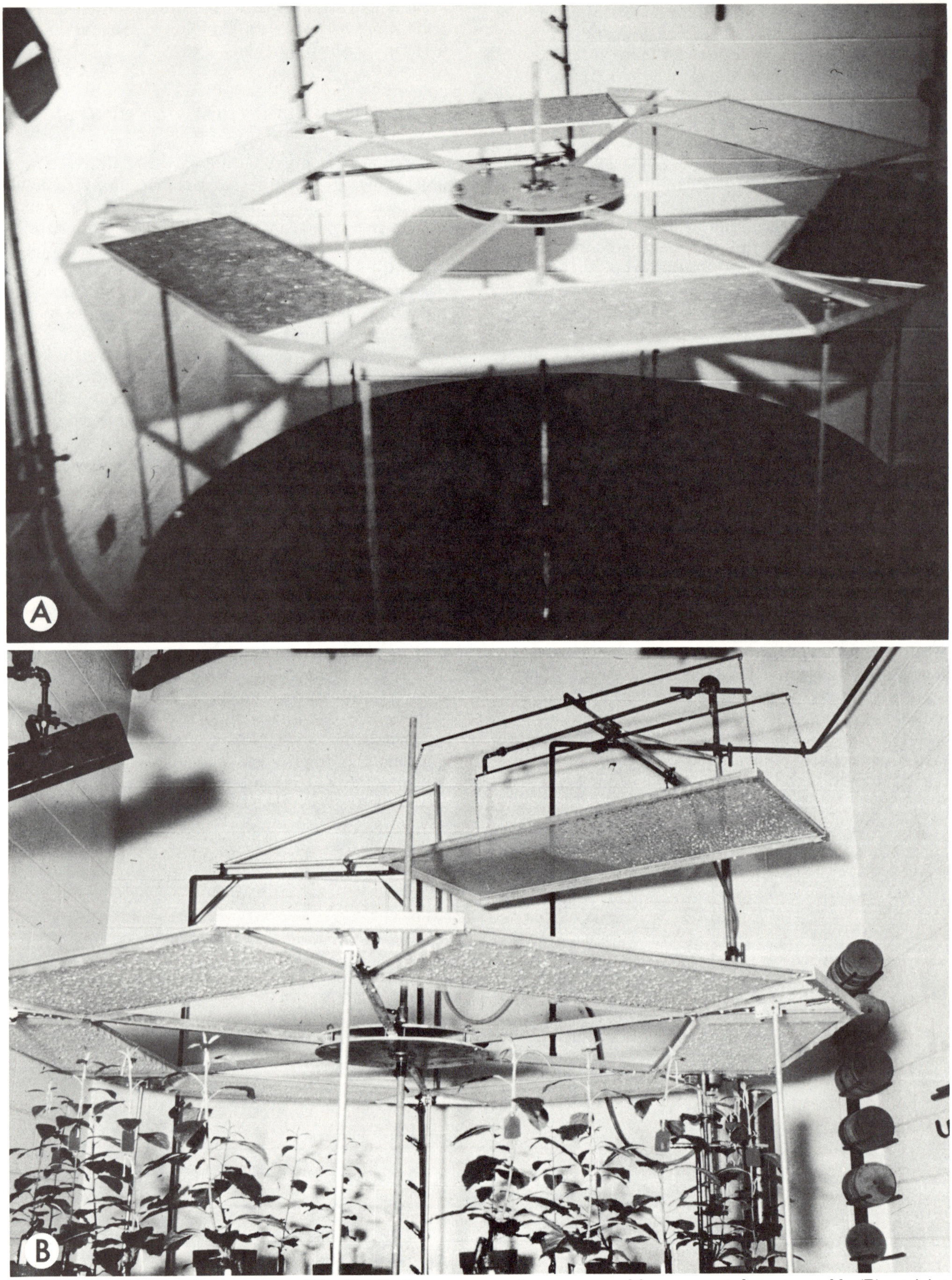

Fig. 2. Fungicide redistribution system: **(A)** redistribution screen panels on a carrousel frame mounted on a turntable; **(B)** precision-sprayed panels and apple trees positioned for rainfall delivery.

before use by placing them horizontally on the carrousel and passing them beneath an overhead washing rain for 1 hr. The screens are then dried and stored until used. After each experiment, the Saran cloth is removed and the frames are refitted with new material.

Redistribution test. *Precision spraying of screens.* In preparation for each redistribution experiment, each screen panel is placed horizontally on the carrousel and precision-sprayed with a fungicidal treatment, including water check, according to procedures described (1) for spraying apple trees. After the sprayed screens have dried at least a day, each screen is placed with the sprayed side up within an individual segment of the carrousel.

Apple tree arrangement. Three unsprayed, single-shooted, growing apple trees are placed on the turntable beneath each redistribution panel with the pots adjusted so that the youngest open leaf on each shoot is about 15 cm (6 in.) beneath the screen. The carrousel is adjusted by means of adjustable support legs so that all panels are in the same horizontal plane. A rain gauge is positioned on a support on the turntable so that the top is level with treetops.

Redistribution rainfall. With all panels in place on the carrousel, the turntable, operating at 2 rpm, is activated. Each panel passes repeatedly beneath an overhead rainfall system described elsewhere (1,2) for fungicide retention studies on trees. A single rainfall nozzle (Spraying Systems No. TX26) directly above the center of the redistribution panels supplies a spray onto a horizontal screen (2), which converts the spray into rainlike droplets that fall onto the redistribution screens. This rainfall carries any dislodged fungicide from the screen onto the unsprayed apple trees beneath.

Following a desired rain, usually 0.6 or 1.3 cm (0.25 or 0.5 in.), the simulated rainfall is stopped and the screens are firmly tapped to dislodge the last droplets hanging to the underside of the screens. The trees are then removed for inoculation. A second lot of unsprayed trees can immediately be placed beneath the redistribution screens and the rainfall process repeated to determine what level of fungicidal residue remains for effective redistribution during the second rainfall.

Tree inoculation. Following 3 hr of drying, the trees are ready for inoculation with a suspension of the apple scab fungus of 70,000 conidia per milliliter. For the greatest inoculating uniformity, the trees are set on the floor in a row (or rows) with the pots built up (with wood blocks of varied thicknesses) so that the tops of all trees are at the same height as determined by a taut string line. By use of a paint sprayer at 1.75 kg/cm^2 (172 kPa; 25 psi) with the nozzle approximately 30 cm (12 in.) directly above the treetops, the inoculum is applied in repeated passes at walk speed. After about 12 or more passes, the beaded droplets on coated glass slides, positioned horizontally on a stand at treetop level, should be 2–3 mm in diameter. The test slides are placed on glass U-tubes on wet filter paper in petri dishes. Once the inoculum on the trees has dried, the trees are placed in the mist chamber at 20°C (68°F) together with the inoculum-bearing petri dishes. A determination on spore germination on glass slides is made in 24 hr. After an infection period of about 30 hr, the trees are returned to the greenhouse for normal maintenance and evaluation as described elsewhere (1–3).

LITERATURE CITED

1. Szkolnik, M. 1978. Techniques involved in greenhouse evaluation of deciduous tree fruit fungicides. Annu. Rev. Phytopathol. 16:103-129.
2. Szkolnik, M. 1978. Evaluation of physical modes of action of fungicides against the apple scab fungus. Proc. Apple Pear Scab Workshop, Kansas City, MO (1976); N.Y. Agric. Exp. Stn. Spec. Rep. 28, pp. 22-27.
3. Szkolnik, M. 1981. Physical modes of action of sterol-inhibiting fungicides against apple diseases. Plant Dis. 65:981-985.
4. Szkolnik, M. 1983. Unique vapor activity by CGA-64251 (Vangard) in the control of powdery mildews roomwide in the greenhouse. Plant Dis. 67:360-366.
5. Szkolnik, M. 1983. Redistribution of fungicide residues in the protection of unsprayed apple foliage against scab. Fungic. Nematic. Tests 38:157.

Testing Procedures to Determine the Volatile Activity of Fungicides for Control of Powdery Mildews in the Greenhouse

DUANE L. COYIER, U.S. Department of Agriculture, Agricultural Research Service, Corvallis, OR 97330

Control of powdery mildew on the multitude of crops grown in greenhouses can often be difficult. Because the environmental conditions are near optimum for growth and sporulation of powdery mildew fungi, frequent treatment is often needed for good control. This is both time-consuming and expensive because fungicides must be applied on 7- to 10-day schedules over an extended period.

Many fungicides that effectively control powdery mildews are volatile at elevated temperatures (1,6). Bergman (2) was apparently the first to recognize the volatility of fungicides when he sprinkled sulfur on wet greenhouse heating pipes to control grape powdery mildew. Coyier and Gallian (4) adapted this technique to modern fungicides and showed that rose powdery mildew could be efficiently controlled in the greenhouse with a minimum effort. The technology has since been applied to control powdery mildews of other commercial crops and to control powdery mildew selectively in research plots where it might interfere with the behavior of other disease interactions.

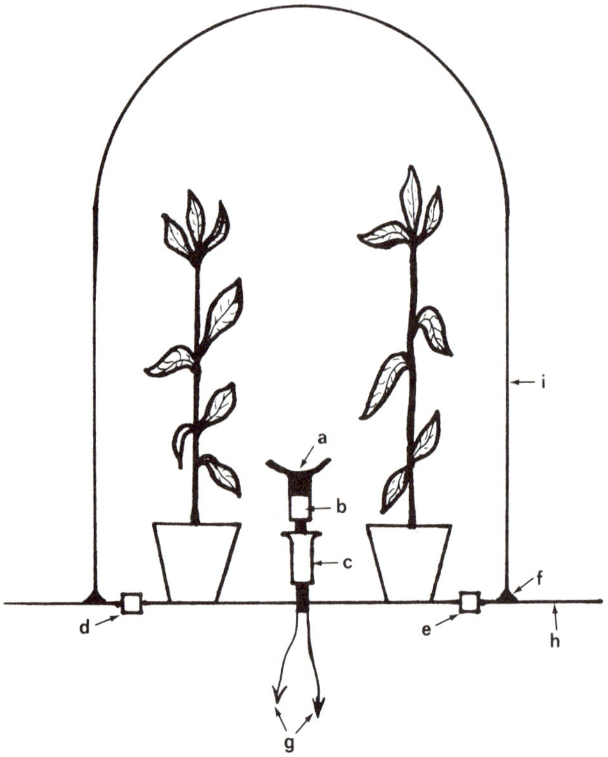

Fig. 1. Glass bell jar suitable for preliminary screening of fungicides for volatile activity against powdery mildews: a = fungicide receptacle, b = 15-W lamp, c = lamp base, d and e = air inlet and outlet for exhausting chamber after fumigation (stoppered during treatment period), f = sealant, g = leads to rheostat, h = stainless steel base, and i = glass bell jar.

MATERIALS AND METHODS

Laboratory screening. *Fumigation chambers.* Preliminary screening of volatile fungicides can be done in small, sealed chambers. The chamber should be of an adequate size to accommodate the plant material and large enough not to interfere seriously with normal respiration of the plant during fungicide exposure. A glass bell jar (Fig. 1) with a volume of 0.1 m^3 was suitable for a 6-hr treatment of four rose plants grown in 12.7-cm (5-in.) plastic pots and/or detached leaf material. The bell jar was placed on a stainless steel surface and sealed with stopcock grease. Any chamber that is of adequate volume, constructed of materials that are nonabsorbent, and easily cleaned between treatments would be suitable for small-scale fumigation tests. Small quantities of fungicide were heated inside the chamber by being dispensed onto a copper receptable fitted over a 15-W electric lamp (Fig. 2). The temperature of the receptacle

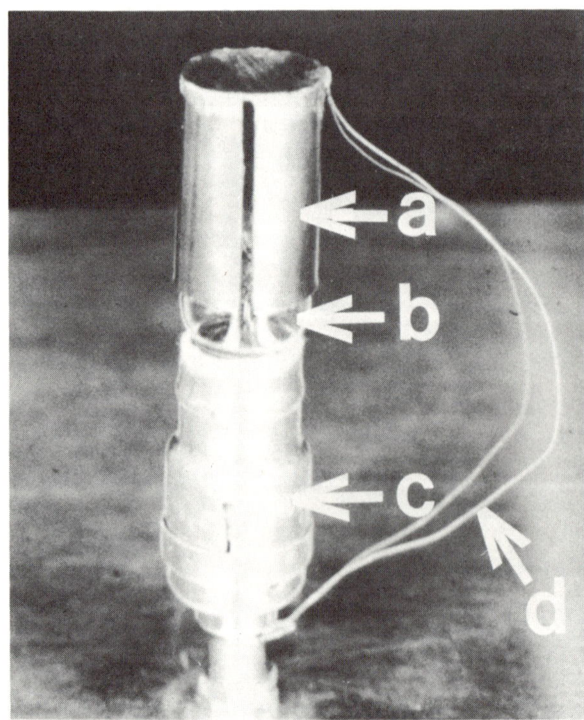

Fig. 2. Detail of fungicide heating unit: a = fungicide receptacle, b = 15-W lamp, c = lamp base, and d = thermocouple wire.

was adjusted by changing the voltage supplied to the lamp with a rheostat. Thermocouples were used to monitor the fungicide and ambient air temperature.

Many factors determine the fungicide dose to which the fungus is exposed during fumigation. The most critical of these is the temperature at which the fungicide is volatilized. Therefore, it is necessary to use a range of temperatures of at least 100, 150, and 200°C in the initial screening process. These temperatures can be further refined in later tests. Ambient air temperature should be maintained within the range of 16–24°C.

Duration of exposure also contributes to total dose. Gallian (5) showed that *Sphaerotheca pannosa* var. *rosae* colonies were killed when fumigated with nuarimol (10 mg a.i./m^3) for 6 hr, but hyphal growth resumed after a 4-hr exposure period. Because the lethal dose may vary among fungicides, a range of fumigation exposure times should be included in preliminary tests. Suitable intervals should include 4, 6, 8, and 10 hr for each fungicide rate tested. When the exposure period is concluded, the test chamber should be exhausted and flushed with clean air for 15–30 min before the plants are removed and returned to the greenhouse.

Although the activity of fungicides in the volatile state varies considerably, it has been determined that most are effective when 5–200 mg a.i./m^3 are heated to a suitable temperature. Therefore, I would recommend that dosages of 5, 10, 100, and 200 mg a.i./m^3 of test chamber airspace be used for preliminary experiments. Sufficient water should be added to wettable powder formulations to produce a thin paste. Liquid formulations can be added directly to the containers in which they will be heated.

Greenhouse fumigation. Any greenhouse that can be adequately sealed during the fumigation period is suitable for testing the volatile activity of fungicides for control of powdery mildews. Automatic ventilation and cooling systems should be inactivated during the test to assure that the fungicide concentration is maintained throughout the treatment period. The chemical should be applied during the nighttime hours when ambient air temperature can be maintained without ventilation and when workers are not in the area. The greenhouse should be entered first by someone equipped with an adequate respirator and proper clothing and vented 30–60 min before reentry by workers.

Although most fungicides disperse rapidly in the gaseous phase, the chemical should be dispensed in several containers in large greenhouses and the air circulated sufficiently to ensure that the fungus is exposed at full dosage for the allotted time. One fungicide heating unit per 1,000 m^3 of greenhouse airspace should provide adequate coverage.

Inoculation. When *S. pannosa* var. *rosae* is the test organism, the inoculum should be grown on roses in the greenhouse and the old conidia removed by directing an airstream over the spore-bearing colonies 24 hr before the conidia are to be used. Recently mature conidia are readily transferred by gently brushing the colonies with a camel's-hair brush over healthy host plants or detached leaf material. A settling tower or other protective device can be used to assure that air currents do not interfere with proper delivery of conidia to the host surface. The inoculated plant materials should be incubated at 21°C and 95–99% RH for 24 hr prior to fungicide exposure.

Data collection. *Detached leaf material.* Nondestructive examination of the effect of fumigation on the fungus can be achieved by use of a microscope equipped with vertical illumination or a high-quality dissecting microscope at 40–80× magnification. If suitable microscopic equipment is unavailable, the fungal hyphae and spores can be readily removed from the leaf surface using the cellophane-tape mounting technique and examined with a standard transmitted light microscope (3).

Observations should begin 24–48 hr after treatment and the results recorded at 24-hr intervals until the untreated controls develop secondary conidia. Because the volatile phase of many fungicides causes many powdery mildew fungi to develop swollen hyphal tips, the incidence of this phenomenon should be noted. In addition, the rate of hyphal growth and colony development following fumigation should also be observed by noting the average time necessary for developing secondary hyphal branching, conidiophore initiation, and the production of secondary conidia. Each treatment should replicated 10 times.

Intact plants. Uninfected rose plants can be inoculated, treated, and the data recorded as described previously for detached leaf material, or infected plants can be tested using multiple fumigation treatments. If the latter is preferred, 10 uniformly infected plants should be selected for each treatment and fumigated four times at weekly intervals. One week following the final fumigation, the plants should be evaluated for the incidence of powdery mildew on newly produced shoots. Either of two methods is acceptable for recording data. In the first, the plants can be rated subjectively on a 0–5 scale, where 0 = no colonies visible to the unaided eye; 1 = few scattered, small discrete colonies; 2 = larger but still discrete colonies; 3 = colonies merging to form larger mildew lesions; 4 = mildew covering half the total leaf surface; and 5 = most leaf and stem surfaces covered by the fungus.

A second and more objective method can be used for larger potted plants or roses grown in ground beds. Young stems that have developed during the treatment period and have four 5-leaflet leaves are selected at random from each treatment and the total number of powdery mildew colonies on the upper and lower surface of each leaf is counted. Twenty-five stems per treatment are adequate for the evaluation of the effect of volatile fungicides. Percentage of infected leaves is subjected to analysis of variance and Duncan's multiple range test.

RESULTS AND DISCUSSION

The procedures described in this chapter have been satisfactory for screening volatile fungicides for control of powdery mildew on roses. The test can also be readily adapted for use on other plants with only slight modifications to accommodate the conditions peculiar to each host.

Relatively large numbers of fungicides can be screened using the preliminary detached leaf methods, and those that show activity can be further tested on intact plants in the laboratory or on larger numbers of plants in research or commercial greenhouses. Phytotoxicity can be determined on numerous cultivars with a minimum number of plants.

Fungicides that were volatile above 105°C and controlled *S. pannosa* on roses include dinocap, fenapanil, fenarimol, nuarimol, oxythioquinox, piperalin, and triadimefon. However, dinocap fumigation caused rose leaves to develop yellow spots not unlike those that result from spray application of this chemical. Excessive fumigation with triadimefon caused phytotoxicity identical to that caused by high-dosage spray treatment. Although bupirimate was ineffective at 105°C, its efficacy was greatly enhanced at 185°C (4). Benomyl and triforine were ineffective against *S. pannosa* on roses when heated to 150°C for 6 hr.

Some of the fungicides that controlled powdery mildew on roses were also effective against powdery mildews that infect other greenhouse-grown crops. Powdery mildews of apple, azalea, begonia, peas, rhododendron, and wheat were controlled by fumigation with fenarimol or oxythioquinox. Volatilization of dinocap controlled powdery mildew on begonia but caused phytotoxicity.

Application of fungicides by volatilization offers several advantages over conventional spray treatment. They include ease of application, reduced exposure of applicator and greenhouse workers, lack of visible and dislodgeable residue on plant surfaces, and reduced total fungicide residue in and on host plants.

LITERATURE CITED

1. Bent, K. J. 1967. Vapour action of fungicides against powdery mildews. Ann. Appl. Biol. 60:251-263.
2. [Bergman, M.] 1852. Gard. Chron. 12:419.
3. Butler, E. E., and Mann, M. P. 1959. Use of cellophane tape for mounting and photographing phytopathogenic fungi. Phytopathology 49:231-232.
4. Coyier, D. L., and Gallian, J. J. 1982. Control of powdery mildew on greenhouse-grown roses by volatilization of fungicides. Plant Dis. 66:842-844.
5. Gallian, J. J. 1983. Effects of volatile fungicides on *Sphaerotheca pannosa* on rose. Ph.D. thesis, Oregon State University, Corvallis.
6. Hislop, E. C. 1967. Observations on the vapour phase activity of some foliage fungicides. Ann. Appl. Biol. 60:265-279.

Greenhouse Evaluation of Seed-Treatment Fungicides for Control of Sugar Beet Seedling Diseases

L. D. LEACH and J. D. MacDONALD, Department of Plant Pathology, University of California, Davis 95616

Damping-off of sugar beet (*Beta vulgaris* L.) seedlings is caused by several fungi. *Pythium ultimum* Trow, *P. aphanidermatum* (Edson) Fitzp., *Rhizoctonia solani* Kühn (teleomorph = *Thanatephorus cucumeris* (Frank) Donk) , and *Aphanomyces cochlioides* Drechs. are all soilborne pathogens commonly associated with seedling disease of sugar beets (1). In addition, *Phoma betae* (Oudem.) Frank (teleomorph *Pleospora bjoerlingii*) is a common seedborne pathogen of sugar beets that may be present on a large percentage of seeds produced in moist climates (2).

Although the ultimate test of seed-treatment fungicides is based on field performance, several factors often complicate evaluation of field results. Various fungi may be associated simultaneously with seedling diseases in the field, and fungicides frequently vary in their activity against different taxonomic groups of fungi. Moreover, excessive dosages of fungicides may have independent adverse effects on seedlings. Information relevant to specificity and phytotoxicity of seed-treatment fungicides can be obtained readily in the laboratory and greenhouse. This preliminary information is necessary for evaluation of fungicides in the field.

MATERIALS AND METHODS

Pathogen isolation and maintenance. Pathogenic isolates of *R. solani* and *Phoma betae* can readily be obtained from infected plant materials or established culture collections and kept in pure culture for experimental use. Once obtained, these fungi can be maintained for long periods in tubes of potato-dextrose agar with no loss of pathogenicity. Pure cultures of *Pythium* and *Aphanomyces*, on the other hand, are difficult to obtain without use of selective media. A modified PV medium (5) works well for culturing *Pythium* spp., whereas Pfender et al (6) recently described a medium for selective isolation of *Aphanomyces* spp. from plant tissues. Cultures of *Pythium* spp. can be maintained for long periods in sealed tubes of Difco cornmeal agar with no appreciable loss of pathogenicity or viability, but this is not the case with *Aphanomyces* spp., and it may be necessary to obtain fresh cultures periodically from the field.

Soil infestation. Because many seed-treatment fungicides are selective in their protection, testing each fungicide in soils infested with a single pathogen is necessary before testing in soils infested with combinations of pathogens or in field soils where more than one pathogen is likely to occur. This procedure allows identification of specificity and aids in selecting effective dosages.

Several methods can be used to infest soils with the various soilborne pathogens. One method that works well with *R. solani* is to grow the pathogen in petri plates of potato-dextrose agar and to mix the cultures in distilled water with a blender. The mixture can then be added to pasteurized soil at the rate of two culture plates per 10–12 kg of moist, friable soil for a heavy infestation. Infested soil should be incubated at ambient temperature for several days with periodic mixing to assure uniform distribution of the inoculum.

Cultures of *Pythium* and *Aphanomyces* spp. also can be added to soil directly. However, cultures of these coenocytic fungi should be diced with a knife rather than mixed in a blender. As an alternative, cultures of *Pythium* can be cultured on a V-8 juice/vermiculite medium (5) and incorporated into soil as washed, colonized vermiculite. Schneider (7) described a method for growing oospore inoculum of *A. cochlioides* that offers the advantage of quantifying and reproducing specific inoculum levels. Soils must be thoroughly mixed after infestation to assure uniform distribution of inoculum.

Application of seed-treatment fungicides. Fungicide treatments should be used at the manufacturer's recommended dosages. If the optimum dosage for use as a seed treatment is unknown, it may be necessary to evaluate several rates. Fungicides should be applied to commercially processed, graded seeds using cultivars adapted to the specific geographic area of interest.

Typically, candidate fungicides are applied to relatively small samples of seeds, and treatment of the seeds involves a correspondingly small amount of fungicide. Often, it is necessary to dilute the fungicides with an appropriate carrier to assure adequate distribution. Fungicides may be applied to seeds as dusts or spray suspensions in a revolving-drum seed treater or as a dust or slurry in a closed container. If applied to seed in liquid form at more than 4% of the seed weight, it may be necessary to dry the seed immediately after treatment.

Evaluation of phytotoxicity. Before evaluating fungicides for their effectiveness against various pathogens, they must first be evaluated for any independent effects on seedling germination or growth at the rates to be used. Treated seeds should be planted in flats of pasteurized soil and each treatment replicated four to six times in a randomized arrangement among flats. Within each row, seeds should be spaced approximately 1/cm and planted to a uniform depth of 2.5 cm. The soil should be kept moist and incubated at 15–20°C.

Emerging seedlings should be counted daily for any evidence of reduced or delayed emergence. Delayed emergence is measured by calculating the mean emergence period (MEP). The MEP for each replication is determined by multiplying the number of seedlings that emerge each day (N) by the corresponding number of days since planting (D) and then dividing the sum of these products by the total number of seedlings that emerge (T) [$MEP = \Sigma(ND)/T$].

In addition to changes in emergence characteristics, the seedlings should be observed for abnormal development. Seedlings should be harvested, washed, blotted dry, and fresh weights determined 20–24 days after planting. With this information, statistical comparisons of total seedling emergence, MEP, and fresh weight can be made (3). A significant delay or reduction in emergence, or reduction in seedling weight, is evidence of phytotoxicity. This information will identify dosages that can be safely used in tests for activity against specific pathogens.

Protection from soilborne pathogens. Seeds treated with candidate fungicides should be planted in infested soils and incubated as described above. In addition to seeds treated with candidate fungicides, nontreated seeds should be planted to provide a check of disease severity and seeds treated with commonly used fungicides should be included to provide a standard of comparison for new materials. For example, in soil infested with *Pythium* or *Aphanomyces* spp., seeds treated with metalaxyl at 0.31 g a.i./kg of seed should be included as an internal check; and in soil infested with *Rhizoctonia* spp., seeds treated with pentachloronitrobenzene at 0.94 g a.i./kg of seed should be included.

Daily counts of seedling emergence should be recorded, with dying seedlings noted, removed from the flats, and cultured in sterile water or on a suitable agar medium to confirm pathogen identity. At the end of the experiment, the records should include the emergence per 100 seed units, the percentage of seedlings damped-off, the surviving seedlings per 100 seed units, and the pathogens identified from infected seedlings.

Protection from seedborne *Phoma betae*. Evaluation of fungicides for activity against *Phoma betae* requires a seed lot that is heavily infested. Although this could be accomplished by inoculating seeds with pycnidiospores of *P. betae*, it is desirable to use a naturally infested seed lot. Seed samples can be obtained from seed production areas where *Phoma* spp. are a problem and assayed for *P. betae* using the water agar method (4). This is a relatively simple and effective assay that involves pouring a shallow layer of 1.2% water agar into 9-cm plastic petri dishes and placing 25–100 seed units (5 per dish) on the agar surface. The plates are then incubated at 20°C for 7 days, after which they are examined for "holdfast-like" structures by inverting the dishes and observing them at 50–100× at the interface between the agar and the petri dish bottom. Culturing seeds without surface disinfestation indicates the level of *P. betae* infestation in the seed lot, whereas the number of seeds with *P. betae* after disinfestation with NaOCl indicates the amount of seed with deep penetration. Comparison of these values provides information about the relative severity of *Phoma* infection (2).

After producing or selecting a seed lot with a high level of *Phoma*, the effectiveness of seed-treatment fungicides can be evaluated by treating samples of the seed and planting them in pasteurized soil in the greenhouse. Nontreated and treated seeds should be planted in flats using a randomized design as described above and incubated at 12–15°C to provide optimum conditions for disease. As infected seedlings are observed, they should be counted and removed at intervals for culturing on water agar to verify that they were killed by *P. betae*. About 14 days after emergence, all seedlings should be harvested, washed free of soil, and examined for evidence of belowground infection by *P. betae*.

Protection against multiple pathogens. Under field conditions, it is common for several pathogens to be simultaneously involved in seedling diseases. Many fungicides are selective in their activity, and it is necessary to determine which combinations of treatment are most effective in soils infested with more than one pathogen. Additional information can be obtained by collecting soils from fields with a history of seedling disease problems and testing the most promising fungicide combinations.

DATA ANALYSIS AND REPORTING

Data should be analyzed statistically by procedures suitable to the experimental design. Methods for analysis of variance and separation of treatment means are described elsewhere (3). The relative efficacy of candidate fungicides is measured as the improvement in emergence, or the relative freedom of seedlings from infection, in comparison with untreated controls.

A number of factors can influence the results of seed-treatment tests. These factors must be described so that other workers can evaluate test results accurately. First among these is the need to identify the cultivar used. Sugar beet cultivars have been developed specifically for different geographic regions according to endemic problems. In addition, the processing grade of the seed should be indicated. The use of processed seed (seed that has been milled to remove superficial tissues) or nonprocessed seed can influence the efficacy of fungicides, as has been demonstrated with *P. betae* (2). Additionally, the planting date and mean temperature during seedling emergence should be reported because temperature can influence the activity of seedling pathogens. Failure to recognize this possibility could lead to erroneous conclusions about fungicide efficacy. A description of the experimental design as well as the identification of statistical tests used in mean separation are also necessary.

Knowledge of the effectiveness of specific fungicides in controlling specific pathogens will aid in selecting the particular seed-treatment fungicides, or combinations thereof, that have the best chance of controlling the seedling pathogens in the field. With this information, field trials can be conducted to test the fungicides under natural conditions, and evaluation of the results should be simplified.

LITERATURE CITED

1. Bennett, C. W., and Leach, L. D. 1971. Diseases and their control. Pages 223-285 in: Advances in Sugar Beet Production: Principles and Practices. R. T. Johnson et al., eds. Iowa State University Press, Ames. 470 pp.
2. Leach, L. D., and MacDonald, J. D. 1976. Seed-borne *Phoma betae* as influenced by area of sugar beet production, seed processing and fungicidal seed treatments. J. Am. Soc. Sugar Beet Technol. 19:4-15.
3. Little, T. M., and Hills, F. J. 1978. Agricultural Experimentation: Design and Analysis. John Wiley, New York.
4. Mangan, A. 1971. A new method for the detection of *Pleospora bjoerlingii* infection of sugar beet seed. Trans. Br. Mycol. Soc. 57:169-172.
5. Mircetich, S. M., and Matheron, M. E. 1976. Phytophthora root and crown rot of cherry trees. Phytopathology 66:549-558.
6. Pfender, W. F., Delwiche, P. A., Grau, C. R., and Hagedorn, D. J. 1984. A medium to enhance recovery of *Aphanomyces* from infected plant tissue. Plant Dis. 68:845-847.
7. Schneider, C. L. 1977. Use of oospore inoculum of *Aphanomyces cochlioides* to initiate blackroot disease in sugarbeet seedlings. J. Am. Soc. Sugar Beet Technol. 20:55-62.

Evaluation of Soil-Applied Fungicides for Control of Fusarium Wilt of Chrysanthemum

ARTHUR W. ENGELHARD and S. S. WOLTZ, IFAS, University of Florida, Gulf Coast Research and Education Center, Bradenton 34203

This method is intended for use in the evaluation of fungicides applied to the soil of potted plants for the control of Fusarium wilt of chrysanthemum. The level of Fusarium wilt symptoms on chrysanthemum is regulated independently by each of three factors: soil pH, form of nitrogen, and chemotherapeutant. The level of disease is increased in the pH range of 5 to 6 and decreased in the range of 6.5 to 7.0 (chrysanthemums do not grow well above pH 7.0). The level of disease is increased when nitrogen is applied in the form of NH_4 [NH_4NO_3, $(NH_4)_2SO_4$] and decreased by the form NO_3 [KNO_3, $NaNO_3$, $Ca(NO_3)_2$]. Similarly, disease is decreased when an effective chemotherapeutant (benomyl 50W) is added. Each of these three factors provides partial control of disease, and the factors are additive, resulting in complete control of symptoms when used together.

TEST PLANTS

Rooted chrysanthemum (*Chrysanthemum morifolium* Ramat.) plants should be set one per 10-cm pot in a well-drained, sterilized soil mix, such as one composed of one-third peat and two-thirds sandy soil. The soil mix may be sterilized by steaming for 30 min at 82° C (180° F) or with methyl bromide at 2.37 kg/m^3 (4 lb/yd^3). The pH of the soil mix should be adjusted to 5.5–6.0 (see below for details). The plants should have a balanced nutritional program that supplies the major and minor elements needed for vigorous plant growth.

PATHOGEN AND HOST

The pathogen suggested for use (because of its common occurrence) is *Fusarium oxysporum* f. sp. *chrysanthemi* Litt., Armst. and Armst. (2). Another fungus, *F. oxysporum* f. sp. *tracheiphilum* race 1 of cowpea (1) is also a known wilt pathogen of chrysanthemum, but it is rarely found. *F. oxysporum* f. sp. *chrysanthemi* sporulates readily when grown on potato-dextrose agar in petri plates under continuous cool-white fluorescent lights at 27° C (80° F). Mycelial growth covers the plates completely within 10 days. Sufficient numbers of spores are produced on 10 plates to infest at least 60 pots. Conidia are harvested by flooding the petri plates with deionized or distilled water, rubbing the agar surface gently with a rubber policeman, and filtering the resulting suspension through cheesecloth. The culture should be derived from a single-spore isolate to increase genetic uniformity and to ensure a pure culture. The spore concentration can be determined using a Levy Corpuscle Counting Chamber (hemacytometer).

Susceptible cultivars that may be used are Delaware, Yellow Delaware, Nob Hill, Southern Comfort, and Yellow Bonnie Jean. Symptomatology and disease reactions were described for 14 chrysanthemum cultivars inoculated with the two recognized formae speciales by Engelhard and Woltz (4).

NUTRITION, pH, AND DISEASE MANAGEMENT

Soil pH and nitrogen source and level significantly affect the level of Fusarium wilt development (5,6,8,9) in chrysanthemum. It is therefore necessary to control them to obtain representative disease development. Nitrate-N reduces disease, whereas ammonium-N increases it. An increase in soil pH obtained by adding liming materials [$Ca(OH)_2$, $CaCO_3$] reduces disease. Applications of zinc and certain combinations of iron, manganese, and zinc increase severity of Fusarium wilt in tomato (7,10) and chrysanthemum (*unpublished data*). Therefore, lime amendments, nitrogen sources, and micronutrient additives should be manipulated in such a way as to ensure disease development in the control and fungicide-treated experimental units. The effect of adding various liming materials and nitrogen sources on the final soil pH and on disease development is discussed in detail by Woltz and Engelhard (9). The interactions of nitrogen source and chemotherapeutants and of soil pH, nitrogen source, and effect of systemic fungicides on Fusarium wilt control are discussed by Engelhard and Woltz (5,6) and Engelhard (3) respectively.

Nutrient management of test plants. The test plants, growing in soil with a pH in the range of 5.5 to 6.0, should be fertilized immediately after potting to encourage vigorous growth. Nitrogen should be applied as ammonium nitrate or ammonium sulfate since disease is encouraged by nitrogen in the form NH_4 compared with NO_3. A suggested nutrient solution is 1.5 g of NH_4NO_3, 1.0 g of $NaH_2PO_4H_2O$, 0.85 g of KCl, and 0.2 g of $MgSO_4 \cdot 7H_2O$ per liter. The solution should be applied at 50 ml/10-cm pot twice weekly. Minor elements should be added to the soil mix or applied in a soil drench, or they should be sprayed on the foliage as needed. An all-NO_3 nitrogen source [3.2 g of $NaNO_3$, 3.8 g of KNO_3, or 4.4 g of $Ca(NO_3)_2$ per liter] should be used as a supplemental treatment to lessen disease severity and to permit the detection of marginally efficacious fungicides.

INOCULATION AND INCUBATION

Plants should be inoculated 11 days after potting (1 day after the second fungicide drench). They should be well established with vigorously growing roots. The roots of a plant are injured by making a vertical cut with a knife into the soil 2 cm from the stem and 2 cm deep on opposite sides of the plant. *F. oxysporum* f. sp. *chrysanthemi* is a wound pathogen, and this procedure facilitates the infection process. The spore suspension is poured into the cuts. Good infection is obtained when at least 10 million spores are applied in a 50-ml suspension to each 10-m pot (200,000 spores per milliliter). Soil temperature is extremely important in the development of Fusarium wilt. It should not be allowed to drop below 23°C (73°F) because pathogenesis is greatly inhibited at lower temperatures. Daytime temperatures should be maintained within the 26–30°C (79–86°F) range. No or few symptoms will develop at lower temperatures.

SYMPTOM DEVELOPMENT

The initial symptoms of Fusarium wilt should appear on control plants within 8 days after inoculation if the proper temperature is maintained. They include a unilateral chlorosis of one or more leaves at or near the stem apex and slight to pronounced curvature of the chlorotic leaves and the stem toward the affected side of the plant. As disease progresses, chlorosis of the affected leaves becomes more general and severely affected leaves wilt. Wilted leaves occur initially on the most severely affected side of the plant, but as disease progresses the entire plant wilts and dies. In the cultivars Delaware and Yellow Delaware, the main vein in affected leaves is visibly darkened from an external viewpoint before any other symptoms appear.

Rating system. Fusarium wilt can be rated on a 0–5 basis as follows: 0 = no visible leaf symptoms; 1 = leaf chlorosis or vascular discoloration or both in one or more leaves; 2 = chlorosis or vascular discoloration or both, plus epinasty of leaf (leaves) or stem or both; 3 = leaf symptoms plus wilting; 4 = leaf symptoms, wilting, and stunted growth of plant; and 5 = dead plant. Percentage of disease control (PDC) is based on the following formula:

$$\text{PDC} = \frac{\text{Mean disease rating in water control} - \text{Mean disease rating in treatment}}{\text{Mean disease rating in water control}} \times 100$$

Additional rating indexes that may be used are plant height, plant weight, and ratings of flowering plants for commercial desirability either as cut flowers or as potted plants.

REPLICATION AND FUNGICIDE APPLICATION

All treatments, including a known standard fungicide and a water control, should be replicated a minimum of four times. Each replicate may consist of a single plant. Fungicide drenches should be applied 3 and 10 days after the plants are potted. The liquid should wet the soil mass. Usually about 50 ml/10-cm pot is required.

Standard fungicide. Benomyl 50W at 0.3 and 1.2 g/L (0.25 and 1.0 lb/100 gal) should be used as a standard treatment. When ammonium nitrate or ammonium sulfate is used as the nitrogen source in the nutritional program, disease should be severe at the rate of 0.3 g/L, whereas control should be at a high level but not complete at the rate of 1.2 g. The possibility of pathogen tolerance should be investigated if the rate of 1.2 g/L fails to give a high level of control. Tolerance is overcome by using a nontolerant strain and maintaining it in sand culture in the absence of benomyl. A possible substitute for benomyl would be propiconazole (CGA 64250, Banner, Tilt).

In research or more advanced testing, the relative effectiveness of a compound in controlling wilt can be evaluated using the integrated fungicide/lime/nitrate regime (5,6) to take advantage of the strong additive effects of the combination. If the soil mix is limed to 6.5–7.0 and an all nitrate-N source (calcium, potassium, or sodium nitrate) is used, complete control of symptoms is obtained with benomyl 50W at 0.3 g/L. This system could be used to evaluate higher rates of less efficacious materials or lower rates of more effective fungicides.

Phytotoxicity. The crop should be carefully inspected at regular intervals for any indication of phytotoxicity. Phytotoxicity symptoms range in severity from death of the plants to marginal chlorosis or necrosis of the leaves, or to plants that have no visible symptoms other than retarded growth. The kind and amount of phytotoxicity should be described and recorded.

Statistical analysis. The data should be analyzed by an appropriate method that may include a least significant difference or Duncan's multiple range rating system.

LITERATURE CITED

1. Armstrong, G. M, and Armstrong, J. K. 1965. Wilt of chrysanthemum caused by race 1 of the cowpea Fusarium. Plant Dis. Rep. 49:673-676.
2. Armstrong, G. M., Armstrong, J. K., and Littrell, R. H. 1970. Wilt of chrysanthemum caused by *Fusarium oxysporum* f. sp. *chrysanthemi*, forma specialis nov. Phytopathology 60:496-498.
3. Engelhard, A. W. 1974. Effect of pH and nitrogen source on the effectiveness of systemic fungicides in controlling Fusarium wilt of chrysanthemums. Proc. Am. Phytopathol. Soc. 1:121.
4. Engelhard, A. W., and Woltz, S. S. 1971. Fusarium wilt of chrysanthemum: Symptomatology and cultivar reactions. Proc. Fla. State Hortic. Soc. 84:351-354.
5. Engelhard, A. W., and Woltz, S. S. 1973. Fusarium wilt of chrysanthemum: Complete control of symptoms with an integrated fungicide-lime-nitrate regime. Phytopathology 63:1256-1259.
6. Engelhard, A. W., and Woltz, S. S. 1973. Fusarium wilt of chrysanthemum: A new cultural-chemical control method. Fla. Flower Grower 10:1-3.
7. Jones, J. P., and Woltz, S. S. 1970. Fusarium wilt of tomato: Interaction of soil liming and micronutrient amendments on disease development. Phytopathology 60:812-813.
8. Littrell, R. H. 1966. Effects of nitrogen nutrition on susceptibility of chrysanthemum to an apparently new biotype of *Fusarium oxysporum*. Plant Dis. Rep. 50.882-884.
9. Woltz, S. S., and Engelhard, A. W. 1973. Fusarium wilt of chrysanthemum: Effect of nitrogen source and lime on disease development. Phytopathology 63:155-157.
10. Woltz, S. S., and Jones, J. P. 1973. Interactions in source of nitrogen fertilizer and liming procedure in the control of Fusarium wilt of tomato. HortScience 8:137-138.

In Vivo Fungicide Screening on Greenhouse-Grown Grasses

P. SANDERS and H. COLE, JR., Department of Plant Pathology, Pennsylvania State University, University Park 16802

Our laboratory has used the greenhouse procedures described herein to evaluate protectant and systemic fungicides for phytotoxicity, control, and duration of efficacy (residual activity) against *Sclerotinia homoeocarpa* F. T. Bennett, *Rhizoctonia solani* Kühn, and *Pythium aphanidermatum* (Edson) Fitz. on Penncross creeping bentgrass (*Agrostis palustris* Huds.) and Pennfine perennial ryegrass (*Lolium perenne* L.). With suitable modification, these procedures may be appropriate for preliminary testing of fungicides for control of other plant pathogens that cause disease on grasses under greenhouse conditions. Testing can be done at any time of the year if a temperature-controlled greenhouse or growth chambers are available. The data obtained are valuable in determining dosages and application methods that may provide maximum disease suppression, longest residual efficacy, and minimum phytotoxicity under field conditions. Such information provides a basis for more efficient field testing of fungicides.

TEST PLANTS

Pennfine perennial ryegrass and Penncross creeping bentgrass are the hosts that are usually used in our greenhouse tests. These species germinate rapidly and are easily managed under greenhouse conditions. As the need arises, other species and cultivars may be used. With ryegrass, fescues, or bluegrasses, the seeding rate and emergence time may vary. Seeding rates are converted from commercial rates (pounds per thousand square feet) to grams per 78 cm^2 (the area of a 10-cm pot). In general, the seeding rate should be two to four times the equivalent commercial rate to allow for rapid establishment of a uniform, dense stand. The time between seeding and availability of grass for use depends on cultivar and species. Penncross bentgrass, ryegrasses, and fescues may be used 3–4 weeks after seeding. Certain bluegrass cultivars, however, may require 5 or more weeks from seeding to use.

Grass is seeded in 10-cm-diameter plastic pots of steam-sterilized greenhouse mix (two parts soil, one part sand, one part peat). After seeding, pots are covered with brown paper to retain moisture and to speed germination. Throughout the experiment, grass is regularly watered and maintained at a cutting height of about 2.5 cm above soil surface. Chemical treatments may be applied 3–4 weeks after seeding.

FUNGICIDE APPLICATION

Fungicides may be tested as foliar sprays, soil drenches, or granular formulations. To obtain a wide spectrum of control effectiveness and phytotoxic manifestation, the test fungicides are applied at a range of dosages both above and below the recommended rate. Fungicide dosages and water diluents are calculated and applied to the pots of grass on an area basis at rates equivalent to the standard dosages (g/93 m^2 or oz/1,000 ft^2) commonly used in turfgrass research and management. An active ingredient or product basis may be employed as appropriate.

The following are some sample calculations:

i. Area of a 10-cm pot:

$$\pi r^2 = 3.14 \times 25 \text{ cm}^2 = 78.5 \text{ cm}^2.$$

ii. Dosage per 10-cm pot (78.5 cm^2):

$$2 \text{ oz}/1{,}000 \text{ ft}^2 = 56.7 \text{ g}/93 \text{ m}^2 = 56{,}700 \text{ mg}/930{,}000 \text{ cm}^2$$

$$\frac{56{,}700 \text{ mg}}{930{,}000 \text{ cm}^2} = \frac{X}{78.5 \text{ cm}^2}$$

$$X = 4.8 \text{ mg}/78.5 \text{ cm}^2.$$

iii. Water dilution rate per 10-cm pot (78.5 cm^2):

$$3 \text{ gal}/1{,}000 \text{ ft}^2 = 11.4 \text{ L}/93 \text{ m}^2 = 11{,}400 \text{ ml}/930{,}000 \text{ cm}^2$$

$$\frac{11{,}400 \text{ ml}}{930{,}000 \text{ cm}^2} = \frac{X}{78.5 \text{ cm}^2}$$

$$X = 1 \text{ ml}/78.5 \text{ cm}^2.$$

Since the method of application can greatly affect both the residual activity of a fungicide and the amount of phytotoxicity produced, any or all of the following methods of application may be used: (i) foliar spray at various water dilution rates, (ii) spray with or without previous wetting of the foliage, (iii) spray followed by irrigation of the foliage, (iv) soil drench, and (v) granular application. Fungicide sprays are applied with an air-pressure, venturi-pickup atomizer sprayer powered by an electric pressure/vacuum pump.

All pots required for a given treatment block are sprayed simultaneously, but care must be taken so that each pot within the treatment block receives equivalent amounts of fungicide. Although almost any pattern of spray that provides uniform coverage may be used, experience has shown that a circular pattern of spray on individual pots results in spray overlap in the center of the pot area. Therefore, a back-and-forth sweeping motion should be used instead.

Drenches are used to determine root uptake and acropetal translocation, and they are applied to pots on an individual basis at the appropriate water dilution rate. Care is taken not to wet the grass foliage during the drenching procedure. Individual pots of grass are drenched by retracting the foliage at the side of the pot adjacent to the rim and pouring the fungicide solution over the soil surface.

Granular fungicides may be applied on an individual pot or treatment block basis. Granulars are "diluted" with sterile sand to assure uniform coverage and applied with a shaker bottle or can.

All pots of grass are treated with the test fungicides simultaneously. A standard fungicide of known effectiveness against the test pathogen may be included for comparison. An untreated check is included in the test to verify the virulence of the test pathogen under the conditions of the experiment. All treatments are replicated at least three times. Factorial experimental designs with arrangement of the replications in a randomized block plan have proved most satisfactory for us.

INOCULUM PREPARATION, INOCULATION, AND INCUBATION

Virulent isolates of the aforementioned fungal pathogens are grown for approximately 1 week on autoclaved rye grain inoculum carrier. The carrier consists of 50 g of rye grain, 1 g of $CaCO_3$, and 75 ml of water placed in 250-ml Erlenmeyer flasks. The flasks are stopped with cotton plugs and autoclaved for 45 min at 15 lb of pressure. Ten to 15 kernels of rye inoculum are placed in the center of the grass area in each pot. After inoculation, each pot is covered with a transparent polyethylene bag to maintain high humidity and placed either on a shaded greenhouse bench or in a temperature-controlled growth chamber. Temperature during incubation is maintained at the optimum for disease development and growth of the test pathogen (21–27°C for *S. homoeocarpa*, 27–30°C for *R. solani*, and 30–36°C for *P. aphanidermatum*). These environmental conditions and the relatively large quantities of virulent inoculum provide a rigorous challenge to the fungicide treatments.

If the purpose is to test immediate suppression of symptoms by the test fungicide, inoculation is usually carried out within 24 hr after application. If, however, information is required on the duration of control provided by the fungicide, all pots of grass are treated simultaneously, and successive weekly inoculations are made. Initial inoculation of some of the treated plants is made within 24 hr after application of the fungicide. Inoculum is applied to a set of previously treated, noninoculated pots of grass at weekly intervals. Sequential weekly inoculations continue until the residual effect of the fungicide is no longer apparent. Testing protectant fungicides for duration of control usually requires two to three weekly inoculations after chemical treatment, but systemic fungicides may require 6–8 weeks of testing after treatment before residual activity completely dissipates.

EVALUATION OF RESULTS

The plastic covers are removed from inoculated pots and fungicide efficacy is evaluated about 7 days after inoculation. This interval is determined by the rate of spread of the pathogen in the untreated pots. When placed in the center of a pot of seedling grass, *S. homoeocarpa, P. aphanidermatum*, and *R. solani* produce a foliar blight that advances radially toward the pot margins. Our procedure is to remove the plastic cover and terminate the test when the grass in the untreated pots is completely blighted. At this time, all inoculated pots are evaluated individually for disease severity. A 0–10 visual rating scale is used that corresponds to diameter (cm) of pot area blighted: 0 = no disease, 1 = 1 cm of blight, 2 = 2 cm of blight, 7 = 7 cm of blight, and 10 = blighting of all foliage in the 10-cm pot.

Data obtained from disease severity evaluations are subjected to analysis of variance and Waller-Duncan *K*-ratio *t*-test.

About 1 week after fungicide application, all treated, noninoculated grass is evaluated for phytotoxicity. Phytotoxic effects, which vary with the plant and the fungicide, may be exhibited as stunting, color changes, thinning, or chemical burn. Phytotoxicity evaluation is descriptive, based on severity of symptoms and type and location of injury. An appropriate visual rating scale is used.

CORRELATION WITH FIELD TEST RESULTS

The procedures described above generally produce reliable, uniform data that are valuable for determining effective dosages and application methods for field use. The variable environmental conditions and differing physiological states of plants encountered in field test situations, however, can influence the performance of fungicides. Greenhouse-grown seedling turf is usually more sensitive to phytotoxic substances than are mature field stands. One exception is clinical injury associated with senescent plants. Generally, however, if injury appears in greenhouse tests, one must be alert for similar effects under field conditions.

Dosages that are effective for disease control in greenhouse pot tests are usually effective in field usage. A possible exception concerns systemic fungicides when thatch and organic matter interfere with movement in and absorption from the soil. In this instance, greenhouse results are normally more optimistic than those obtained under field conditions.

Chemicals that act either totally or primarily to elicit antifungal host responses may not be adequately screened using greenhouse procedures, except for purposes of determining activity in greenhouse production situations. If a compound interacts primarily with the host resistance mechanism, it is advisable to test in particular use settings.

Despite the foregoing limitations, preliminary greenhouse testing of fungicides can be invaluable in selecting dosages and application techniques for future field testing, assuring more efficient use of time and space in fungicide field evaluation.

Testing Chemical Sprays on Blossoms of Deciduous Fruit Trees

MICHAEL SZKOLNIK, Department of Plant Pathology, New York State Agricultural Experiment Station, Geneva 14456, and K. D. HICKEY, Department of Plant Pathology, Pennsylvania State University, Fruit Research Laboratory, Biglerville 17307

Blossoms of deciduous fruit trees can be used to considerable advantage in the evaluation of formulated chemical sprays for control of plant diseases, particularly under controlled conditions in the greenhouse. Procedures covered here are for two blossom blight diseases of commercial importance: brown rot blossom blight, a fungal disease of stone fruits (*Prunus*) caused by *Monilinia fructicola* (Wint.) Honey, and fire blight, a bacterial disease of pome fruits caused by *Erwinia amylovora* (Burrill) Winslow et al.

BROWN ROT BLOSSOM BLIGHT ON *PRUNUS*

Plant Material

The procedure for evaluation of fungicides against brown rot blossom blight has, with revisions, been used successfully in the greenhouse for several decades (3). An important prerequisite in the use of this procedure is long-term advance planning in the development of a reservoir of suitable blossoming trees. As with any evaluative study, the plant material should be very suitable from the standpoint of uniformity of good vigor and blossoming. Specifically, newly acquired nursery stock of peach, cherry, or prune will not produce suitable numbers of blossoms until the second or third year from planting. Most commercially available sweet and sour cherry, peach, and prune cultivars are adaptable to greenhouse culture and blossom blight studies.

Establishment of a good blossom research program should begin with potting of nursery-stock stone fruit trees of good vigor and uniformity, having a caliper of at least 1.6 cm, planted in 20-cm clay pots in a mixture of 1:1:2 parts of good field soil, clean sand, and peat moss, respectively. These can be maintained in the greenhouse or outdoors during the growing season with appropriate watering, weekly application of complete analyses (including minor elements) soluble fertilizer, and timely spraying for diseases, insects, and mites. Following fall leaf drop, the trees should be stored in a cold indoor environment with temperatures high enough to prevent cold damage to sensitive cultivars.

About 6–8 weeks before normal orchard bud swell, the trees should be stored in a refrigerated room at 1 or 2°C to favor longer-lasting dormancy. The overall usability of trees for greenhouse blossom blight tests is some 4 or 5 months, approximately February through June in the northern hemisphere. At 2°C, the fruit buds may develop in storage to such an extent that by late June or early July further storage would result in abnormal blossom development unsuitable for fungicidal evaluation in the greenhouse.

As tree groups are removed from storage to the greenhouse for research, they should first be repotted by removing each tree from the pot, scraping off about 2.5 cm of the outer soil and roots, then replacing the tree in the same pot with a new supply of a mixture of soil, sand, and peat moss added and tamped in. At this time, some pruning may be in order to remove dead, abnormal, and long rangy branches. The trees should be kept in a warm (22°C) greenhouse to force blossoming. Trees may bloom quite uniformly or may have to be programmed to a fairly uniform bloom by spray time. Advanced trees should be held back as they reach about 60–75% bloom in a cooler greenhouse or in a holding room (with lights) at about 10°C; slower blooming trees can be advanced in warmer greenhouses. When all the trees reach about 75–95% bloom, they should be brought together for exposure in a warm greenhouse for at least 1 day preceding sprays. The best time to use the trees for fungicide evaluation is at early full-bloom stage when the blossoms are most susceptible to infection. If this is delayed to late bloom, the researcher may find variability in infection and confusion in differentiating between actual brown rot blossom blight and chemical or other damage as blossoms (especially ones not pollinated) wither and drop.

Procedures

The procedures given here are for (i) protective and (ii) eradicative (after-infection or curative) modes of action (6,7) in the control of brown rot blossom blight on potted trees; and (iii) for the residual protective action of fungicides tested on detached blossoms in plastic containers or excised branches with blossoms in a mist chamber (1).

Protective action. *Spray application.* On the day tests are initiated on blossoming trees, all unopened floral units should be removed first because they usually do not become infected and may affect the actual control data later. The trees should be randomized into treatment groups of similar blossom count, preferably more than 100 blossoms per treatment. Dilute fungicidal sprays are applied to these groups by paint sprayer at 172 kPa (75 kg/cm^2). The sprayer should be shaken often to minimize settling of fungicides, and the sprays should be applied from all sides and the top of each tree. The trees, including a water-sprayed treatment, should be allowed to dry about 3 hr before inoculation or exposure of trees to artificial rainfall. Rainfall exposure is desirable for fungicide redistribution and a test on fungicide residue erosion that occurs with natural rains in the orchard. Artificial rainfall is achieved either by placement of trees on a round turntable directly beneath one overhead, solid-cone nozzle (e.g., Spraying Systems TeeJet 8004) or by positioning the trees on the floor in a circle directly beneath the path of an overhead nozzle on an automatically revolving boom. A suitable rainfall would be one of 1.3 cm measured 100 cm below the height-adjustable nozzle at about mid-tree level.

Blossom inoculation. About 3 hr after the fungicidal spray or shortly after completion of the rainfall, the trees are inoculated. The inoculum source should be prepared at least 3 days before the spray test by inoculation of canned peach or apricot halves or slices with *M. fructicola* (see Szkolnik, "Preparation of inoculum for brown rot of stone fruit and apple scab," *this volume*) (4). Conidia are atomized from the fruit with water at 172 kPa into a beaker and then poured into a blender, where they are agitated at high speed for about 1 min to break up spore chains that would otherwise interfere with accurate standardization of suspension. At times, a drop of Triton X-100 or Tween 20 surfactant added to the agitated spores facilitates chain breakup. The spores are then diluted in distilled water (without additives) and standardized by hemacytometer to a count of 60,000/ml.

The inoculum is applied by paint sprayer at 172 kPa with periodic shaking of the sprayer to maintain uniformity. As with spraying, the trees should be inoculated from all sides and top to ensure wetting of all blossoms. At this time the inoculum spray is misted onto coated glass slides into droplets about 2–3 mm in diameter. The slides are then placed on glass rods on wet filter paper in a petri dish for a microscopic examination in about 24 hr on adequacy of spore germination. The trees and covered dish with slides are then kept in a mist chamber at 23°C to maintain wetness of all blossoms and favorable infection conditions for 30 hr, after which they are removed for normal maintenance in the greenhouse.

Recording and analyzing data. Blight infections appear as tan to brown, slightly sunken areas on the shuck. Two or three daily counts may be necessary, starting with the appearance of observable symptoms as early as the third day following inoculation. The first count should include blossoms with blight symptoms and those without; this provides a total blossom count. Infected blossoms are removed. Subsequent counts and removal of infected blossoms eventually provide a cumulative count of infected blossoms per treatment that, with a known total blossom count, will provide the means for determination of fungicidal efficacy. To improve accuracy, the trees may be returned to the mist chamber at 23°C for about 16 hr (overnight) about 5 days after inoculation. Counts made right after this exposure are simplified by the aerial sporulating signs of the pathogen.

At times, confusion arises as to whether certain blossoms are actually infected by *M. fructicola* or are objects of fungicide phytotoxicity or blossom senility, particularly unfertilized blossoms following petal fall. Verification of brown rot can be made by placement of blossoms in question on a flat aluminum wire screen in a covered plastic container into which is added first a volume of 10% sucrose solution or of a 1:3 dilution of a canned peach or apricot syrup with just enough water to wet the wire screen without submerging it. Typical *M. fructicola* mycelium and sporulation within 24 hr will verify blossom blight infection.

Relative fungicidal efficacy for protection against brown rot blossom blight is based on percentage of blossoms infected in comparison with other treatments and water-sprayed checks. It would be desirable to run three experiments with the same treatments. Statistical analyses of the data in any one experiment may require handling of each tree of the treatment as a replicate; in multiple repeat experiments, each test may be handled as a treatment replicate.

Each time blight data are taken it is advisable to record other observations, such as brown rot spotting on leaves and phytotoxicity to petals, shuck, and leaves.

Postexperiment tree handling. Following completion of each experiment, the trees should be well maintained either in the greenhouse and later outdoors (when there is no further danger of tree freezing) through appropriate watering, fertilizing, and pest control favorable for good fruit bud initiation. When suitably handled, the potted stone fruit trees should be very satisfactory for brown rot blossom blight research (into both biology and control measures) for many years.

After-infection activity. *Inoculation.* Evaluation of fungicides on stone fruit blossoms for fungicidal after-infection (eradicative, curative) mode of action should be conducted at the early full-bloom stage after first removing unopen floral units. The trees should be inoculated and kept in the mist chamber for infection according to the procedure described above.

Spray application. At a desired after-infection interval, usually 18, 24, or 30 hr, the required number of trees is randomly removed from the mist chamber and the fungicidal treatments applied within a minimal time lapse. Spraying by paint sprayer is the same as described above. For each after-infection time period in a given experiment, the treatment group should include its own water-sprayed check treatment and inoculum-atomized glass slides for accurate information on the level of spore germination and amount of blossom infection in the mist chamber for that time period.

Recording and analyzing data. Taking and handling of blight data are the same as discussed previously in the section on protective action.

Data from detached blossoms or excised branches. *Detached blossoms.* The use of detached blossoms in securing blight control data from bloom sprays in the orchard or the greenhouse is an alternative to intact blossoms on the trees (1). It may prove useful in orchard spray research when cold or dry conditions prevent suitable blossom blight infections and negate chances of recording data in the field.

Dry blossoms from sprayed orchard or greenhouse trees are collected well before the start of petal fall, placed in plastic boxes with minimal handling or rubbing, and stored at 4°C pending use. Infection chambers are prepared by placement of about 1.0 cm of sterilized sand in the bottom of a transparent plastic box, measuring about 25 × 30 × 13 cm, with a lid. The sand is thoroughly moistened but not wet. Ten blossoms of each spray treatment are embedded stem down in a row in the sand. This is repeated for other treated blossoms, with rows in the same chamber spaced 5 cm apart. This random arrangement of rows of blossoms is continued in other plastic boxes to provide five replicate rows for each treatment and check. Appropriate identifying designations are made at each row on the side of the box.

Mass inoculation of all blossoms in each box is by hand atomizer (e.g., DeVilbiss Atomizer No. 152) affixed to an air line at 60 kg/cm^2 (138 kPa). The inoculum consists of 80,000 *M. fructicola* conidia per milliliter of suspension prepared by diluting spores obtained from inoculated canned peach or apricot pieces (see Szkolnik, *this volume*) similar to inoculation procedures described for brown rot blossom blight above. The tops are replaced and the boxes kept at 22°C. The high relative humidity resulting from the moist sand helps keep blossoms wet and prone to infection.

Infections, appearing as brown spotting to general browning of petals, can usually be determined 72 hr after inoculation. A suitable method of examination is by assignment of infection ratings of 1–5 for each blossom infection, with 1 being uninfected and 5 being a completely infected blossom; other numbers represent intermediate levels of infection.

With five replicates per treatment, the data can be treated statistically for either total percentage of infection per treatment or for differential degrees of infection.

Sometimes there is difficulty in differentiating between actual brown rot infection and other causes for petal browning, especially marginal burn caused by handling, phytotoxicity, or blossom senility. Also, chemical vapor from certain treatments may reach high enough concentrations within the closed boxes to affect the development and sporulation of the pathogen in blossoms of other treatments.

Excised branches with blossoms. It is possible to conduct fungicide evaluations for brown rot blossom blight on cut branches collected at the dormant stage or as the blossoms open from bearing trees in the orchard. An important advantage of the dormant samples is that these can be gathered and stored for use over a span of 4–5 months. The procedure for handling cut branches from producing stone fruit trees is similar to that described for pear branches in this chapter. This method has been used successfully at the New York State Agricultural Experiment Station by Szkolnik and W. F. S. Schwabe (Fruit and Fruit Technology Research Institute, Stellenbosch, South Africa).

Samples of excised branches with blossoms may also be collected at blossoming time from trees treated with fungicides in field experiments or from untreated trees. Five to 10 blossoming shoots should be collected from each replicate orchard-sprayed tree, inoculated, and given an infection period in the mist chamber at 23°C and 100% relative humidity for 30 hr as described above for potted trees. Untreated branches may be sprayed in the laboratory, dried, inoculated, and infected as described.

FIRE BLIGHT ON PEAR

Plant Material

Numerous chemicals have been effectively evaluated on pear (*Pyrus communis* L.) blossoms in the greenhouse for efficacy against *E. amylovora.* The procedure can be used over the approximately 5 months that pear blossoms are available in good condition for greenhouse studies. Suitable blossoms may be those produced on cut branches from orchard bearing trees or on potted pear trees adapted for laboratory-greenhouse research (2,3,5,8). Cut branches have the advantage of being readily available in orchards during winter dormancy until the orchard bloom period; nursery stock trees may take 2 years from time of potting before satisfactory numbers of blossoms are produced on them. Among the pear varieties found acceptable for blossom blight research, Bartlett and Bosc proved to be most suitable.

Cut branches from bearing orchard. Cut branches from the orchard should be promptly stored at 1–2°C with the cut ends immersed in containers of water. Branches are withdrawn from storage according to need for each experiment. They are then cut to smaller branches of manageable size (to minimize danger of tipping over) and placed in crocks or heavy containers containing 1% dextrose. The random selection of cut branches should represent a good assortment of fruit buds that will provide more than 100 blossoms per treatment. During the forcing of blossoms in the greenhouse at approximately 21–24°C, the base of the branches should be pruned diagonally about 2 cm every 3 days or so with the cuts made in the water so as not to introduce air into the vascular system and to preclude occlusion of the vascular system by yeasts or other organisms that might thrive in the sugar-charged substrate.

At times it may be advisable to replace the substrate completely with the 1% dextrose solution if microbial contamination is great. The time needed for development of full bloom varies. Dormant branches collected in the orchard in February, stored, and brought into the greenhouse early (e.g., early March) require more time until flowering (about 5 weeks) than those brought to the greenhouse in May, which may bloom in 2 weeks.

Potted nursery stock. Nursery stock trees, preferably Bartlett and Bosc, should be uniformly of good vigor and caliper (not below 1.3 cm). They should be potted in 18-cm clay pots and maintained much as described above for potted stone fruit trees.

Blossom Spray Procedure

Appropriate cut pear branches or potted trees are forced to bloom in the greenhouse at temperatures between 21 and 24°C. When there is a tendency toward irregular bloom, the branches and trees can be programmed toward bloom equality in colder and warmer rooms as described for peach blossoms. They are brought together in one greenhouse room when they all attain nearly full bloom at least 1 day before spray application.

On the spray date, all unopened blossoms should be removed because these usually do not show infection during the course of the test. Infection of pear blossoms usually occurs in the calyx cup of open blossoms.

Dilute bactericide sprays are applied from all sides and the top of each tree or branch bouquet with a paint sprayer at 172 kPa as described above for blossom blight of stone fruit. Thoroughness of spraying to wet the calyx cup of all blossoms is important. Each test should include water-sprayed checks and streptomycin at 50 and 100 mg a.i./L as standards.

Although one may apply artificial overhead rain for study of erosion of spray residues, as in the case of brown rot blossom blight tests, this step is usually omitted in fire blight control research because of the scarcity of promising candidate antibacterial chemicals available from the chemical industry.

Inoculation and Infection

The inoculation source, prepared in advance, is a 3-day nutrient broth culture of *E. amylovora* of proven pathogenicity. A suitable suspension contains 1 or 2 ml of this culture per liter of distilled water, without nutrient additive. Pyrax ABB clay, if available, may be added at 4 g/L to enhance blossom abrasion during inoculation. Inoculation is made with a paint sprayer at 210 kg/cm^2 (480 kPa) with emphasis on uniformity of application to cover all blossoms. The trees are kept wet in a mist chamber at 24°C for 36 hr for infection, then returned to the greenhouse.

Recording and Analyzing Data

As infections progress, they are first detected as water-soaked tissue that darkens to brown and black. When the latter symptoms reach a maximum level in water-sprayed checks (about 6 days after inoculation), data can be recorded. Intensity of infection may vary because some treatments may be more restrictive than others on the progress of infection. The data for each treatment should include total number of blossoms with and without infection and total number of blossom clusters with and without infection. These infections should be further defined to include numbers of blossoms with infection visible only in the calyx cup, infection progressing to the receptacle, and infection progressing down the stem to the fruit spur or twig.

The data can be analyzed using each branch or tree as a replicate in any given experiment. Where several tests are conducted with the same treatments, the individual tests may serve as replicates.

Perhaps the data most likely to be used to assess the comparative efficacies of treatments would be the percentage of blossoms infected. The accumulated data above would also provide the researcher with other options in comparing treatments for least cluster infection and relative intensity of infection. Some clues might be sought to help discern between bacteriostatic and bactericidal action. Observations on phytotoxicity to blossoms and foliage should be recorded.

LITERATURE CITED

1. Gilpatrick, J. D., and Smith, C. A. 1980. Fungicide control of tart cherry diseases. Fungic. Nematic. Tests 35:30-31.
2. Hamilton, J. M., and Szkolnik, M. 1957. Omadine, a promising new organic fungicide for the control of blossom blight of pears. Plant Dis. Rep. 41:301-302.
3. Hamilton, J. M., Szkolnik, M., and Nevill, J. R. 1964. Greenhouse evaluation of fruit fungicides in 1963. Plant Dis. Rep. 48:295-299.
4. Nevill, J. R., Szkolnik, M., Gilpatrick, J. D., and Ogawa, J. M. 1978. Mass production of conidia of brown rot fungi on canned fruit pieces. Plant Dis. Rep. 62:966-969.
5. Szkolnik, M. 1973. Pear fire blight control. Fungic. Nematic. Tests 29:30.
6. Szkolnik, M. 1978. Techniques involved in greenhouse evaluation of deciduous tree fruit fungicides. Annu. Rev. Phytopathol. 16:103-129.
7. Szkolnik, M. 1981. Protective and eradicative fungicidal action in the control of cherry brown rot blossom blight. Fungic. Nematic. Tests 36:30.
8. Szkolnik, M., and Hamilton, J. M. 1966. Tree fruit disease and fungicide research in 1965. Proc. N.Y. State Hortic. Soc. 111:163-174.

Methods for Field Evaluation of Fungicides for Control of Foliar and Fruit Diseases of Apple

KENNETH D. HICKEY, Department of Plant Pathology, The Pennsylvania State University, Fruit Research Laboratory, Biglerville 17307; KEITH S. YODER, Department of Plant Pathology, Physiology, and Weed Science, Virginia Polytechnic Institute and State University, Fruit Research Laboratory, Winchester 22601; and ELDON I. ZEHR, Department of Plant Pathology and Physiology, Clemson University, Clemson, SC 29631

Evaluation of fungicides for efficacy against major pathogens of apple (*Malus domestica* Borkh.) begins with primary screening of many chemical compounds from chemical manufacturers in laboratory, greenhouse, and field experiments. For those that pass the primary tests, the acceptability for usage in commercial orchards is determined through continued greenhouse and field trials under varied conditions of culture, environment, and application methods. Tests must also be conducted to determine phytotoxicity, acceptable residues in food, safety to applicators, and environmental impact.

The methods described here for the evaluation of fungicides on leaves and fruits of apple under field conditions have been used successfully by a number of investigators (2–5,7–9,11,12) for many years. These methods are particularly suited for areas of the eastern United States with moderate temperatures and moderate to high incidence of important apple pathogens. Economically important diseases and pathogens that are commonly found in commercial orchards and for which fungicides are used include apple scab (*Venturia inaequalis* (Cke.) Wint.); apple powdery mildew (*Podosphaera leucotricha* (Ell. & Ev.) Salm.); cedar apple rust (*Gymnosporangium juniperi-virginianae* Schw.); quince rust (*G. clavipes* Cke. & Pk.); sooty blotch (*Gloeodes pomigena* (Schw.) Colby); flyspeck (*Zygophiala jamaicensis* Mason); black rot (*Botryosphaeria obtusa* (Schw.) Shoemaker); bitter rot (*Glomerella cingulata* (Ston.) Spauld. & Schrenk); and bot rot, or white rot (*B. dothidea* (Moug. ex Fr.) Ces. & de Not.). Other diseases may be important in specific regions.

MATERIALS AND PROCEDURES

Test site. A test site should be in an area with climatic conditions favorable for the development of disease. The site must accommodate legal limitations on test chemicals, and it should be located where cultural practices can be managed carefully and where inoculum levels are high enough to show differences in fungicide efficacy. The soil should be well drained with sufficient depth, structure, and fertility to assure moderate to vigorous growth of the trees. Sites with several years of cropping history of grass, field crops, or row crops generally have less potential for soilborne pathogens that cause root rot or collar rot in the test trees. Good air drainage minimizes injury from low temperatures and spring frost and also promotes rapid drying of fungicide spray suspensions. Windy conditions on hilltop sites can interfere with proper application and cause inadequate coverage or interplot contamination of test compounds. A ground slope of 5% or less facilitates the movement of equipment.

Test plants. Cultivars of apple to be used for fungicide evaluation should be of importance in the test region and preferably in other major producing areas. Since they vary widely in their susceptibility to major diseases, only those with moderate to high susceptibility should be used. Cultivars with low to moderate susceptibility, however, have been successfully used under high disease pressure, because lesions may be abundant and discrete. The cultivars McIntosh, Cortland, Stayman Winesap, Rome Beauty, and several red sports of Delicious are moderately to highly susceptible to scab. Cultivars susceptible to powdery mildew and suitable as test plants include Jonathan, Rome Beauty, Cortland, Idared, Granny Smith, Jonamac, Macoun, Gravenstein, Monroe, Baldwin, and Stayman Winesap. Delicious, one of the most widely planted cultivars, is susceptible to scab, black rot, bitter rot, sooty blotch, flyspeck, and quince rust. Many of the red sports planted in the United States, however, are immune to cedar apple rust and only slightly susceptible to quince rust and powdery mildew. When they are used for studies on sooty blotch and flyspeck, their red color sometimes makes detection difficult in cases of light infections. Golden Delicious is excellent for evaluating fungicides for control of these diseases because of its light color. It is susceptible to cedar apple rust, scab, and the fruit rots and is very sensitive to fruit surface russeting, a form of phytotoxicity. Jonathan fruit is highly susceptible to cedar apple rust, Brooks spot, fruit rots, and fruit russet caused by powdery mildew.

Standard, semidwarf, or dwarf trees may be used, depending on the desired tree size. The preferred rootstock is M7 or M26, but MM106, MM111, or apple seedling rootstocks are also used. Trees grown on MM106 are generally of the most desirable size and productivity, but because of their high susceptibility to collar rot (*Phytophthora* spp.) and tomato ringspot virus, loss of such trees may be high.

Trees should be planted in rows at spacings recommended for the particular rootstock-cultivar combination or spaced at distances suitable for the application equipment. Orchard blocks four rows wide provide maximum flexibility in the movement of spray equipment. Within rows, trees may be arranged either singly, in clusters of one each of three cultivars per tree site, or in short sections containing three to six different cultivars spaced 4 m apart within each section. To allow for easy movement of spray equipment down and across rows, the distance between rows should be 10 m; within rows, single trees, clusters, or short sections of rows should be separated by 8–9 m. Trees grown on M26 rootstock and supported by a single- or double-wire trellis for each short section along the rows are suitable where several cultivars are desirable. Excellent results have been obtained with a combination of Rome Beauty, Delicious, and Golden Delicious cultivars. McIntosh, Cortland, Jonathan, or other cultivars may also be used where desirable. Another method to obtain two or three cultivars per plot involves grafting one or two additional cultivars onto each tree in the second year after planting. Trees with equal parts of each cultivar can be grown and maintained.

If possible, rows should be perpendicular to the prevailing wind, to help minimize spray drift to adjacent plots. The middles between rows should be maintained in fescue or other grass sod to minimize soil erosion and facilitate spraying during wet periods. Control of vegetation under the trees can be obtained by occasional cultivation with a mechanical tree hoe or by the use of herbicides.

Four- or five-year-old trees generally are of sufficient size and bear enough fruit to be useful for fungicide evaluation. Experiments for the control of powdery mildew, scab, frogeye leaf spot, or cedar apple rust may be conducted on younger trees under some conditions. Care should be taken, however, to avoid overcropping or other major stresses on young trees before the root system is well established. The trees should be sufficiently pruned and fertilized annually and the orchard floor mowed to optimize tree shape and density, annual bearing, and general health.

Inoculum. A high level of virulent inoculum is essential in evaluating fungicide efficacy. Natural sources of inoculum are preferred to artificially introduced sources. However, with some test fungi, inoculation may be necessary to assure a satisfactory test. In such cases, placement of infected leaves or stems in the test site is preferred to application of spore suspensions. Inoculum of most of the major apple pathogens can be maintained at moderate levels by leaving a number of untreated trees distributed throughout the test site, in addition to the untreated check treatment.

Apple scab. Additional ascospore inoculum of the pathogen that causes apple scab can be introduced by placing under each test tree about 0.06 m^3 of overwintered apple leaves infected the previous season. Leaves collected from abandoned McIntosh orchards at about the green tip phenologic stage are an excellent source of inoculum. Early-season infections of apple scab are best obtained by spraying the trees during one or more infection periods with conidia of *V. inaequalis*. The conidia can be obtained from wick cultures or from infected potted trees in greenhouses (11). Lesions that develop from early primary infections are very important sources of inoculum for secondary spread throughout the test block.

Apple powdery mildew. The fungus *P. leucotricha* is an obligate parasite that overwinters in infected buds on vegetative shoots (terminals) produced during the previous growing season. Fruiting or nonfruiting spurs (clusters) may also be sites of overwintering, and most infected buds are located near the apex of the shoot. Infections on the first developing leaves result from conidia produced on opening infected buds. Because the opening of infected buds is slightly delayed, the first infections occur around the tight-cluster stage of healthy bud development. Secondary spread results from inoculum produced by both primary and secondary infections. Powdery mildew infections occur at temperatures between 4 and 30°C when the relative humidity is between 70 and 100%. Leaves are susceptible to infection for only a few days after they emerge but can be infected at any time if they are injured. Fruit infections occur while the fruit is young—between the pink bud stage and petal-fall or a few days later. The time of infection on developing buds, where the fungus overwinters, is not well understood, but presumably infection occurs from midseason until the apical buds on the shoots have fully formed.

Cedar apple rust and quince rust. Galls of the pathogen that causes cedar apple rust can be collected from red cedar (*Juniperus virginiana*) when the telia horns first appear and can be stored at 0.5°C until they are placed in the tree canopy. An effective method of inoculating is to soak the galls in water for 30 min and then place three to five into a small wire basket suspended 60 cm above the center of a tree. Galls should be placed in baskets above each tree one to three times when 12–24-hr periods of rain and temperatures of 12–18°C are forecast, from the late pink stage through the blossoming stage.

The fungus causing quince rust produces small cankers on branches of red cedar, which may be used as an inoculum source. Because quince rust infections on fruit occur only during a relatively short period, between the pink stage and petal-fall, predictable results are more difficult to obtain for this disease than they are for cedar apple rust.

Fruit rots and blotch. Inoculum for the pathogens causing sooty blotch, flyspeck, and various fruit rots, such as black rot, bitter rot, and bot rot, is usually present on branches and twigs of the test trees or on nearby woody plants, and supplemental inoculum is seldom required. The fungi causing fruit decay often grow and sporulate on fire blight cankers or other necrotic woody tissues produced the previous season. Pieces of diseased tissue placed above the center of each tree provide ample inoculum for fungicide evaluation. Generally, five or six pieces (10–15 cm long) placed in trees at the pink stage are sufficient.

In humid southern areas, fruit decay, sooty blotch, and flyspeck are especially important. As a group, they are often referred to as summer diseases, but infections by *B. obtusa* can occur as early as flowering time and in some localities can cause substantial losses of fruit early in the season. Fungicide tests for control of black rot should be designed according to the behavior of the pathogen in the locality when the test is being done.

Experimental design. All field tests of fungicides should be designed to facilitate proper statistical analyses of the data. Randomized complete block designs and completely randomized plot designs are the most commonly used. The number of replicates of each treatment can vary from three to six, according to subjective estimates of the number required for adequate statistical comparison. The number of replicates also depends on the number of trees available, the uniformity of trees, the inoculum level, and local environment in the orchard. Single-tree replicates are satisfactory and are commonly used for evaluating experimental compounds that are in short supply. If little is known about the fungicides being tested, the major variable among treatments is the test compound or rate. Two or three rates are commonly selected, ranging from 25% below to 25% above the optimum found in previous tests. In tests in which treatments are applied with an air-blast sprayer, larger blocks are needed, with spacing similar to that in commercial orchards. In such tests, each replicate should consist of three or more consecutive trees along the row; adjacent rows on each side serve as barriers to reduce drift between plots.

Experiments may have more than one objective, and the design should facilitate achieving these goals. Usually one replicate is not treated with fungicides and serves as a control to determine disease severity in the test area. Additionally, one treatment should be a commonly recommended fungicide, to serve as a standard for comparison. In tests designed for control of particular diseases, such as apple scab or powdery mildew, control of other diseases that may interfere with the test pathogen is often desirable. In such cases, a fungicide that is effective against an unwanted pathogen can be used across all plots, or combining it with the test compound may be desirable.

Fungicide application and equipment. Uniform coverage of all susceptible tissue as it develops sequentially during the growing season is essential for evaluating seasonal efficacy of fungicides. The frequency of applications may vary, depending on the rate of tree growth and the characteristics of particular fungicides, but generally it is five to eight days during the period from the green tip stage through the blossoming stage and 12–16 days in postbloom sprays. The method of application should be appropriate for the objectives of the test, the orchard layout, and the plot design. Where trees are small to medium in size or the amount of experimental compound is limited, dilute aqueous suspensions mixed in 75–200-L lots are applied with a hydraulic sprayer equipped with a single-nozzle spray gun or a multinozzle boom operated at a pressure of 2,800–3,500 kPa. To assure complete coverage, the spray mixture should be applied until runoff occurs. Although this method does not mimic the commercial application of concentrate spray mixtures with air-blast sprayers, it is useful for establishing the efficacy of uniform fungicide deposits on plant surfaces. If an evaluation of fungicide efficacy under the conditions of commercial orchards is desired, concentrate sprays should be applied with an air-blast sprayer.

Careful attention to proper operation and calibration of the sprayer, along with matching the size and type of the sprayer with tree size and density, results in more accurate evaluation of fungicide performance. The amount of spray mixture per treated area should be adjusted to match tree size or sprayer type. As a general guide, the amount of dilute spray mixture should be 2,800–3,700 L/ha for standard orchards (with trees 5.5–6.0 m high planted in rows 10.6 m apart) and about 60–75% of this amount for dwarf and semidwarf trees. The amount of fungicide per hectare may be concentrated from two to 10 times and applied as concentrated sprays at equivalent rates with a low-volume sprayer.

DATA COLLECTION AND REPORTING

Observation of disease. Disease incidence and severity on leaves and fruit should be recorded for each plot (or tree) on one or more dates during the growing season. If disease progress curves are needed, three or more counts are needed between the appearance of the first symptoms and the time when the disease ceases to spread. The most valuable single count to measure seasonal control of fungicide treatments on leaves is at midsummer, after the growth of vegetative shoots (terminals) has ceased. Disease incidence is determined by observing all leaves on 10–20 vegetative shoots per tree; the shoots are selected at random from all sides of the tree. Shoots that are uniform in length should be selected, because leaf development along the shoots should occur at about the same time and should span the entire growing period. Short shoots whose growth terminates early in the season should be avoided. All leaves on the shoot must be observed, because apple scab, powdery mildew, and cedar apple rust rarely occur on the same portion of the shoot. For each leaf, records of the date when it emerged and of disease incidence are desirable for correlating the time of infection with weather conditions and spray application. The incidence of apple scab during the prebloom period is best determined from observations of the flowering clusters. Observations should be made on all leaves (usually eight to 12) and young fruit on 25–50 clusters per tree about two to three weeks after petal-fall and before infected leaves or fruit fall.

A level of disease severity should be established for untreated trees or for treatments that are relatively ineffective. Severity can be determined by the number of lesions per infected leaf or fruit or by a rating system such as that reported by Horsfall and Barratt (6). In this method, the amount of surface area showing symptoms is estimated by means of 12 unevenly divided categories, with narrow limits between categories on both ends of the scale and wide increments in the middle. The method is particularly effective where treatments allow only a few lesions to develop or where treatments used in several applications have strong suppressive action on lesions of powdery mildew or apple scab.

Determination of the efficacy of fungicides for controlling the primary phase (overwintering) of apple powdery mildew requires observations during the growing season following treatment. Infection of buds and subsequent overwintering can be measured the following spring, at the pink stage to the early bloom stage, by the number of opening buds that are infected. One or all of three methods may be used to record the number of infected buds. In the first method, the total number of shoots with primary infection that show growth of the fungus as the leaves emerge is determined for each tree. In the second method, on large trees, the percentage of shoots having infected buds is determined by observation of 100 shoots (selected at random) per tree or by a count of all infections observed in a period of 3–5 min. In the third method, the percentage of dead or infected buds on 20 shoots per tree is determined by observation of the 10 most distal buds, starting with the apical bud. This can be accomplished by excising shoots and forcing them to open in a greenhouse or laboratory during late winter or as growth begins in the spring. Mycelium and spores of the fungus are apparent on infected buds as they open. Because infection cannot be assured in dead buds that remain unopened, they should be counted and recorded separately from infected buds.

Disease incidence and severity on fruit should be recorded by observation of 100 fruits per tree. Differences among fungicide treatments used for control of apple scab and cedar apple rust are generally less apparent on fruit than on leaves. Sooty blotch and flyspeck may appear on fruit from midsummer to near harvest. Observations made at harvest are adequate when infections appear late in the season. Since active fungicide residue on the fruit is closely correlated with level of control, a series of counts, beginning when the disease appears and continuing to harvest, is necessary to measure fungicide efficacy if persistence or longevity of activity is being studied. Infections by the decay-producing fungi are often apparent at harvest but may be more pronounced after regular cold storage (0–1°C) for a period of one to six months.

Determinations of phytotoxicity. Observations for leaf necrosis or chlorosis should be made several times during the growing season. Additionally, any phytotoxicity in floral parts or young fruit that may affect fruit set should be recorded. Evaluation of fruit for russeting that may affect fruit set should also be recorded. Evaluation of fruit for russeting or reduction in color intensity can be made by random observation of 25–100 fruits per tree on unharvested trees late in summer, but harvested fruit is preferred. The data should be expressed as the percentage of the surface that is affected, based on standard rating systems such as the Horsfall-Barratt method (6) or other recognized standard grading systems for fruit. Effect on crop yield may be measured by total weight per tree, area, or size and weight of individual fruits. Caution must be exercised in interpreting the results, since many factors, such as previous crop load per tree, cropping method, or unrelated pest control programs, can affect tree performance. To obtain reliable yield data, observations should be made on the same trees treated with the same treatments for a minimum of three years. More reliable data are obtained from trees 10 or more years old on replicated plots containing five or more trees per plot.

Data analyses and reporting. Data should be compiled from field observations, and a percentage of infected leaves or fruit should be determined for each treatment. To help in the interpretation of the results, the data should be statistically analyzed by conventional methods of analysis of variance to determine variance among treatments. The percentages may be transformed by appropriate methods before analysis if transformation is required for suitable statistical treatment. Significance between treatments should be determined by Duncan's multiple range test (1), the Waller-Duncan Bayesian least significant differences test (10), or other suitable methods (see chapter 3). Qualified statisticians can be of invaluable assistance in choosing the appropriate statistical test and can otherwise help to interpret the results of experiments.

Reports of field trials of fungicides should include 1) general conditions under which the test was conducted, including the type and number of trees, the inoculum level, and weather conditions; 2) the name of the test compound, its formulation, and the name and percentage of each active ingredient; 3) the method of application, the rate, the concentration (dilute or concentrate), the spray volume per tree or per area, the equipment that was used, and the number of applications; 4) the number of replicates, mean disease incidence, details of when and how the data were secured, the level of control that was obtained, and the relative effectiveness compared with standard treatments; and 5) the type and degree of phytotoxicity.

LITERATURE CITED

1. Duncan, D. B. 1955. Multiple-range and multiple *F*-test. Biometrics 11:1-42.
2. Hickey, K. D. 1984. The management of apple powdery mildew in commercial orchards. Mountaineer Grower 456(4):4-10.
3. Hickey, K. D., Garretson, M., and May, J. 1984. Efficacy of SAT and RSAT treatments of Bravo and Difolatan for disease control, 1983. Fungic. Nematic. Tests 39:12.
4. Hickey, K. D., and Travis, J. W. 1984. Management of resistant strains of orchard pathogens. Pa. Fruit News 63(4):60-62.
5. Hickey, K. D., and Yoder, K. S. 1981. Field performance of sterol-inhibiting fungicides against apple powdery mildew in the mid-Atlantic apple growing region. Plant Dis. 65:1002-1006.
6. Horsfall, J. G., and Barratt, R. W. 1945. An improved grading system for measuring plant diseases. (Abstr.) Phytopathology 35:655.
7. Lewis, F. H. 1980. Control of deciduous tree fruit diseases: A success story. Plant Dis. 64:258-263.
8. Lewis, F. H., and Hickey, K. D. 1972. Fungicide usage on deciduous fruit trees. Annu. Rev. Phytopathol. 10:399-428.
9. Mills, W. D. 1944. Efficient use of sulfur dusts and sprays during rain to control apple scab. N.Y. State Coll. Agric. Bull. 630, War Emerg. Bull. 114:1-4.
10. Waller, R. A., and Duncan, D. B. 1969. A Bayes rule for the symmetric multiple comparison problem. J. Am. Stat. Assoc. 64:1484-1499.
11. Williams, E. B. 1976. Handling the apple scab organism in the laboratory and greenhouse. Pages 16-18 in: Proc. Apple and Pear Scab Workshop, Kansas City, MO. A. L. Jones and J. D. Gilpatrick, eds. Cornell University, Ithaca, NY.
12. Yoder, K. S., and Hickey, K. D. 1981. Sterol-inhibiting fungicides for control of certain diseases of apple in the Cumberland-Shenandoah region. Plant Dis. 65:998-1001.
13. Yoder, K. S., and Hickey, K. D. 1983. Control of apple powdery mildew in the mid-Atlantic region. Plant Dis. 67:245-248.

Field Evaluation of Fungicides for Control of Diseases on Tart Cherries

A. L. JONES and G. R. EHRET, Department of Botany and Plant Pathology and Pesticide Research Center, Michigan State University, East Lansing 48824

The tart cherry (*Prunus cerasus* L.) produced commercially in the United States and Canada is the Montmorency. Montmorency cherry production is concentrated around the Great Lakes states, with Michigan as the center. Michigan ranks first among the states in tart cherry production and accounts for nearly 84% of the crop, with an annual value of $27–57 million. Utah is a second center of production. Timely and accurate evaluation of fungicides is important to the prosperity and continued growth of the tart cherry industry in North America.

The Montmorency cherry is susceptible to several common fungal diseases. The most important of these diseases are cherry leaf spot, caused by *Coccomyces hiemalis* Higgins (=*Blumeriella jaapii* (Rehm) Arx), and brown rot, caused by *Monilinia fructicola* (Wint.) Honey. A severe outbreak of leaf spot (sometimes called shot hole) can reduce bud survival, decrease fruit set, and result in eventual tree loss. An epidemic of brown rot has the potential to destroy an entire crop almost overnight. Of lesser economic importance are European brown rot, caused by *M. laxa* (Aderh. & Ruhl.) Honey, which is found sporadically throughout the Great Lakes region; powdery mildew, caused by *Podosphaera oxyacanthae* (DC.) d By., which is an important problem in Utah and an occasional problem in the Great Lakes region; and fruit rot, caused by *Alternaria* spp., which occurs on mature to overripe fruit damaged by wind, birds, mechanical equipment, or hail. Descriptions of each disease have been published elsewhere (12).

Brown rot and leaf spot are controlled primarily with protective fungicides. Fungicide sprays for brown rot are timed based on tree growth stages during the bloom period and on a calendar basis during the preharvest period. Up to four applications are made for blossom blight at white-bud, bloom, petal-fall, and shuck-split stages. During the preharvest period, two to three applications are repeated at intervals of 7–10 days to protect the maturing fruit. Leaf spot control starts at petal-fall, and sprays are applied every 10–14 days until harvest. A postharvest spray is applied in years of severe disease pressure to prevent early defoliation of the trees. In practice, six to eight fungicide applications are made each year to control both diseases. In years with above-normal rainfall and epidemic brown rot, two to three additional sprays are applied. Postinfection sprays of cycloheximide (Acti-dione) were used on an emergency basis to control leaf spot in the 1960s. Although the ergosterol biosynthesis-inhibiting fungicides have good postinfection activity against leaf spot (3), none of these compounds is currently registered for this purpose.

This chapter describes methods for evaluating the effectiveness of fungicides for disease control on tart cherries. Because the objectives of each investigator may vary, some modifications of these methods are described to emphasize the scope of research that can be conducted. Researchers new to this area should also review Lewis (13).

METHODS

Test Site

A good test site for tart cherries should have uniform, deep, well-drained soil of moderate fertility. Because Montmorency blooms relatively early in the spring, the orchard site should be located sufficiently above the surrounding area to allow for drainage of cold air to lower levels. Also, a site influenced by a large body of water such as Lake Michigan or Lake Ontario tends to bloom later, reducing the probability of crop reduction because of late freezes.

When a new orchard is planted, first consideration should be given to sites in the vicinity of commercial orchards with consistent cropping histories. Sites recently cleared of oak and other forest trees should be avoided because of the danger of infection by *Armillaria mellea*. Chokecherries should be eradicated from all adjacent fencerows and woodlots to avoid problems with X-disease. The soil should be checked for nematodes and fumigated if necessary.

Besides choosing a site large enough to allow for suitable experimental plot designs, it is wise to select a somewhat isolated location when applying unlabeled materials for which the toxicology is incomplete. Consideration should be given to preventing drift to surrounding areas and to the containment of runoff that might contaminate water sources. Planting a windbreak of tall, closely spaced trees on the west side of the site cuts the wind, improves conditions for spraying, and protects fruit from wind-whip. The test site should also be secure from trespassers and human traffic that might disrupt an experiment.

Because research plots are frequently located near wooded areas, trees and fruit may need to be protected from wildlife. Deer can be a major problem when a new orchard is established. They can be discouraged from browsing on young trees by hanging unwashed human hair balls covered by Styrofoam cups, bars of deodorant soap, or small bags of tankage from each tree in the orchard. Where deer pressure is heavy, a high-tensile-strength electric fence should be erected with at least five wires spaced in a vertical or slanted configuration. Raccoons can be prevented from breaking limbs and destroying fruit by livetrapping and relocation. Bird damage is alleviated by spraying repellent chemicals on the fruit as they color and mature.

Test Plants

Montmorency cherry trees can be used for testing purposes in the fourth growing season if they are grown under good cultural conditions, including trickle irrigation. Only virus-free trees should be planted, and the orchard should be isolated from older blocks infected with ring spot or prune dwarf viruses. Mazzard rootstock is preferable to Mahaleb rootstock on all but sandy, well-drained soils because it tolerates poorly drained soil better. If soils are too wet, trees on Mahaleb rootstock will die after 5–10 years. Drainage in small pockets of wet soils underlain by a clay "lens" can be improved by installing tile drains. Subsoiling is of limited effectiveness for improving drainage, because the clay layers reform after a few years.

Annual pruning is important in the test orchard to permit thorough spray coverage, stimulate new fruiting wood in the lower portion of the tree, and develop a sound framework for long-term production. Failure to prune young trees properly often leads to limb breakage and loss of experimental units once the trees come into production. Judicious pruning makes it possible to maintain twice the normal number of trees in the early years of a test plot and then remove alternate trees when crowding occurs. This technique allows earlier and more intensive testing in the orchard by doubling both fruit and foliage production and by maintaining a more favorable microclimate for disease development. Doubling the number of trees planted per acre also allows the use of buffer trees to trap spray drift and serve as sources of inoculum.

A tree spacing of 20 ft (6.1 m) between rows is adequate for handgun spraying with intensive pruning. But airblast spraying is more convenient at wider spacings with at least 40 ft (12.2 m) for turning areas on the ends. For best sunlight exposure, tree rows should be oriented north to south when topography and other considerations permit.

Inoculum

Before tests are begun in the fourth or fifth growing season, it is often necessary to build up inoculum of cherry leaf spot in the orchard. This can be done either in late autumn or in spring before petal-fall by spreading infected leaves from another orchard under the test trees. Once the pathogen is established in the orchard, unsprayed buffer rows around the periphery of the orchard and check trees in the interior generally provide sufficient inoculum for good leaf spot pressure. Because early defoliation can lead to severe weakening and death of unsprayed trees, buffer trees should receive one or two fungicide applications in years of severe disease to prevent complete defoliation. To preserve leafborne inoculum, it is best to avoid fertilizing at leaf-fall in autumn and cultivating or mowing in the spring before fruit set.

Brown rot inoculum can be introduced into a test block by techniques similar to those for leaf spot. Brown rot mummies or infected fruit gathered from problem orchards are placed in baskets or plastic mesh bags within the test trees before bloom or fruit ripening. Once brown rot is established, avoid pruning out overwintering mummies and twig infections so that a high level of natural inoculum can be maintained for the next growing season.

If natural inoculum is not available, fruit on buffer trees can be artificially inoculated by syringe with conidia produced on canned fruit halves (14). Inoculations should be made when the fruit start to color, to allow more time for inoculum to build up and spread to treated trees. Also, if fruit are inoculated too early, they will fall from the tree before the fungus sporulates. Inoculum can also be harvested from naturally infected fruit, but care must be taken to avoid unwanted pathogens such as *Rhizopus* spp. We have found that inoculating the fruit on buffer trees with pure cultures of brown rot gives more consistent results than placing naturally infected fruit in baskets in the sprayed trees.

Monitoring Leaf Spot Infection Periods

To accurately assess the merits of individual fungicides, it is essential to know when infection periods occur and how severe they are. Several methods are available for this purpose.

A model developed at Michigan State University to identify cherry leaf spot infection periods (2) generates an "environmental favorability index" value from 0 to 100. An index value of 14 represents the minimal conditions of temperature and wetness necessary for infection. Index values can be manually estimated from a computer-designed nomogram (3) or from an equation (2). To use the nomogram or equation, the mean air temperature (°C) and hours of leaf wetness for each wet period must be known. These parameters should be recorded with weather monitoring equipment located at or very near the fungicide test site. Typical weather monitoring equipment would be a 7-day recording hygrothermograph (Bendix Co., Baltimore, MD 21204) and a 7-day recording leaf wetness meter (De Wit Instruments, Roden, The Netherlands, or Belfort Instrument Co., Baltimore, MD 21224).

A Reuter-Stokes predictor (Reuter-Stokes Instruments, Inc., Twinsburg, OH 44087) programmed with the predictive model for cherry leaf spot can facilitate monitoring of infection periods (6,9–11). The computer continuously monitors leaf wetness and dry-bulb and wet-bulb temperatures. Weather and infection data are stored in the predictor's memory bank and can be retrieved through a keyboard display system and with a plug-in printer. The relative favorableness of the weather for leaf spot infection is displayed as "none," "low," "moderate," or "high." Predictions and important weather data retrieved from the predictor should be included in research reports. The unit saves considerable time and eliminates human errors in making routine calculations.

Because predictions are based only on the weather, inoculum levels in the orchard should be monitored with spore traps to determine whether inoculum is limiting when weather is favorable for infection. We have used two types of spore traps to monitor the release of ascospores. The first type is the Rotorod sampler (Ted Brown Associates, Los Altos Hills, CA 94002). This trap is relatively inexpensive, and data on spore discharge during each wet period can be calculated relatively quickly because the spore-trapping surface is easy to examine. Samplers must be resupplied with fresh rods before and soon after each wet period so that the trapping surface is relatively free of dust; spore counting is much easier if spores are not mixed with large quantities of dust. If the rods cannot be routinely changed, retractable collection heads work well. Samplers should be activated during wet periods with a moisture sensor (15). The second type is the Burkard spore trap (Burkard Scientific Limited, Rickmansworth, Hertfordshire, England), which is considerably more expensive than the Rotorod sampler but gives more detailed information because it samples the air continuously for 7 days. Burkard traps are particularly suited to epidemiologic studies because the results can be analyzed on an hourly basis.

Leaf spot infection periods can be verified by placing unsprayed, actively growing, potted cherry trees (trap plants) in the test plot before a wet period and removing them after. Inoculum levels should be high near the trap plants. After exposure, trap plants should be held in a cold frame for 2–3 weeks, when infection can be rated. If suitable trap plants are not available, new leaves on shoots can be labeled with their date of emergence and examined for infection after about 2 weeks (2).

Checking for Resistant Strains

If disease develops on trees sprayed with fungicides that normally give excellent control, samples should be collected and tested to see if resistant strains have developed. Before samples are collected, the trees should be sprayed with a normally effective rate of the suspect fungicide to determine whether spread to new leaves or healthy fruit can be stopped (6). If the spread of infection is apparently not retarded by the fungicide, the samples should be tested in the laboratory for possible resistant strains.

The best leaf or fruit samples for testing are those with active lesions that are sporulating or can be induced to sporulate in the laboratory. For comparison, samples should be collected from trees that have never been sprayed with the fungicide in question and also from unsprayed trees in the test block. If resistance testing must be delayed, leaves with cherry leaf spot may be stored 24–48 hr under refrigeration or frozen in sealed bags for longer storage. If fungicide-resistant *M. fructicola* is suspected between growing seasons, infected fruit mummies can be collected in winter or spring and induced to sporulate in the laboratory for poison agar tests. Fungicide resistance of *C. hiemalis* can be tested in the spring by collecting fallen leaves under suspected treatment trees and ejecting the ascospores onto water agar plates in spore discharge towers. Ascospores can then be transferred to plates of media amended with fungicides. Testing methods vary with fungicides (7,8).

Handling Fungicides

Researchers working with experimental fungicides are faced with the problem of handling and storing chemicals for which the toxicology is incomplete and the environmental impact unknown. To ensure worker safety and to act in an environmentally sound manner, all fungicide researchers should have a well-designed storage and handling system.

In 1979, we constructed a pesticide handling and storage facility at our field laboratory to minimize chemical vapors and odors and to provide a safe work area for handling chemicals for research studies (1). Certain features of the design of this facility have proved important to the safe handling of experimental fungicides. A ventilation system controlled by a time clock automatically exhausts air in the storage room before work hours. Forced-air ventilation of storage cabinets has largely eliminated puffs of chemical dust that occur when doors of unvented facilities are opened. Ten to 15 air changes per hour are required to remove objectionable odors and vapors from the room. A stainless steel transfer hood is easy to clean and a convenient place to measure small quantities of fungicides without cross-contamination or worker exposure. Spilled chemicals are flushed from the hood to a gravel-filled, clay-lined pit. Storing weighing materials, safety materials, and pesticides in locked cabinets increases the accuracy and security of the fungicide testing program.

To protect the environment and avoid costly disposal problems, a few simple guidelines should be followed in handling unused fungicides. Researchers should accept only the minimum amount of chemical needed to complete the experiment and should return any unused chemicals to the manufacturer. Any unregistered chemicals that have accumulated at the end of the spraying season should be sent to a chemical disposal company if they cannot be returned to the manufacturer. Excess registered fungicides may be given to commercial cherry growers for use according to label directions. Tanks of fungicides mixed for orchard application should be emptied as completely as possible in the spraying process and should be rinsed over a containment pit.

Application Equipment

Fungicides are generally applied at high pressure with either a handgun or an air-blast sprayer to the point of runoff. Ideally, these sprayers should be custom-built to conform to each researcher's requirements according to experimental plot design and tree size. However, a beginning plant pathologist may successfully use a small commercial unit when trees are small and the test program is not extensive. As the testing program develops, the researcher will require more elaborate equipment with increased capacity to cover larger trees and expanded acreage.

We tested 26 spray mixtures in 1983 on cherry trees from 8 to 20 ft (2.4–6.1 m) in height. Our equipment for handgun applications consists of four 50-gal (189-L) stainless steel tanks mounted on a trailer frame with a 20-gal/min (76-L/min) pump (John Bean, FMC Corp., Jonesboro, AR 72401) designed to operate up to 500 psi (3,500 kPa) and powered by a 12-hp gasoline engine. Sprays are applied with 50 ft (15.2 in.) of half-inch (1.27-cm) polyvinyl chloride spray hose attached to a brass spray gun 25 in. (63.5 cm) long that delivers 10–12 gal (38–45 L) per minute. The four spray tanks have mechanical agitation and are interconnected by a system of ball and gate valves that allows the applicator to spray a different mixture from each tank; each tank may also serve as a nurse tank for rapid filling and rinsing when treatments requiring less than 50 gal (189 L) are applied. In addition, an expanded steel platform and a guardrail are provided on the sprayer so that the applicator may ride between treatments.

Our air-blast sprayer was designed and built at Michigan State University using a trailer frame on which we mounted a John Bean airhead capable of delivering 30,000 ft^3 (849.5 m^3) of air per minute with a 20-gal/min (76-L/min) pump similar to the one previously described. The airhead and pump are driven by a shaft powered by a 60-hp tractor with an independent power takeoff. Four conical, 30-gal (114-L) plastic mixing tanks were attached to the sprayer frame, and electric agitators consisting of the propeller and motor unit from a commercial trolling motor were suspended at the base of each tank. Tank agitators were wired to a central control box accessible to the tractor operator with three-way toggle switches for high and low speed and the off position. Also wired to the control box were two solenoid valves that operated the left and right nozzle banks and a gate valve that was combined with an electric motor and microswitch to remotely dump unused chemicals. Finally, the sprayer was equipped with John Bean flip-over spray nozzles for convenience in switching spray concentrations and quick-change hose couplings to accept a spray gun as a backup for the dilute sprayer.

Because most commercial tart cherry growers use concentrate air-blast units, fungicide researchers should consider setting up at least one test block specifically for air-blast studies. Air-blast test plots should consist of groups of trees, not individual trees. Buffer rows are required to minimize spray drift from one plot to another. Planting of windbreaks may also be helpful. Minimum plot size is three-row plots with three trees per row. Data are recorded from the center tree in the plot. Calibration of air-blast units must be checked regularly to ensure uniform and repeatable spray coverage. As with dilute spraying with handguns, applicators need to be protected either by a spray cab or by a respirator and protective clothing.

All sprays should be applied under good drying conditions and at moderate temperatures for the best fungicidal activity and avoidance of phytotoxicity. Treatments applied before high temperatures and in slow-drying locations may injure foliage and fruit. Fungicide applications made before a heavy rain with little drying time may be ineffective because of removal of the spray deposits.

Data Collection

For accurate data collection, treatments in the test block should be clearly marked with field stakes and plot plans recorded in a field book. We use data collection forms of our own design with spaces provided to record the experiment number, day, year, replication, treatments, and number of terminals, fruit, and leaves to be examined. Data recorded on these forms can be keypunched directly onto computer cards or into a microcomputer.

Leaf spot ratings should be made two or three times during the growing season, generally once before harvest and one or two times after harvest. The date the first symptoms are observed on unsprayed trees is also noted.

When rating leaf spot, it is best to have two persons examining the foliage and one person recording. Each observer should examine one terminal shoot at a time and then report the results to the recorder. Data are usually taken on 20 terminals per tree. Terminals at about chest height are selected at random from four sides of the tree. The total number of nodes, leaves, and infected leaves are counted on each terminal. The number of lesions on each leaf may also be determined. Lesion counts may help to separate treatments when disease pressure is low or when most treatments are effective.

Defoliation from leaf spot infection is determined by subtracting the number of remaining leaves from the number of nodes. Defoliation estimates for leaf spot are less precise when sour cherry yellows is present in the orchard. It is important to use virus-free trees in establishing new orchards.

An alternative and more quantitative method for counting leaf spot involves tagging the base of 10–20 terminal shoots per tree and the last unfolded leaf on each shoot. The date, total number of leaves, and number of diseased leaves are recorded when shoots are tagged; subsequent counts at regular intervals reveal the advance of the disease. Fungicides can then be ranked according to their efficacy in preventing defoliation, leaf infection, and sporulation.

Powdery mildew data may be collected at various times during the growing season, but the greatest differences between treatments are usually seen at about the time terminal growth stops. We evaluate powdery mildew on 20 shoots per tree. Because mildew is most prevalent on actively growing shoots in the center of the tree, shoots from this part of the tree are included among the 20 terminals.

Brown rot may be rated on the tree at harvest and 1–3 weeks after harvest by counting the number of infected fruit in 200–500 fruit samples selected at random. To have enough ripe fruit, it is necessary to apply two or three sprays of a bird repellent on a 10-day interval. Fruit can also be picked from each replication and held in cold storage for rating at regular intervals.

In 15 years of testing fungicides on tart cherries, we have not had sufficient brown rot blossom blight in any test to evaluate the effectiveness of fungicides for preventing natural infections. To evaluate blossom blight control activity, sprayed blossoms are brought into the laboratory, placed in plastic boxes, and inoculated with a suspension containing 10^5 conidia per milliliter. After 4–5 days at room temperature, the blossoms are rated on a 0–5 scale for discoloration from brown rot infection (5).

As an aid in timing data collection, it is important to gather weather monitoring data from leaf wetness meters, computers, and hygrothermographs along with spore-trapping data. By establishing when infection periods have occurred, the researcher can pinpoint the optimal time for making disease ratings.

It is often necessary to collect fruit samples for residue analysis. Residue samples generally consist of 3–5 lb (1.4–2.3 kg) of fruit collected as a composite from all the replicates. To avoid problems with contamination, unsprayed trees should be sampled before treated trees are sampled. Each sample should be identified as to the treatment, formulation, cooperator, test location, crop, variety, and date harvested. Samples are usually frozen immediately after sampling for later shipment in dry ice to a residue-testing laboratory. Most chemical companies require that a comprehensive information form accompany each sample when it is shipped.

Fruit quality data may be required for fungicides nearing registration. Samples weighing 2–3 lb (0.9–1.4 kg) are collected randomly around the perimeter and from the center of each tree. Fruit size is estimated by determining the number of cherries per pound of fruit or by determining the weight of 100 cherries. The percentage of soluble solids in the juice is determined with a refractometer. Usually, juice from each of 25 cherries per replication is tested. The suitability of the fruit for pitting can be determined with a simple hand-operated cherry pitter. When possible, fruit quality should be measured in cooperation with a food science department equipped to determine the processing characteristics of the fruit in accordance with commercial standards. These tests may include canning and freezing the fruit in small lots and determining consumer acceptability through the use of taste panels.

Data Analysis

When data collection is complete, the results should be summarized and then analyzed with a suitable statistical method. Disease control data are frequently compiled as percentages, with each treatment compared with the others. Duncan's multiple range test or a similar range test should be used to detect significant differences among treatment means. A statistical analysis is essential so that both the researcher and the reader may judge the validity of the results.

Phytotoxicity

Trees in the test plot should be examined visually on a regular basis for evidence of phytotoxicity from the fungicide treatments. Chemical damage to the foliage may appear as necrotic spots or a general yellowing of the leaves with or without necrotic areas. Fruit injuries may appear as circular spots at the points where fruit touch or as rings on the bottom where spray materials have accumulated.

Factors contributing to spray injury can often be assessed by noting the location and extent of the injury within the tree, by reviewing the weather conditions during and after application, and by checking the calibration of the spray equipment. Injury may be localized to fruit or foliage in slow-drying locations in the interior of trees or may be generally distributed throughout the tree. In any case, unsprayed check trees must be observed to determine whether the problem is due to the fungicide or to some other factor.

Rating phytotoxicity can be difficult, particularly if the symptoms are not distributed uniformly. In general, visual ratings are satisfactory for estimating the extent of the problem. Ratings may be expressed as a percentage or on a 0–5 or 0–10 scale. Sometimes a rating system similar to the one described earlier for leaf spot is appropriate. Where factors such as high temperatures or chemical concentration are believed to contribute to phytotoxicity to fruit, the role of these factors can be studied in growth chamber experiments (4).

With the increasing popularity of sterol-inhibiting fungicides, researchers must be prepared to identify and evaluate new types of phytotoxicity. Some members of this family of fungicides have both fungicidal and growth regulator-type activities. On other tree fruit crops the sterol inhibitors at high rates have shortened internodes, thickened leaves, reduced leaf size, and darkened leaf color. Fruit shape, size, and finish have also been influenced at certain rates of application. As a result, traditional ways of evaluating phytotoxicity may need to be revised to detect and evaluate subtle physiologic changes in the growth of tart cherry trees.

LITERATURE CITED

1. Ehret, G. R., Zehr, C. L., and Jones, A. L. 1981. A field-based handling and storage facility for experimental pesticides. Plant Dis. 65:959–961.
2. Eisensmith, S. P., and Jones, A. L. 1981. A model for detecting infection periods of *Coccomyces hiemalis* on sour cherry. Phytopathology 71:728–732.
3. Eisensmith, S. P., and Jones, A. L. 1981. Infection model for timing fungicide applications to control cherry leaf spot. Plant Dis. 65:955–958.
4. Jones, A. L. 1973. Phytotoxicities of dodine and azinphosmethyl to cherry fruit. Plant Dis. Rep. 57:428–431.
5. Jones, A. L. 1975. Control of brown rot of cherry with a new hydantoin fungicide and with selected fungicide mixtures. Plant Dis. Rep. 59:127–130.
6. Jones, A. L., Comstock, R., and Switalski, D. 1977. Control of tolerant and sensitive leaf spot in 1976. Fungic. Nematic. Tests 32:38.
7. Jones, A. L., and Ehret, G. R. 1976. Isolation and characterization of benomyl-tolerant strains of *Monilinia fructicola*. Plant Dis. Rep. 60:765–769.
8. Jones, A. L., and Ehret, G. R. 1980. Resistance of *Coccomyces hiemalis* to benzimidazole fungicides. Plant Dis. 64:767–769.
9. Jones, A. L., Ehret, G. R., and Comstock, R. E. 1983. Evaluation of a protective Difolatan schedule and four after-infection schedules. Fungic. Nematic. Tests 38:114.
10. Jones, A. L., Fisher, P. D., Seem, R. C., Kroon, J. C., and Van DeMotter, P. J. 1984. Development and commercialization of an in-field microcomputer delivery system for weather-driven predictive models. Plant Dis. 68:458–463.
11. Jones, A. L., Lillevik, S. L., Fisher, P. D., and Stebbins, T. C. 1980. A microcomputer-based instrument to predict primary apple scab infection periods. Plant Dis. 64:69–72.
12. Jones, A. L., and Sutton, T. B. 1984. Diseases of tree fruits. North Central Region Ext. Bull. 45. 60 pp.
13. Lewis, F. H. 1978. Methods of testing fungicides on tart cherries (*Prunus cerasus* L.). Pages 51–53 in: Methods for Evaluating Plant Fungicides, Nematicides, and Bactericides. Am. Phytopathol. Soc., St. Paul, MN.
14. Nevill, J. R., Szkolnik, M., Gilpatrick, J. D., and Ogawa, J. M. 1978. Mass production of conidia of brown rot fungi on canned fruit pieces. Plant Dis. Rep. 62:966–969.
15. Small, C. G. 1978. A moisture-activated electronic instrument for use in field studies of plant diseases. Plant Dis. Rep. 62:1039–1043.

Field Evaluation of Fungicides for Control of Peach Leaf Curl

I. C. MacSWAN, Department of Botany and Plant Pathology, and H. L. DOOLEY, U.S. Department of Agriculture Horticultural Crops Research Lab, Oregon State University, Corvallis 97331

Peach leaf curl is caused by the fungus *Taphrina deformans* (Berk.) Tul. This organism attacks most if not all cultivars of peach (*Prunus persica* (L.) Batsch) and many cultivars of nectarine (*P. persica* (L.) *nucipersica* (Suckow) C. K. Schneid.) (7). In our tests, we used the cultivars Rochester, Elberta, and Early Improved Elberta, because they are susceptible under the conditions present in the Willamette Valley, Oregon.

The exact relationship between climatic conditions and infection and disease development is still in doubt. Most researchers (1–3,5,6), however, agree that infection and disease development progress best in wet, cool weather when the buds begin to swell. Most areas of the country receiving 20 in. (50.8 cm) or more of rain have more disease than do areas with less rainfall and warmer temperatures.

Peach leaf curl is a short-cycle disease. The spores infect the leaves in swelling buds. Mycelium grows within a leaf and produces ascospores, which are discharged before the infected leaf drops (1–3,5). The ascospores lodge on the bark and twigs, where they may germinate to form bud conidia several times before new bud swelling and infection occur.

Weather plays an important role. Research indicates that rain and dew may wash ascospores and bud conidia onto the swollen leaf buds to cause infection and complete the life cycle (1–3,5). Cool weather slows leaf development and thus allows more time for infection (1–3). During warm weather, the infection pegs may die or the leaf may mature so rapidly that the fungus is outgrown and no disease occurs.

A single application of fungicide can control peach leaf curl (1,4,5). The fungicide must be present throughout the period of bud swelling, opening, and leaf elongation. Although a single spray application has been proven to control the disease (1,2,5), two applications are recommended to assure acceptable control in the Willamette Valley, where the disease can be extremely severe. Applications in that area should be made in mid-December and early January.

The objective of this method is to evaluate fungicides used to control peach leaf curl caused by *T. deformans*. The test should be conducted at a site suitable for peach growth, where the disease is endemic.

PROCEDURE

A randomized block or other valid statistical design should be used, with a minimum of four single-tree replications per treatment. Treatments should be arranged according to Snedecor's table of randomly assorted digits (8) or its equivalent.

Spray applications may be made with a handgun sprayer, at pressures of 300–350 psi (2,068–2,413 kPa). A backpack duster for single-tree applications has been used successfully. If speed sprayers or dusters are used, however, blocks of trees should be treated, rather than single trees, to reduce drift to adjacent plots. To minimize drift, spray or dust applications should be made when there is little or no wind.

Treatments, including an unsprayed check and a treatment with a standard fungicide, should consist of a minimum of four single-tree replicates. In areas where the disease is severe, use of a less effective fungicide (e.g., Bordeaux mixture 8-8-100) on the check trees may be preferable to leaving them untreated. Untreated trees are used to determine the severity of disease. A standard fungicide treatment, such as Bordeaux mixture 12-12-100 (5.44 kg, 5.44 kg, 379 L), is used to determine whether the timing of application was adequate for control. The performance of test fungicides may be compared with that of the standard fungicide, or disease incidence in the treated trees may be compared with that in the untreated control trees; in some cases, both comparisons may be made.

DATA COLLECTION

The number of diseased and healthy leaf terminals should be recorded after the leaves have emerged and unfolded in early spring but before defoliation. A leaf terminal consists of the rosette of leaves located at the tip of a branch. The mean percentage of diseased leaf terminals should be calculated, and the total number of leaf terminals that were evaluated should be reported. For dwarf trees, a minimum of 50 leaf terminals per tree should be observed; for full-size trees, 100 leaf terminals per tree. The severity of any phytotoxicity as well as a description of injury should be reported. Phytotoxicity may appear as wood or twig burning.

RESULTS

Although 95–98% disease control is possible with an effective fungicide and proper timing, a product being evaluated is considered effective if at least 85–90% disease control is obtained, under the conditions present in the Willamette Valley.

REPORTING

The following information is helpful if reported: 1) the name of the product and its EPA registration number or experimental permit number, 2) the active ingredient or ingredients and the percentage of each, 3) the application rate, 4) the method of application, 5) the number of trees per treatment, 6) the number of replications, 7) the application date or dates, 8) the total number of leaf terminals evaluated per treatment, 9) the percentage of diseased leaf terminals, 10) the date of bud swelling, and 11) the percentage and type of phytotoxicity, if it occurs.

LITERATURE CITED

1. Anderson, H. W. 1956. Diseases of Fruit Crops. McGraw-Hill Book Co., Inc., New York.
2. Fitzpatrick, R. E. 1934. The life history and parasitism of *Taphrina deformans*. Sci. Agric. 14:305-306.
3. Fitzpatrick, R. E. 1935. Further studies on the parasitism of *Taphrina deformans*. Sci. Agric. 15:341-344.
4. MacSwan, I. C. 1969. Peach: Leaf curl (*Taphrina deformans*). Fungic. Nematic. Tests 25:48.
5. Mix, A. J. 1935. The life history of *Taphrina deformans*. Phytopathology 25:41-66.
6. Mix, A. J. 1936. The genus *Taphrina*. I. An annotated bibliography. Kans. Agric. Exp. Stn. Bull. 34:113-149.
7. Ritchie, D. F., and Werner, D. J. 1981. Susceptibility and inheritance of susceptibility to peach leaf curl in peach and nectarine cultivars. Plant Dis. 65:731-734.
8. Snedecor, G. W., and Cochran, W. G. 1967. Statistical Methods, 6th ed. Iowa State University Press, Ames.

Testing Fungicides and Bactericides for Control of Foliar and Fruit Pathogens of Peach in the Eastern United States

ELDON I. ZEHR, Department of Plant Pathology and Physiology, Clemson University, Clemson, SC 29631; DAVID F. RITCHIE, Department of Plant Pathology, North Carolina State University, Raleigh 27650; and CHARLES R. DRAKE, Department of Plant Pathology, Physiology and Weed Science, Virginia Polytechnic Institute and State University, Blacksburg 24061

In the early stages of development of fungicides and bactericides for control of peach (*Prunus persica* (L.) Batsch) diseases, field tests are conducted in small plots at industry and university experiment station research farms. The methods and equipment used at various locations often have developed independently and may differ considerably. Although we do not argue here for standardized or uniform procedures, reviewing published information can be difficult because of the wide variation in methods used. Data from fruit pesticide tests at research stations would be easier to use and understand if investigators used similar techniques and procedures. Our purpose here is to discuss certain principles of fungicide testing on peaches in the eastern United States and to suggest some criteria that we have found to be important for reliable evaluations of fungicide performance.

Pathogens of importance and their behavior in the climate of the United States west of the Rocky Mountains sometimes differ significantly from those in the East. Readers who wish to use this chapter as a guide for fungicide testing in the West should be aware of these important differences and should make appropriate adjustments for them. The discussions by Ogawa, Manji, and MacSwan (*this volume*) and MacSwan and Dooley (*this volume*) should be considered carefully. Previous discussions by Ogawa et al (10), Dooley and MacSwan (2), and Lewis (7), as well as reports published annually by the New Fungicide and Nematicide Data Committee of The American Phytopathological Society in *Fungicide and Nematicide Tests*, are also helpful.

DISEASES OF IMPORTANCE IN THE EAST

Leaf Curl

Peach leaf curl, caused by *Taphrina deformans* (Berk.) Tul., develops early in the season, and fungicides to control it usually are applied at times when tissues susceptible to other diseases have not yet developed. Because this pathogen behaves similarly in the East and the West, procedures described by MacSwan and Dooley (*this volume*) apply in both areas. Therefore, details of procedures for studying peach leaf curl are not discussed here.

Blossom Blight and Brown Rot

Two fungi cause blossom blight and brown rot in the United States. *Monilinia laxa* (Aderh. & Ruhl.) Honey causes significant disease losses in the western United States but is uncommon east of the Rocky Mountains. Procedures for testing fungicides for control of *M. laxa* are described by Ogawa, Manji, and MacSwan (*this volume*). *M. fructicola* (Wint.) Honey is important in all parts of the United States where stone fruits are grown except in some semiarid climates. Brown rot has the potential to cause such significant crop losses of peaches in the eastern United States that control of *M. fructicola* is of primary interest for testing new fungicides.

Peaches are especially susceptible to *M. fructicola* during the flowering period, 2–3 weeks before harvest, and postharvest. Immature peach fruit generally are resistant to infection, but injured fruit, those that fail to develop, and those removed during thinning may become infected. Latent or incipient infections during the period of maturation have been reported in some areas. Although the flowering and preharvest periods are of particular interest for brown rot control, it is desirable that effective fungicides also be registered for use during the intervening period to suppress sporulation on infected flowers, control infection of abscised fruits on the ground, and suppress incipient infections if necessary.

Scab

Peach scab caused by *Cladosporium carpophilum* Thuem. is especially important beginning with abscission of the calyx and for 6–8 weeks thereafter. Some infections probably occur after that time if inoculum continues to be available. Fungicide sprays for scab control usually begin at "shuck-split" (a colloquial term that refers to the splitting of the calyx by the developing fruit) and continue for a period that varies in length according to time of fruit maturity and residual activity of the fungicide. Some fungicides that are applied during bloom may affect scab development if the chemical has both good residual activity and efficacy for scab control.

Bacterial Spot

The bacterial spot disease caused by *Xanthomonas campestris* pv. *pruni* (Smith) Dye affects leaves and fruit of susceptible cultivars of peach. Fungicides in use for control of fungal diseases on peach have little effect on bacterial spot. Bactericides for suppression of bacterial spot usually are tested independently of fungicide tests. Applications for effective control usually begin at petal-fall and continue until a few weeks before harvest. The availability of moisture strongly influences infection. Spray intervals and length of seasonal applications depend on characteristics of the bactericides tested and weather conditions during the period of susceptibility.

Other Diseases

Some additional diseases of local or regional importance may be of interest in some fungicide tests. Rust, caused by *Tranzschelia discolor* (Fckl.) Tranz. & Litv., is important in states bordering the Gulf of Mexico and occasionally appears in areas further north. Powdery mildew, perhaps caused by *Sphaerotheca pannosa* (Wallr.:Fr.) Lév. or by fungi of other genera, and rusty spot (a related problem of uncertain etiology) appear sporadically in the mid-Atlantic region. Anthracnose, caused by *Glomerella cingulata* (Ston.) Spauld. & Schrenk, is an important problem in the Piedmont region of the Carolinas. Peach tree gummosis, caused by certain strains of *Botryosphaeria dothidea* (Moug. ex Fr.) Ces. & de Not. and perhaps other *Botryosphaeria* spp., is spreading in the South and may require more fungicide tests for control in the near future. Perennial canker caused by *Cytospora* spp. is of widespread importance and may be amenable to fungicidal control. Investigators should be alert for the appearance of these diseases in test plots and be careful to collect information on fungicide efficacy when possible. Such information is often scarce or absent in published literature.

PLANNING THE ORCHARD

Planning an orchard is one of the most important tasks for long-term success of fungicide testing. Factors that should be considered include precautions to avoid spring frost damage, selection of soils that do not accommodate root disease problems that sometimes decimate peach orchards, type of fungicide application equipment to be used, diseases to be tested, sources of inoculum, numbers of replications, number of trees per replicate, and criteria to be evaluated (e.g., whether yield is to be measured as well as efficacy).

Potential for Frost Damage

In most of the eastern United States, spring frosts during or after bloom are often responsible for partial or total loss of the fruit crop. Such occurrences are very damaging to fungicide testing programs. Careful site selection can do much to alleviate potential frost or freeze damage, although the risk rarely can be eliminated.

When selecting a site, obtain the spring minimum temperature records for several consecutive years, if possible. If the records indicate that frequent frost damage is likely, the site should be avoided. If temperature records cannot be obtained easily, experiences of fruit growers in the area can be helpful. Sites always should have good air drainage, which helps to avoid the accumulation of stagnant, cold air in the orchard that may cause cold injury. Sites near large bodies of water are especially desirable because of the moderating influence of the water on air temperature during periods of cold weather. When establishing test orchards, it is always desirable to plant frost-sensitive cultivars (e.g., Sunhigh or Loring) at or near the crest of the hill and the most frost-tolerant cultivars (e.g., Cresthaven or Redhaven) at the base of the slope.

Overhead irrigation and wind machines have been used successfully to avoid frost injury to peach flowers and fruit. If a frost-prone site must be used, installation of frost-protection equipment before planting may be considered. This equipment is expensive, and use of overhead irrigation may saturate the soil with water. In addition, limb breakage may be severe if the trees are not trained or pruned properly, or if too much water is applied. Provisions must be made for excellent soil drainage. If fungicides for blossom diseases are to be studied, the effects of irrigation on fungicide protection also must be considered.

Fruit buds may be killed during the dormant season during rapid drops in temperature to −23°C or lower on sites with excessive wind exposure. A rapid drop in temperature is usually accompanied by strong winds that aggravate the freeze injury. Wind machines and other precautionary measures do not prevent the killing of dormant buds. If the site under consideration has a history of freezing injury for 2 out of 5 years, it is too risky to use, and a more favorable site should be chosen.

Soil

Well-drained soil is essential for long-term survival of peach orchards. Soils that remain saturated with water for more than 24 hr usually result in injury to peach roots. Root rots or crown rots often appear in such soil. Sandy soils, however, may be prone to nematode problems. For long-term survival of trees, some provisions for nematode control must be considered in sandy soils where nematodes thrive.

Another consideration is the cropping history of the site. If fruit trees or forest trees have grown in the site, researchers must be alert for potential root problems induced by *Clitocybe tabescens* or *Armillaria mellea*. So common are these problems in old forest sites in the Southeast that fruit growers have named them "oak root rot."

The best orchard sites are those on sloping land with good internal soil drainage and sufficient loam or clay content to discourage nematode activity. The potential for root rots is reduced if the site has been in grassland or row crops for many years before orchard establishment.

Application Equipment

The type of fungicide application equipment has an important influence on orchard layout and design. Widely spaced, single-tree replicates are satisfactory for most application equipment, but commercial air-blast sprayers are an important exception. There are good reasons for using air-blast sprayers. They are the type of equipment used in most orchards to apply fungicides. When the machines are carefully calibrated and operated, fungicide deposit is uniform and hazards of accidental exposure to experimental materials may be reduced. Fungicide drift to nearby trees is a major problem that prohibits the use of air-blast sprayers in many test sites. Using air-blast sprayers usually requires the establishment of large blocks of trees with spacing similar to that in commercial orchards. To compensate for fungicide drift, each plot usually must be several rows wide and six to eight trees long. Data on fungicide performance must be obtained from trees near the center of each plot.

Because large blocks of trees are expensive to maintain and harvest, most investigators use smaller application equipment that permits small orchards to be maintained for experimental use. With a handgun or similar application equipment, fungicide drift can be minimized or eliminated by spacing trees somewhat farther apart than in commercial orchards and by pruning the trees to maintain relatively small size. Unless the location is such that drift is not a problem, spacing at 25–30 ft (7.6–9.1 m) is desirable to avoid drift and facilitate movement of equipment through the orchard.

Researchers might also consider separating each replicate by an unsprayed tree or trees or by trees that receive an effective fungicide treatment, depending on the objectives of the test (14). Treatments each year could be shifted among the available trees so that tests are not conducted on a given tree two years in succession. This approach must be used with caution. Large numbers of unsprayed trees in a test orchard may overwhelm effective fungicides with large numbers of available spores (14). On the other hand, spraying the trees with an effective fungicide may be inconvenient, and drift of the fungicide to test trees may be a problem.

Diseases and Inoculum

Diseases to be tested and sources of the pathogens involved should influence orchard establishment. For some diseases (e.g., leaf curl and scab), the source of inoculum often is in the tree to be sprayed or those nearby. For others, such as brown rot, the source may be some distance from the test trees.

If the pathogen overwinters or resides in the tree to be sprayed, the amount of inoculum available may depend considerably on the amount of disease in the tree and the fungicides applied to the tree the preceding growing season. For such diseases, it is wise to use alternate years for fungicide tests to encourage both uniformity of inoculum and sufficient inoculum for a satisfactory test. During intervening seasons when it is not being used for testing, the orchard may receive only minimal fungicide applications, or it may be sprayed with fungicides that have little effectiveness for the pathogens involved.

When the test trees are not important sources of inoculum, as may be the case for *M. fructicola*, the orchard may be placed near sources of inoculum such as wild plums (*Prunus angustifolia* or other *Prunus*), or unsprayed peaches, plums, or other hosts may be planted nearby. Such sources of inoculum help to encourage blossom blight development in test orchards and provide a natural environment that resembles many commercial peach orchards. This procedure is especially useful for blossom blight because peach flowers are difficult to inoculate successfully in the field.

Replication and Criteria for Evaluation

The number and size of replicates are important factors for determining the size of the test orchard. Number and size of replicates required depend on the diseases to be tested, criteria used for evaluation, and method of application.

For small plots that are to be sprayed with a handgun, a single tree is often sufficient for each replication, but many investigators prefer to use more than one tree per replicate. Obviously, many more trees per replicate will be needed if aircraft application is to be used. The number of trees required per replicate usually depends on the number needed to achieve uniform, accurate fungicide deposition with the application equipment being used.

The number of replicates to be used also is variable. If the inoculum is uniformly distributed in the test plot, and efficacy, phytotoxicity, and fruit quality are the criteria to be evaluated, four replicates often are sufficient. More replicates will be needed when inoculum is not uniform or when yield is an important criterion. Young trees often produce relatively uniform fruit yield, but variability increases with age. Peach trees are prone to limb breakage, nonuniform ripening, and uneven fruit loads when frosts reduce the crop, and they are difficult to prune and thin uniformly. Each of these factors affects fruit yield; consequently, quantity of fruit per tree becomes increasingly variable as trees age. However, these variables become less important if more than one tree per replicate is used. If yield data are to be collected, 8–10 single-tree replicates probably are needed, whereas 4–6 may be sufficient with two or more trees per replicate. Therefore, the kind of data to be collected is important for both the size of the orchard and tree spacing in the orchard.

One design that has considerable utility and flexibility is to plant four to six closely spaced trees (3–4 m apart) in the row and separate these trees and adjacent rows by enough open space to avoid fungicide drift problems. These closely spaced trees may constitute one replicate of one treatment. All trees probably should be of the same cultivar to facilitate pest control practices that might become difficult for different cultivars that ripen at different times. This design allows air-blast as well as handgun application as needed. Fruit yield may be obtained if desired, and one tree in the group may be left unsprayed as a source of inoculum for the test, depending on objectives.

Careful planning before trees are planted is very important for long-term success. Careful consideration of the overall objectives and procedures improves versatility and suitability to accommodate unforeseen needs that usually develop as time progresses. Because peach trees are expensive to plant and maintain, experimental orchards often must be of limited size. When funds are limiting, a priority should be assigned to each objective and the orchard should be designed to accommodate the more urgent needs.

ORCHARD MAINTENANCE

Routine orchard maintenance (fertilization, nematode control, weed control, etc.) may consist of practices common to commercial orchards in the region. Insecticides may be mixed with test fungicides, especially when compatibility information is needed. However, most insecticides are hazardous to use, and mixing an insecticide with an experimental fungicide and applying it to trees increase the risk of accidental exposure. For this reason many researchers apply insecticides separately to the entire orchard, using equipment that is less risky from the standpoint of accidental exposure.

Because the application of experimental fungicides requires much traffic in the orchard, grasses or natural vegetation usually is permitted to grow between rows of trees, and herbicides are used to control vegetation under trees. Undisturbed soil with leaf litter under the trees encourages the development of apothecia from mummies on the ground (1; E. I. Zehr, *personal observation*). This practice may help to increase the inoculum for *M. fructicola* in springtime.

If control of anthracnose is an objective of the fungicide test program, then hosts of *G. cingulata*, such as lupine, vetch, or clover (11) may be used as cover crops in the orchard. Legumes, however, should be used cautiously to avoid excessive nitrogen availability at inappropriate times that may affect fruit quality.

PLOT DESIGN

Fruit orchards are established with the assumption that they can be used for experimental purposes over and over again. As discussed previously, the orchard layout should be sufficiently flexible to accommodate experiments of different purposes, replications, and numbers of treatments.

Because few of the diseases that are studied in foliar fungicide tests are greatly affected by soil-related factors, a randomized complete block design often functions well for randomly assigning treatments in the test orchard. If soil-applied fungicides are to be tested, or if bacterial spot is one of the diseases to be studied, soil-related factors, such as internal soil drainage, may become important conditioning factors for fungicide or bactericide performance. Other designs, such as the Latin square, should be considered in these instances. Researchers should consult the excellent discussion of this topic by Nelson (8). If sources of inoculum for aerially disseminated pathogens exist or have been established near the test orchard, an appropriate plot design to accommodate varying distances from the inoculum source should be chosen.

When choosing the number of replications or deciding on an appropriate experimental design, researchers should seek the advice of qualified statisticians. Their assistance can be very helpful in planning experiments to achieve the best possible results.

APPLICATION OF TEST MATERIAL

Careful application of test materials is absolutely essential for the success of any fungicide test. The most carefully planned experiments may be of no value if test materials are not applied properly.

Uniform application is one of the most important requirements. To accomplish uniform application, many researchers use the so-called "dilute" fungicide spray in which the spray is applied to the point of runoff, usually with a handgun or wand applicator. The primary advantage of this method is that it avoids excessive fungicide deposit on all portions of the tree, because excess material runs off of the sprayed surface. Careful application so that all surfaces become wet results in a reasonably uniform deposit of fungicide over the entire tree.

The dilute method of application has two important disadvantages. Sprayers in most commercial orchards now use only a fraction of the volumes of water required for runoff to occur. Because the dilute method thus differs from commercial practice, certain limitations apply when results of experiments are extrapolated to commercial orchard situations. A second disadvantage is the difficulty of establishing the optimum rate of fungicide to apply in commercial orchards. Pesticide rates on fungicide labels usually are given as rate per acre rather than as rate per given volume of water carrier. When trees are sprayed to the point of runoff, much less spray material is needed early in the season than later when the trees are in full leaf. In addition, considerable fungicide is wasted when runoff occurs. Rates per acre, therefore, often must be established on the basis of the experimenter's estimation of the amount of material required to achieve acceptable disease control in orchards that differ greatly in tree size, spacing, and amount of surface area to be protected.

Orchard equipment that more closely imitates commercial practice is available and is being used in many test orchards. This equipment is useful for fungicide testing, but it requires careful attention to ensure that fungicide deposits are uniformly applied on all surfaces. Sometimes trees must be pruned more carefully to permit easy passage of the spray through the tree. Wind velocity during application also affects uniform deposit considerably. Often the equipment is expensive and may require a larger land area or more trees than hand-held spray guns that are used for dilute sprays. If financial resources are small, costs of application and orchard maintenance may be limiting factors.

Changing temperature or moisture conditions sometimes influence fungicide performance. Because various fungicide treatments in a given experiment rarely can be applied simultaneously, the size of experiments usually should be limited to the number of treatments that can be applied within a few hours to avoid large differences in temperature and moisture conditions between the first and last treatments applied. Some researchers have designed equipment that enables rapid application of multiple treatments (7). Others use small-plot sprayers or commercial orchard equipment for application (2,6,10,13,15). The kinds of equipment used are not critical to the outcome of experiments if the equipment delivers uniform application of the fungicide in fine droplets that adhere to the plant surface. However, reports should record details of the equipment used and the conditions under which fungicides were applied so that results can be compared with those reported by other investigators.

INOCULATION

Availability of effective inoculum is essential for satisfactory completion of a test of fungicide efficacy. In general, when possible a natural source of inoculum is preferred to artificial inoculation because it usually gives a more realistic test. If artificial inoculation is used, best results may be obtained if the inoculation imitates a natural setting, such as placing diseased plant material in an orchard or in individual trees.

For peach leaf curl and scab, sprays that were applied during the previous season may determine whether sufficient inoculum is available for a satisfactory test. Application of captan, sulfur, or benomyl during the growing season controls scab infections of twigs as well as fruit and reduces the amount of inoculum for the following season. Captan also may influence leaf curl development during the following spring (9); therefore, tests for leaf curl in orchards where captan was used may not have enough disease development. Inoculum increases may be encouraged by omitting fungicide applications the preceding season or by spraying with fungicides that are ineffective for scab or leaf curl and by using cultivars that are especially susceptible to leaf curl (e.g., Redskin or Loring).

Effective inoculum for *M. fructicola* or *X. campestris* pv. *pruni* is less easily encouraged in test orchards, because favorable weather conditions are very important for successful inoculation. Blossom blight caused by *M. fructicola* is sporadic in most locations and is probably best encouraged by maintaining unsprayed *Prunus* spp. nearby. Overhead sprinklers may be necessary to provide sufficient periods of wetting when dry weather prevails during flowering time. Artificial inoculation of peach flowers with *M. fructicola* is difficult and not very reliable, but inoculation with *X. campestris* pv. *pruni* can be effective. Sprays of inoculum under high pressure provide microscopic wounds that encourage successful inoculation. Overhead irrigation after inoculation, if available, may help to increase the number of infections.

When inoculum levels of *X. campestris* pv. *pruni* are low or not uniformly distributed, bactericide may be applied two or three times starting at shuck-split. Two to 3 weeks later, the trees are inoculated with *X. campestris* pv. *pruni*, and the bactericide spray schedule is continued. The bacteria are grown in liquid culture (0.8% Difco nutrient broth, 0.5% yeast extract, and 0.5% dextrose) for 24 hr. One liter of broth culture is added to 50 L of water in the sprayer, and a stream of high-pressure spray at 2,070–2,415 kPa (300–350 psi) is applied across the upper third of the trees. Efficacy is evaluated at and/or after harvest for infections that occur below the area of inoculation. By this method, bactericides for preventing the spread of *X. campestris* pv. *pruni* can be evaluated in a manner that simulates natural conditions.

Zehr (15) routinely inoculates two fruits on each tree with a culture of *M. fructicola* about 10–14 days before harvest. Sporulation on the inoculated fruits begins approximately 3 days later, and secondary spread of conidia provides a relatively uniform supply of inoculum in the experimental orchard. This method simulates conditions that develop in commercial peach orchards and is reliable if rain falls between the time of inoculation and harvest. It has not been effective when the intervening period has been dry, even when overhead irrigation is supplied. Ogawa et al (10) suggest applying conidia to fruit with an air-blast sprayer or overhead sprinklers.

If weather conditions have been dry and brown rot has not developed in the test plot, it is still possible to obtain efficacy data by using postharvest inoculation. Preharvest fungicides are applied in the field, and a representative sample of 25–50 firm-ripe peaches is harvested for each replicate. Overripe fruit should be avoided because secondary microorganisms cause rapid decay. The fruit are placed by replicate on a sheet of 1- to 2-mil plastic in a shaded location, such as the floor of a packhouse. Care should be taken to place the stem end of the fruit down to avoid possible infection through the point at which the stem was attached. Using a 1- to 2-L hand sprayer, a low concentration of *M. fructicola* conidia (25,000/ml) grown on inoculated fruit is misted over the fruit surface. The fruit are then covered with a second sheet of clear plastic. Numbers of infected fruit are recorded after 5–7 days. This method provides a means of obtaining efficacy data when brown rot does not develop in the orchard, assures that all treatments receive uniform inoculum, and provides a way to determine residual fungicide activity by harvesting and inoculating fruit at intervals after the last fungicide application. A disadvantage of the method is that it fails to detect efficacy of different seasonal spray schedules and does not detect eradicative activity of a fungicide.

Sometimes cultivar selection is important to encourage disease development in test plots. Peach cultivars with large, showy flowers (e.g., Loring, Redskin, and Redglobe) often are more susceptible to blossom blight than other cultivars. Sunhigh, Suwanee, Blake, certain clingstone, and other cultivars are very susceptible to bacterial spot. Rio-Oso-Gem is especially susceptible to anthracnose because the suture of the fruit is not entirely closed, and infections often develop on the suture.

For some diseases that occur sporadically (e.g., powdery mildew and rust), little can be done to encourage disease development. These diseases usually do not cause serious economic losses, but it is desirable to know the relative effectiveness of fungicides for controlling them (4). Up to 74% fruit infection has been observed in some orchards in Virginia (3). Investigators should be alert for the appearance of minor diseases in test plots so that data on fungicide control can be obtained.

Time of harvest may have important influences on the appearance of fruit decay. Brown rot usually becomes well established by the time peaches are in the firm-ripe stage, but Rhizopus decay begins after peaches have begun to soften. When data on control of Rhizopus decay are needed, it is helpful to permit the earliest maturing fruit to ripen on the tree or fall to the ground and become infected with *Rhizopus* spp. before fruit are collected for storage.

EVALUATION OF FUNGICIDE AND BACTERICIDE PERFORMANCE

Blossom Blight

Evaluating the efficacy of fungicides for blossom blight control sometimes is subjective. When blossom blight is severe, the percentage of infected flowers on 10–20 randomly selected twigs per tree may be used to estimate the percentage of infection. More commonly, only a small percentage of flowers become infected. These relatively small numbers may be important because twig infections that subsequently develop may cause death of the twigs and subsequent loss of many more flowers. Furthermore, many conidia are produced on infected flowers and twigs.

If tree growth has been controlled by pruning, counting all the infected flowers or twigs on each tree is feasible. This procedure becomes easier as twig lesions become more developed. Blighted twigs are easily observed, and twig lesions characteristically exude gum, which allows them to be easily identified. However, the counting must be done before the trees are in full foliage because infections then become difficult to observe.

Ritchie and Bennett (13) reported that certain fungicides may affect the amount of sporulation on infected peach blossoms and twig lesions. This attribute may be very important for reducing the number of spores available for later infections in the orchard. Suppression of sporulation may be tested in the greenhouse using detached flowering twigs that are first sprayed with a spore suspension of *M. fructicola* and later sprayed with the test fungicide. Washings of conidia from the infected flowers may be collected and studied for effects of fungicides on numbers of conidia, percentage germination, and infectivity. Similar collections may be made from infected flowers in the field, but numbers of conidia may be influenced by weather conditions in the field as well as by the fungicide applied.

Scab

Although *C. carpophilum* initiates infections early in fruit development, symptoms do not appear before pit-hardening. On immature fruit, the small, olivaceous lesions may be confused with superficial spots resulting from other causes. Scab lesions are most easily identified and counted when fruit are mature.

Numbers of fruit to be examined may vary depending on the purposes of the experiment. Numbers of lesions per infected fruit sometimes may be evaluated numerically or by the Horsfall-Barratt system (5). Usually, the number of lesions per fruit is ignored because almost complete disease control can be achieved with several fungicides in commercial use. Many investigators, therefore, count only infected fruit, and one well-defined scab lesion is sufficient to count the fruit as infected. For most experimental work, examination of 50–100 mature fruits picked at random per replicate is sufficient for evaluation.

Bacterial Spot

X. campestris pv. *pruni* infects both leaves and fruit of susceptible peach cultivars. Infected leaves eventually abscise, and defoliation resulting from bacterial spot can be evaluated at the time of fruit maturity or sometimes before. Percentage defoliation can be estimated by determining the percentage defoliation on 10–20 randomly selected shoots per tree.

Infection of fruit is more difficult to study. A few lesions per fruit do not detract substantially from fruit quality, but often the number of lesions per fruit is high. Because no commercial bactericide is very effective for controlling bacterial spot, an experimental chemical that reduces the number of lesions per fruit has potential for commercial use. Investigators often use an evaluation procedure that lists both numbers of fruits infected and an index based on the number of lesions per fruit (6,12).

One of us (Ritchie) uses the following procedure for evaluating bacterial spot infection. For foliage infection, the percentage defoliation is estimated, and then the severity of infection on the foliage remaining on the tree is estimated on a scale of 0 to 5, where 0 = no symptoms observed, 1 = trace (a few infected leaves observed), 2 = up to 5% of foliage exhibiting symptoms, 3 = 6–15% of foliage with symptoms, 4 = 16–40% of foliage affected, and 5 = more than 40% of foliage affected. Fruit infection can be evaluated more quantitatively by examining a specific number of randomly selected fruit per replication. Thus, a percentage of fruit infected can be determined, and a rating scale similar to that used to estimate foliage infection can be used to estimate the area of the fruit surface affected.

Fruit Decay

Control of fruit decay at maturity is one of the most important evaluations in most fungicide tests. Ripening fruits also are especially sensitive to chemical injury, and researchers should be particularly alert for unusual ripening patterns, surface injury, or effects on the color or shape of fruit at that time.

Ripening peach fruit are susceptible to infection by several fungi. Infections that begin in the field sometimes do not become visible until after harvest; therefore, fungicide sprays commonly are applied until one or two days before harvest of the fruit. To evaluate the residual effectiveness of such applications, it is customary to store fruit 4–7 days or sometimes longer at temperatures favorable for symptom development. A common procedure is to examine 100 firm-ripe peaches per replicate and record evidence of scab, bacterial spot, and fruit decay. Then 100 decay-free fruits per replicate are packed in suitable containers for postharvest storage. Because fungi may spread rapidly from one fruit to another during storage, fruit should be stored so that they do not touch. Fruit trays often are used for this purpose. The trays of fruit usually are packed in closed, but not airtight, containers for storage so that relative humidity is favorable for diseases.

STATISTICAL ANALYSIS

Data collected in well-designed experiments should always be analyzed statistically—not for the purpose of making ironclad judgments about performance or to satisfy the editors of a journal, but to assist the investigator in interpreting the data. Computers are very useful for this purpose, but the statistical test to be used should be considered very carefully. Qualified statisticians available at many universities and other locations can contribute immensely when determining the appropriate test to use, offer suggestions for improving experimental designs, and help to interpret the results of tests. Because replicates in experiments completed in the natural environment can vary greatly, statistical tests sometimes are essential for determining the relative similarities or differences among treatments.

LITERATURE CITED

1. Byrde, R. J. W., and Willetts, H. J. 1977. The Brown Rot Fungi of Fruit: Their Biology and Control. Pergamon Press, Elmsford, NY. Page 53.
2. Dooley, H. L., and MacSwan, I. C. 1978. Field evaluation of fungicides for control of peach leaf curl. Pages 57–58 in: Methods for Evaluating Plant Fungicides, Nematicides, and Bactericides. E. I. Zehr, ed. Am. Phytopathol. Soc., St. Paul, MN.
3. Drake, C. R. 1970. Powdery mildew of peach in Virginia. Plant Dis. Rep. 54:686-688.
4. Drake, C. R. 1974. Powdery mildew control on peach. Fungic. Nematic. Tests 30:43-44.

5. Horsfall, J. G., and Barratt, R. W. 1945. An improved grading system for measuring plant diseases. (Abstr.) Phytopathology 35:655.
6. Latham, A. J., and Carlton, C. C. 1983. Evaluation of fungicides for control of peach diseases, 1982. Fungic. Nematic. Tests 38:120.
7. Lewis, F. H. 1978. Method of testing fungicides on tart cherries (*Prunus cerasus* L.). Pages 51–53 in: Methods for Evaluating Plant Fungicides, Nematicides, and Bactericides. E. I. Zehr, ed. Am. Phytopathol. Soc., St. Paul, MN.
8. Nelson, L. A. 1978. Use of statistics in planning, data analysis, and interpretation of fungicide and nematicide tests. Pages 2–14 in: Methods for Evaluating Plant Fungicides, Nematicides, and Bactericides. E. I. Zehr, ed. Am. Phytopathol. Soc., St. Paul, MN.
9. Northover, J. 1978. Prevention of peach leaf curl, caused by *Taphrina deformans*, with preharvest and pre-leaf fall fungicide applications. Plant Dis. Rep. 62:706-709.
10. Ogawa, J. M., Manji, B. T., English, H., Sall, M. A., Yates, W. E., Chiarappa, L., and Rough, D. 1978. Laboratory and field test procedures for evaluation of fungicides for control of brown rot diseases of stone fruits. Pages 54–57 in: Methods for Evaluating Plant Fungicides, Nematicides, and Bactericides. E. I. Zehr, ed. Am. Phytopathol. Soc., St. Paul, MN.
11. Petersen, D. H. 1955. Additional hosts of *Glomerella cingulata* in the families of Gramineae and Leguminosae. Plant Dis. Rep. 39:576-577.
12. Powell, D. 1967. Peach: Bacterial spot (*Xanthomonas pruni*). Fungic. Nematic. Tests 22:49.
13. Ritchie, D., and Bennett, M. 1984. Sporulation inhibition on blossom blight cankers and control of fruit brown rot, 1983. Fungic. Nematic. Tests 39:45.
14. Shoemaker, P. B. 1974. Fungicide testing: Some epidemiological and statistical considerations. Fungic. Nematic. Tests 29:1-3.
15. Zehr, E. I. 1979. Tests of new fungicides for control of peach diseases, 1978. Fungic. Nematic. Tests 34:42-43.

Evaluating Spray Materials to Control Fire Blight: Laboratory, Greenhouse, and Field Techniques

STEVEN V. BEER and JOHN L. NORELLI, Department of Plant Pathology, Cornell University, Ithaca, NY 14853 and Geneva, NY 14456

Fire blight of pear (*Pyrus communis* L.) and apple (*Malus pumila* Miller), caused by *Erwinia amylovora* (Burrill) Winslow et al, can lead to significant crop losses in the season in which infection takes place and in subsequent years. The current season's crop is reduced in direct proportion to the percentage of blossoms infected. If fire blight lesions extend into older tree parts, structural damage reduces fruit production for at least several seasons. In severe cases, particularly in young and high-density orchards, trees may be killed or rendered permanently unproductive (2,6,20,40).

Infections that originate in primary blossoms are considered most important, because they reduce yields, may kill major limbs, and provide inoculum for later secondary blossom and shoot infection (10,20,40). Specific measures aimed at reducing blossom blight incidence have been more successful and cost-effective than measures designed to control vegetative shoot infection, and recommendations for controlling the blossom blight phase of the disease with chemical sprays have been emphasized in most areas (2).

Successful development of new spray materials (chemicals or biologicals) to control blossom infection of apple or pear requires strong evidence that the new materials are efficacious under natural orchard conditions. However, testing candidate materials in commercial orchards is expensive and often has been disappointing because of the sporadic occurrence of the disease. Often insufficient or nonuniform infection develops in commercial orchards, and valid conclusions cannot be drawn concerning the efficacy of test materials. Orchard testing is expensive because of tree and crop loss from nonefficacious treatments and because fruit treated with materials not approved by regulatory authorities cannot be used or sold. However, the high costs of orchard testing can be justified if there is evidence from preliminary evaluations that experimental materials are effective in controlling fire blight. Evidence sufficient to justify orchard testing of an experimental material might be based on its previous efficacy in orchards in other geographic areas or on reliable laboratory or greenhouse tests of the candidate materials.

Although the development of chemical bactericides has been emphasized in the past, recent research indicates that biological control techniques have potential for controlling fire blight (12). The testing procedures described in this chapter for evaluating materials for the control of fire blight are generally applicable to both chemical and biological materials. The laboratory and seedling tests to be described have been successful in identifying materials that warrant further evaluation under orchard conditions (35). The research orchard testing procedures have been described by Beer (7) and have given consistent results in 13 of 14 seasons of testing in research orchards in New York State.

Chemical and biological materials for the control of fire blight (or other plant diseases) must be evaluated under conditions consistent with government regulations. The most comprehensive regulations applicable to operations conducted in the United States have been published in the *Federal Register* (36). These regulations are revised periodically, and researchers are advised to keep abreast of the revisions.

In most of the techniques we describe, host plants are artificially inoculated to ensure the development of sufficient disease to evaluate the efficacy of candidate materials. Each test requires the use of specialized plant material, which with sufficient planning can be obtained in all pome fruit-growing areas.

SELECTING, TESTING, AND PREPARING INOCULUM

Pathogen

The success of evaluations of materials for the control of fire blight depends on uniform disease pressure throughout the test plot. This can be accomplished by artificial inoculation of test plant material with *E. amylovora* at an appropriate time. Any pathogenic strain of *E. amylovora* derived from a single cell or single colony to assure genetic purity and reproducibility may be used, provided it is sensitive in vitro to the materials to be tested. The strain also should be sensitive to streptomycin so that the activity of new materials can be compared with this antibiotic, which is the accepted material for fire blight control in many areas (2). In the case of orchard testing, if the strain used for artificial inoculation is sensitive to streptomycin, fire blight infection in adjacent orchards can be controlled by application of the antibiotic. The pathogenicity and sensitivity to streptomycin of the candidate strain must be tested before the strain is used.

Streptomycin sensitivity may be tested by dilution plating the strain on nutrient agar (Difco Laboratories, Detroit, MI) that has been amended with 0, 10, and 100 mg/L of streptomycin (Sigma Chemical Inc., St. Louis, MO) (9), followed by incubation for several days at 25–27°C. Alternatively, "sensitivity disks" containing known amounts of streptomycin (Difco Laboratories,

Detroit, MI; Baltimore Biological Laboratories, Baltimore, MD) may be deposited on freshly prepared lawns of the candidate strains according to the directions of the supplier. A strain of *E. amylovora* (or other bacterium) known to be resistant to 100 mg/L of streptomycin should be included in the test as a check. Such strains have been isolated from orchards in California, Washington, and Oregon (17,32). Only strains of *E. amylovora* that fail to grow on medium containing 10 mg/L of streptomycin sulfate should be used for artificial inoculation.

In certain experiments, it may be desirable to use a strain of *E. amylovora* that is resistant to a specific antibiotic to facilitate its recovery from treated trees. If populations of *E. amylovora* are to be determined at some time after inoculation, the use of an antibiotic-resistant strain has the advantage of allowing the use of simple media, amended with the antibiotic, rather than more complex media selective or differential for the pathogen. In addition, the use of an antibiotic-resistant strain permits differentiation of the strain used as inoculum from other strains of *E. amylovora* that may be present in the test block.

Several antibiotics may be used to mark strains of *E. amylovora*. Researchers should choose one that is not currently used for fire blight control (i.e., *not* streptomycin or oxytetracycline) and one whose resistance is encoded by chromosomal rather than plasmid-borne genes (30). Rifampicin and nalidixic acid have been used successfully in New York. Procedures for selecting antibiotic-resistant strains have been described (24,30).

Once selected, the resistant strain should be compared with the sensitive parent for similarity with respect to growth rate in culture, plating efficiency, stability of the antibiotic resistance marker, and pathogenicity toward apple or pear. Failure to test the strain may result in the use of a strain that cannot be recovered quantitatively or that differs widely from the parent strain in ability to induce disease.

The pathogenicity of the selected *E. amylovora* strain may be checked by several techniques. Succulent shoots of greenhouse-grown apple or pear seedlings or cultivars, or susceptible ornamental species such as *Cotoneaster* and *Pyracantha* may be inoculated with a hypodermic needle and syringe (3). Etiolated apple or pear seedlings (39) or young apple seedlings as described below also may be used. Several candidate strains and a nonpathogen (as a check) should be included in pathogenicity tests. Because differential virulence toward apple cultivars exists among strains of *E. amylovora* (33), the virulence of test strains toward the particular apple cultivars to be inoculated should be determined.

Production of Inoculum

Two considerations must be kept in mind in preparing inoculum for use in tests of bactericides to ensure uniform and reproducible tests: first, the inoculum should be of known viable cell concentration, expressed in colony-forming units (cfu) per milliliter, and second, all cells should be in a similar physiologic state. Other factors being equal, the amount of infection that develops in response to artificial inoculation is proportional to the number of viable cells of *E. amylovora* applied (4,8). If the inoculum preparation contains too few viable cells, infection will be insufficient to test candidate materials properly. If inoculum is too concentrated, so much infection may develop that test and check materials may be overwhelmed.

The effectiveness of inoculum is determined in part by the physiologic state of the cells it contains. Log-phase cells are more effective than stationary-phase cells (S. V. Beer, *unpublished*). The lag time (time between application and the beginning of active growth) is much shorter with log-phase cultures. In addition, it is possible that some cells recovered from a stationary-phase culture by plating on a rich medium would not survive inoculation procedures and would not be effective in inducing disease. Thus, for maximum disease-inducing effect, log-phase cultures should be used.

Regardless of which type of culturing is used to prepare inoculum, great care must be taken to ensure that the cells within the culture are in a uniform growth stage. This is particularly important when experiments are to be repeated. Agitated broth culture, rather than stationary broth culture or growth in petri plates, provides better conditions for the development of uniform physiologic states of the bacteria. In addition, the phase of growth of a culture can be monitored readily in broth cultures by measuring culture turbidity (30).

If broth cultures are used to produce inoculum, the cultures should be started with a standardized amount of seed culture (e.g., initial turbidity, set quantity of seed culture). The cultures should be grown under standard conditions of media composition, agitation, and temperature and for a standard length of time. The viable cell concentration should be determined by standard dilution plating procedures (30). By knowing the relationships between growth rate, turbidity, and viable cell concentration, the researcher can dilute the culture to the approximate desired viable cell concentration. A suitable diluent should be used to preserve the viability of cells in the inoculum; buffers, dilute media, or proteinaceous additives have proved effective with *E. amylovora* (S. V. Beer, *unpublished*) and other bacteria (30). Chilling inoculum suspensions diluted in 0.05*M* potassium phosphate buffer (pH 6.5) in an ice bath has preserved the viability of suspensions of *E. amylovora* for at least 4 hr.

Inoculum also may be produced by culturing *E. amylovora* in petri dishes. If this method is used, the cultures should be started from small numbers of cells that are spread uniformly over fresh plates. Plates should be incubated for a limited time (12–24 hr) to maximize the proportion of cells present that are in (or near) log-phase. Bacteria should be recovered from the plate by adding a suitable diluent and scraping with a sterile rubber policeman. Further dilutions should be made to arrive at the desired concentration for application, as described above.

Media

Any of several general complete media may be used to culture *E. amylovora* for inoculum, including medium 523 of Kado and Heskett (27), nutrient yeast extract glucose broth (23), and Luria broth (29). If an antibotic-resistant strain is used as inoculum, appropriate antibiotic concentrations should be included to maintain selection pressure for antibiotic resistance.

Media solidified with agar (1.5–2.0% w/v) for use in estimating viable cell populations may be those mentioned above or any of the several selective or differential media that have been described for *E. amylovora* by Miller and Schroth (31), Crosse and Goodman (18), or Ishimaru and Klos (25). If an antibiotic-resistant strain is used, the antibiotic should be included at an appropriate level.

Advance Preparation of Inoculum

Because the ideal time for inoculating orchard trees generally cannot be predicted more than 1 or 2 days in advance, techniques for preparing inoculum of known concentration on short notice should be used. The following procedure (43) has been used successfully in New York.

The selected *E. amylovora* strain is incubated at 25–29°C in a suitable broth medium with vigorous mixing to provide good aeration. After incubation, the bacteria are handled as aseptically as possible. All containers should be autoclaved or rinsed with 70% (v/v) ethanol and allowed to dry before use. *E. amylovora* cells are harvested during the late log-phase of growth (18–22 hr) by centrifugation at 10,000 *g* for about 10 min at 4°C. The supernatant liquid is decanted carefully, and the bacteria-containing pellets are resuspended in a milk suspending medium at 10–25% of the original medium volume. Suspending medium is prepared by rehydrating 200 g of instant nonfat dry milk (Carnation Co., Los Angeles, CA) in 1 L of distilled water. After the milk has been vigorously shaken to dissolve it, it is filtered through glass wool and then pasteurized in an autoclave for 13 min at 115°C (75.8 kPa, 11 psi). The resuspended pellets are pooled and mixed to equalize concentration, dispensed into plastic containers of 10- to 500-ml capacity, capped, and frozen at −20°C.

After at least 1 day in the freezer, samples of the inoculum are tested for viable cell concentration expressed as cfu/ml. Immediately after thawing at room temperature, the inoculum preparation and dilutions thereof are maintained in an ice bath. Dilutions of the inoculum concentrate are made in chilled 50-m*M* potassium phosphate buffer (pH 6.5) and plated in triplicate on a suitable agar medium (30). Seeded plates are incubated (inverted) at 27–29°C for 1–3 days (depending on the medium used) before counting.

Inoculum should be maintained at −20°C until used. Large quantities may be transported to the field in insulated chests with dry ice (solid carbon dioxide). For lesser quantities or for short trips, mixtures of ice and sodium chloride are effective for keeping inoculum frozen. Inoculum is prepared in the field by thawing containers in air or lukewarm water with shaking. After thawing, inoculum concentrates should be maintained in an ice bath to retard changes in cell numbers.

The water used to dilute inoculum to field concentration should not contain residual chlorine or other components that are lethal to the bacteria. If water quality is questionable, a sample should be tested, in advance, for its effect on the viability of *E. amylovora*. Buffered distilled or deionized water, or buffered tap water with a final potassium phosphate buffer concentration of 5 m*M*, pH 6.5, may be used. No more concentrate should be diluted to the application concentration than can be used in 1 hr.

IMMATURE PEAR FRUIT TISSUE TEST

Many in vitro and greenhouse evaluations of chemicals for fire blight control have been unreliable for selecting chemicals that are effective in the orchard. The growth of *E. amylovora* in vitro may be inhibited by chemicals, such as maneb and mancozeb (26), that do not control fire blight in the orchard. Tests conducted in the greenhouse in which chemicals are applied to vegetative apple shoots followed by inoculation with *E. amylovora* are not useful for screening chemicals for the control of blossom blight (22).

The immature pear fruit tissue test (35) is an inexpensive, rapid, and reliable laboratory method for preliminary evaluation of experimental chemicals. Evaluation of test chemicals is based on the effect of increasing chemical concentration on the development of fire blight symptoms on immature pear fruit tissue. The test is responsive to low concentrations of test chemicals and can be conducted with small quantities of test chemicals. The method is more reliable for predicting orchard activity than traditional in vitro techniques, but it cannot be used to determine the exact concentration of experimental chemical to be used in the orchard. Its major limitation is that it requires a supply of immature pear fruit, which can be stored for only 4–5 months after harvest.

Immature pear fruits have been used also to evaluate the effects of bacterial antagonists of *E. amylovora* in studies aimed at developing biological control methods for fire blight (11,12). For evaluating the efficacy of potential biological control agents, pear fruits are first treated with strains of the antagonists to be tested; 2–3 hr later, the fruits are inoculated with *E. amylovora*. Effectiveness is rated on the basis of the number and length of time that fruits remain free of fire blight symptoms. Although the remainder of this section deals with the evaluation of chemicals for the control of fire blight, the principles are the same for the evaluation of biological control agents.

Pear Fruits

Immature pear fruits are harvested 4–8 weeks before normal harvest, when fruits are 25–50 mm in diameter. Mature fruit cannot be substituted for immature fruit. Only fruit of highly susceptible pear cultivars such as Bartlett or Bosc should be used. Immature apple fruit of the Twenty Ounce and Idared cultivars were not a useful substitute for pear fruit because they are less susceptible to *E. amylovora*. Although fruits of other apple cultivars have not been tested, we suspect that they also would not be useful. After prolonged storage (more than 5 months), immature pear fruit lose susceptibility and can no longer be used as test material. The useful life of stored fruit varies with the quality of storage conditions. Commercial grade fruit storage facilities (0–1°C) are superior to laboratory refrigeration.

Test Procedure and Selection of Chemical Treatments

Immature pear fruits are cut into cubes of approximately 1 cm^3, placed in solutions or suspensions of test chemicals for 4–5 min, and drained. The treated cubes are placed equidistantly on moist filter paper in a petri dish, and a 10-μl drop of a suspension of *E. amylovora* containing 10^9–10^{10} cfu/ml is placed on top of each cube.

The inoculated cubes are incubated at 28° C for 48 hr and then evaluated for disease symptoms on the following scale: 0 = no symptoms; 1 = slight ooze or water-soaking; 2 = much ooze and discoloration. The mean rating obtained for five cubes in one petri dish constitutes one replicate.

Chemicals are tested at four concentrations at fivefold increments; each concentration is replicated four times. Each chemical is tested at one concentration above the expected rate required for fire blight control in the orchard and at three concentrations below that rate. For example, streptomcyin is tested at 2, 10, 50, and 250 mg/L. The expected rate required for fire blight control in the orchard for new chemicals can be inferred from rates that have been established for similar types of compounds and in some cases from in vitro activity. All tests should include streptomycin and a material such as mancozeb, which has in vitro activity against the growth of *E. amylovora* but no activity in controlling fire blight in the orchard, as check treatments.

Interpretation of Data

A decrease in disease rating with increasing chemical concentration indicates that the test material is effective in the immature pear fruit tissue test. Disease ratings are plotted against the $\log_{10}$ of the chemical concentration, and the least squares estimated regression line is calculated. From the regression line, the concentration for a disease rating of 1 is calculated. If the calculated concentration is less than the manufacturer's suggested rate of application, further testing of the material for control of fire blight is considered justified. Illustrations of the procedure and more details on the analysis of test data have been published (35).

APPLE SEEDLING TEST

The apple seedling test permits preliminary evaluation of experimental chemicals with young apple seedlings growing in a controlled environment chamber or greenhouse (35). As with the immature pear fruit tissue test, evaluation of test chemicals is based on the effect of increasing chemical concentration on the development of fire blight symptoms. The major advantage of this procedure is that it is not limited by seasonal availability of plant material. However, the test requires more work and time for preparation of test materials, and it has not been reliable in predicting the activity of some synthesized organic compounds (35). The apple seedling test has not been used to evaluate potential biological control agents.

Test Procedure and Selection of Chemical Treatments

Apple seeds are stratified, germinated, and allowed to develop until the seed coat can be removed easily. Four seedlings are placed into holes (about 5 mm in diameter) in the lid of a plastic petri dish so that the roots are submerged in water and the cotyledons project above the lid. Seedlings may be grown in the greenhouse or in a controlled environment chamber under a combination of fluorescent and incandescent lamps (14-hr photoperiod) until at least two and preferably three true leaves develop. The seedlings are sprayed to runoff with solutions or suspensions of test chemicals and incubated for 24 hr at approximately 25°C. Seedlings are then inoculated by cutting the youngest unfolded leaf with scissors that have been dipped in a suspension of *E. amylovora* containing 10^9–10^{10} cfu/ml.

Four days after inoculation, the proportion of diseased plants per dish is recorded. Plants are considered diseased if discoloration of the inoculated leaf has progressed into the vascular tissue of the petiole. One dish of plants constitutes one replicate. Chemicals are tested at four concentrations at fivefold increments, and each concentration is replicated four times. Two concentrations are selected above the expected rate required for fire blight control in the orchard, and two concentrations are selected below that rate. For example, streptomycin is applied at 10, 50, 250, and 1,250 mg/L. For new chemicals of unknown activity, the expected rates required for fire blight control in the orchard must be inferred from rates that have been established for similar types of compounds or from in vitro inhibition data. Streptomycin and a material that has no activity in controlling fire blight in the orchard should be included in all tests as check treatments.

Interpretation of Data

A significant decrease in the proportion of diseased seedlings with increasing chemical concentration indicates that the chemical is effective in the apple seedling test. A regression line describing the change in the proportion of infected plants as a function of the $\log_{10}$ of the chemical concentration is calculated using a least squares method. The null hypothesis that the slope of the regression line is equal to 0 is tested using an F statistic (41). Rejection of the null hypothesis indicates that further testing of the chemical for fire blight control is justified.

EVALUATING SPRAY MATERIALS IN THE RESEARCH ORCHARD

Evaluating spray materials for the control of fire blight blossom infection in an orchard used exclusively for fire blight research allows materials to be tested under natural host and environmental conditions. By using rather simple artificial inoculation techniques, uniform disease incidence can be ensured throughout test plots to facilitate judging the efficacy of spray materials in the orchard. Because evaluations in research orchards are costly and time-consuming, only materials that reasonably can be expected to be efficacious against fire blight should be tested there. Thus, materials to be tested in the research orchard should have been efficacious against fire blight in at least one of the laboratory or greenhouse tests described above or in previous orchard tests, perhaps in other geographic areas. Research orchard evaluations normally would not be used for initial testing of new materials.

The detailed description of research orchard testing procedures that follows is based on techniques developed and used over a 14-year period in several orchards in western New York State. These procedures have been used to evaluate several registered materials for fire blight control as well as experimental chemicals (4,5,34) and biological control materials (12).

Techniques similar to those described below may be used to evaluate the efficacy of potential chemical or biological control agents under greenhouse conditions. For such tests, container-grown apple or pear trees that have been trained to induce blossom production are used (S. V. Beer, J. L. Norelli, and J. Barnard, *unpublished*).

Test Site

The research orchard test site should be suitable for the growth of apple or pear or both with respect to soil and climate. Ideally, the test site should be located in an area apart from areas of commercial pome fruit production so that the anticipated extensive disease on the test site does not threaten commercial plantings. This isolation is not required, however, if the precautions described below are taken. If the test site is dry during bloom, overhead irrigation facilities should be available to establish higher moisture conditions that favor development of the disease.

Test Trees

Any fire blight-susceptible apple (3) or pear (37) cultivar can be used. Combination plantings of several cultivars are desirable to ensure cross-pollination. If pollination occurs, infected blossoms are more likely to remain on clusters and lead to cluster infection. Idared, Rhode Island Greening, and Twenty Ounce apple cultivars and Bartlett and Bosc pear cultivars have been used with success in New York. If less susceptible apple cultivars such as McIntosh and Empire are used, more inoculum is required to produce a given level of infection, but damage to tree structure is less severe.

Trees propagated on size-controlling rootstocks are most desirable because they bloom at a young age and their size allows the researcher easy access. For apple, the clonal rootstocks M 7 and MM 106 are suggested. Rootstocks M 9 and M 26 should not be used, because their own high susceptibility may lead to early tree death (19,28). Trees propagated on seedling understocks may be used, but they take longer to flower and are more difficult to manage and manipulate during tests.

Trees should not be more than 3 m (10 ft) high and should be spaced more widely than they normally are for commercial fruit production. Wider spacing facilitates maneuvering of orchard machinery and reduces the effects of spray drift. If individual plots include sets of contiguous trees (see experimental design below), tree spacing within rows may be comparable to that used in commercial practice.

Table 1. Effect of *Erwinia amylovora* concentration on fire blight blossom infection[a] in two apple cultivars[b]

Inoculum concentration[c]	Cultivar	Growing season			
		1981	1982	1983	1984
10^7	Idared	77.8	58.4	5.4	93.7
	Empire	56.5	5.1	ND[d]	81.0
10^5	Idared	26.4	30.1	ND	ND
	Empire	28.4	0.4	ND	ND
Mantissa[e]		1.0	0.5	4.0	6.0

[a] Data are percentages of infected clusters.
[b] All tests were conducted in the same research orchard at the New York State Agricultural Experiment Station, Geneva.
[c] Nominal inoculum concentration.
[d] ND = not determined.
[e] Mantissa describing precise inoculum concentration in each year.

Test Plot Management

Recommended commercial orchard maintenance practices (fertilization, cultivation, pruning, and insect, disease, rodent, and weed control) should be used for the research orchard. Where feasible, a sod cover crop should be maintained to facilitate access to the test site when soil is wet. Test plots should be kept free of natural fire blight (6,38) so that inoculum levels will be uniform throughout the test orchard.

All pome fruit plantings within 400 m (0.25 mile) of the research orchard should be protected from fire blight by a rigorous protective spray program. Such plantings should be patrolled frequently while infection is developing in the test orchard, and appropriate action should be taken if disease develops outside the test orchard (6).

Inoculation Techniques

Appropriately diluted inoculum may be applied to blossoms by various techniques. For orchard tests, application of inoculum by low-pressure sprayer or compressed gas-powered atomizer is suggested. Applying inoculum to individual blossom clusters to the point of runoff with a cone-jet tipped wand (Spraying Systems Inc., Wheaton, IL) has been successful (5). For large-scale tests, inoculum may be pressurized (21–28 kPa, 30–40 psi) with a power takeoff-driven nylon roller pump (Model C-6100, Hypro Inc., St. Paul, MN). Low pressure is used to reduce contamination of adjacent trees or orchards. For smaller scale tests, containers of inoculum can be pressurized with air supplied by a hand pump or portable power compressors. Specially modified atomizers (Model 127, DeVilbis Company, Somerset, PA) equipped with a control valve and a reservoir that holds 0.2–0.5 L and connected to tanks of compressed nitrogen, air, or carbon dioxide also may be used. In areas isolated from other pome fruit trees, inoculum might be applied by air-blast sprayer, but limited survival of *E. amylovora* in small droplets has been reported (42). The container used for inoculum should be cleaned thoroughly to prevent inactivation of the pathogen by residues of previously used compounds.

Weather conditions after inoculation, blossom age, and cultivar susceptibility influence the amount of infection that results from a given inoculum level (1; S. V. Beer, *unpublished*). Because weather conditions cannot be predicted reliably, choosing an inoculum concentration that normally results in a relatively large amount of infection is preferable. The percentages of cluster infection that resulted from disease production tests on apple over a 4-year period in western New York are listed in Table 1. An inoculum concentration of 1×10^7 cells per milliliter is suggested for initial trials. The most desirable inoculum level depends on individual orchard and climatic conditions.

Application of Test Materials

Test materials may be applied to test trees by several means. Dilute applications may be made with a hand-operated backpack sprayer, compressed air sprayers, hydraulic handgun at 1,725–3,100 kPa (250–450 psi) to the point of runoff, or air-blast sprayer calibrated to wet trees completely without runoff. Concentrate sprays may be applied with a low-volume sprayer such as the Kinkelder model P50 (Marwald, Ltd., Burlington, Ontario) at a rate between 560 and 1,120 L/ha (60–120 gal/acre) (5). If low-volume sprayers are used, accurate calibration of the sprayer delivery system is particularly important and should be checked with water before spray is applied (21).

Steps should be taken to minimize spray drift between treated trees. This is particularly important when air-blast sprayers are used or windy conditions occur at the time of spraying. If air-blast sprayers are used, untreated buffer trees should be used between treatment areas. If small trees are used in tests, large plastic sheets can be used to cover adjacent trees during treatment.

Timing of Inoculation and Spray Application

The relative timing of spray material application and inoculation determines the type of evaluation obtained. Trees may be sprayed before inoculation to test the protective action of a spray material or after inoculation to test eradicative action. In preliminary tests of a new material, sprays are frequently applied both before and after inoculation to quickly indicate whether more extensive and costly testing is justified.

Materials to be tested for protective action should be applied 1 or 2 days before inoculation and when 40–70% of the blossoms are fully open (15). Because not all of the open blossoms that will be inoculated subsequently have been sprayed, complete control of the disease cannot be expected with nonsystemic materials. Clusters bearing only unopened blossoms should be removed to reduce the number of experimental units that are not treated but that otherwise would be evaluated later.

Materials to be tested for eradicative action should be applied 1–3 days after inoculation but before 20% petal-fall. Inoculation should be done when at least 50–70% of the blossoms are open. Separate check plots and plots treated with a standard material must be included for each day of inoculation, because weather conditions and blossom age affect the amount of disease that develops.

In combined tests of protective and eradicative activity, protective sprays are applied at 25–50% bloom. Twenty-four hours later, or at about 50–70% bloom, blossoms are inoculated. Eradicative sprays are applied 1–3 days after inoculation.

Experimental Design

Several considerations influence the choice of experimental design and the amount of replication needed. The advice of a statistician should be obtained before orchard tests begin. Points to be considered include sources of variability in the test plot, sources of experimental error in the test procedure, and the questions to be addressed in the experiment. The objectives should be clearly defined and the treatments carefully selected. Based on these considerations, the simplest experimental design that will provide the precision required should be chosen. The statistical tests to be used for analysis should be outlined before the experiment is begun.

As variability in the test plot, experimental error, and the degree of resolution desired between test treatments increase, the number of replicates required for a valid test increases. More replicates may be required to evaluate materials of moderate to high efficacy, for which small differences between treatments are anticipated, than for preliminary tests of materials of undefined efficacy. In preliminary tests of new materials, we have routinely used five replicates, which has permitted separating treatments that result in approximately 40% differences in percentage of disease control.

Individual blossom clusters generally are the unit of observation. The larger the number of blossom clusters observed per replicate, the more accurately the amount of disease can be estimated. However, as the number of blossom clusters inoculated per tree increases, experimental error can also increase because it becomes more difficult to control inoculation of each cluster and to make thorough observations of treated clusters. In general, we treat, inoculate, and observe 50–100 blossom clusters per replicate. If the number of blossom clusters is limited or if the precision needed is great, individual blossoms (within clusters) can be examined for symptoms. This is rarely done, however, because of the time and expertise required. Examination of individual blossoms must be timed properly to distinguish direct infections (from artificial inoculation) from those that develop via the spur base and peduncle from adjacent infected blossoms.

In general, an individual tree or a set of contiguous trees in the orchard is the unit of replication. We have found that treating individual branches within trees has several advantages: a higher level of precision can be maintained in blossom cluster inoculation and disease evaluation within a limited portion of the tree; major tree losses due to fire blight can be avoided more easily; and spray drift between trees can be reduced. If specific branches are chosen for treatment and inoculation, they must be labeled carefully so that they can be identified later for observation when more leaves will be present.

The randomized complete block is the experimental design most often used in orchard tests because it compensates for variation within the test site. For example, an orchard planted on a slope could have a gradient of physical factors, such as soil type and air drainage, that affect tree growth and disease development. If trees within the orchard are not uniform, patterns in the sources of variability should be identified. The experimental design should be blocked to minimize variability within the blocks and maximize variability between blocks. Blocks need not be determined by physical location. For example, trees in an orchard may be blocked according to yield or vigor. To minimize experimental error, trees within a block should be treated as uniformly as possible. For example, if inoculation is done by two people, each should work within blocks and not between blocks. Similarly, if more than one person evaluates symptom development, each evaluator should work within a block and not between blocks. If there are no identifiable sources of variation within the test orchard, a completely randomized design should be used.

Control (Check) Treatments

A check (no chemical or biological) treatment should be included in all tests. Check plots should be treated with water at the same time that test plots are treated with materials to be evaluated. Water-treated plots are preferred to unsprayed plots because the carrier water used for spray application may redistribute inoculum and affect the development of infection. Check plots provide an index of the maximum amount of infection to be expected for a noneffective material and are used to judge the efficacy of test materials.

A standard treatment that is expected to provide good disease control should be included in all tests. A commercial formulation of streptomycin sulfate (such as Agri-Strep Type D, Merck & Co., Rahway, NJ) is suggested unless streptomycin-resistant strains of *E. amylovora* are endemic to the area. By including streptomycin-treated plots in the test, the researcher can compare candidate materials with an accepted standard material for the control of fire blight. Streptomycin should be tested at a concentration of 100 mg/L, the recommended rate for fire blight control (2), and at 50 mg/L. These treatments allow the researcher to evaluate the severity of the test conditions and to detect partial activity of new test materials. Under severe test conditions, 100 mg/L of streptomycin should provide significantly better fire blight control than 50 mg/L (34).

Data Collection

The efficacy of spray materials is based on the percentage of blighted blossom clusters on treated trees relative to the percentage of blighted clusters on check (water-sprayed) trees. Control should be evaluated after the pathogen has progressed into the cluster base and cluster leaves show symptoms (wilting or black or brown necrosis, or both) but before individual cluster infections coalesce and become indistinguishable. Under western New York conditions, the proper stage for evaluation is reached 20–30 days after inoculation. At that time, uninfected fruit are 1–2 cm (0.4–0.8 in.) in diameter.

Clusters must be examined extremely carefully to determine if they are infected. Because vegetative shoots or clusters sometimes show symptoms at the same time as blossom clusters, all infections must be examined closely to determine the source of infection. Blossom clusters can be identified by the presence of infected or healthy fruit or peduncle scars on the cluster base. Noninfected blossom clusters without fruit are less obvious than infected or fruiting clusters. Although most blossoms that become infected remain attached to the cluster base, some abscise before *E. amylovora* enters the base. Clusters with symptomatic cluster leaves or necrotic spur bases have been considered infected; those with only necrotic peduncle scars have been considered healthy.

All treated and inoculated clusters (up to a maximum of 400) on each test tree should be examined. On trees bearing more than 400 clusters, representative limbs of the tree are examined. Whole limbs in each tree quadrant should be selected at random, and up to 100 blossom clusters on the selected limbs are examined.

Variation in the judgment of different people collecting data can be a significant source of experimental error. Variation can occur because of differences in the expertise of the evaluators or in the criteria used to evaluate disease. Before collecting data, all evaluators should examine several clusters together, and the criteria for disease should be defined clearly. If a blocked experimental design is used, individual evaluators should work within a single block.

Phytotoxicity

Trees treated with candidate materials should be examined frequently for evidence of phytotoxicity. Within 2 days of application, sprayed blossoms should be examined for evidence of petal burning or necrosis. Later, foliage and fruit should be examined for evidence of chlorosis, necrosis, russeting, or abnormal morphology. In all cases, symptoms on trees treated with candidate materials should be compared with water-sprayed check trees.

Data Analysis

Test data should be analyzed statistically by procedures suitable for the experimental design. The proportion of infected clusters per tree (a decimal between 0 and 1.0) is calculated from the counts of healthy and infected clusters. The percentage data may need to be transformed to the arc sine square root proportion before statistical analysis if proportions of infection are near 0 or 1.0. An analysis of variance is usually performed on the data. Depending on the experimental design, treatment means are compared by partitioning variance among the degrees of freedom for treatment in the analysis of variance or by an appropriate multiple comparison procedure (14,16).

Although statistical analysis is performed on data relating to infected clusters, the effect of each treatment can be interpreted more easily if these data are transformed to percentage of disease control (PDC) using the following formula:

$$PDC = \frac{(DI_{ck} - DI_{tr})}{DI_{ck}} \times 100\,,$$

where DI_{ck} is mean disease incidence in water-sprayed check plots and DI_{tr} is mean disease incidence in treated plots. The effect of the transformation is to relate the efficacy of a candidate material to that of water. When PDC = 100, infection is not present in treated plots; when PDC = 0, treated plots have the same level of infection as the water-sprayed check plots.

Reporting Test Results

Investigators should publicize the results of tests of candidate materials. Reports to manufacturers affect future development of candidate materials and may be used in preparing applications for registration. Test reports are also useful in making recommendations to growers and to researchers working on fire blight or other bacterial diseases. Information that should be included in a complete report has been listed elsewhere (7).

EVALUATING MATERIALS IN COMMERCIAL ORCHARDS

Final evaluations of potential new materials for the control of fire blight should be conducted in commercial orchards, without the benefit of artificial inoculation, and should use application techniques that are used in commercial practice. Only materials with demonstrated efficacy in the previously described tests involving artificial inoculation justify evaluation in commercial orchards.

The key to successful testing in commercial orchards is the identification, in advance, of orchards that are likely to sustain sufficient natural fire blight infection to permit evaluation of test materials. Identifying such orchards is difficult but should be based on history of fire blight, especially in the previous season. Several other orchard characteristics, such as cultivars, age, nutrition, and soil type (13), also help to assess the probability that particular sites will be worthwhile. The researcher must provide for the possibility that disease may be insufficient in a selected plot by arranging test plots in at least four different orchards over a period of two seasons.

The experimental design for use in commercial orchard tests depends on the orchard layout, numbers of trees available for treatment with test materials, approvals by regulatory agencies, and materials available. The same considerations used in research orchard testing with respect to site variability also apply to commercial orchard tests. Application equipment should be checked in advance to ensure that it is properly calibrated. The applicator should be thoroughly familiar with application technique, timing of applications, and layout of treatments.

DISCUSSION

Evaluating materials for their efficacy in controlling fire blight is a difficult and painstaking process. The sporadic occurrence of the disease under natural conditions, the fact that only one test is possible during each growing season, the perennial nature of host plants, and the limited period of susceptibility all contribute to the difficulties. Although the ultimate test of efficacy involves performance in commercial orchards, initial testing of candidate materials under natural conditions can be justified only rarely. However, the multistep process of evaluations described in this chapter facilitates efficient and cost-effective testing of materials. Thus, candidate materials are first tested in small-scale laboratory or greenhouse studies. Only materials that perform well in these tests warrant more costly and involved testing in a research orchard. And only materials that perform well in the research orchard are finally tested under commercial orchard conditions.

Many of the problems associated with orchard testing of bactericides can be overcome by conducting tests in a research orchard. Although establishing and maintaining an orchard exclusively for fire blight research is costly and time-consuming, the yield of usable data that can be derived from it far outweighs the costs. In addition, by using an orchard exclusively for fire blight work, researchers have complete control over management practices that may influence disease incidence or severity, and they need not be overly concerned with potential damage to trees or the marketability of the crop.

The test procedures described here provide the necessary procedural information for the orchard testing of bactericides. However, each researcher must adapt these guidelines to the situation at hand. For example, the concentration of bacterial inoculum should be adjusted to individual test conditions that affect inoculum survival and disease development, such as forecasted weather, tree vigor and susceptibility, percent bloom, and time of day when inoculation will be done.

Because the exact date of bloom and its duration cannot be predicted accurately, prior planning and organization are critical to successful orchard testing. Test treatments and experimental design should be planned well in advance of bloom. Orchard plots should be laid out as soon as tree development allows blossom clusters to be distinguished from vegetative clusters. Equipment and supplies should be assembled and checked before bloom. Before testing starts, all personnel involved should become familiar with the orchard layout, test procedures, and their specific responsibilities.

Results from evaluations of spray materials to control fire blight often have been disappointing and ambiguous because of the sporadic occurrence of the disease. This problem can be overcome by artificially inoculating test material and by carefully selecting plant material for testing. If these steps are taken to ensure sufficient and uniform infection, laboratory, greenhouse, and field tests will provide the necessary data to evaluate reliably the ability of new chemical and biological materials to control fire blight.

LITERATURE CITED

1. Aldwinckle, H. S., and Beer, S. V. 1976. Nutrient status of apple blossoms and their susceptibility to fire blight. Ann. Appl. Biol. 82:159-163.
2. Aldwinckle, H. S., and Beer, S. V. 1979. Fire blight and its control. Hortic. Rev. 1:423-474.
3. Aldwinckle, H. S., and Preczewski, J. L. 1976. Reaction of terminal shoots of apple cultivars to invasion by *Erwinia amylovora*. Phytopathology 66:1439-1444.
4. Beer, S. V. 1974. Production of fire blight of apple under field conditions. (Abstr.) Phytopathology 64:578.
5. Beer, S. V. 1976. Fire blight control with streptomycin sprays and adjuvants at different application volumes. Plant Dis. Rep. 60:541-544.
6. Beer, S. V. 1976. Fire blight—Its nature and control. N.Y. State Coll. Agric. Life Sci. Info. Bull. 100.
7. Beer, S. V. 1978. Techniques for field evaluation of spray materials to control fire blight of apple and pear blossoms. Pages 46–50 in: Methods for Evaluating Plant Fungicides, Nematicides, and Bactericides. E. I. Zehr, ed. Am. Phytopathol. Soc., St. Paul, MN. 141 pp.
8. Beer, S. V., and Norelli, J. L. 1975. Factors affecting fire blight infection and *Erwinia amylovora* populations in pome-fruit blossoms. (Abstr.) Proc. Am. Phytopathol. Soc. 2:95.
9. Beer, S. V., and Norelli, J. L. 1976. Streptomycin-resistant *Erwinia amylovora* not found in western New York pear and apple orchards. Plant Dis. Rep. 60:624-626.
10. Beer, S. V., and Opgenorth, D. C. 1976. *Erwinia amylovora* on fire blight canker surfaces and blossoms in relation to disease occurrence. Phytopathology 66:317-322.
11. Beer, S. V., and Rundle, J. R. 1983. Suppression of *Erwinia amylovora* by *Erwinia herbicola* in immature pear fruits. (Abstr.) Phytopathology 73:1346.
12. Beer, S. V., Rundle, J. R., and Norelli, J. L. 1984. Recent progress in the development of biological controls for fire blight—A review. Acta Hortic. 151:195-202.
13. Beer, S. V., Schwager, S. J., Norelli, J. L., Aldwinckle, H. S., and Burr, T. J. 1984. Towards a practical warning system for fire blight blossom infection. Acta Hortic. 151:23-36.
14. Byron-Jones, J., and Finney, D. J. 1983. On an error in "Instructions to Authors." HortScience 18:279-282.
15. Chapman, P. J., and Catlin, G. 1976. Growth stages in fruit trees—From dormant to fruit set. N.Y. Food Life Sci. Bull. 58.
16. Chew, V. 1977. Comparisons among treatment means in an analysis of variance. USDA, ARS H-6. 64 pp.
17. Coyier, D. L., and Covey, R. P. 1975. Tolerance of *Erwinia amylovora* to streptomycin sulfate in Oregon and

Washington. Plant Dis. Rep. 59:849-852.
18. Crosse, J. E., and Goodman, R. N. 1973. A selective medium for and a definitive colony characteristic of *Erwinia amylovora.* Phytopathology 63:1425-1426.
19. Cummins, J. N., and Aldwinckle, H. S. 1973. Fire blight susceptibility of fruiting trees of some apple rootstock clones. HortScience 8:176-178.
20. Eden-Green, S. J., and Billing, E. 1974. Fireblight. Rev. Plant Pathol. 53:353-365.
21. Fisher, R. W., and Hikichi, W. 1971. Orchard sprayers—A guide for Ontario growers. Ont. Dep. Agric. Food. Publ. 373.
22. Franek, F. R., and Klos, E. J. 1963. The performance of several new chemicals for the control of fire blight of pear. Plant Dis. Rep. 47:348-351.
23. Gantotti, B. V., Kindle, K. L., and Beer, S. V. 1981. Transfer of the drug-resistance transposon Tn5 to *Erwinia herbicola* and the induction of insertion mutations. Curr. Microbiol. 6:377-381.
24. Gerhardt, P. 1981. Manual of Methods for General Bacteriology. Am. Soc. Microbiol., Washington, DC. 524 pp.
25. Ishimaru, C., and Klos, E. J. 1984. New medium for detecting *Erwinia amylovora* and its use in epidemiological studies. Phytopathology 74:1342-1345.
26. Jones, A. L. 1964. Bactericidal action of several organic fungicides against *Erwinia amylovora* and fire blight disease control. Plant Dis. Rep. 48:182-186.
27. Kado, C. I., and Heskett, M. G. 1970. Selective media for isolation of *Agrobacterium, Corynebacterium, Erwinia, Pseudomonas,* and *Xanthomonas.* Phytopathology 60:969-976.
28. Keil, H. L., and van der Zwet, T. 1975. Fire blight susceptibility of dwarfing apple rootstocks. Fruit Var. J. 29:30-33.
29. Maniatis, T., Fritsch, E. F., and Sambrook, J. 1982. Molecular Cloning—A Laboratory Manual. Cold Spring Harbor Laboratory, Cold Spring Harbor, NY. 545 pp.
30. Meynell, G. G., and Meynell, E. 1970. Theory and Practice in Experimental Bacteriology. 2nd ed. Cambridge University Press, Cambridge.
31. Miller, T. D., and Schroth, M. N. 1972. Monitoring the epiphytic population of *Erwinia amylovora* on pear with a selective medium. Phytopathology 62:1175-1182.
32. Moller, W. J., Schroth, M. N., and Thomson, S. V. 1981. The scenario of fire blight and streptomycin resistance. Plant Dis. 65:563-568.
33. Norelli, J. L., Aldwinckle, H. S., and Beer, S. V. 1984. Differential host × pathogen interactions among cultivars of apple and strains of *Erwinia amylovora.* Phytopathology 74:136-139.
34. Norelli, J., and Gilpatrick, J. D. 1981. Chemical control of fire blight of apple during bloom, 1980. Fungic. Nematic. Tests 36:13.
35. Norelli, J. L., and Gilpatrick, J. D. 1982. Techniques for screening chemicals for fire blight control. Plant Dis. 66:1162-1165.
36. Office of Science and Technology Policy. 1984. Proposal for a coordinated framework for regulation of biotechnology; Notice. Federal Register (U.S.) 49(252):50856-50907.
37. Oitto, W. A., van der Zwet, T., and Brooks, H. F. 1970. Rating of pear cultivars for resistance to fire blight. HortScience 5:474-476.
38. Riedl, H., Burr, T. J., and Stiles, W. C. 1984. 1984 Tree-fruit production recommendations. Coop. Ext. N.Y. State Coll. Agric. Life Sci. 76 pp.
39. Ritchie, D. F., and Klos, E. J. 1974. A laboratory method of testing pathogenicity of suspected *Erwinia amylovora* isolates. Plant Dis. Rep. 58:181-183.
40. Schroth, M. N., Thomson, S. V., Hildebrand, D. C., and Moller, W. J. 1974. Epidemiology and control of fire blight. Annu. Rev. Phytopathol. 12:389-412.
41. Snedecor, G. W., and Cochran, W. G. 1967. Statistical Methods. 6th ed. Iowa State College Press, Ames. 593 pp.
42. Southey, R. F. W., and Harper, G. J. 1971. The survival of *Erwinia amylovora* in airborne particles: Tests in the laboratory and in open air. J. Appl. Bacteriol. 34:547-556.
43. Weiss, F. A. 1957. Maintenance and preservation of cultures. Pages 99-119 in: Manual of Microbiological Methods. M. J. Pelczar, Jr., ed. McGraw-Hill, New York.

Evaluating Fungicides for Control of Black Knot of Plums

D. A. ROSENBERGER, Department of Plant Pathology, New York State Agricultural Experiment Station, Hudson Valley Laboratory, Highland 12528, and A. L. JONES, Department of Botany and Plant Pathology and Pesticide Research Center, Michigan State University, East Lansing 48824

Black knot, caused by *Dibotryon morbosum* (Schw.) Theiss. & Syd., is a disease of plum and prune trees (*Prunus salicina* and *P. domestica*, respectively) in the eastern half of North America. The symptoms, life history, early chemical control measures, and naturally occurring mycoparasites of the pathogen have been extensively investigated by Koch (3–7).

Infections occur at nodes on twigs and spurs of the current season's growth (6,10). The fungus causes hyperplasia of the host tissue, the bark erupts during late summer or early the following spring, and a large stroma composed of fungal and hyperplastic host tissue develops on the infected branch (15). The black knot stroma is at first light brown, then turns a velvety green during the period of conidia production about 1 year after infection, and finally turns black and produces ascospores the second year after infection. Sometimes, however, perithecia are produced on 1-year-old infections (6,7,10,13).

Ascospores are released during rain periods and are spread by wind and rain. The period of ascospore discharge may begin at budburst and continue until active growth of terminal shoots stops (10,15). Most black knots probably develop from ascospore infections, because conidia rarely cause infection (2,7,15). No function has been determined for chlamydospores (4,16).

Black knot is usually controlled through a combination of sanitation and fungicide programs. Koch (3) showed that removing knots during pruning reduced infection by 80%, and pruning plus sprays of Bordeaux mixture improved control to 95%. Other fungicide tests have been reported from Canada, Michigan, New York, and Pennsylvania (2,8,10,11,13).

This chapter describes methods that have been used for field testing of fungicides for control of black knot. Fungicides tested for black knot control may also be effective against other plum pathogens such as *Monilinia fructicola* (Wint.) Honey and *Coccomyces prunophorae* Higgins, the fungi causing brown rot and leaf spot, respectively. In field tests designed to evaluate black knot control, relatively little additional work may be required to also evaluate brown rot and leaf spot control. However, specific methods for evaluating other plum diseases are not included here.

TEST PROCEDURES

Selecting a Test Site

The potentially devastating effects of severe black knot infection dictate that tests of fungicides for controlling the black knot pathogen be conducted only on sites where tree loss is acceptable. Unsprayed control trees and trees sprayed with ineffective fungicides may become so severely infected in the course of a single trial that they are destroyed either when the knots are pruned out at the end of the experiment or by the debilitating effects of knots left in the trees. Although *D. morbosum* infects only the current season's growth, scaffold limbs may become infected by invasion from fruiting spurs. The severity of black knot infection in test trees varies with the inoculum pressure and weather conditions during the test. Under some conditions, even unsprayed control trees are only lightly infected, but under severe conditions significant loss of bearing wood may occur in even the best treatments (11).

The test site should also be isolated from other plantings of susceptible host plants to avoid unwanted spread of the disease. No fungicide program has proved to be totally effective against black knot when disease pressure is high. Thus, plantings of plums immediately adjacent to the test site may sustain some black knot damage even if they are protected with fungicides. Black knot may also infect cherry and peach, but infections observed in peach were less severe than in plums, and black knot in cherry may be caused by a different race of the pathogen (2,7,10).

Trees selected for fungicide testing should be of the same age and cultivar. Cultivars of plum and prune vary in susceptibility to black knot; Stanley prune and Damson plum are the most susceptible (14). Fungicide test data will be most directly applicable to commercial agriculture if *P. domestica* 'Stanley' is used. Stanley prunes can also be used for evaluating fungicides for the control of brown rot fruit infection.

Trees should be at least 3 years old because younger trees usually do not have enough terminals and fruiting spurs (i.e., susceptible infection sites) to allow appropriate evaluation of fungicides. Smaller trees may be suitable if used in multiple-tree plots or with many replications. The smaller the tree, the more severe the damage from black knot infection is likely to be because small trees have more fruiting spurs on or near the trunk. Where fungicides are to be evaluated for brown rot also, 5- to 7-year-old trees are required for adequate fruiting.

If trees are planted specifically for fungicide testing, the planting distance between trees and rows should be adequate to minimize interplot interference caused by drift. In areas where tomato ringspot virus (TmRSV) is prevalent in tree fruits, trees should be propagated on Marianna 2624 rootstock rather than on the TmRSV-susceptible Myrobalan rootstock (9,12). Trees on Myrobalan rootstock may collapse suddenly or decline slowly if they become infected by TmRSV. Differences in tree vigor may affect susceptibility to black knot.

Black knot fungicide tests require at least 12–13 months for completion. Data collection is simplified if evaluations can be done at the end of the second growing season. Whenever possible, the test site should be reserved for the test for 2 years.

Introducing Inoculum

A source of healthy black knot inoculum should be identified before a field fungicide trial is initiated. Existing knots in test trees or in abandoned orchards may not be satisfactory because ascospore production in many knots is severely reduced by mycoparasites (2,4). Use of benzimidazole fungicides in commercial orchards may also inhibit ascospore production (10). Knots selected for use as inoculum should have uniformly raised, rather glossy, rounded "caps" over the ascospore-bearing locules on the surface of the stroma. Thus, the healthy stroma is coal black with minute convex bumps over the entire surface. Black knot stromata with a dull or gray-white surface, sunken "caps," or otherwise lacking the raised locule "caps" generally fail to discharge significant numbers of ascospores.

The ascospore-producing potential of mature knots can be tested by soaking excised knots in water for 10–20 min and then placing knots in a spore-collecting tower (1) for 20 min. Knots collected during the peak ascospore discharge period can be brought from the field and tested immediately, but knots collected earlier in the season may require several days at room temperature before ascospores are mature enough to discharge. Four or five healthy, medium-sized knots should produce macroscopically visible accumulations of spores on slides in the discharge tower. The quality of knots for ascospore production can also be evaluated by sectioning locules with a razor blade and examining the sections under a microscope.

Unless trees in the test orchard are already uniformly infected with healthy knots, knots should be collected from another source and suspended in mesh bags or in wire baskets from the top of each test tree. Knots from wild black cherry (*P. serotina*) should be avoided because ascospores from cherry may fail to infect plums (6,15). One or two healthy, medium-sized knots per tree should provide adequate inoculum pressure, but care should be taken to use similar sized knots in all trees to minimize variations in disease pressure.

The natural ascospore discharge cycle can best be duplicated in the test orchard if previously collected knots are replaced with freshly collected knots several times during the season. Excised knots may cease discharging ascospores earlier than knots still attached to living trees. Knots collected and refrigerated for use later in the season may prolong the ascospore discharge season beyond its normal duration.

Monitoring Ascospore Discharge

To accurately assess the merits of individual fungicides, it is essential to know when infection periods occurred. Although our knowledge of the pathogen's requirements for wetness and temperature is not sufficient to predict black knot infection periods, it is possible to estimate when infection most likely occurred by monitoring ascospore discharge from knots in the orchard. One method for monitoring ascospore release is to place glass slides covered with petroleum jelly directly above the knots (10). The slides are changed and examined for ascospores after each rainfall. Rotorod samplers (Ted Brown Associates, Los Altos Hills, CA 94002) can also be used very effectively to monitor ascospore release (G. R. Ehret and A. L. Jones, *unpublished*).

Applying Fungicides

Treatments should be assigned to test plots in a randomized block design with a minimum of four replicates. Fungicides are usually applied at dilute rates using a handgun, but air-blast application and low-volume sprays can be used if tree spacing is adequate to minimize drift.

Fungicides are usually applied on a 7- to 10-day schedule, but shorter spray intervals may be needed during periods of rapid tree growth or unusually heavy rainfall. When to begin and end the black knot spray season should be determined by testing knots for ascospore release in a spore tower. Usually, fungicide protection is needed from bloom until 4–6 weeks after petal-fall, but in some localities (and perhaps in some seasons) ascospores are released before bloom (10,15).

Conditions Favoring Disease Development

Infection may occur during wetting periods as short as 6 hours at 21°C, but little or no infection occurs when temperatures are below 11°C (15). Koch (7) reported that periods of high humidity following wetting periods may also contribute to disease development.

DATA COLLECTION

Incipient black knots may sometimes be visible as distinct swellings or even as erupted knots by September following May–June infections. However, evaluations must usually be delayed at least 12 months from the time of infection. It is not clear whether the duration of the presymptomatic period is a function of the cultivar, environment, time of infection, or black knot isolate. To be certain that slowly developing infections are not overlooked, plots evaluated at the end of the first season should be rechecked for new infections after bloom the following spring. Although all infections will be apparent 12–14 months after infections are initiated, evaluations are easier if they can be delayed until after leaf fall at the end of the second season. Because of the long incubation period, care must be taken to prune trees uniformly so that differences in infection between trees are not obscured. If possible, trees should not be pruned until after the data are recorded.

Black knot fungicide plots can sometimes be evaluated by counting the number of knots per tree or the number on 20 terminals per tree (8,10,13). If terminals are used, terminals of similar length and position should be selected on all trees. If trees are severely infected, counting individual infections may prove difficult or impossible because infections merge or overlap on opposite sides of the same limb. In tests where some trees are severely infected, evaluations may be made by measuring infected and healthy portions of limbs and expressing infection as the proportion of infected to total limb length (11). If entire trees are not measured, similar portions of each tree should be measured. Measurements based on the weight or size of excised knots are less useful because such measures are confounded by the amount of host tissue removed with the knot.

Observations of Fungicide Side Effects

Foliage and fruit should be carefully checked throughout the spray season, and plums from all treatments should be compared at harvest to determine whether any fungicides caused phytotoxic or growth-regulator effects. If trees were very similar in size and productive capacity at the start of the trial, yield data collected during the second season may also provide a measure of black knot severity (11).

DISCUSSION

Because black knot is so destructive and slow to develop symptoms, testing fungicides for control of black knot is not likely to be done on a regular basis. The suggestions and guidelines provided here should help to assure a successful test when the opportunity for such a test arises.

Much additional information is needed on the biology of black knot. For example, are there different races with different host ranges? Do conidia or chlamydospores play a role in survival or disease spread? What conditions favor mycoparasites, and what conditions are required for infection by black knot? Carefully designed experiments can often provide biological information about the pathogen in addition to information about chemical control.

LITERATURE CITED

1. Gilpatrick, J. D., Smith, C. A., and Flowers, D. R. 1972. A method of collecting ascospores of *Venturia inaequalis* for spore germination studies. Plant Dis. Rep. 56:39-42.
2. Gourley, C. O. 1962. A comparison of growth, life cycle, and control of *Dibotryon morbosum* (Sch.) Th. & Syd. on peach and plum in Nova Scotia. Can. J. Plant Sci. 42:122-129.
3. Koch, L. W. 1933. Investigations on black knot of plums and cherries. I. Development and discharge of spores and experiments in control. Sci. Agric. 13:576-590.
4. Koch, L. W. 1934. Studies on the overwintering of certain fungi parasitic and saprophytic on fruit trees. Can. J. Res. 11:190-206.
5. Koch, L. W. 1934. Investigations on black knot on plums and cherries. II. The occurrence and significance of certain fungi found in association with *Dibotryon morbosum* (Sch.) T. & S. Sci. Agric. 15:80-95.
6. Koch, L. W. 1935. Investigations on the black knot of plums and cherries. III. Symptomatology, life history, and cultural studies of *Dibotryon morbosum* (Sch.) T. & S. Sci. Agric. 15:411-423.
7. Koch, L. W. 1935. Investigations on the black knot of plums and cherries. IV. Studies in pathogenicity and pathological histology. Sci. Agric. 15:729-743.
8. Lewis, F. H., Smith, D. H., and Hickey, K. D. 1978. Fungicides for the control of black knot and brown rot of plums. Pa. Fruit News 57(9):13-16.
9. Mircetich, S. M., and Hoy, J. W. 1981. Brownline of prune trees, a disease associated with tomato ringspot virus infection of Myrobalan and peach rootstocks. Phytopathology 71:30-35.
10. Ritchie, D. F., Klos, E. J., and Yoder, K. S. 1975. Epidemiology of black knot of 'Stanley' plums and its control with systemic fungicides. Plant Dis. Rep. 59:499-503.
11. Rosenberger, D. A., and Gerling, W. D. 1984. Effect of black knot incidence on yield of Stanley prune trees and economic benefits of fungicide protection. Plant Dis. 68:1060-1064.
12. Rosenberger, D. A., Gonsalves, D., and Cummins, J. N. 1983. Decline of apple, prune, and peach trees caused by a nematode-transmitted virus. Proc. N.Y. State Hortic. Soc. 128:81-85.
13. Rosenberger, D. A., and Meyer, F. W. 1979. Fungicide evaluation for control of black knot and brown rot of prunes, 1978. Fungic. Nematic. Tests 34:44.
14. Smith, D. H., and Lewis, F. H. 1966. Some factors involved in outbreaks of black knot on plums. (Abstr.). Phytopathology 56:586.
15. Smith, D. H., Lewis, F. H., and Wainwright, S. H. 1970. Epidemiology of the black knot disease of plums. Phytopathology 60:1441-1444.
16. Wainwright, S. H., and Lewis, F. H. 1970. Developmental morphology of the black knot pathogen on plum. Phytopathology 60:1238-1244.

Evaluating Canker Disease Severity in Deciduous Fruit Trees

P. F. BERTRAND, The University of Georgia Cooperative Extension Service, Tifton 31793, and HARLEY ENGLISH, Department of Plant Pathology, University of California, Davis 95616

Evaluating disease severity in perennial plants gradually decimated by a slow-moving canker disease can be difficult. Cytospora canker caused by *Cytospora leucostoma* (Pers.) Sacc., as it occurs on French prunes in California, is one such disease. *C. leucostoma* is most virulent in weakened or predisposed trees (1). Work was undertaken in 1972–1974 to determine the factors that contribute to the development of Cytospora canker in French prunes. A prerequisite of this work was to make an objective evaluation of disease severity in an orchard so that disease severity and other correlative factors could be compared for a wide range of orchards of various ages.

A disease index for evaluating severity of Cytospora canker in French prunes was developed. With proper modification, this method could be used to evaluate many other canker diseases.

METHODS

The severity of Cytospora canker in an orchard was estimated by counting the number of separate fruiting cankers on a tree and giving each canker a weighted value based on its location, but not its size, within the tree. Cankers on small branches and terminals were given a value of 1; those on large branches and secondary scaffolds, 3; and those on the major, or primary, scaffolds, 9 (Fig. 1). Each canker was placed in one of these three categories. Canker location rather than size was used because commercial surgical removal of diseased tissues nearly always involves branch removal and not detailed scraping out of cankers. Large continuous cankers extending from minor to major branches were given a single value according to the most important limb on which they occurred; thus, a single continuous fruiting canker from a terminal to a major scaffold was given a value of 9.

Fig. 1. Dormant French prune tree showing how canker ratings were weighted according to location. Cankers on small branches (**A**) were assigned a value of 1; on large branches (**B**), 3; and on main scaffolds (**C**), 9.

The sum of the weighted canker values for all trees evaluated, divided by the number of trees evaluated, gives the average canker value per tree. The amount of Cytospora canker in a susceptible orchard tends to increase yearly because of canker enlargement and new infection. Dividing the average canker value by an age factor minimizes the effect of age on the disease index and allows comparison of orchards of different ages. In this work the age factor used was orchard age minus 6 years, because Cytospora canker generally is not a problem in California until the trees begin to bear crops, which is generally after about 6 years' growth. In comparing orchards of different ages, it must be recognized that dividing by the age factor results only in approximations because it assumes, perhaps falsely, equal yearly increases in the amount of disease.

Disease counts were made from December to February, because cankers were most easily seen in the absence of leaves. The disease index was calculated by the following formula:

$$\mathrm{DI} = \frac{\left[\dfrac{\text{sum of weighted canker values}}{\text{number of trees evaluated}}\right]}{\text{age factor}} \times 100\,,$$

where DI = disease index. Multiplying by 100 adjusts the DI ratings to values greater than 1.

In our studies, initial sampling was done in a 20-year-old, 16.2-ha (40-acre) orchard on nonuniform soil. Management practices do not vary within this orchard. A second set of samples was taken from single plots in 13 separate orchards of various ages and management regimes located throughout the northern California prune-growing regions. Data gathered in these plots demonstrated that nutritional and soil physical factors could accurately predict the observed disease index calculated by this method (2).

DISCUSSION

This disease evaluation method is best suited to canker diseases that develop indiscriminately throughout the framework of the tree and colonize the tree slowly over a period of time and to diseases that result in readily visible signs or symptoms such as fruiting structures, killed limbs, or peeling surfaces. The method could be used for canker diseases that require surgery to evaluate, but the process would be very slow, and the surgery would interfere with repeat observations.

Canker diseases often work their way into tree trunks. A value for trunk cankers would have to be assigned in these cases. Trunk cankers were not encountered in our test sites, although Cytospora canker is not rare in trunks of dead or nearly dead prune trees. A value of 27 was to have been assigned to trunk cankers had they been found.

An orchard age factor must be determined for each disease. Age alone may in many cases be the best factor. In other situations, some adjustments of age may be more useful, as in our work.

Canker diseases that preferentially invade the trunks or major scaffolds because of some condition unique to these tree parts may not lend themselves to evaluation by this method. The invasion of tree shaker injuries on the trunks of prune trees by *Ceratocystis fimbriata* Ell. & Halst. is a situation of this type (11). Using the described method to evaluate cankers developing in winter-injured tissues can give misleading results if the winter injuries preferentially occur on trunks or major scaffolds, as they often do. Cytospora canker of stone fruit in the colder parts of the United States (5,6,10) and Botryosphaeria canker in apples caused by *Botryosphaeria dothidea* (Moug.:Fr.) Ces. & de Not. (7) are often associated with winter-injured tissues and can present problems of this sort.

We have not used the disease index described herein to evaluate pesticide or other treatments used for the control of canker diseases, but we believe it could be used satisfactorily for this purpose. The disease index would appear to be especially useful in an experiment designed to evaluate the use of spray treatments (fungicides or sun protection paints) or soil treatments (fungicides, nematicides, or fertilizers) extending over several years in orchards of different ages. It would also be useful in cases where sites of infection are not specific or are unknown.

In some cases the sites of infection by canker-causing fungi are very specific. The invasion of pruning wounds of apricots and grapes by *Eutypa armeniacae* Hansf. & Carter or the invasion of apple leaf scars by *Nectria galligena* Bres. are examples of such cases (11). The long-term effects of these diseases can be evaluated with our disease index. However, evaluating fungicide treatments in these two and other similar cases can be very simple and straightforward. The method of assaying infections of pruning stubs on apricot and grape has been described elsewhere (8,9). European apple cankers on year-old shoots caused by invasion of leaf scars by *N. galligena* the previous fall have very characteristic symptoms (3,4). These can be counted simply without resorting to an elaborate disease rating system.

We have outlined a useful method for evaluating canker diseases in deciduous fruit trees. It is best suited to easily visible diseases that occur at random throughout the framework of the tree. It is not well suited to diseases that preferentially occur on one part of the tree such as the trunk. The method can be used to evaluate control procedures. However, much simpler ways to evaluate control practices are possible for some canker diseases.

LITERATURE CITED

1. Bertrand, P. F., and English, W. H. 1976. Virulence and seasonal activity of *Cytospora leucostoma* and *C. cincta* in French prune trees in California. Plant Dis. Rep. 60:106-110.
2. Bertrand, P. F., English, H., and Carlson, R. M. 1976. Relationship of soil physical and fertility properties to the occurrence of Cytospora canker in French prune orchards. Phytopathology 66:1321-1324.
3. Dubin, H. J., and English, H. 1974. Factors affecting apple leaf scar infection by *Nectria galligena* conidia. Phytopathology 64:1201-1203.
4. English, W. H., Dubin, H. J., and Moller, W. J. 1972. European canker of apple in California. Univ. Calif. Agric. Ext. Serv. AST-n85. 4 pp.
5. Helton, A. W. 1961. Effect of simulated freeze-cracking on invasion of dry-ice-injured stems of Stanley prune trees by naturally disseminated Cytospora inoculum. Plant Dis. Rep. 46:45-47.
6. Hildebrand, E. M. 1947. Perennial peach canker and the canker complex in New York, with methods of control. Cornell Univ. Agric. Exp. Stn. Memo. 276. 61 pp.
7. Lewis, G. D. 1956. Botryosphaeria canker and fruit rot of apple in New York. Plant Dis. Rep. 40:228.
8. Moller, W. J., and Kasimatis, A. N. 1978. Dieback of grapevines caused by *Eutypa armeniacae*. Plant Dis. Rep. 62:254-258.
9. Ramos, D. E., Moller, W. J., and English, H. 1975. Susceptibility of apricot tree pruning wounds to infection by *Eutypa armeniacae*. Phytopathology 65:1359-1364.
10. Rolfs, R. M. 1910. Winter killing of twigs, canker and sunscald of peach trees. Mo. State Exp. Stn. Bull. 17. 101 pp.
11. Wilson, E. E., and Ogawa, J. M. 1979. Fungal, bacterial and certain nonparasitic diseases of fruit and nut crops in California. University of California, Berkeley. 109 pp.

Methods for Field Evaluation of Fungicides for Control of Cytospora Canker of Stone Fruits

J. W. TRAVIS, Department of Plant Pathology, The Pennsylvania State University, University Park 16802; and K. D. HICKEY, Department of Plant Pathology, The Pennsylvania State University, Fruit Research Laboratory, Biglerville 17307

Perennial peach canker is a major problem on peaches (*Prunus persica* (L.) Batsch) (6,8,24,31,36) and other stone fruit (4,11,15,28) in the United States. It is caused by either of two fungi, *Leucostoma cincta* (Pers. ex Fr.) Höhn. (syn. *Cytospora cincta* (Pers.) Fr.) or *L. persoonii* (Nits.) Höhn. (syn. *C. leucostoma* (Pers.) Fr.). The fungi infect the woody parts of the tree through dead twigs and buds, leaf scars, winter-injured wood, and injuries such as pruning cuts (34,35,36,42). One of the first visible symptoms of infection is the exudation of gum. Then, as the diseased area enlarges, the tissue collapses, and a canker is formed. The bark remains intact at first but later sloughs off. The canker enlarges more longitudinally than laterally, and it expands each year, by first invading periderm tissue that was produced in response to the infection and then extending into nearby cambial tissue (1,21).

PRELIMINARY CONSIDERATIONS

Several experimental methods may be used to evaluate the efficacy of fungicide treatments for managing Cytospora canker in the field. Selection of the appropriate methods depends on the objectives of the experiment, the site of infection, the season of the year, tree vigor, the location of canker on the tree, and the potential for interference of additional stresses on the tree. Before specific tests are initiated, the important sites of infection on the tree must be determined. Leaf scar infections are considered of primary importance in Canada (43) and Illinois (31), but in Pennsylvania they have not been found to be important infection sites (36). Another consideration is the cultivar to be tested. Although cultivars differ in susceptibility (7,38), the level of inherent resistance is generally believed to be insignificant and can be misleading (33,40). Many researchers have chosen to work with only one cultivar unless they are specifically looking for differences between cultivars in response to treatments.

EXPERIMENTAL OBJECTIVES

The first objective to be determined is whether control measures are to be directed toward prevention or eradication of the canker, because the methods used for each vary greatly.

To prevent canker, protective sprays are applied to the infection site before infection; thus spray timing is difficult, because the tissue is exposed to infection immediately after wounding. It is not clear whether the tissue is susceptible to infection immediately after wounding or whether some period of time elapses after wounding before the tissue becomes susceptible (27). Recent work by Biggs (1) indicates that a few days may be required before wounded tissue is susceptible to infection. In the field, infection is assumed to be possible soon after wounding but dependent on the release of spores by rainfall. Preventative treatments are usually for protecting pruning wounds before rain occurs and are applied when most of the pruning is done. Proper timing of preventative treatments for other types of wounding is further complicated by the imprecise occurrence of wounding from cold injury, twig dieback, and leaf abscission.

Methods for the testing of canker eradication treatments are not as difficult as those for prevention, because they usually involve either direct application of the test chemical to the canker (25) or surgical removal of diseased tissue and subsequent application of the fungicide (37).

HOST-PATHOGEN INTERACTION

The interaction of host and pathogen dictates some of the experimental parameters. Canker elongation occurs primarily from the onset of dormancy until growth begins again in the spring (22). Tests investigating the suppression of canker expansion are limited to the dormant season. Two periods of maximum infection of pruning cuts and dead twigs occur, during the two to three weeks just before budbreak and again in the fall as dormancy begins (23). To study the effect of chemicals in preventing infections of new wounds, chemical treatments should be applied during these periods of potentially maximum infection (23). Maximum formation of periderm after wounding occurs during the rapid growth period in the spring (38,39); therefore, surgical removal of diseased tissue followed by chemical treatments is best done at that time. The location of the wound on the tree is critical to the potential for wound healing. Younger, more vigorous wood produces more periderm than older, less vigorous wood (40); for this reason, treatments should be limited to wood of similar age and vigor. Finally, care must be taken to select uniform trees. The trees' general state of health has a profound effect on wound healing and therefore on the success or failure of the treatments. Trees that are stressed by winter injury, nematodes, weed competition, overcropping, or drought naturally respond very differently in a test than healthy, unstressed trees do (36).

TEST PROCEDURES

Orchard test site and trees. Because Cytospora canker is a destructive disease, tests are usually conducted in growers' orchards where the disease already exists. Care must be taken in selecting a test orchard. A "good" test site is as typical of the local fruit-growing area as possible. Tree uniformity is the main consideration. Trees should be of the same cultivar and uniform in age and vigor; their actual age and vigor depend on the objectives of the experiment. The particular cultivar that is selected is of little importance, because all peach and nectarine cultivars are susceptible to infection by *Cytospora.* If weak and dying trees are scattered throughout the orchard, they must be marked and excluded from the experiment. In eradication tests involving individual canker treatment, cankers should be examined and marked before testing, because some are not suitable for surgical removal.

Recommended cultural and pest management practices should be followed in the test area. The soil type and the slope of the site should be as uniform as possible. As is often observed, trees in low-lying sections of orchards often have more canker; this greater incidence of the disease may be due to more cold damage, poorer soil drainage, or slower drying conditions. Enough canker must be present in the orchard to supply inoculum; if not enough is naturally present, inoculation is necessary. The orchard should be large enough to adequately test, replicate, and buffer between the treatments.

Methods of wounding. In some cases, wounding of trees before inoculation may be desirable. By controlling the wounding before inoculation or infection, the experimenter controls the time, method, and extent of wounding and provides more uniformity across the experiment. Several wounding methods have been used, including cutting wounds (11,24), crushing wounds (8,11,26), and a combination of freezing and crushing wounds (10,33). Although all are effective, their differences make some more suitable than others in certain experimental situations. Surgery, although intended for removal of diseased areas, also results in wounding of the tissue. Surgical wounds have been the center of considerable effort in research on protection from infecton (24,37).

Methods of culturing and inoculum production. Royse (31) reviewed the requirements for culturing of *Cytospora* and production of inoculum. Briefly, isolates of *C. cincta* and *C. leucostoma* may vary in growth characteristics and pathogenicity in culture (4,6,12). Isolates vary from nonvirulent to extremely virulent (14,46). Although isolates are not host-specific, those from stone fruit trees are more virulent to stone fruit trees than to forest trees (9). Conidial germination of *C. leucostoma* has been shown to be dependent on a carbon source (30). Optimum conidia germination has been observed from cultures grown on natural peach gum, mannitol, sucrose, sorbitol, and maltose.

Methods of inoculation. Artificially wounded tissue is often inoculated to ensure uniform inoculum type and uniform infection pressure and to determine the time of infection. The most common method of inoculation has been placement of inoculum grown on agar into the wound. Pieces of inoculum are either placed in a hole bored under a bark flap (3) or are spread across the surface of an area of crushed tissue. Spore suspensions of *Cytospora* may also be applied (11). In many cases, natural infection from the orchard is sufficient to infect the wounded surface, and it may be difficult to prevent. Natural infection by *Cytospora* may be preferable to infection from inoculation because of local differences in virulence between isolates (4), in cultivars (15), and in suscept sources (9).

Protection of wounds after inoculation. In most instances, wounds are not protected with physical barriers after treatment or inoculation. On occasion, however, wounds have been covered to prevent further infection or invasion by insects. Helton used petroleum (11), plastic (10), and elastic tape (16) as protectants against pathogen invasion; Breth et al (3) placed cages of mesh screen over wounded and inoculated areas to prevent infestation by the lesser peach tree borer.

Treatment of wounds. Chemical treatments can be administered to wounds that occur naturally or wounds inflicted in the normal course of orchard operation.

Leaf scars. Leaf scars have been found to be a major site of infection (33,34,43) in some fruit-growing regions. Fungicide treatments have been applied with success (29) to prevent leaf scar infection. Treatments are applied soon after leaf drop with an air-blast sprayer or by dilute application.

Dead twigs. Dead twigs occur commonly in peach orchards, usually on the inside shaded areas of trees. The shading effect of the upper canopy causes gradual decline and weakening of the twigs on the inside and lower scaffold limbs. These twigs often die and become infected. The infection moves down a twig and into a main scaffold limb, resulting in a canker. The canker initially has a major effect on the tree's productivity and eventually on its longevity. Chemical treatments have not been very successful in preventing infection of dead twigs, because of the problem of application timing. Buds on twigs die at different times over a long period of time, and the protection of dead buds is therefore difficult.

Pruning wounds. Wounds made as a result of pruning are a major site of infection. University extension programs have long recommended preventative sprays to protect pruning cuts. The fungicide should be applied as a preventative spray soon after the cut is made and before rain occurs (5). Eradicant fungicides are applied after rain (20).

Methods of fungicide application. The objective of the experiment determines the method and timing of fungicide application. Fungicides have been applied as concentrated sprays with an air-blast sprayer (18,19,29,32), as dilute sprays with a spray gun (29), as trunk drenches (17), by direct application with a paintbrush (8,13,37), and by injection (13).

Evaluation. Because host response is intimately involved in evaluation of the effectiveness of a treatment, time must be allowed for tree response before evaluation. Canker treatments, therefore, require several months to several years before final evaluations can be made. Researchers have used the presence of gum and collapsing bark as symptoms of infection by *Cytospora*. Boothby (2) and Wisniewski (44) have shown that once-infected stone fruit trees excrete more gum than wounded trees whose wounds are not infected. However, the gumming response is not specific to *Cytospora*; many other pathogens may also initiate it (45). Isolation and identification of *Cytospora* is a positive diagnosis of infection. Another method of evaluation is based on the response or lack of response by the tree after treatment. Tissue that becomes infected does not form periderm as healthy, uninfected tissue does. The following evaluation system, based on the formation of periderm tissue in the absence of infection (37), assigns numerical values for the formation of periderm: 0, no periderm (extensive diseased tissue); 1, incomplete periderm at the margin of the treatment area (some diseased tissue at the margin); 2, complete periderm along the margin of the treatment area (no diseased tissue).

LITERATURE CITED

1. Biggs, A. 1984. Boundary-zone formation in peach bark in response to wounds and *Cytospora leucostoma* infection. Can. J. Bot. 62:2814-2821.
2. Boothby, D. 1983. Gummosis of stone-fruit trees and their fruits. J. Sci. Food Agric. 34:1-7.
3. Breth, D. I., Travis, J. W., and Hull, L. A. 1985. The relationship between Cytospora canker and the lesser peach tree borer. Pa. Fruit News 64(1):16, 20.
4. Chiarappa, L. 1960. Distribution and mode of spread of Cytospora canker in an orchard of the President Plum variety in California. Plant Dis. Rep. 44:612-617.
5. Crassweller, R. M., Daum, D. R., Tetrault, R. C., Travis, J. W., Collison, C. H., Greene, G. M., Hickey, K. D., Hock, W. K., Hull, L. A., and Jaffee, B. 1985. Tree Fruit Production Guide. The Pennsylvania State University, University Park. 83 pp.
6. Daines, R. H. 1967. Cytospora problem of peach and the replant problem. Pages 42-45 in: Proc. Canker Complex and Decline of Peach Trees. North Carolina State University, Raleigh.
7. Gairola, C., and Powell, D. 1970. Cytospora peach canker in Illinois. Plant Dis. Rep. 54:832-835.
8. Harder, H. H., and Luepschen, N. S. 1974. Chemical control of peach canker. Colo. Agric. Exp. Stn. Prog. Rep. 74-6.
9. Helton, A. W. 1961. First year effects of 10 selected *Cytospora* isolates on 20 fruit and forest tree species and varieties. Plant Dis. Rep. 45:500-504.
10. Helton, A. W. 1962. Effect of Bordeaux spray and plastic covering in preventing invasion of low-temperature-injured stems of prune trees by *Cytospora* fungi. Plant Dis. Rep. 46:77-79.
11. Helton, A. W. 1962. Prevention of *Cytospora* invasion of injured stems of prune trees with wound treatments. Phytopathology 52:1061-1064.
12. Helton, A. W. 1970. Effect of culture age on virulence of artificial *Cytospora* infections in *Prunus domestica*. Phytopathology 60:1694-1695.
13. Helton, A. W., Kochan, W. J., and Dostal, H. C. 1964. Controlling Cytospora canker and other conditions in prunes with spray-applied systemics—A preliminary report. Trans. 70th Annu. Meet. Idaho State Hortic. Soc., pp. 48-54.
14. Helton, A. W., and Konicek, D. E. 1962. An optimum environment for the culturing of *Cytospora* isolates from stone fruits. I. Temperature. Mycopathol. Mycol. Appl. 16:19-26.
15. Helton, A. W., and Moissey, J. S. 1955. *Cytospora* damage in Idaho prune orchards. Plant Dis. Rep. 39:931-943.
16. Helton, A. W., and Rohrbach, K. G. 1967. Chemotherapy of Cytospora canker disease in peach trees. Phytopathology 57:442-446.
17. Helton, A. W., Smith, W. T., and Williams, R. E. 1965. Cycloheximide thiosemicarbazone as a systemic fungicide for orchard trees. Trans. 71st Annu. Meet. Idaho State Hortic. Soc., pp. 29-36.
18. Hickey, K. D., and Parker, K. G. 1963. Peach: Perennial canker (*Valsa cincta* and *V. leucostoma*). Fungic. Nematic. Tests 19:50-51.
19. Hickey, K. D., and Parker, K. G. 1964. Peach: Perennial canker (*Valsa cincta* and *V. leucostoma*). Fungic. Nematic. Tests 20:45-46.
20. Hickey, K. D., and Travis, J. W. 1984. Effect of preventative and post-infection fungicide treatments and the time of pruning on incidence of *Cytospora* infection on pruning wounds on peach. Pages 65-70 in: Proc. 1984 Stone Fruit Decline Conf., Kearneysville, WV. U.S. Dep. Agric. Agric. Res. Serv., Washington, DC.
21. Hildebrand, E. M. 1947. Perennial peach canker and the canker complex in New York, with methods of control. Cornell Univ. Agric. Exp. Stn. Mem. 276:1-61.
22. Jones, A. C., and Luepschen, N. S. 1971. Seasonal development of Cytospora canker on peach in Colorado. Plant Dis. Rep. 55:314-317.
23. Jones, A. C., Luepschen, N. S., Mahannah, S. M., and Smith, L. R. 1970. Peach gummosis studies: Spore dissemination and seasonal development. Colo. Agric. Exp. Stn. Prog. Rep. 70-8.
24. Luepschen, N. S. 1965. Protecting peach trees from Cytospora canker. Colo. Agric. Exp. Stn. Prog. Rep. 142.
25. Luepschen, N. S. 1976. Use of benomyl sprays for suppressing Cytospora canker on artificially inoculated peach trees. Plant Dis. Rep. 60:477-479.
26. Luepschen, N. S., Jones, A. C., Rohrbach, K. G., and Dickens, L. E. 1970. Peach varietal susceptibility to Cytospora canker. Colo. Agric. Exp. Stn. Prog. Rep. 70-5.
27. Luepschen, N. S., and Rohrbach, K. G. 1969. Cytospora canker of peach trees: Spore availability and wound susceptibility. Plant Dis. Rep. 53:869-872.
28. Lukezic, F. L., DeVay, J. E., and English, H. 1960. Occurrence of Cytospora canker in stone fruit trees in California. (Abstr.) Phytopathology 50:84-85.
29. Northover, J. 1976. Protection of peach shoots against species of *Leucostoma* with benomyl and captafol. Phytopathology 66:1125-1128.
30. Rohrbach, K. G., and Luepschen, N. S. 1968. Environmental and nutritional factors affecting pycnidiospore germination of *Cytospora leucostoma*. Phytopathology 58:1134-1138.
31. Royse, D. J. 1978. Epidemiology and control of perennial canker of peach. Ph.D. thesis, University of Illinois, Urbana. 73 pp.
32. Royse, D. J., and Ries, S. M. 1978. The influence of fungi isolated from peach twigs on the pathogenicity of *Cytospora cincta*. Phytopathology 68:603-607.
33. Scorza, R., and Pusey, P. L. 1984. A wound-freezing inoculation technique for evaluating resistance to *Cytospora leucostoma* in young peach trees. Phytopathology 74:569-572.
34. Tekauz, A. 1968. Peach canker studies. M.S. thesis, University of Toronto. 99 pp.
35. Tekauz, A. 1972. The role of leaf scar and pruning cut infections in the etiology and epidemiology of peach canker caused by *Leucostoma* species. Ph.D. thesis, University of Toronto. 161 pp.
36. Travis, J. W. 1984. Peach integrated crop management orchard survey. III. The prevalence and importance of Cytospora canker. Pa. Fruit News 63(4):54-55.
37. Travis, J. W., and Hickey, K. D. 1984. Efficacy of surgical removal and fungicide wound treatments on eradication of Cytospora canker on peach. Pages 60-64 in: Proc. 1984 Stone Fruit Decline Conf., Kearneysville, WV. U.S. Dep. Agric. Agric. Res. Serv., Washington, DC.
38. Weaver, G. M. 1963. A relationship between the rate of leaf abscission and perennial canker in peach varieties. Can. J. Plant Sci. 43:365-369.
39. Wensley, R. N. 1966. Rate of healing and its relation to canker of peach. Can. J. Plant Sci. 46:257-264.
40. Wensley, R. N. 1970. Innate resistance of peach to perennial

canker (*Leucostoma cincta*) (Fr.) v. Hohnel (*Valsa cincta*). Can. J. Plant Sci. 50:339-343.

41. Willison, R. S. 1933. Peach canker investigations. I. Some notes on incidence, contributing factors, and control measures. Sci. Agric. 14:32-47.
42. Willison, R. S. 1936. Peach canker investigations. II. Infection studies. Can. J. Res. 14:27-44.
43. Willison, R. S. 1937. Peach canker investigations. III. Further notes on incidence, contribution factors, and related phenomenon. Can. J. Res. 15:324-339.
44. Wisniewski, M., Bogle, A. L., and Wilson, C. L. 1984. Histopathology of canker development on peach trees after inoculation with *Cytospora leucostoma*. Can. J. Bot. 62:2804-2813.
45. Wormald, H. 1938. Bacterial diseases of stone-fruit trees in Britain. 7. The organisms causing bacterial disease of sweet cherry. J. Pomol. Hortic. Sci. 16:280-290.
46. Wysong, D. S., and Dickens, L. E. 1962. Variation in virulence of *Valsa leucostoma*. Plant Dis. Rep. 46:274-276.

Field Test Procedures for Evaluation of Fungicides to Control *Monilinia laxa* on Stone Fruits

J. M. OGAWA and B. T. MANJI, Department of Plant Pathology, University of California, Davis 95616; and I. C. MacSWAN, Department of Plant Pathology, Oregon State University, Corvallis 97331

Brown rot disease, caused by *Monilinia laxa* (Aderh. & Ruhl.) Honey, is common on almonds, apricots, sweet cherries, and prunes in California (3,9,14) and on prunes, sweet cherries, and peaches in Oregon. The disease has been reported on apricots in Wisconsin and sweet cherries in New York. The blossom blight phase is more common than the fruit rot phase on these crops. For the design of a fungicide test plot in an orchard, consideration must be given to the many cultivars planted for each crop as well as to the requirement of more than one cultivar for pollination of almonds and sweet cherries. Field test procedures for *M. laxa* differ from those published for *M. fructicola* (8).

Disease cycle. The disease cycle for brown rot blossom blight on almonds in California (Fig. 1) demonstrates the relationship between crop phenology and disease development. Disease cycles for apricots and prunes are similar to that for almonds. During the dormant season, the fungus overwinters in mummies, infected peduncles, and blighted blossoms and twigs. Between the middle of November and early December, the combination of rain, dew, fog, and cooler temperatures fulfills the necessary conditions for emergence of sporodochia. Numbers of sporodochia increase until the blossoming period, which starts in the middle of February for almonds, late February for apricots, early March for prunes, and late March for sweet cherries. Free moisture and temperatures over 13°C are necessary for blossom infections.

The primary infection courts on blossoms of almond and sweet cherry are pistils, anthers, and petals. All parts of apricot and prune blossoms, including the sepals, are susceptible to infection. Green fruits of almonds, apricots, prunes, and sweet cherries are somewhat resistant to new infections. However, green-fruit infections originating from infected flower parts may result in decay at shuck-fall. Quiescent infections that occur during bloom may also result in preharvest decay.

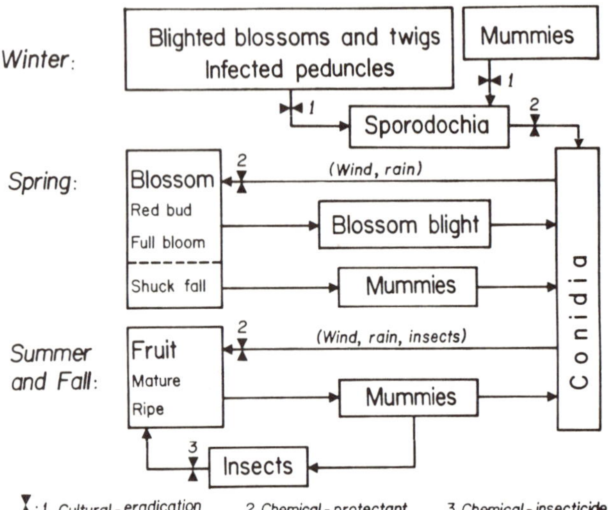

Fig. 1. Disease cycle of *Monilinia laxa* on almonds.

Inoculum sources for infection of mature fruit include blighted blossoms and green-fruit mummies. Insects can act as vectors, spreading spores to ripening fruits with split pits or with injuries caused by insects or by mechanical means.

Mummies are formed from infected fruit. Apricot and prune mummies tend to drop during the decay process, whereas almond mummies tend to remain attached. Harvesting operations with mechanical shakers do not dislodge almond mummies; most of them drop to the ground when the trees are shaken during the winter to prevent the overwintering of navel orangeworms in mummified nuts. Unfortunately, complete elimination of mummies is impossible.

Current control measures are directed at reducing the inoculum source with an eradicant fungicide (to control the emergence of sporodochia) before the blossoming period and protecting the opening blossom parts with protectant fungicides (11). Preharvest sprays to control *M. laxa* fruit rot have not been effective for almonds, apricots, prunes, and sweet cherries, although fruit rot caused by *M. fructicola* on peaches and nectarines can be reduced by preharvest fungicide applications.

Susceptibility of almond, apricot, and prune cultivars to *M. laxa* infections. Although 40 almond cultivars are grown in California, eight cultivars account for over 90% of almond plantings. These eight cultivars differ in their blossoming time (4), in the proportion in which they are planted (2), and in susceptibility to blossom blight. Table 1 indicates that most orchards are planted largely with Nonpareil with adjacent rows of other early (Ne Plus Ultra) or late (Mission) cultivars. Recently, some orchards have been planted with one early cultivar and two that bloom at an intermediate time.

Of the 13 apricot cultivars grown in California (2), Blenheim-Royal (41%) and Tilton (41%) account for most of the 11,000 ha. All apricot cultivars are susceptible to infection by *M. laxa*, and Flaming Gold, Modesto, Derby Royal, and Perfection are highly susceptible.

California prune acreage (2) is planted primarily with the cultivar French (98%). Plantings of Imperial cultivar amount to slightly over 1% of prune acreage. Both cultivars are susceptible to brown rot blossom blight.

Laboratory tests. Fungicides to be tested in the field

are first evaluated in vitro in laboratory tests. The purpose of in vitro tests is to establish the baseline sensitivity of *M. laxa* isolates to a fungicide (that is, to establish, for a given fungicide, the lowest concentration that inhibits growth of the fungus). The *M. laxa* isolates to be tested are collected from the major crop-growing areas. At least 100 samples having blossom or twig blight with sporulating *M. laxa* are collected. Single-spore isolates are obtained from each sample.

The effects of a fungicide on spore germination and mycelial growth are determined by placing spores or mycelium on agar medium amended with the fungicide. Technical-grade fungicide (or the proprietary compound, if the technical grade is not available) is first dissolved in water or some other solvent and is then added to sterile, cooled potato-dextrose agar. If preliminary experiments show that the fungicide is not inactivated by heat, it may be added to the medium before autoclaving. Spores of *M. laxa* for the tests can easily be produced on oatmeal agar in the presence of continuous light or on slices of canned apricots (5).

Inhibition of spore germination and mycelial growth on test samples in vitro should be calculated in relation to germination and growth on fungicide-free controls. ED_{50} values (fungicide concentrations resulting in 50% growth inhibition) for spore germination are determined from linear regression equations using logarithms for each concentration and probits for percentage of growth inhibition (1). Inhibition of mycelial growth should be reported as minimum inhibitory concentration (mic), which is the minimum fungicide concentration necessary for complete inhibition of mycelial growth. ED_{50} and mic values enable researchers to compare the activity of different fungicides against *M. laxa*.

Laboratory in vivo tests of fungicide spray may be conducted with blossoms on detached twigs, which are placed in water-saturated sand in enclosed containers. Such tests have provided satisfactory data for evaluation of systemic fungicides, such as benomyl and thiophanate-methyl, but are not suitable for testing protectants, such as captan or fixed coppers. To provide uniform infection, spore concentrations of 1×10^6 conidia per milliliter are required.

Contamination by *Rhizopus, Botrytis,* and *Sclerotinia* may occur if the incubation temperature is 20°C, but incubation at lower temperatures is not satisfactory for quick analyses. The incidence of petal necrosis and blossom blight and the sporulation of *M. laxa* are recorded two to five days after inoculation. Positive identification of *M. laxa* is required if contamination with *M. fructicola* (12) is a possibility.

Field tests. The criteria for selection of an orchard for evaluation of fungicides should include 1) a previous history of brown rot blossom and twig blight in the orchard and surrounding areas, 2) the presence of sporodochia (inoculum source) during the winter before the test, and 3) the presence of susceptible cultivars in the orchard. An orchard should be chosen in which at least 5–10% of the blossoms were infected by *M. laxa* in the year preceding the test. Trees in the orchard should be examined in the winter to determine whether sporodochial development on twigs is sufficient to ensure that enough inoculum will be present during bloom.

Highly susceptible trees grown for the purpose of testing experimental fungicides should be used for the first small-scale field studies. Suitable stone fruit cultivars include Drake almond, Blenheim apricot, and French prune. To establish the approximate concentrations of an active ingredient required for achieving disease control without phytotoxicity, at least six to eight single-tree replications should be sprayed with each of the fungicides to be evaluated. Preliminary tests are intended to evaluate an experimental fungicide's protectant activity, and not necessarily its systemic activity. Sprays are applied to almonds at the pink bud stage, to apricots at the red bud stage, and to prunes at the green bud stage. An additional spray is applied at about 80% bloom for all tree crops. For apricots and prunes, the second spray may be required earlier than 80% bloom, when the petals have started to separate, thus exposing the anthers. For all crops, sprays should be timed to provide protectant activity before rain or free moisture from fog or dews.

Tests during the second year should provide more definitive data on proper timing of sprays. A single concentration of experimental fungicides should be compared to a currently recommended standard fungicide. In addition, winter applications are essential to determine whether a fungicide has any suppressive action on the development of sporodochia (10). One should determine whether a fungicide can be combined with dormant oils, which could improve the fungicide's penetration of twigs to reduce the number of sprays required for disease control.

An experimental-use permit is required for harvesting and selling any crop that has been treated with experimental fungicides. After the permit has been approved, large-scale tests may be conducted in commercial orchards. The fungicides should be applied by both ground rigs and aircraft. If an experimental fungicide provides effective disease control without phytotoxicity in commercial tests, an application for full registration for use of the fungicide is in order. The application must include data on fungicide residues on the crop after maximum use patterns. Residue data are preferably collected during the first year of field tests. Because of the special requirements of spray application and sample collection for residue analyses, a test plot reserved strictly for obtaining residue data may be convenient.

Collection of data. Procedures to evaluate the efficacy of experimental fungicides are the same for apricots and prunes. The number of blighted twigs per quadrant (a quarter portion of a tree) on wood produced the preceding year (fruiting wood) may be counted, or

Table 1. Susceptibility of eight almond cultivars grown in California to infection by *Monilinia laxa*

Bloom time	Cultivar	Production area, (%)[a]	Susceptibility[b]
Early	Ne Plus Ultra	6	+++
	Peerless	2	+++
Intermediate	Nonpareil	54	++
	Merced	6	++
	Carmel	8	++
	Price	3	+
Late	Mission (Texas)	12	+[c]
	Thompson	2	++

[a]Percentage of 187,379 ha in California.

[b]+++, most susceptible; +, least susceptible. Less common cultivars such as Drake, IXL, Jordanolo, and Harpareil would be classified ++++.

[c]Rated ++++ in some orchards.

the total number of blighted twigs per tree may be counted. If twig blight is severe, examination of 50–100 segments (10–20 cm long) of fruiting wood per quadrant is sufficient; if disease incidence is low, the number of blighted twigs on the whole tree is counted. The latter technique could provide biased results if tree size and vigor are not uniform.

Counts of individual blighted blossoms in the absence of twig blight are difficult to obtain and unreliable for apricots and prunes, because, although blossoms turn brown and *Monilinia* may sporulate on blossom parts, developing fruit may later appear to be healthy. If the incidence of disease is extremely low, counting twigs with blighted blossoms that show gumming at the base may be necessary. Symptoms of blossom blight on apricots and prunes could be caused by *M. fructicola* as well as *M. laxa*. To distinguish the two species, approximately 20 twigs from different trees should be collected. Isolations from the samples on potato-dextrose agar medium reveal which *Monilinia* species is present (3,12).

On almonds, fungicide efficacy is usually evaluated by counting blighted twigs as for apricots and prunes. Additionally, blossom blight may be evaluated by counting 100–200 individual blossoms per quadrant.

In all counting procedures, the incidence of disease in a few unsprayed trees is evaluated first, to determine the minimum number of twigs or blossoms to observe. Then the most effective treatments are evaluated to determine the maximum numbers to observe for possible separation between all treatments.

Data on fruit rot have not been evaluated for almonds and apricots because of erratic occurrence. Quiescent infections of young prune fruits usually result in decay that can be evaluated, but meaningful data on the efficacy of fungicides against decay on mature and ripe fruit have not been collected.

DISCUSSION

The methods presented here have been used to evaluate eradicant fungicides (13), protectant fungicides (6), and systemic fungicides (7,9–11) that were later registered for commercial use. However, new research problems create a need for new techniques to evaluate fungicides. For example, recently developed chemicals have demonstrated genestatic effects, which have not received appropriate evaluations in total disease control. Furthermore, control data are lacking for fungicides in combination with adjuvants (spreaders and stickers), other fungicides, insecticides, and even foliage nutrients. Other factors receiving little attention are the exact timing of treatments with respect to the growth stage of the host and environmental conditions (relating conditions such as periods of rain, fog, dew, sunshine, and relative humidity to requirements for the next cover treatment). In fact, information on the half-life of fungicide residue and the minimum dosage required is not available. Then, with the development of new fungicides that may render hosts resistant to pathogens, a technique to show host resistance is needed.

In testing fungicides in a semiarid climate, such as that found in California, all new fungicides should be evaluated for disease control under conditions present in commercial orchards. These conditions could include different planting patterns and irrigation systems as well as different disease pressures. Treatments should be sufficiently replicated to detect small differences in disease incidences.

The most successful approach to evaluation of experimental fungicides involves considerations of 1) the type of activity—eradicant, protectant, suppressant (10), or genestatic; 2) the efficacy of treatments applied from the ground or from the air under different environmental conditions; 3) the ease with which the required number of sprays or dusts can be applied with the required timing; and 4) the cost-effectiveness of the new treatment compared to that of the standard treatment. Once a fungicide has been registered, the most appropriate approach is for growers to leave untreated portions in their orchards, where the efficacy of the treatment can be evaluated. Growers are often reluctant to do this, but without such a program, assessment of fungicide efficacy on cultivars grown in varied environments will continue to be uncertain.

LITERATURE CITED

1. Bliss, C. J. 1935. The calculation of the dosage-mortality curve. Ann. Appl. Biol. 22:134-167.
2. California Crop and Lifestock Reporting Service. 1982. California Fruit and Nut Acreage. Department of Food and Agriculture, Sacramento. 28 pp.
3. Hewitt, W. B., and Leach, L. D. 1939. Brown-rot *Sclerotinias* occurring in California and their distribution on stone fruits. Phytopathology 29:337-351.
4. Micke, W., and Kester, D., eds. 1978. Almond Orchard Management. Priced Publ. 4092. University of California, Division of Agricultural Sciences, Oakland. 150 pp.
5. Nevill, J. R., Szkolnik, M., Gilpatrick, J. D., and Ogawa, J. M. 1978. Mass production of conidia of brown rot fungi on canned fruit pieces. Plant Dis. Rep. 62:966-969.
6. Ogawa, J. M., Hall, D. H., and Koepsell, P. A. 1967. Spread of pathogens within crops as affected by life cycle and environment. Pages 247-267 in: Airborne Microbes. Symp. Soc. Gen. Microbiol., London.
7. Ogawa, J. M., Manji, B. T., and Bose, E. 1968. Efficacy of fungicide 1991 in reducing fruit rot of stone fruit. Plant Dis. Rep. 52:722-726.
8. Ogawa, J. M., Manji, B. T., English, H., Sall, M. A., Yates, W. E., Chiarappa, L., and Rough, D. 1978. Laboratory and field test procedures for evaluation of fungicides for control of brown rot diseases of stone fruits. E. I. Zehr, ed. Pages 54-58 in: Methods for Evaluating Plant Fungicides, Nematicides, and Bactericides. E. I. Zehr, ed. American Phytopathological Society, St. Paul, MN.
9. Ogawa, J. M., Moller, W. J., Manji, B. T., Rough, D., and Koike, S. T. 1980. Brown rot of stone fruits. Leafl. 2206. University of California, Division of Agricultural Sciences, Oakland.
10. Ramsdell, D. C., and Ogawa, J. M. 1973. Reduction of *Monilinia laxa* inoculum potential in almond orchards resulting from dormant benomyl sprays. Phytopathology 63:830-836.
11. Ramsdell, D. C., and Ogawa, J. M. 1973. Systemic activity of methyl 2-benzimidazolecarbamate (MBC) in almond blossoms following prebloom sprays of benomyl + MBC. Phytopathology 63:959-964.
12. Sonoda, R. M., Ogawa, J. M., and Manji, B. T. 1982. Use of interactions of cultures to distinguish *Monilinia laxa* from *M. fructicola*. Plant Dis. 66:325-326.
13. Wilson, E. E. 1942. Experiments with arsenite sprays to eradicate *Sclerotinia laxa* in stone fruit trees as means of controlling the brown rot disease in blossoms. J. Agric. Res. 64:561-594.
14. Wilson, E. E., and Ogawa, J. M. 1979. Fungal, Bacterial, and Certain Nonparasitic Diseases of Fruit and Nut Crops in California. Priced Publ. 4090. University of California, Division of Agricultural Sciences, Oakland. 190 pp.

Field Test Procedures for Evaluation of Fungicides to Control Shot Hole Disease on Stone Fruits

J. M. OGAWA, B. T. MANJI, and L. M. HIGHBERG, Department of Plant Pathology, University of California, Davis 95616; DON ROUGH, Farm Advisor, San Joaquin County, Stockton, CA 95207; and MARIO VIVEROS, Farm Advisor, Kern County, Bakersfield, CA 93303

Shot hole disease, caused by *Stigmina carpophila* (Lév.) Ellis (syn. *Coryneum beijerinckii* Oud.), is common in the Pacific Coast states and is a major disease of peaches, nectarines, apricots, and almonds in California (7,8,10). It occurs rarely on sweet cherries, plums, and prunes. Field test procedures for evaluation of fungicides for control of shot hole are based on the disease cycle of the fungal pathogen, differences in susceptibility of hosts and host parts, cultural practices in the orchard, types of spray equipment, and the fungicides to be tested. The test procedures reported here have been developed over a 30-year period.

Identification and isolation of *S. carpophila.* Although shot hole can be recognized by lesions produced on various plant parts, positive identification of the disease is difficult unless the fungus is sporulating. The fungus produces sporodochia on leaves, twigs, and blossoms. Microscopic examination of the sporodochia reveals typical *Stigmina* spores, which are elongate to ovoid and three- to eight-septate, have greenish yellow central cells and hyaline end cells, and are characteristically constricted at the septa.

In the absence of sporulation, isolations are necessary for identification of the causal agent. The most important requirement for successful isolation of *Stigmina* is the collection of fresh samples of newly infected leaves, twigs, or fruit. Isolation must be made in the early stages of infection, to consistently isolate the fungus without contamination from other organisms. The plant material is surface-sterilized in 400 ppm chlorine for 3 min and air-dried. Small pieces of tissue (1–2 mm) are then placed on lactic acid potato-dextrose agar and incubated at 20°C. The medium on which the fungus is cultured is important for positive identification. Spores produced on lactic acid potato-dextrose agar usually have three septa; on cornmeal agar, the number of septa is variable. To confirm the pathogenicity of a *Stigmina* isolate, spores can be placed on surface-sterilized almond or apricot leaves, which are highly susceptible to the pathogen, and incubated in a moisture chamber at 20°C.

Disease cycle. The disease cycle is similar on different stone fruit hosts. In the fall, the fungus attacks the leaves, twigs, or buds. Free moisture is required for infection. The incidence of infection is low at temperatures below 10°C, but infections are numerous at temperatures near 20°C. In California, infections continue during the late fall and winter. The fungus overwinters as individual spores (3) on the surface of host parts and in infected buds (9). When new leaves or blossoms emerge in the spring, new infections occur. Infections on young leaves usually produce lesions, which become necrotic and drop out, and thus create shotholing. Blossom infections produce local necrotic lesions on the sepals and occasionally result in blossom blight.

Spores that develop on lesions are dispersed in water droplets and by wind-blown rains. Germinating spores produce germ tubes, which originate from one or two of four to eight cells and cause infection by direct penetration of host tissue. If infection does not occur from the initial penetration by the germ tube, nongerminated cells of spores can germinate later and cause infections, extending the viability of the spore and providing another chance for infection.

In the presence of moisture, new spores are readily formed in lesions and provide a source of viable spores as fruits form and develop throughout the season. New infections may occur with each rain. The susceptibility of host parts to infection in relation to temperature and length of the period of free moisture has not been clearly defined. Protective treatments should be applied before wet periods to obtain effective control.

Meteorological observations. The effectiveness of fungicides as protectants, eradicants, or suppressants is influenced by spray coverage of susceptible host parts, the presence of free moisture, and temperature. Continuous-recording rain gauges and hygrothermographs located in test plots are helpful for determining relationships between fungicide effectiveness and weather conditions. For example, information on the length of the period of free moisture at certain temperatures could be useful for forecasting the occurrence of shot hole. Data on relative humidity and temperature provide information on the deposition of dew on susceptible host parts. This information could be used in determining the timing of fungicide treatments. Data on environmental conditions from the local weather bureau are useful for making decisions on whether to spray, but not for establishing critical data on epidemiology of the disease.

Susceptibility of hosts and host parts and evaluation of disease. Methods for evaluation of shot hole on the various stone fruit crops differ since host parts of each crop differ in susceptibility. On peach and nectarine, twig blight causes the most injury, as it affects the current year's crop through loss of fruiting wood and foliage; on apricot, bud blight and fruit infections are the most important aspects; on almond, leaf infections appear to be the most important.

All cultivars of peaches and nectarines are susceptible to infection. The most serious damage occurs during the fall and winter, when dormant buds and twigs become infected. Disease severity is evaluated before the blossoming period by counting blighted buds, twig lesions, or blighted twigs. Blossom, fruit, and leaf infections are not common on peaches, so that data on them are difficult to obtain.

All apricot cultivars are extremely susceptible. In the fall and winter, bud infections are common, but twig infections are rare. In the spring, fruit and leaf infections are abundant. Susceptibility of apricot blossoms is not known. Immature fruits appear more susceptible than mature fruits. Disease evaluations are made on dormant bud infections before bloom and on fruit infections at the pit-hardening stage or even when fruits are mature or ripe. Leaf infections can be difficult to evaluate, because any injury to leaves can induce reddening and necrosis, which could be confused with symptoms of shot hole. Infections caused by the brown rot fungi (*Monilinia fructicola* (Wint.) Honey or *M. laxa* (Aderh. & Ruhl.) Honey) and the gray mold fungus (*Botrytis cinerea* Pers. ex Fr.) and certain spray treatments (such as coppers) cause lesions similar to those caused by the shot hole fungus.

All almond cultivars are susceptible and, like peaches, nectarines, and apricots, require control measures. Bud infections during the winter are rare, and twig infections occur only in the presence of high levels of inoculum. Twig blight from bud or stem infection is rare, but shoot dieback can occur and can be mistaken for bacterial blast or brown rot. Occasional blossom infections in the spring can be mistaken for infections from the organisms that cause bacterial blast or brown rot. Newly formed leaves are highly susceptible and have a tendendy to drop when infected. Disease evaluations, therefore, are best made on leaf or fruit infections. Leaf samples are collected as soon as symptoms develop; 50–100 leaves per quadrant are collected at random (the number of leaves is based on the incidence of disease). More than one collection is appropriate during years with intermittent rains. Counting of fallen leaves on the orchard floor could also provide some basis for evaluation, but one must be careful not to count leaves that fell from other causes, such as potassium deficiency. Evaluation of infections on the upper surface of the fruit, before the hulls start to dry, provides a good measure of disease severity during the season.

Selection of a test plot. The most important criterion for selecting a test plot for fungicide evaluation is the presence of high levels of inoculum canker of *S. carpophila*. An amount sufficient for a test plot can be expected to be present in an orchard if the incidence of the disease was high the previous year. Where a commercial orchard must be used, the difficulty lies in obtaining permission from the grower to leave an unsprayed check. Although more disease may occur on some cultivars than on others, a test plot should be located where the incidence of disease is uniform. Furthermore, researchers should be prepared to conduct an experiment in the same test plot for more than one year.

For tests on peaches, nectarines, and apricots, one would look for an orchard that has received poor care or has not been sprayed because the grower was unaware of the disease. In California, test plots from which data on twig blight can be obtained are difficult to find for peaches and nectarines, because disease levels are so low in most commercial orchards (2). Data on bud blight on apricots are also difficult to obtain, but one may detect trends through evaluation of fruit infections. For almonds, however, suitable test plots can be found quite easily. For experimental purposes, a separate block of trees with peaches or nectarines and apricots on experimental field stations would be helpful in testing nonregistered fungicides for control of shot hole disease.

Spray equipment and plot design. Initial fungicide screening can be accomplished with handgun equipment, which is adequate for test plots with eight to 10 single-tree replications. Ground or aircraft applications are evaluated in subsequent trials. For ground air-blast applications, the minimum size for a test plot is 5 × 5 trees replicated at least five times. In such a test plot, spray drift is unlikely to cover more than two rows, so that the middle row can be used for disease evaluation. Aircraft applications (4,6) require plots of at least 5 × 20 trees, and the aircraft should fly between rows. Fixed-wing aircraft traveling 90 mph provide adequate coverage and disease control on dormant, defoliated trees. When foliage is present, a helicopter traveling at 20 mph provides coverage of both upper and lower leaf surfaces (6). Fungicide applications through overhead sprinklers provide effective disease control, but most orchards now use low ground sprinklers or drip irrigation (1).

Statistical analysis of data. Researchers who are not familiar with the design of field plots should consult a statistician before a test plot is set up. Methods for disease evaluation (data from twigs, leaves, or fruit) and for crop yield determination should be decided at that time. Selecting a uniform orchard near peak production is better than trying to work in an orchard that is past its prime. In crop yield data, the method of harvest and the size of subsamples can be important, especially for almonds. For example, almond subsamples (usually 2 kg) must be separated into hulls, shells, and meats for individual weighing. Meats must be dried before their weights are recorded. In California, data on crop losses from shot hole disease have been studied for almonds (3), but not for other crops.

DISCUSSION

The methods commonly used to evaluate fungicides for shot hole disease use small-scale field plots, instead of laboratory tests, for initial screening. With new fungicides, periodic evaluations are necessary to determine redistribution and long-term residual from rainfall, unless one has facilities to simulate rain by overhead sprinkler irrigation. Laboratory tests showed thiabendazole (5) to be active against the shot hole fungus, but it has not been recommended. Chemicals that have been determined to be effective for control of shot hole disease in field tests are copper compounds (both Bordeaux mixture and fixed coppers), ziram, captan, captafol, dichlone, and (recently) chlorothalonil and iprodione (2,7). Timing of sprays to protect susceptible host parts is extremely critical to disease control. In addition, disease severity in the previous year and existing weather conditions (rain and temperature) influence whether a chemical significantly reduces disease incidence. To make a fair evaluation of a fungicide's effectiveness under low disease incidence, a large-scale field test in a commercial orchard is necessary.

Fungicides should be evaluated for their effect on crop yield as well as for their effectiveness for disease control. Final conclusions concerning fungicide effectiveness for control of shot hole should be based on differences in crop yield. The cost of disease control should be compared to the benefits received, if any, from the increase in yield. For shot hole disease, this information should be compiled for at least three to five years.

Researchers should be aware that certain fungicides may not be appropriate for disease control on commodities that may eventually be fed to lactating animals or to animals for slaughter.

In the final analyses of experiments conducted over the years, results of tests vary considerably between orchards with low disease incidence and those with high disease incidence.

LITERATURE CITED

1. Aldrich, T., Moller, W. J., and Schubach, H. 1974. Shot hole disease control in almonds—By injecting fungicides into overhead sprinklers. Calif. Agric. 28(10):11.
2. English, H., and Davis, J. R. 1962. Efficacy of fall applications of copper and organic fungicides for the control of Coryneum blight of peach in California. Plant Dis. Rep. 46:688-691.
3. Highberg, L. M. 1983. Shot hole disease of almond: Studies on yield reduction in relation to disease incidence and survival of inoculum in association with dormant bud. M.S. thesis, Department of Plant Pathology, University of California, Davis.
4. Kilgore, W. W., Yates, W. E., and Ogawa, J. M. 1964. Evaluation of concentrate and dilute ground air-carrier and aircraft spray coverages. Hilgardia 35:527-536.
5. Luepschen, N. S., Rohrbach, K. G., Core, C. M., and Gilbert, H. A. 1968. Coryneum blight infection and control studies. Colo. Agric. Exp. Stn. Prog. Rep. 68-2.
6. Ogawa, J. M., Yates, W. E., and Kilgore, W. W. 1964. Susceptibility of almond leaf to Coryneum blight, and evaluation of helicopter spray applications for disease control. Hilgardia 35:537-543.
7. Ogawa, J. M., Teviotdale, B. L., Rough, D., Manji, B. T., Highberg, L. M., Viveros, M., and Yoshikawa, F. 1983. Shot hole of stone fruits. Leafl. 21363. University of California, Division of Agricultural Sciences, Oakland. 4 pp.
8. Smith, R. E. 1907. California peach blight. Calif. Univ. Agric. Exp. Stn. Bull. 191:73-100. (out of print)
9. Wilson, E. E. 1937. The shot-hole disease of stone fruit trees. Calif. Univ. Agric. Exp. Stn. Bull. 608:3-40. (out of print)
10. Wilson, E. E., and Ogawa, J. M. 1979. Fungal, Bacterial, and Certain Nonparasitic Diseases of Fruit and Nut Crops in California. University of California, Division of Agricultural Sciences, Oakland. 190 pp.

Field Methods for Evaluating Fungicides for Control of Greasy Spot, Melanose, and Scab on Citrus

J. O. WHITESIDE, University of Florida, Institute of Food and Agricultural Sciences, Citrus Research and Education Center, Lake Alfred 33850

The number of fungicide applications that can be used to control greasy spot (caused by *Mycosphaerella citri* Whiteside), melanose (caused by *Diaporthe citri* Wolf), and scab (caused by *Elsinoë fawcettii* Bitanc. & Jenkins) in Florida citrus groves is economically limited. Any candidate material that would need to be applied more than twice annually to control any one of these diseases probably could not compete with materials that are now recommended (1). Greasy spot is usually controlled satisfactorily by one summer spray treatment of copper fungicide or, if the disease pressure is not too heavy, of 1% oil-water emulsion (2,11). One application of copper fungicide in late April or early May usually gives adequate control of melanose (9). Scab can often be controlled satisfactorily by one spray of captafol at bloom (10); an earlier treatment of captafol in late winter, just before new shoots emerge, is needed only if the disease was severe the previous year.

The number of times that citrus trees can be sprayed each year is limited by the expense of the relatively high spray volumes required to cover the characteristically dense canopy. Because the capital outlay is high, each spraying unit is expected to handle a large acreage, and spraying often has to begin before the most appropriate time epidemiologically. Therefore, only materials that have long residual protective action, inoculum-reducing ability, or both are satisfactory. The screening procedures for materials for control of greasy spot, melanose, and scab are necessarily severe. Initially, a test material is applied only once, at a time when it is likely to have maximum effect. Because the major infection periods for the three diseases differ, materials are tested against each disease in separate experiments.

Trees are sprayed to runoff with single-nozzle handguns at 2,758 kPa (400 psi). Volume of spray applied averages from 30 L (8 gal) per tree for 3-m (10-ft) trees to 76 L (20 gal) per tree for 6-m (20-ft) trees.

GREASY SPOT CONTROL

Test Sites and Cultivars

Tests for greasy spot control are made on bearing grapefruit (*Citrus paradisi* Macf.) and sweet orange (*C. sinensis* (L.) Osbeck) trees that have had moderate to severe greasy spot in previous years. Grapefruit trees are used to test the effectiveness of spray materials against greasy spot rind blotch, because this facet of the disease is much more conspicuous on grapefruit than on sweet orange.

The most serious aspect of leaf infection is the defoliation it causes during the winter after infection. Excessive defoliation before the spring growth has emerged causes the new growth flush to be weak and consequently reduces the yield of the next crop of fruit. Because more premature defoliation usually occurs from the spring growth flush than from summer flushes, disease ratings are best obtained from the previous spring flush.

Organism

Ascospores of *M. citri*, released from perithecia produced on partially decomposed fallen citrus leaves, are the major source of inoculum. Conidia of the anamorphic (*Stenella*) state of this fungus are produced sparsely on transient mycelium that develops on the surface of leaves and fruit following ascospore or conidia germinations, but they are of minor epidemiologic importance. Host penetration occurs only through stomata, and many stomata per unit area must be penetrated for greasy spot to develop. Development of mycelial growth on the leaf surface considerably increases the chances of this happening.

The supply of infected fallen leaves reaches a peak in May following heavy leaf drop through winter and spring. Perithecial development and maturation are hastened by more frequent rainstorms in May and June. Ascospore release (mostly by rain) usually peaks in June or July (11). During August or September, ascospore numbers decline rapidly because of advancing decomposition of the leaf litter, perithecial exhaustion, and the fact that fewer leaves drop from citrus trees during the summer to replenish the inoculum supply (7). The combined high temperatures and prolonged high humidity that occur almost nightly in Florida from June through September favor development of the external mycelial growth and infection (3). Because ascospores are fewer after July, however, more infection occurs in June and July than in August or September.

Normal cultural practices are followed in the test groves. Routine disk harrowing in late spring buries much of the leaf litter, but enough litter remains under the tree canopy itself to cause potentially severe attacks.

Procedure

The first tests with a new material are made either on whole trees, if they are less than 3 m (10 ft) high, or on 1.8-m- (6-ft-) wide sectors of the canopy of large trees. On very large trees, several such sectors can be used on each tree if a pressure-retaining hand sprayer is used to apply the treatments to each sector. Confining all of the plots of one replication to either the north or the south side of the tree is important, however, because defoliation from greasy spot occurs earlier on the shaded north side of the canopy. The single-tree or canopy-sector plots are replicated at least six times. After a material has been determined to be effective against *M. citri* in small-tree or tree-sector plots, it is tested on large trees using one- or two-tree plots replicated at least six times in a randomized block design.

The risk of interference from drifting spray droplets is slight on calm days, because such droplets are unlikely to settle on the underside of the leaves. Fungicides must be deposited on the lower leaf surface (to which the stomata are confined) to prevent infection. Therefore, unsprayed trees are not left as buffers between plots unless the space between the canopies of neighboring trees is less than 1.8 m (6 ft). As a precaution, however, sprays are not applied to nonbuffered trees if the wind speed exceeds 8 km/hr (5 mph).

The test materials are applied only once any time between mid-June and late July. Special care is taken to wet the underside of the leaves thoroughly. Unsprayed check trees and standard treatments of 1% oil-water emulsion alone and a copper fungicide are included in every test. No other fungicides or oil sprays are applied to the trees during the year of test. Promising materials are later tested in combination with spray oil (meeting FC 435-66 specifications of Florida Pesticide Law, Rules and Regulations), which is commonly applied in the summer to control several pests as well as greasy spot.

Data Collection

Greasy spot has a long incubation period, and symptoms do not normally appear before November or December. By this time, identifying shoots of the spring growth can be difficult. Spring growth can be confused with late growth of the previous year and even with subsequent summer growth of the current year. Therefore, shoots of the spring flush must be labeled in May soon after they have fully expanded so that they can be identified the following winter. For this purpose, a copper-wired white 1.5 × 9.0-cm plastic tag is tied around the base of each shoot to be sampled. The number of shoots thus labeled on each canopy sector or tree plot varies from 20 to 40. To aid retrieval, the tags are placed in groups at the same locations on each plot. On a canopy-sector plot, they are all placed in a group near the center of the sector. When whole trees are used, tags are placed at the same two or four compass points on each tree.

Procedures used to assess greasy spot involve cutting off the labeled shoots and examining the leaves closely for symptoms. In some early studies (2), greasy spot severity was based only on the percentage of the area of the leaf blade with disease symptoms. This method requires that leaves be examined before substantial defoliation begins in unsprayed checks and poorer treatments. This usually means assessing disease severity long before the spring growth emerges. More recently (11), the fact that greasy spot is harmful because it causes premature leaf abscission has been taken into consideration. Data now include the amount of defoliation that occurs by the time the new shoots begin to appear the following spring. All leaves on each shoot are counted the previous summer before greasy spot appears, and a final count is made in February or March. The mean percentage of defoliation is calculated from these data. Leaves that do not abscise are examined for greasy spot to provide an additional rating based on the percentage of remaining leaves that are diseased.

Materials for controlling greasy spot rind blotch are tested mostly in older grapefruit groves because, for unknown reasons, greasy spot rind blotch seldom appears on the fruit of young grapefruit trees. About 80–100 fruits are picked at random from the lower part of the canopy on each one- or two-tree plot. The sample is washed to remove sooty mold and other obscuring deposits and graded for presence of greasy spot rind blotch. Sometimes the disease is also rated as a percentage of fruit surface showing symptoms. These washed fruit are also examined for any evidence of phytotoxicity caused by the spray materials.

All percentage data are arc sine transformed to degrees of an angle before being subjected to an analysis of variance and Duncan's multiple range test.

MELANOSE CONTROL

Test Sites and Cultivars

All citrus cultivars are susceptible to melanose, but tests usually are made with grapefruit, on which the disease has a greater commercial impact. The severity of melanose is determined partly by the amount of recent twig dieback in the tree canopy. Old trees usually carry more dead twigs than young trees. In Florida, melanose severity on fruit rind varies considerably from year to year, and sufficient disease pressure for fungicide screening purposes cannot be assured even by using overhead sprinkler irrigation.

The use of older trees to increase the chances of heavy infection creates problems of spray drift, because of tree height. Buffer trees are needed between plots, which increases the number of trees required for fungicide evaluations. To reduce the costly wastage of fruit that inevitably occurs when large trees are used for testing nonregistered materials, a bioassay procedure using sprayed fruit from small trees is used for primary screening.

Organism

In Florida, *D. citri* produces airborne ascospores and water-liberated and splash-dispersed pycnidiospores. Only the latter, however, are responsible for serious melanose development on fruit rind. Rind blemish occurs when rain washes large numbers of pycnidiospores onto the fruit from dead twigs above. No pycnidia occur on the melanose pustules themselves.

Melanose tends to be unevenly distributed. Fruit on the lower part and inside of the canopy carry more melanose than those located at the top and outside. The amount of infection depends on the occurrence of long wetting periods before the rind becomes immune to infection about 12 weeks after petal-fall (9). At 25°C, a minimum of 10–12 hr of continuous wetting is required for infection, but at 15°C, the minimum time increases to 18–24 hr (5).

Usually, more infection occurs in May or June than in March or April (9), partly because of differences in the number of days of rain in these months and partly because of temperature. Also, precipitation after April occurs mostly as afternoon and evening thunderstorms that are often conducive to overnight wetting periods. In contrast, rainfall in March and April is usually associated with fast-moving cold fronts and rapid drying. The probability of infection after cold-front precipitation is reduced further by the rapid drop in temperature that generally follows.

For fungicide screening tests, a record of probable infection days is maintained. This record is based on data provided by a continuously recording rain gauge and hygrothermograph and is done by inspecting fruit for raindrops in early morning and by periodically examining fruit for early symptoms of melanose.

Procedures—Primary Screening

When young citrus fruit are picked and inoculated with *D. citri*, a melanose reaction becomes visible within 7 days, which is before substantial degeneration of the fruit begins. This phenomenon can be used to advantage to detect activity of fungicidal residues on the fruit surface (8).

Young bearing grapefruit trees are sprayed with test materials about 4 weeks after petal-fall. Fruits, with 4–6 cm of stem attached, are later clipped at random from the tree. Care must be taken to avoid removing or smudging the spray residue; the fruits are held only by their stems. The fruits are placed on a 3-cm-high hardware cloth (mesh size 12 mm) platform in a plastic wash basin measuring 32 × 62 cm and 14 cm deep. The stem of each fruit is inserted through the mesh, and the fruits are placed close enough together to prevent movement and possible change in their orientation. Within 2 hr of picking, the basins are returned to the laboratory and filled with demineralized water to the level of the platform.

To produce the pycnidiospores that are required as inoculum, *D. citri* is cultured on autoclaved 1- to 2-yr-old twigs from grapefruit trees. After pycnidia develop, the twigs are air-dried for 2 days and then placed in a damp chamber to induce exudation of spore tendrils. The tendrils are picked off with a needle and immersed in 1% orange juice in distilled water. Pycnidiospores disperse rapidly but sink and adhere to the bottom of the container. Repeated dislodging with a camel's-hair brush is needed to maintain a uniform suspension. The suspension is standardized at 5×10^5 pycnidiospores per milliliter. This concentration is high enough to form a continuous necrotic spot over the inoculated area, provided no fungicidal residue is present in this area.

Two small drops (each about 15 μl) of inoculum are deposited with a syringe about 10 mm apart near the uppermost point of each fruit. The basin is covered as soon as possible with a sheet of polyethylene to prevent the drops from evaporating. Water is essential for survival of conidia as well as for germination and infection. The temperature is maintained at 25–28°C, and the cover is removed after 2 days.

Procedures—Grove Evaluation

Test materials are applied to one- to four-tree plots replicated at least six times in a randomized block design. A single line of unsprayed buffer trees must separate the test trees to avoid spray drift effects. No other fungicides are applied to the trees during the year of test. Rust mites are controlled by timely spraying with acaricides before the fruit russeting they cause can obscure symptoms of melanose.

Basic copper sulfate, 1 g of metallic copper per liter (0.8 lb/100 gal), is included as a standard treatment in all tests. Sometimes it is applied also at half this rate, particularly when different copper fungicides are to be evaluated (5).

Fungicides are tested as postbloom fruit-protectant sprays. For this purpose, the test materials are applied in late April or early May—about 4 weeks after petal-fall and before the 6- to 8-week period when infection is most likely to occur.

Data Collection—Primary Screening

In assessing the results of the inoculation tests on detached fruit, it is assumed that a fungicidal residue was present where the drop of inoculum was applied if no symptoms (brown discoloration of the rind surface) develop within 7 days. The first evaluation is made within 7 days of spraying the tree. To determine the relative persistence of the test materials on the fruit surface, further samples are bioassayed weekly for the next 2 or 3 weeks.

Data Collection—Grove Evaluation

Because melanose pustules are more obvious after the rind has colored, collection of fruit for disease assessment is usually delayed until December or later. The sample consists of 80–100 fruits picked at random only from the lower part of the canopy, where melanose is likely to be more severe. Fruits are washed to remove superficial deposits and graded as either essentially melanose-free (up to 100 small pustules per fruit are allowed if they are widely scattered and none exceeds 1 mm in diameter) or as diseased (pustules sufficiently conspicuous and numerous to detract from fruit appearance). Data are presented as the percentage of melanose-free fruit. An analysis of variance and Duncan's multiple range test are done after the percentage values are transformed to degrees of an angle.

SCAB CONTROL

Fungicides for scab control are screened in the field in two stages (4). A test material is first included in small-scale experiments to determine whether it has sufficient activity against the scab fungus to justify further interest. These tests are made on small, nonbearing trees and are based on the ability of a material to reduce the amount of inoculum at source or to become redistributed or both, thereby protecting shoots that emerge after treatment. Materials that pass the primary screening tests are then applied to bearing trees to determine their effectiveness in preventing fruit infection.

Test Site and Cultivars—Primary Screening

Primary tests are made on trees of a rough lemon (*C. jambhiri* Lush.) variant (author's accession No. 27) that is highly susceptible to scab but resistant to the Alternaria leaf spot that makes most rough lemons less suitable for work on scab. Although other scab-susceptible citrus species such as sour orange (*C. aurantium* L.) could be used, this rough lemon variant is preferred because it is so vigorous that the trees recover quickly when scab pressure is reduced. Thus, the trees can be used again for further tests after only a short recovery period.

The site is irrigated with overhead sprinklers to encourage scab attack when rainfall is inadequate. After completion of a test, the canopy is pruned periodically to keep tree size within bounds and to promote abundant new growth that can become infected with scab. The trees are not used again for further tests until disease uniformity among trees has been restored.

Test Site and Cultivars—Grove Evaluation

Bearing trees of the commercially important Temple (*C. temple* Hort. ex Y. Tanaka) cultivar are used for grove testing of promising candidate materials. Temple fruit react to infection by producing conspicuous blister-type scab lesions. Temple trees are, therefore, more suitable for testing fungicides for scab control than are citrus cultivars such as grapefruit, which produce a flat, scurfy type of lesion that can be confused with wind scar. Temple trees less than 3.6 m (12 ft) high are used for fungicide screening tests, making buffer trees against spray drift unnecessary.

Organism

The scab fungus overwinters in infected living leaves, twigs, and fruit. Successive crops of conidia are produced from old scab lesions when wetted by rain or dew. The perfect state of the fungus has not been observed in Florida. Conidia are mostly water-liberated and splash-dispersed. The disease therefore tends to be unevenly distributed throughout the tree canopy, with the severity of fruit infection depending on proximity to the inoculum source. Sometimes, another type of conidium is produced that can become airborne. Such conidia can withstand limited desiccation after dispersal and germinate in dew the following night (6). However, although these conidia may be important in long-distance dissemination, they are too few, in themselves, to cause severe symptoms on the fruit.

Shoot and leaf infection is not of major economic importance except on susceptible rootstock varieties such as rough lemon and sour orange in nurseries. Shoot infection is, however, of epidemiologic importance, because it builds up inoculum to infect fruit. Fungicides are required to inhibit conidia production on scab lesions while new shoots and young fruit are susceptible to infection. Critical spraying times for scab control are just before shoots emerge in early spring and before major fruit set. Leaves are susceptible to infection for only a few days after emergence, but fruit rind remains susceptible for about 3 months after petal-fall.

Conidia of *E. fawcettii* require only a short period of wetting to germinate and penetrate the host; even at 15°C, 3–4 hr suffice (6). In this respect, scab differs from melanose. It can be promoted by cold front-induced precipitation and relatively short periods of overhead sprinkler irrigation. Furthermore, only a short period of wetting is required for conidia development. Operation of overhead sprinklers for 4–6 hr is long enough to promote conidia formation and dispersal and infection.

Procedure—Primary Screening

Each material is applied to at least five trees arranged in a randomized block design. The materials are sprayed either in late winter just before spring shoot growth is expected or during the summer at times when no shoot extension is present. If one sees in summer tests that the next flush of growth may be long delayed or uneven, the trees are pruned lightly to promote new shoot growth. If rainfall is inadequate, scab infection is assured by running the overhead sprinkler irrigation periodically for 12–16 hr during the early phases of shoot expansion. Standard treatments of captafol are included in every test.

Procedure—Grove Evaluation

Treatments are applied to two- or three-tree plots of Temple trees replicated at least six times in a randomized block design. Sprays are applied shortly before major petal-fall. Unsprayed check treatments are included in each test, but no buffer trees are left between plots.

Data Collection—Primary Screening

The degree of control is based on the number of new shoots that show scab symptoms. After the new shoots have fully expanded, 50 are selected at random from the lower half of the canopy of each tree. Turgid shoots must be examined, because wilting obscures symptoms. No attempt is made to index disease severity on each shoot; the shoot is considered diseased if even one lesion is visible. Data are expressed as the percentage of diseased shoots.

Data Collection—Grove Evaluations

Scab is assessed only on the fruit. Because scab symptoms are conspicuous on immature Temple fruit, data can be obtained before normal picking time. Usually, however, the crop is allowed to mature before sampling. Washing fruit samples before examination is not necessary, because scab lesions show clearly through any surface deposits. Fruit can therefore be graded as healthy or infected in the grove at harvest. Disease assessment is based on a sample of 100 fruits picked randomly around the middle of the canopy. Percentages of diseased fruit are transformed to degrees of an angle and then subjected to analysis of variance and Duncan's multiple range test.

DISCUSSION

Greasy Spot

Coefficients of variation in screening tests for greasy spot control have fallen mostly between 15 and 25%. Although greater precision might be achieved by increasing the sample size and number of replications, the current procedures are considered precise enough for screening and ranking. Standard treatments of copper fungicides have consistently been effective (2,11) and have provided a satisfactory reference for interpreting test results. Oil sprays, however, have been less reliable as standards because of their variable effectiveness.

Defoliation data provide the best means of expressing greasy spot severity. Nevertheless, greasy spot sometimes develops too late or is not severe enough to cause substantial defoliation before the end of winter. In these cases, useful comparative data can still be obtained by rating the disease according to the percentage of leaves or of leaf area affected.

Spray oil is applied almost routinely to Florida citrus

groves in the summer to control various pests as well as greasy spot. Mostly, fungicides are used as a supplement to oil to help insure control. Testing candidate materials for compatibility with oil is, therefore, an essential part of the testing procedure.

A candidate material must have certain qualities to be effective against greasy spot. Long residual action is of primary importance. The material must also kill on contact any mycelial growth of *M. citri* already present on the leaf and fruit surface.

Melanose

The bioassay test used for primary screening of experimental fungicides for control of melanose is intended to rapidly eliminate candidate materials that have little or no fungitoxicity against *D. citri*. It is not sensitive enough to detect minor differences in treatment efficacy of different materials. Even in conventional field plot experiments, the level of precision in comparing different treatments for melanose control has not been high. Coefficients of variation in fungicide screening tests against melanose have ranged from 18 to 28% (8). Although greater precision is preferred, practical considerations impose limitations. The amount of spraying that can be done in 1 day restricts the number of trees per plot and the number of replications that can be covered. More replication and larger fruit samples would be required to detect differences of lower magnitude statistically.

Relatively few materials have equaled copper fungicides in their ability to control melanose, presumably because few other materials are sufficiently stable to provide long-term residual protection. The practice of applying test materials only once during the 3-month period of rind susceptibility highlights the residual qualities of chemical treatments.

Scab

In Temple groves, scab severity varies from tree to tree, resulting in coefficients of variation approaching 50% in some tests. This problem has been partly solved by using the areas in Temple groves that showed the heaviest, most uniform infection the previous year, and by increasing the number of replications.

Including the highly effective standard, captafol, in each test has assisted considerably in interpreting results from candidate materials. Copper fungicides and ferbam are no longer used as standards because their performance is inconsistent and often poor (4,10).

The possibility that windblown conidia may cause interplot interference must be considered in tests for scab control. Results could be misinterpreted if new shoots or young fruit become infected with conidia from external sources, particularly if substantial secondary infection follows. In sprayed plots, conidia from external sources could obscure eradicative effects on conidia from old scab lesions. For this reason, unsprayed buffer trees are omitted. Spray drift problems are avoided by using relatively small trees. Interplot interference is also reduced by including in the grove tests only those materials that seemed highly promising in the primary screening tests. Unsprayed check plots are necessary, however, to measure the variability in scab pressure that can occur over the experimental area and the considerable differences in disease severity that occur from year to year.

LITERATURE CITED

1. Florida Cooperative Extension Service. 1986. Florida Citrus Spray Guide. Fla. Coop. Ext. Serv. Circ. 393-L. University of Florida, Gainesville.
2. Whiteside, J. O. 1973. Evaluation of spray materials for the control of greasy spot of citrus. Plant Dis. Rep. 57:691-694.
3. Whiteside, J. O. 1974. Environmental factors affecting infection of citrus leaves by *Mycosphaerella citri*. Phytopathology 64:115-120.
4. Whiteside, J. O. 1974. Evaluation of fungicides for citrus scab control. Proc. Fla. State Hortic. Soc. 87:9-14.
5. Whiteside, J. O. 1975. Evaluation of fungicides for the control of melanose on grapefruit in Florida. Plant Dis. Rep. 59:656-660.
6. Whiteside, J. O. 1975. Biological characteristics of *Elsinoë fawcetti* pertaining to the epidemiology of sour orange scab. Phytopathology 65:1170-1177.
7. Whiteside, J. O. 1977. Behavior and control of greasy spot in Florida citrus groves. Proc. Int. Soc. Citriculture 3:981-986.
8. Whiteside, J. O. 1977. Method for rapid screening of fungicides as candidates for citrus melanose control. Proc. Fla. State Hortic. Soc. 90:67-70.
9. Whiteside, J. O. 1980. Timing of fungicide spray treatments for citrus melanose control. Proc. Fla. State Hortic. Soc. 93:21-24.
10. Whiteside, J. O. 1981. Evolution of current methods for citrus scab control. Proc. Fla. State Hortic. Soc. 94:5-8.
11. Whiteside, J. O. 1982. Timing of single-spray treatments for optimal control of greasy spot on grapefruit leaves and fruit. Plant Dis. 66:687-690.

Evaluating Fungicides for Control of Phytophthora Crown and Root Rot Diseases of Tree Crops

L. W. TIMMER, University of Florida, Institute of Food and Agricultural Sciences, Citrus Research and Education Center, Lake Alfred 33850, and M. ELLIS, Department of Plant Pathology, The Ohio State University, Ohio Agricultural Research and Development Center, Wooster 44691

PHYTOPHTHORA DISEASES OF CITRUS

Phytophthora spp. cause serious foot rot and root rot problems in most citrus-producing areas of the world. Root rot is primarily a problem in seedbeds and nurseries on highly and moderately susceptible rootstocks. It is a problem in citrus orchards only on susceptible rootstocks under conditions favorable for disease development. The most serious problem is foot rot or gummosis, which generally occurs above the bud union on susceptible scion cultivars such as sweet orange (*Citrus sinensis* (L.) Osbeck) and grapefruit (*C. paradisi* Macf.). Crown rot below the soil surface usually occurs only on more susceptible rootstocks. In most citrus areas, the species of *Phytophthora* associated with foot rot and root rot are *P. parasitica* Dast. and *P. citrophthora* (R. E. Sm. & E. H. Sm.) Leonian. Phytophthora diseases are endemic problems in most citrus orchards.

Control of *Phytophthora* spp. in seedbeds and nurseries has depended on soil fumigation and sanitary practices to prevent reinvasion of treated areas. Control in the field has been attained by use of resistant rootstocks and by following cultural practices such as high budding, avoiding trunk wounds, and keeping the areas around trunks clean and dry.

At one time, the use of fungicides was limited to painting the trunks of young trees with copper fungicides or captan before banking trees with soil or wrapping trees with insulating materials for freeze protection. Also, foot rot lesions on trees in orchards were often treated surgically and the affected area painted with copper fungicides. More recently, the advent of protective and systemic fungicides highly effective against *Phytophthora* spp. has given the citrus grower additional options for controlling root rot and foot rot. As these materials became available, methods were developed to evaluate their efficacy and longevity. These methods are described herein.

Methods of Inoculation and Fungicide Application

Test site and test plants. Fungicides for control of root rot usually are evaluated most efficiently in containers in the greenhouse or in field seedbeds or nurseries using highly susceptible sweet orange seedlings. In commercial nurseries, more resistant rootstock seedlings such as Carrizo citrange can be used, if *Phytophthora* populations are high and conditions are favorable for disease development.

For fungicide evaluations for foot rot control, 1- to 2-year-old sweet orange or grapefruit trees budded on susceptible or moderately resistant rootstocks are useful test material. The scion cultivar may be less susceptible if budded on a highly resistant rootstock. Older trees are less susceptible to infection by *Phytophthora* spp., lesions expand more slowly, and fungicide treatments of large trees seem to be less effective (3).

Inoculation methods. *Preparation of inoculum.* For mycelial inoculum, *Phytophthora* spp. are grown on V-8 juice agar (10). Plugs of mycelium from agar plates have been used for trunk inoculation. Where larger quantities of inoculum are required for infestation of soil, *Phytophthora* spp. can be grown in autoclaved vermiculite saturated with dilute V-8 juice.

For production of chlamydospores for infestation of soil, *P. parasitica* can be grown in bottles containing clarified V-8 juice (14). Bottles are incubated horizontally until thick mycelial mats are formed. Bottles are then placed upright, and 100 ml of sterile water is added. Chlamydospores are formed in abundance on the submerged mat after incubation for 2–3 weeks. Chlamydospores are separated from the mycelium by blending and low-speed centrifugation.

Zoospores for trunk or root dip inoculation of seedlings are produced by first growing the fungus in V-8 juice broth for 2–3 days. The mycelial mat is rinsed several times in sterile distilled water and then incubated in a shallow layer of sterile water for 2–3 days. Zoospore formation can be induced by chilling sporangia at 6°C for 15–20 min and allowing them to warm to room temperature (12).

Inoculation procedure. For tests of fungicide efficacy against root rot, infested soil is needed. Naturally infested nursery soil can be used at the site or placed in containers for greenhouse evaluations. Alternatively, soil or potting mixtures can be infested by adding chlamydospores (6) or by mixing mycelial inoculum grown on vermiculite with the potting medium. Levels of infestation can be determined by soil dilution plating on selective media. The $P_{10}VP$ medium plus hymexazol developed by Tsao and Guy (15) and the PARP medium (8) plus 50 mg of hymexazol per liter are highly effective for determining population levels of *P. parasitica*. Various levels of infestation can be established by diluting infested soil with sterile soil.

For some types of studies, it may be desirable to inoculate roots directly. This can be best accomplished by dipping roots in zoospore suspensions (3).

Uniform inoculation of trunks of young trees to evaluate the effectiveness of fungicides against foot rot is difficult. Commonly, mycelial plugs are placed under the bark, and the wounded area is wrapped or covered to prevent drying. It is generally necessary to maintain moisture around the inoculation site for at least 1–2 weeks to assure foot rot development. This can be accomplished by packing moist cotton around the inoculation site or by mounding soil up against the trunk (11).

An effective means of inoculation was described by Whiteside (17). A watertight collar is formed around the trunk using Tygon tubing sealed with wax and petroleum jelly. The trunk is wounded by making a vertical cut in the bark, and a zoospore suspension is added to the watertight collar. Moisture is maintained around the wound for 6 days by adding water or by placing wet absorbent cotton within the collar.

Fungicide application. The selection of a method of application depends on the nature of the fungicide and the purpose it is designed to serve. Soil drenches have been used effectively for systemic fungicides that are translocated acropetally (3,4,6,13). Foliar sprays have been effective only with systemic fungicides like fosetyl Al that are translocated basipetally (3,4,6).

One of the most effective ways to control foot rot has been to apply protective or systemic materials as trunk paints (3,4,12,13). Captafol, some copper fungicides, and metalaxyl have provided protection against infection for up to a year when applied as trunk paints (12,13). Systemic materials may be applied after infection to test their curative ability. Curative effects have been demonstrated with metalaxyl applied as a trunk paint or soil drench (3,4,6,13). Protective fungicides may be applied after affected bark has been surgically removed to prevent further lesion development, but this approach to foot rot control has not been highly effective (13).

Applications of fungicides as soil surface sprays or through irrigation systems have not been investigated thoroughly but if effective would be simple and inexpensive ways to apply fungicides.

Evaluating Fungicide Efficacy

Disease incidence. The ability of a fungicide to reduce disease occurrence under natural conditions is difficult to evaluate because of the large number of trees needed. However, the ability of protective fungicides to prevent foot rot is ultimately the only measure of their effectiveness. Artificial inoculations are inappropriate because the wounds needed for successful infection breach the barrier of protective fungicide applied as trunk paint. The bioassays described below help circumvent this problem and allow detection of the distribution and persistence of fungicides on the bark surface. However, fungicides that survive initial screening and testing by other means should be evaluated in large-scale field plots under natural conditions.

Disease severity. Disease ratings and other measures of the effects of the disease are generally more useful indicators of fungicide efficacy than disease incidence. Visual ratings of root rot and shoot symptoms are useful for evaluating fungicides for root rot control. These ratings can be combined with quantitative measurements of root and top dry weights for comparisons of the benefits of fungicide application. The split root systems described by Munnecke (9) for evaluating systemic effects of fungicides for avocado root rot control may also be useful for citrus.

Fungicides for control of foot rot lesions can be evaluated by ratings of lesion severity or lesion activity (presence or absence of gum). More precise methods involve measuring lesion width or the lesion area by tracing the outline onto paper or plastic sheets and using a planimeter to determine the area.

Bioassays. A bioassay to measure fungicidal activity against *Phytophthora* spp. (12) has been used to evaluate protective and systemic fungicides (3,4,13). In this method, zoospores, produced as described above for inoculm production, are applied to tissue that has been treated with a fungicide. A 30-μl droplet of a zoospore suspension of about 5×10^4 zoospores per milliliter is placed on treated pieces of tissue held in a petri dish on moistened filter paper. After a 24-hr incubation at 25–30°C, the tissue pieces are inverted onto PVP selective media (cornmeal agar plus 10 mg of pimaricin, 100 mg of pentachloronitrobenzene, and 200 mg of vancomycin per milliliter). The area of the resultant colony minus the tissue piece is calculated after 72 hr of incubation at 30°C. Data are expressed as percent reduction in colony area compared with an untreated control. The method was first used to determine the longevity of fungicide residues on the bark of trees treated with trunk paints. It also proved effective for detecting systemic activity of fungicides in disks of trunk bark and cross sections of twigs. It is less effective for detecting activity in leaf disks and cross sections of roots, because of problems with bacterial contamination.

Population determination. At times it may be desirable to determine the effects of fungicide applications on populations of *Phytophthora* spp. in soil. This is readily accomplished by adding 10 g of soil to 90 ml of 0.25% agar and plating 1.0 ml of the slurry on each of 10 plates of the PVPH medium of Tsao and Guy (15) or the selective medium of Kannwischer and Mitchell (8) plus 50 mg of hymexazol per liter. Plates are incubated for 2–3 days at 30°C in the dark, and the colonies are counted after the slurry has been washed from the plates. Propagules may range from 1–5 per gram in lightly infested field soils to 300–600 per gram in heavily infested nursery soils under optimal conditions for disease development.

Perspective

The need for fungicides to control the Phytophthora diseases of citrus is limited. A great deal can be accomplished through proper use of resistant rootstocks and cultural practices to prevent disease. Currently available fungicides are highly effective against *Phytophthora* spp. and at times are being used as a substitute for good management practices. However, repeated applications of some of these fungicides, such as metalaxyl, may well result in the appearance of resistant isolates of *Phytophthora* spp. Effective fungicides are needed to control root rot in seedbeds and nurseries, but widespread use of such materials is likely to result in loss of these fungicides as rapidly as they are discovered.

Most current research indicates that little can be gained by trying to control Phytophthora foot rot and root rot in mature, producing orchards by fungicide applications. Direct application of systemic fungicides to active foot rot lesions is effective and can reduce tree losses but is labor-intensive. The benefits to be derived from soil surface applications of fungicides in bearing orchards solely to control feeder root rot on susceptible rootstocks are unknown. The effectiveness of fungicide applications as soil surface sprays or through irrigation systems for control of foot rot on young field trees has not been determined. Direct application of the fungicide to the site of infection, i.e., the trunk or crown of the tree, is likely to be most effective. Rapid, economical methods to make such applications are needed.

PHYTOPHTHORA DISEASES OF APPLE

Collar and crown rot of apple are caused by *P. cactorum* (Leb. & Cohn) Schroet., but several other *Phytophthora* spp. have been associated with the disease (7). In collar rot, the fungus invades at the soil line on the tree trunk, moves laterally and longitudinally, and can eventually girdle and kill the tree. In crown rot, invasion occurs below ground in the region where major roots emerge from the lower trunk, and the infection extends distally along the primary roots (7).

Until recently, the use of fungicides to control collar and crown rot was not effective. Recently introduced fungicides such as metalaxyl (Ridomil 2E and other trade names), which are highly effective against pythiaceous fungi as well as being systemic, could provide effective chemical control. However, the use of fungicide should not be considered the major means of controlling collar or crown rot. The development and use of resistant rootstocks, combined with proper site selection (i.e., well-drained soil), remain the most economical and successful means of control.

There is little information on established procedures for evaluating fungicides for control of crown and collar rot of apple. The procedures described here were used in Ohio to study the in vitro effects of fungicides on mycelial growth, sporangium production, and zoospore germination of *P. cactorum*; upward movement of fungicide in apple trees following soil drenches; and efficacy of fungicide for disease control in the field. Although these procedures were used specifically to study metalaxyl (5), they should be useful in evaluating the efficacy of other fungicides on other fungi on other tree crops.

In Vitro Laboratory Procedures

Preliminary testing in vitro may help to eliminate compounds that show little or no activity against the fungus. However, poor or no activity in in vitro tests is not proof that the fungicide is ineffective. For example, metalaxyl incorporated into a synthetic medium at 1 μg/ml prevented growth of mycelia of *P. cactorum*, but when fosetyl Al (aluminum tris (-O-ethyl phosphonate)) (Aliette) was incorporated into the same medium, it had no effect on mycelial growth at 1,000 μg/ml (M. Ellis, *unpublished*). Based on in vitro tests, fosetyl Al appeared ineffective, and further testing would have been stopped. Even though fosetyl Al performs poorly in in vitro tests, it appears to be effective in controlling pythiaceous fungi.

Fungal isolates. Isolates should be obtained from infected apple trees. Stock cultures are maintained on dilute V-8 juice agar (10) in screw-top test tubes or bottles. To assure pathogenicity, apple fruit or seedlings should be inoculated with the test isolate and the fungus reisolated from diseased tissue before inoculations in the greenhouse or field.

Effect of fungicide on mycelial growth. The required amount of fungicide is incorporated as a suspension or solution in a solvent into lima bean agar (16), after the medium has been autoclaved and cooled to approximately 45–50°C. Fifteen milliliters of medium containing the desired concentration is placed in petri dishes. At least 10 dishes are used per concentration tested, with medium containing no fungicide as the control. Lima bean agar disks 5 mm in diameter from the margin of 5-day-old cultures of the fungus are placed in the center of each plate. Plates are incubated at 24°C under constant light, and linear growth of the fungus on each plate is measured daily for 7 days.

Effect of fungicide on sporangium formation. Three colonized agar disks are placed in each of three petri dishes 52 mm in diameter with 5 ml of the test solutions. A solution without fungicide serves as the control. Dishes are incubated in the light for 3 days. Sporangia present in four different microscope fields (×160) on each agar disk are counted and recorded daily.

Effect of fungicide on zoospore germination. Zoospores are produced by placing three colonized disks in a petri plate containing 15 ml of lima bean broth. Plates are incubated in light for 3 days at 24°C. Broth cultures are washed with Chen-Zentmyer salt solution (2) to induce sporangium production. The lima bean broth is replaced with salt solution four times at 1-hr intervals. Abundant sporangia are produced overnight. To induce zoospore production, cultures with sporangia are placed at 4°C for 30 min, then returned to room temperature (24°C). Zoospore discharge begins within 25 min. Zoospore concentrations are determined with a hemacytometer and adjusted by dilution with 0.001 M phosphate buffer.

The effect of fungicide on zoospore viability is measured by pipetting 0.5 ml of a suspension containing 10^5 zoospores per milliliter into 4.5 ml of solution containing the desired fungicide concentration. Three replicate plates per concentration are used. After 2 and 4 hr, 200 zoospores in each plate are examined at ×160 and recorded as germinated, not germinated, or lysed.

All in vitro tests dealing with fungal response to a range of fungicide concentrations are analyzed by regression analyses.

Greenhouse and Field Procedures

In greenhouse and field tests to study fungicide movement or efficacy for disease control, the commercial formulation of the fungicide is used rather than analytical grade. A susceptible rootstock is selected for inoculations. The most common dwarfing apple rootstocks vary in susceptibility to collar and crown rot. East Malling (M) rootstocks M-4 and M-9 are quite resistant, and the Canadian rootstock Ottawa-3 has M-9-type resistance. Of the others, M-25, M-26, Malling-Merton (MM) 103, MM-104, MM-106, MM-107, MM-109, MM-110, MM-111, MM-113, and MM-115 are quite susceptible; M-2, M-7, and MM-112 are intermediate in susceptibility.

A qualitative assay for root uptake and upward translocation of fungicide. Three- to 4-year-old trees on a susceptible rootstock with a mean trunk diameter of about 3 cm at 15 cm above the soil line are grown in 8-L plastic pots in the greenhouse. Trees are drenched with 500 ml of several fungicide concentrations. Contact of the fungicide with the tree trunk is avoided by adding the solution to the soil slowly. Three trees (replications) per concentration are arranged in a randomized block design on the greenhouse bench.

Seventy-two hours after drenching, wedge-shaped sections (each 1 cm wide, 7 mm long, and 5 mm deep) are cut from each tree at 5, 10, and 15 cm above the crown (soil line). Wedges are cut transversely with the xylem and contain phloem (bark) and secondary xylem. Wedges are then placed phloem side down on pentachloronitrobenzene-benomyl-neomycin-chloramphenicol (PBNC) medium (5), and an agar disk colonized by *P. cactorum* is placed on the upper surface of each wedge. All disks are in contact with secondary xylem. Fungal growth on the agar plug and wood surrounding the plug is observed daily. Fungal growth on agar plugs and the wedge indicates that no fungicide is present in the tree. Absence of fungal growth indicates that sufficient fungicide is present to inhibit fungal growth. This technique was successful in demonstrating root uptake and upward translocation of metalaxyl in apple trees (5) and can be used in field studies.

Field procedures to evaluate the preventive activity of fungicides. The test should be established on a site infested with *P. cactorum* using 1- to 2-year-old trees on a susceptible rootstock. Susceptible roots must be used. Black plastic collars (20 cm tall and 20 cm in diameter) are placed around the base of each tree and embedded 10 cm into the soil to maintain inoculum around the base of the tree and to flood the tree bases with water during dry periods to ensure good conditions for disease development.

Half of the trees in each treatment are inoculated with oospores of *P. cactorum*. Hemp seed broth is prepared by autoclaving 30 g of hemp seed in 1 L of distilled water for 30 min, filtering through four layers of cheesecloth, and autoclaving a second time. An agar plug from the edge of a 7-day-old culture is added to a 125-ml flask containing 25 ml of hemp seed broth. Oospores are produced after 2–3 weeks in the dark at 24°C. Oospores are harvested by grinding cultures in an Omni mixer for 60 sec at high speed and freezing for 24 hr at −20°C to kill mycelial fragments and sporangia (1). After thawing, cultures are reground for 15 sec to suspend the oospores. Oospores are counted with a hemacytometer, and the concentration is adjusted by dilution with distilled water.

Trees are inoculated before fungicide treatment by pouring 10 ml of a suspension containing 10,000 oospores per milliliter around the base of the tree inside the plastic collar. Inoculation may be repeated each spring. One week after inoculation, 1 L of fungicide solution is poured around the base of the tree inside the plastic collar. Fungicides are applied twice annually, during early spring (prebloom) and fall (October). About 24 trees per treatment are used in a split-plot design. If broadcast treatments are desired, the area under the tree canopy should be calculated and the desired amount of fungicide per treated acre applied in a specific amount of carrier (usually water). Drench application around the trunk of the tree has been the most successful means of application (M. Ellis, *unpublished*).

The development of collar rot symptoms is recorded annually. As trees develop symptoms, isolations are made from the base of diseased trees to verify the presence of *P. cactorum*. When this technique was used in Ohio, initial collar rot symptoms developed in 2 years, and the test was continued for 4 years (M. Ellis, *unpublished*).

Field procedures to evaluate the postinfection, curative activity of fungicides. Naturally infected trees in established orchards should be selected on the basis of visible symptoms. Typical collar or crown rot symptoms include yellowing or chlorosis of leaves in the spring (in the fall, this discoloration often changes to a purplish or bronze color); reduced terminal growth; reduced fruit size and yield; and a visible canker or discolored, rotted bark at and below the soil line or crown on most trees. Ten single-tree replicates as uniform as possible in symptom development and size should be selected per treatment. Trees with severe symptoms that cannot recover (i.e., completely girdled around the trunk) should not be selected. Isolations should be made from each tree before treatment to verify the presence of *Phytophthora* spp. Symptoms typical of collar and crown rot are similar to those of other root disorders.

Evaluations of the efficacy of the fungicide are based on postapplication symptom remission of treated trees compared to untreated controls. Symptom remission can be determined by measuring terminal growth and trunk circumference before treatment and each year after growth has ceased in the fall. Trunk circumference should be measured at the same location on each tree. Terminal growth should be measured on at least 10 terminals per tree. Differences in color of foliage are harder to quantitate but can be recorded using a Gardner automatic color difference meter (Gardner Laboratory, Inc., P.O. Box 5728, Bethesda, MD 20014) (5).

LITERATURE CITED

1. Banihashemi, Z., and Mitchell, J. E. 1976. Factors affecting oospore germination in *Phytophthora cactorum*, the incitant of apple collar rot. Phytopathology 66:443-448.
2. Chen, S.-W., and Zentmyer, G. A. 1970. Production of sporangia by *Phytophthora cinnamomi* in axenic culture. Mycologia 62:397-402.
3. Davis, R. M. 1981. Phytophthora foot rot control with the systemic fungicides metalaxyl and fosetyl aluminum. Proc. Int. Soc. Citriculture 1:349-351.
4. Davis, R. M. 1982. Control of Phytophthora root and foot rot of citrus with systemic fungicides metalaxyl and phosethyl aluminum. Plant Dis. 66:218-220.
5. Ellis, M. A., Grove, G. G., and Ferree, D. C. 1982. Effects of metalaxyl on *Phytophthora cactorum* and collar rot of apple. Phytopathology 72:1431-1434.
6. Farih, A., Menge, J. A., Tsao, P. H., and Ohr, H. D. 1981. Metalaxyl and efosite aluminum for contol of Phytophthora gummosis and root rot on citrus. Plant Dis. 65:654-657.
7. Jeffers, S. N., Aldwinckle, H. S., Burr, T. J., and Arneson, P. A. 1982. *Phytophthora* and *Pythium* species associated with crown rot in New York apple orchards. Phytopathology 72:533-538.
8. Kannwischer, M. E., and Mitchell, D. J. 1981. Relationships of numbers of spores of *Phytophthora parasitica* var. *nicotianae* to infection and mortality of tobacco. Phytopathology 71:69-73.
9. Munnecke, D. E. 1982. Apparent movement of Aliette or Ridomil in *Persea indica* and its effect on root rot. (Abstr.) Phytopathology 72:970.
10. Schmitthenner, A. F. 1973. Isolation and identification methods for *Phytophthora* and *Pythium*. Proc. First Woody

Ornamental Disease Workshop, Univ. Mo., Columbia. 128 pp.
11. Timmer, L. W. 1975. Comparative susceptibility of Star Ruby grapefruit to Phytophthora foot rot. J. Rio Grande Valley Hortic. Soc. 29:59-64.
12. Timmer, L. W. 1977. Preventive and curative trunk treatments for control of Phytophthora foot rot of citrus. Phytopathology 67:1149-1154.
13. Timmer, L. W. 1979. Preventive and systemic activity of experimental fungicides against *Phytophthora parasitica* on citrus. Plant Dis. Rep. 63:324-327.
14. Tsao, P. H. 1971. Chlamydospore formation in sporangium-free liquid cultures of *Phytophthora parasitica*. Phytopathology 61:1412-1413.
15. Tsao, P. H., and Guy, S. O. 1977. Inhibition of *Mortierella* and *Pythium* in a *Phytophthora*-isolation medium containing hymexazol. Phytopathology 67:796-801.
16. Tuite, J. 1969. Plant Pathological Methods, Fungi and Bacteria. Burgess Publishing Co., Minneapolis, MN. 329 pp.
17. Whiteside, J. O. 1971. Some factors affecting the occurrence and development of foot rot on citrus trees. Phytopathology 61:1233-1238.

Field Evaluation of Fungicides for Control of Diseases of Grapes

ROGER C. PEARSON, Department of Plant Pathology, New York State Agricultural Experiment Station, Cornell University, Geneva 14456

The widespread use of fungicides to control plant diseases had significant beginnings in the vineyards of Bordeaux, France, where in the 1880s Millardet first applied a mixture of copper sulfate and lime to control grape downy mildew (10). Since then, there has been much interest in chemicals that control diseases of this most widely planted fruit crop in the world.

Three of the most destructive fungal diseases of grapes—powdery mildew (caused by *Uncinula necator* (Schw.) Burr.), downy mildew (caused by *Plasmopara viticola* (Berk. & Curt.) Berl. & de Toni), and black rot (caused by *Guignardia bidwellii* (Ell.) Viala & Ravaz)—had their origins on native American species of *Vitis*. Nevertheless, interest in controlling these diseases was stimulated by their introduction into Europe during the 1800s, when large crop losses occurred on the highly susceptible European grape, *V. vinifera* L. In eastern North America, each of these diseases can be very destructive on native American bunch grapes (*V. labruscana* Bailey) and *V. vinifera* and its hybrids. Because of climatic conditions, downy mildew and black rot do not occur in *V. vinifera* plantings in California.

This chapter addresses methods for field evaluation of fungicides to control the three diseases mentioned above. Botrytis bunch rot (caused by *Botrytis cinerea* Pers.:Fr.), Phomopsis cane and leaf spot (caused by *Phomopsis viticola* (Sacc.) Sacc.), and Eutypa dieback (caused by *Eutypa lata* (Pers.:Fr.) Tul.) are also covered.

GENERAL INFORMATION

Because cultivars vary greatly in susceptibility to specific diseases, it is possible to select specific cultivars that are susceptible to one or two diseases of interest and less susceptible to others. Vineyard location, slope, and bordering vegetation can have a profound effect on microclimate and hence disease occurrence. A history of disease in a specific vineyard location may be essential for adequate disease pressure.

If a vineyard is to be planted for the purpose of testing fungicides, the vines should come from certified, virus-indexed material. Several virus and viruslike diseases, including fan leaf, yellow mosaic, vein necrosis, leafroll, and stem pitting, can cause foliar symptoms that may mask or be confused with symptoms of chemical phytotoxicity (2). Furthermore, viruses that cause abortion of fruit render vines unsuitable for the evaluation of diseases that affect fruit or for the evaluation of effects on yield.

Diseases that depend on free moisture (rainfall or heavy dews) for spread and infection may not develop consistently throughout the season or from one year to the next. One technique to extend wetting periods or even create them is the use of an over-the-trellis misting system. Not only can a misting system encourage infection periods and disease spread, but also redistribution, tenacity, weathering capabilities, and postinfection activity of fungicides can be studied under semicontrolled field conditions.

The main reason for conducting a fungicide trial is to obtain efficacy data. However, many other types of data that reflect the effects of disease or chemical on the vine can be measured, including yield, vigor, quality, phytotoxicity, and enological effects.

Yield

If yield data are to be collected from a trial, the decision must be made at the time of pruning or meaningful yield measurements will not be possible at harvest. Viticulturists frequently employ a technique called balanced pruning (7), which helps to standardize the cropping potential of weak vines and vigorous vines and allows yield differences caused by external treatment effects to be measured. This method requires weighing the current season's prunings (a measure of the previous season's vigor) and adjusting the number of buds to be left based on an appropriate formula for the cultivar in question. For example, a 30 + 10 system is common for Concord (*V. labruscana*) vines: 30 buds are retained for the first pound of prunings and 10 buds are retained for each additional pound of prunings.

The components of yield that should be recorded are number of clusters per vine, cluster weight, and crop weight per vine. These values can then be used to calculate yield per acre. An additional measure of vine productivity is obtained by combining yield with sugar content of the fruit to arrive at a figure termed "soluble solids per vine," the product of percent soluble solids/100 and crop weight (12).

Vigor

Vigor or vine growth can be recorded during the growing season by measuring shoot length. However, a more accurate and more meaningful method involves weighing the prunings of current-season wood during the normal dormant pruning period. To maintain proper vigor and provide carbohydrate reserves for winter survival of buds and canes, especially in cold climates, it is generally necessary to cluster-thin certain cultivars, French hybrid cultivars in particular. Vines are thinned to one cluster per shoot or some other uniform factor.

Quality

Several factors of fruit quality can be measured, e.g. sugar content, total acidity, and anthocyanin content of skins (nonwhite cultivars). Sugar content can be measured with a hand-held refractometer, and results are generally reported as percent soluble solids or degrees Brix. Determining total acidity (g/100 ml) and anthocyanin content of the skins ($\mu g/cm^2$ or mg/kg) requires more elaborate laboratory techniques (15). Wine quality is usually determined by a "blind" tasting panel. Fruit quality standards differ depending on the ultimate use of the crop, i.e., fresh market, wine, processing, or raisins.

Phytotoxicity

Observations on phytotoxicity caused by an experimental compound are just as important as efficacy data. Furthermore, the commercial use of the crop may determine the significance of the injury. Surface blemishes that may be insignificant for the wine or juice market may render the crop unsalable for the fresh fruit or cannery market.

A phytotoxic response may be related to several host factors, the most important of which is the cultivar. For example, the American cultivar Concord is very sensitive to sulfur regardless of environmental conditions, and some hybrids of *V. vinifera* × *Vitis* spp. (American) are also highly susceptible to sulfur injury. Most *V. vinifera* cultivars, however, are tolerant of sulfur except at temperatures above 38°C.

Some compounds are phytotoxic only if applied at specific stages of vine growth or under specific environmental conditions. Young, unfolding or expanding leaves are frequently highly susceptible to injury from chemicals, e.g. dinocap. Berries sprayed during the early stages of their development are often susceptible to injury that may result in scarring, russeting, or even cracking. Environmental conditions may trigger phytotoxic responses, as in the case of sulfur mentioned previously. Coppers generally cause more foliar injury in the latter half of the growing season when heavy dews solubilize copper ions in residues accumulated during the season, and folpet may cause injury under conditions of high temperature and drought.

Data on phytotoxicity, such as severity of foliar or fruit injury and node position of affected leaves, should be accompanied by data on stage of vine growth, defined according to an accepted system (1,4,7) and weather at the time of the spray application (temperature, humidity, wind speed and direction, and cloud cover). The pH of the spray water, formulation of material used, adjuvants used, and other pesticides in the spray mixture should also be noted. Weather conditions following spray applications, especially rainfall and temperature, should be monitored. If possible, observations on several cultivars should be made. A photographic record of the injury, accompanied by detailed notes, generally proves invaluable for future reference.

Enological Effects

Although most experimental compounds are evaluated for their effects on fungal diseases, when working with wine grapes the researcher must also be concerned with possible side effects on another group of fungi, the yeasts. Before a new fungicide is released for wide-scale testing, its effect on fermentation must be determined. Many commercially used fungicides do affect fermentation, and wineries may prohibit their use within a specific number of days before harvest. Fungicides may delay or inhibit the onset of fermentation. Even a delay of 12–24 hr can substantially disrupt the crushing schedule of wineries. These side effects make some compounds unacceptable to the wine industry that may be acceptable to the fresh fruit or juice industry. Lastly, because anything that is sprayed on the vine may eventually affect the quality of the finished bottle of wine, tasting panels must determine if there are any off-flavors.

Plot Design and Plot Size

Various plot designs have been used for fungicide evaluation in vineyards; the most common is the randomized complete block. The number of replications usually ranges from four to six, and the plot size is usually determined by the purpose of the trial. If eradicative fungicides are being evaluated, large plots should be used to lessen the chance of recontamination by inoculum from adjacent plots or from outside the trial area. However, if protective fungicides with low vapor activity are being evaluated, smaller plots may be acceptable. In studies I and my colleagues have done, plot size has ranged from one-half vine to several rows. For instance, we have used half-vine plots in bunch rot studies with protective fungicides to evaluate timing of applications. Vines were sprayed by hand with backpack sprayers, applying the spray evenly on both sides of the trellis. With this procedure it is advisable to leave half of the vine as an unsprayed check. Single- or double-vine plots with unsprayed buffer vines on either side in the row have also been used to evaluate fungicide efficacy against both foliar and fruit diseases. Vines in these plots are usually sprayed with a handgun or CO_2 pressurized backpack sprayers.

Trellising systems in eastern North America frequently use a post at every third or fourth vine in the row. The vines between these posts make convenient plots for spraying with a handgun or with an over-the-row hooded boom sprayer applying dilute sprays. Data are usually taken from the center one or two vines, leaving the vines on each end as sprayed buffers. If vines are plentiful or if vine size varies considerably, three-panel plots consisting of 9–12 vines per plot may be more suitable for spraying with over-the-row hooded boom sprayers. The two end panels generally serve as buffers, and data are taken from the center panel.

If commercial application equipment is used to apply the test compound, large plots should be used. If a hydraulic pump vertical boom sprayer is to be used, three-row plots with a buffer row between them are suitable. The three rows are sprayed from both sides, but data are recorded only from the center row. Three buffer rows separate data rows in this design, and the effects of drift are minimal. Depending on row length, several treatments may be placed along the row. In general, 15–20 treatment vines per plot along the row are sufficient for efficacy data. When air-blast sprayers, with their high drift factor, are used for fungicide trials, three-row plots with a buffer row between them may be used, but five-row plots should be used if treatments are applied to young vines or to vines early in the season when the canopy is sparse. Data are taken from the center row in both situations. The number of vines along the row to be included in each plot generally ranges from 25 to 30. Plots for aerial application are the most difficult to manage, and large vineyards must be chosen where ample buffer rows can be used to avoid contamination by drift. Unfortunately, replication of aerial plots is usually impractical.

Most fungicide trials are conducted in standard vineyards with commercial planting distances between vines and rows. However, where land is at a premium and the cost of maintaining a traditional trellised vineyard planting is prohibitive, compact grape plantings using less than commercial planting distances and no trellis system have been used successfully. In these small plots, three or four rooted cuttings of either the same or mixed cultivars are planted at each site. Even own-rooted *V. vinifera* vines can be used in this type of planting, because the vines are generally maintained for no longer than 3 years before they are replanted. Groups of vines are planted on centers of 1.5–2 m, trained to individual stakes, and sprayed by hand. The vines are not fruited, and only data on leaf infection are collected. Because of the small area involved, this type of planting is especially amenable to sprinkler irrigation for promotion of disease development. Studies of downy mildew, black rot, and powdery mildew have been conducted in these "mini-vineyards."

Rating Systems

Disease may be quantified by incidence, where the percentage or proportion of infected vines, shoots, leaves, clusters, or berries is recorded, or by severity, where the amount of disease per unit, e.g. leaf area infected, is reported. Researchers have used many rating systems to determine the severity of disease. In addition to rating disease on a scale of 0 to 100%, rating numbers have been assigned to specific percentage categories; e.g., 0 to 5, where 0 = no infection, 1 = 1–5%, 2 = 6–25%, 3 = 26–50%, 4 = 51–75%, and 5 = 76–100% infection, or any grouping of percentage categories desired (3). One commonly used rating system is the Horsfall-Barratt (HB) system (5) of 1 to 12, where percentage ranges at the extremes of the scale are much narrower than at the center of the scale. The human eye can supposedly distinguish the difference between healthy and diseased tissue at the extremes of the scale more accurately than in the center of the range (6). To convert the HB rating numbers to percentages, Elanco conversion tables (13) are consulted, or the conversions can be made quickly with the aid of a computer program. The rating numbers themselves are not used for statistical analysis; instead they are converted to percentages and transformed so the errors are more normally distributed around the means, and the transformed data are analyzed.

These rating systems are most easily applied to disease in a single plane such as a leaf surface, but with practice they can be used for three-dimensional surfaces such as berries, clusters, shoots, and in some cases, depending on the growth habit of the cultivar, the entire vine canopy surface area.

Environmental Monitoring

Monitoring the environmental conditions under which a test is conducted is an integral part of conducting a fungicide trial. Data on temperature, rainfall, and duration of leaf wetness, as well as spore trapping to monitor inoculum release, can be indispensable in interpreting results of fungicide trials. Information on inoculum availability and occurrence of infection periods can help explain the failure of some fungicides and the success of others. In addition, this information is essential if postinfection studies are conducted. Information on weather events surrounding a fungicide application can also document conditions under which phytotoxicity may occur. If continuously recording weather instruments are not available, daily minimum and maximum temperatures as well as daily rainfall totals should be recorded. Also, notes on temperature, wind conditions, and cloud cover should be taken at the time the applications are made.

POWDERY MILDEW

The powdery mildew fungus infects all green parts of the vine. Berries infected in the early stages of development may abort or the epidermis may stop growing, resulting in cracked fruit when the berry contents expand. Uncontrolled, this disease can reduce vine vigor, yield, bud hardiness, and crop quality (12).

Most cultivars of *V. vinifera* are highly susceptible to powdery mildew, whereas most cultivars of American species and their hybrids are less susceptible. In New York State, one of the most susceptible hybrid cultivars is Rosette. It is also quite tolerant of low winter temperatures, making it an especially suitable test cultivar in cool climates. Other suitable hybrid cultivars include Aurore, Chancellor, De Chaunac, and Vidal 256. Delaware and Concord are suitable American cultivars for data on foliage or rachis infection but poor for data on berry infection.

One of the least labor-intensive approaches for evaluating fungicide effectiveness is estimating the vine canopy surface area infected. This information is generally most reliable from those American cultivars whose shoots have a pendulous growth habit, creating a "curtain" of foliage, and data are recorded from both sides of the trellis. Cultivars whose shoots have an upright growth habit are less suitable for this measurement. A more critical evaluation of foliage infection involves assessing the infected surface area on individual leaves, usually the adaxial side, using the HB system. The number of leaves analyzed per plot generally ranges from 50 to 100. Disease incidence based on percentage of leaves infected can be calculated from the same data set.

The severity of cluster infection can be obtained by estimating the surface area of the cluster (50–100 clusters per plot) with infected berries using the HB system or another rating system. Powdery mildew on the rachis of clusters can be rated for severity most easily by using a simplified rating scale of 0 to 5. If information on the percentage of infected berries is desired, all berries are removed from the cluster (10–20 clusters per plot) and categorized as healthy or diseased.

Because several foliar diseases occur on grapes in the eastern United States, it is not always possible to screen fungicides for a specific disease without controlling other destructive diseases. This is usually accomplished by adding fungicides that have specific activity and are unlikely to interfere with the experiment. Where downy mildew is a problem, captan or metalaxyl can be sprayed over all treatments, leaving powdery mildew development relatively unchecked. Ferbam is an effective fungicide for controlling black rot in vineyards where powdery mildew trials are being conducted, and vinclozolin or iprodione can be used to control Botrytis bunch rot in tight-clustered cultivars prone to this disease. Fungicide screening trials for control of powdery mildew usually include commercial standard fungicides such as sulfur, dinocap, or triadimefon for comparison with experimental compounds.

DOWNY MILDEW

The downy mildew fungus attacks all green parts of the vine and from prebloom to harvest has the potential of causing complete crop loss. The fungus overwinters as oospores in fallen leaves. Because primary infection and subsequent secondary infection depend on rainfall, the disease is a major problem in central and eastern North American vineyards. Infection periods can be identified when temperature and leaf wetness duration are known.

Most cultivars of *V. vinifera* are highly susceptible to downy mildew, but cultivars of American *Vitis* spp. and their hybrids vary greatly in susceptibility. The American cultivars Delaware, Niagara, and Catawba are quite susceptible, and the hybrid Chancellor is even more susceptible, making it a very useful test cultivar.

One way to increase disease inoculum in a vineyard is to spread cold-pressed pomace from a winery on the vineyard floor after harvest. Because grape seeds remain viable after this process, as opposed to those in hot-pressed pomace, grape seedlings cover the vineyard floor the following spring. Because they are close to the source of overwintering downy mildew inoculum on the vineyard floor, the seedlings frequently become infected before the vines on the trellis. Sporangia from these seedlings are readily spread throughout the vineyard.

Thorough spray coverage of the undersides of leaves is essential for downy mildew control, because infection and sporulation occur through the stomata. Furthermore, only leaves mature enough to have functional stomata are susceptible to infection. Hence, in postinfection studies, shoots should be tagged at the youngest most fully expanded leaf, and data should be taken only from leaves proximal to the tag.

The effectiveness of fungicides can be rated on shoots, leaves, or clusters. The percentage of shoots infected can be estimated by simply examining 25–50 randomly selected shoots per vine. The severity of leaf infection can be determined by rating leaf area infected (using the HB system) either from 50–100 randomly selected mature leaves per plot or from all mature leaves on 6–10 randomly selected shoots per plot. If disease development is light, the number of lesions per leaf can be determined on an equivalent number of leaves. Clusters infected before bloom are frequently destroyed, so a determination of the percentage of clusters infected is generally adequate. If single-vine plots are used, all clusters on each vine should be evaluated. When plots include numerous vines, 50–100 clusters per plot should be adequate for evaluation.

The incidence and severity of fruit infection occurring after fruit set should be determined at midseason and again at harvest. The area of each cluster that is infected may be rated either by the HB scale or a modified scale of 0 to 5 as mentioned earlier. At harvest, the percentage of berries infected per cluster (10–20 clusters per plot) can be calculated by removing all berries from each cluster and separating them into healthy and diseased categories. A purple-berry cultivar such as Chancellor can simplify harvest evaluations of berry infection, because the infected berries are pink or bright red and remain firm, unlike the healthy purple berries, which become soft when ripe.

Fungicide screening trials for control of downy mildew generally include captan or mancozeb as protective standards and metalaxyl as a systemic and postinfection standard. Fungicides useful for controlling unwanted diseases and not affecting downy mildew in test vineyards include sulfur, dinocap, benomyl, or triadimefon for powdery mildew; benomyl or triadimefon for black rot; and benomyl, vinclozolin, or iprodione for Botrytis bunch rot.

BLACK ROT

Black rot is a highly destructive disease of grapes grown in warm, humid climates. The fungus overwinters in mummified berries from which primary inoculum in the form of ascospores or conidia is released with spring rains. Conidia from pycnidia formed in infected tissue provide inoculum for secondary spread. Spore dispersal and infection depend on rainfall, so identification of an infection period is possible when the amount of rainfall, duration of leaf wetness, and mean temperature during the wetting period are known.

To introduce the disease or increase inoculum in test plantings, infected mummified berries can be collected and placed under the trellis. The longevity of mummies, and thus their inoculum-producing potential, is promoted by using weed control methods other than cultivation in test plantings.

Most cultivars of *V. vinifera*, many hybrids, and the major American bunch grape cultivars Concord, Catawba, and Niagara are susceptible to the disease. Catawba and Aurore (a highly susceptible hybrid cultivar) have been widely used for fungicide testing under eastern North American conditions.

Infection of either foliage or fruit is suitable for evaluating fungicide effectiveness. If disease pressure is light, the percentage of leaves infected and the number of lesions per leaf on 50–100 leaves per plot provide adequate information on incidence and severity. The leaf sample may be collected at random, or all the leaves on 6–10 randomly selected primary shoots per plot can be evaluated. When disease pressure is heavy, lesions may coalesce and petiole infections may kill leaves. Under such conditions, a rating of leaf area infected, using the HB scale or another rating scale, is most suitable. Fruit infection should be measured at vêraison (fruit softening and color change), before other rot organisms attack berries and confuse the evaluation. The incidence of disease, based on the percentage of clusters infected (50–100 clusters per plot), may not be discriminating enough to compare effectiveness of different fungicides. Therefore, severity of infection based on the number of infected berries per cluster, usually 10–20 clusters per plot, or on an estimate of the percentage of the surface area of the cluster infected (50–100 clusters per plot) should be included.

If data on postinfection activity of fungicides under field conditions are desired, a special effort must be made not only to identify infection periods, but also to identify the leaves that were susceptible at the time of the infection period. Leaves are susceptible to infection by *G. bidwellii* only for about a week after they unfold (14), so only those leaves susceptible during a specific infection period should be used for data collection. Tags should be applied to shoots identifying leaves that are to be used for analysis. Five or six tagged shoots per vine per plot generally give sufficient data. Vines in postinfection screening trials should receive no more than one spray per season, because redistribution of previously applied fungicides may affect disease development.

Fungicide screening trials generally include ferbam or mancozeb as standard protective fungicides and triadimefon as a standard postinfection fungicide. Fungicides useful for controlling unwanted diseases in black rot trials include sulfur or dinocap for powdery mildew, metalaxyl for downy mildew, and vinclozolin or iprodione for Botrytis bunch rot.

BOTRYTIS BUNCH ROT AND CANE BLIGHT

Botrytis bunch rot and cane blight is quite variable in the damage it causes, depending on the cultivar, cultural practices, and microclimate. Because this fungus is a saprophyte on dead plant material, inoculum can be considered present in the vineyard at all times. The cane blight and leaf infection phases of the disease occur only sporadically, and in general the fungus must first become established on a food base, e.g. dead bud scales at the base of a shoot, or pollen and other debris on a leaf. Cool, foggy weather, but not necessarily rainfall, is favorable for infection.

Fruit may be infected from bloom (8) to harvest. Cultivars with tight, compact clusters and berries with "thin skins" are generally highly susceptible to bunch rot. Fruit that have cracked because of natural internal pressures, powdery mildew, or insects are readily infected by *B. cinerea*. Clusters can be made more compact for experimental purposes by cluster-thinning vines before bloom. Daminozide (Alar) applied before bloom increases berry set, which also results in compact clusters. This treatment can make relatively resistant cultivars like Concord highly susceptible to bunch rot. In New York State, cultivars that are naturally susceptible to Botrytis bunch rot, and hence good research subjects, include Chardonnay and the hybrids Aurore and Vignoles (Ravat 51).

Most researchers time sprays for control of Botrytis bunch rot on the basis of vine phenology rather than calendar date or infection periods, which are difficult to identify. In Europe, most *Botrytis* control programs are based on a four-spray schedule beginning at bloom followed by sprays at bunch closing, véraison, and 2 weeks later. Trials in California usually emphasize bloom sprays, whereas trials in New York State generally compare the European program of four sprays with a newly developed two-spray program beginning at approximately 5% sugar (11). Because berry surfaces are difficult to wet, fungicides applied after bloom are generally most effective if a spreader-sticker is added to them.

Determinations of fungicide effectiveness are generally based solely on an evaluation of fruit infection. Data on both incidence and severity of cluster infection are desirable. Disease evaluation is usually restricted to the basal first and second clusters on a shoot, because they are generally larger and more mature. A sample of 50–100 clusters is generally adequate in large plots, but clusters on all shoots should be examined in small plots of one or two vines. In addition to determining disease incidence by calculating the percentage of clusters infected, the severity of the attack is determined by estimating the cluster area infected, usually using a rating system such as the 0 to 5 system described earlier.

Fungicides used as standards in Botrytis bunch rot trials vary depending on geographic region. Benomyl is a common standard fungicide where resistance to it is not a problem. Other possible standards are dicloran, captan, or folpet. The dicarboximide fungicides vinclozolin and iprodione are the standards in areas where resistance to them has not developed. Fungicides useful for controlling unwanted diseases in Botrytis bunch rot trials include dinocap or triadimefon for powdery mildew, mancozeb or metalaxyl for downy mildew, and mancozeb or triadimefon for black rot.

PHOMOPSIS CANE AND LEAF SPOT

Phomopsis viticola can cause lesions on shoots, which may become weakened and therefore more prone to wind damage. Infection of the rachis can result in shriveled berries. Severely infected leaves turn yellow and drop from the vine, and infections of ripe fruit result in crop loss because fruit drop from the cluster.

The fungus overwinters as pycnidia in previously infected bark. Cool spring rains release conidia, which generally infect the first four to six basal internodes of young shoots. Because conidia are dispersed by rain splash or windblown rain, the most severely infected shoots are those located beneath older wood containing inoculum. In general, if shoots grow 15–20 cm after budbreak without becoming infected, they will be free of infection for the remainder of the season. Because rain may not always fall during this stage of growth, an over-the-trellis misting system can be useful for encouraging disease development at the desirable time. Cultivars differ considerably in susceptibility, and under New York conditions, Concord and the hybrid Chelois are good for fungicide screening.

Phomopsis cane and leaf spot is controlled by applications of dormant sprays (sodium arsenite or dinitro amine herbicides) to eradicate the fungus or, more commonly, by protective fungicides applied at 2–7 cm of growth and, if needed, again at 10–15 cm of growth. These protective sprays are generally most effective if a spreader-sticker is added to the tank to overcome the difficulty of wetting the young, hairy shoots.

Fungicide effectiveness is most easily evaluated by observing *Phomopsis* lesions on the basal four to six internodes of canes in the autumn after leaf fall. When infection is light, the number of lesions per internode or the percentage of infected internodes can be reported. When infection is severe, a rating system for estimating the number of lesions per internode or the percentage of the surface area infected per internode is suitable. If one- or two-vine plots are used, all canes should be examined. If more vines are used, 50–100 canes per plot should be examined. As mentioned previously, canes originating low in the canopy or underneath old wood are most likely to be infected and should be selected for data collection. If data on fruit infection are desired, the percentage of infected berries per cluster can be calculated from a sample of 25–50 clusters per plot. Infected berries are easily dislodged from the cluster, making collection of these data difficult. A cloth placed on the ground under the vine canopy can give an indication of fruit loss caused by shatter. The fruit is collected, weighed, and added to the total harvest weight, from which a yield loss figure can be calculated.

EUTYPA DIEBACK

Eutypa dieback is a very destructive disease that causes cankers on trunks and cordons (arms). Ascospores from perithecia produced on dead wood are the source of inoculum for this pruning wound pathogen. The strategy of fungicide protection is to treat pruning cuts as soon after pruning as practical but before weather events trigger ascospore release.

Because 4–5 years elapse between the time of pruning wound infection and expression of foliar symptoms, most fungicide trials are evaluated before symptom expression. Fungicide efficacy is based on whether or not the fungus can be isolated from wood near treated pruning wounds. Trials are generally conducted as follows. Fresh pruning cuts are made during the dormant season in wood 2 years of age or older. Pruning cut surfaces are treated with the candidate fungicide, and the following day the pruning cuts are artificially inoculated with a suspension of ascospores (100–2,000 spores per wound). After 12–18 months, the portion of the vine containing the treated wound is removed from the vine, and an attempt is made to isolate the fungus from the xylem tissue (9).

Inoculum for artificial inoculation is collected from dead wood containing perithecia. The wood should be stored dry either at room temperature or in a cold room to prevent release of ascospores. Ascospores are collected by soaking wood laden with stromata in water for 1 hr, then blotting it with a towel to remove excess water. Sections of stromata are then embedded in clay on the inside lid of a petri dish, allowing spores to discharge into water in the bottom of the dish, or pieces of wood are placed in a spore settling tower with the airstream directed into a deep-well slide containing water. Spores are collected after 1–2 hr, counted with a hemacytometer, and diluted to the desired concentration.

Treatments in a trial to evaluate effectiveness of pruning wound protectants are generally arranged randomly among vines or in a randomized complete block design. Either individual vines or a group of vines containing equivalent numbers of treated pruning cuts may be considered a plot. Results are presented as the percentage of wounds infected or as the number of infected wounds observed versus expected in a chi-square analysis.

Benomyl is generally used as a standard treatment in fungicide trials for control of Eutypa dieback. Foliar sprays applied during the growing season do not generally interfere with Eutypa dieback trials established during the dormant season.

DISCUSSION

In Europe, there has been an attempt to standardize methods of conducting fungicide trials on grapes and especially methods for evaluating fungicide effectiveness for specific diseases (3). However, general agreement on procedures to follow has not been fully implemented between countries. Nevertheless, within some countries there are standardized procedures to follow for collecting fungicide efficacy data required to support registration. Currently, there are no standardized procedures for testing grape fungicides in the United States.

Standardizing procedures so that results from different parts of the world could be interpreted and compared with relative ease and dependability is an admirable goal. Before such a goal can be reached, however, various factors including cultivars, cultural practices, weather conditions, application equipment, and disease complexes must be considered. Standardization is most likely to succeed on a regional basis where conditions are more homogeneous. In the United States, for example, it may be possible to standardize procedures for the West Coast or the Great Lakes region. Nevertheless, even when standardized procedures are followed, ultimate use strategies must be tailored to local conditions.

LITERATURE CITED

1. Baggiolini, M. 1952. Les stades repères dans le développement annuel de la vigne et leur utilisation pratique. Rev. Romande Agric. Vitic. Arboric. 8:4-6.
2. Bovey, R., Gärtel, W., Hewitt, W. B., Martelli, G. P., and Vuittenez, A. 1980. Virus and Virus-like Disorders of Grapevines. Editions Payot. Lausanne, Switzerland. 181 pp.
3. Bulit, J. 1980. Normalisation de l'évaluation des dégâts provoqués par les maladies cryptogamiques sur la vigne. Bull. OIV 587:3-16.
4. Eichhorn, K. W., and Lorenz, D. H. 1977. Phänologische Entwicklungsstadien der Rebe (Phenological developmental stages of the grapevine). Nachrichtenbl. Dtsch. Pflanzenschutzdienstes (Braunschweig) 29:119-120.
5. Horsfall, J. G., and Barratt, R. W. 1945. An improved grading system for measuring plant diseases. (Abstr.) Phytopathology 35:655.
6. Horsfall, J. G., and Cowling, E. B. 1978. Pathometry: The measurement of plant disease. Pages 119-136 in: Plant Disease, An Advanced Treatise. Vol. II. How Disease Develops in Populations. J. G. Horsfall and E. B. Cowling, eds. Academic Press, New York.
7. Jordan, T. D., Pool, R. M., Zabadal, T. J., and Tomkins, J. P. 1981. Cultural practices for commercial vineyards. Misc. Bull. 111. N.Y. State Coll. Agric. Life Sci., Cornell Univ. 69 pp.
8. McClellan, W. D., and Hewitt, W. B. 1973. Early Botrytis rot of grapes: Time of infection and latency of *Botrytis cinerea* Pers. in *Vitis vinifera* L. Phytopathology 63:1151-1157.
9. Moller, W. J., and Kasimatis, A. N. 1980. Protection of grapevine pruning wounds from Eutypa dieback. Plant

Dis. 64:278-280.
10. Parris, G. K. 1968. A Chronology of Plant Pathology. Johnson and Sons, Starkville, MS. 167 pp.
11. Pearson, R. C., and Riegel, D. G. 1983. Control of Botrytis bunch rot of ripening grapes: Timing applications of the dicarboximide fungicides. Am. J. Enol. Vitic. 34:167-172.
12. Pool, R. M., Pearson, R. C., Welser, M. J., Lakso, A. N., and Seem, R. C. 1984. Influence of powdery mildew on yield and growth of Rosette grapevines. Plant Dis. 68:590-593.
13. Redman, C. E., King, E. P., and Brown, I. F., Jr. 1967. Elanco conversion tables for Barratt-Horsfall rating numbers. Eli Lilly and Co., Indianapolis, IN. 110 pp.
14. Spotts, R. A. 1977. Effect of leaf wetness duration and temperature on the infectivity of *Guignardia bidwellii* on grape leaves. Phytopathology 67:1378-1381.
15. Winkler, A. J., Cook, J. A., Kliewer, W. M., and Lider, L. A. 1974. General Viticulture. University of California Press, Berkeley. 710 pp.

Evaluating Fungicides for the Control of Common Leaf Spot, Gray Mold, and Leather Rot of Strawberry

A. O. PAULUS, Plant Pathology Department, University of California, Riverside 92521, and G. G. GROVE and M. A. ELLIS, Department of Plant Pathology, Ohio State University, Ohio Agricultural Research and Development Center, Wooster 44691

Common leaf spot, caused by *Mycosphaerella fragariae* (Tul.) Lindau (anamorph: *Ramularia tulasnei* Sacc.), and gray mold, caused by *Botrytis cinerea* Pers.:Fr., are strawberry diseases that cause consistent economic concern wherever strawberries are grown in the United States. Leather rot, caused by *Phytophthora cactorum* (Leb. & Cohn) Schroet., occurs more erratically but can cause devastating fruit losses when the proper environmental conditions are prevalent for development of the disease. During the 1981 growing season, an estimated 20–30% of the Ohio strawberry crop was lost to leather rot (4).

The number of fungicide applications needed during the bloom and fruiting period varies in different parts of the United States because of the length of the picking season. Strawberry growers in the northern part of the country tend to have a picking season that lasts for only a few weeks, whereas those in California and Florida may have a period of 2–6 months or longer. Residue problems may also affect whether a fungicide can be used in a particular region; California growers may pick as often as every 3 days and therefore prefer fungicides that have a zero tolerance.

All three diseases are favored by prolonged periods of high humidity or above-normal rainfall. In areas where humidity or rainfall is not consistently adequate for disease development, it may be necessary to sprinkle- or mist-irrigate the fungicide test plot (7). Automatic time clocks, activated by either a battery or electric line, are available from several companies. Timing intervals, day or night operation, etc. must be determined by the environmental strawberry-growing conditions in each area.

The highest yield of fruit is usually obtained from strawberry plants grown on a well-drained sandy loam soil. Recommended commercial strawberry maintenance practices (fertilization, cultivation, pruning, insect and weed control) should be used for the test site. A single or double row of strawberry plants per bed may be best for ease of picking during the fruiting season. Plot size depends on available labor during the fruiting season and number of fungicides to be tested. Successful results have been obtained with as few as 25 plants per replicate. Plots should be replicated. We suggest that a completely randomized block statistical design be used. Local statisticians can help select the best design for each plot area.

COMMON LEAF SPOT

Epidemics of leaf spot can kill entire leaves, resulting in reduced yields, a lower grade of harvested berries, and a loss in vigor of affected plants (2,6). The disease is usually prevalent in the fall and spring. It is first noticeable as small, purplish, circular spots on the upper surface of the leaflets. The center of the spot becomes gray to white, is surrounded by a broad, reddish purple margin, and is generally 0.32–0.64 cm (0.125–0.25 in.) in diameter. The leaf spot fungus may also attack leaf stems, fruit stalks, and fruit calyxes. On the stems and stalks, the lesions tend to be more elongated than circular. Infection takes place through the lower leaf surface, and the fungus invades through the stomata. Infection of strawberry fruit by this fungus is not common.

Test Plants

Strawberry cultivars vary in resistance to the leaf spot fungus (9). A susceptible cultivar should be selected for fungicide trials. Because different races of the fungus exist in different areas of the United States, a strawberry cultivar resistant in one area may be susceptible in another area.

Prolonged wet periods are the key to development of a severe epidemic of common leaf spot on strawberry. A temperature range of 10–18°C is satisfactory for good disease development and is one of the reasons leaf spot can be severe during the spring.

Inoculum for Field Plots

A researcher beginning with a newly planted plot area should select, if possible, naturally infected plants from the nursery. These infected plants usually provide an adequate disease epidemic for the current year's trials.

If inoculum is not present on the plants in the plot, the fungus is usually collected from the field in the imperfect or *Ramularia* form, but the black perithecia (*Mycosphaerella* form) have been found in lesions on overwintered leaves in early spring. Leaves bearing fruited lesions are washed under running water, and a clipped camel's-hair brush is used to remove old spores. Leaves are allowed to dry and then are placed in a moist chamber at room temperature. The following day a new crop of spores is produced on the lesions. Distilled water is used to make the spore suspension to be sprayed on the underside of leaves (5).

Another way to develop inoculum for adequate disease in the plot is to collect infected leaves from the field (9) and place them on moist filter paper in petri plates at 10–15°C. After sporulation, transfer the spores to potato-dextrose agar (PDA). Single-spore these isolates, transfer them to PDA in petri plates, and incubate at 25°C. After 3 days, wash the developed conidia into beakers with distilled demineralized water and use in the inoculations. A concentration of 300,000 conidia per milliliter is usually best for adequate disease development. Wet the foliage of the test plants before applying the conidia to the underside of the leaves. Repeated inoculations may be required or desired to establish an epiphytotic severe enough to distinguish differences among fungicides.

Fungicide Application

Dipping strawberry plants that are naturally infected with common leaf spot in a benomyl suspension before planting has reduced the effects of the disease and, for cultivars with some tolerance, has eliminated the need for additional fungicide sprays in southern California (10). But where growers use susceptible cultivars and the environment is more conducive to disease development, additional foliar sprays are needed for control during the growing season.

A foliar spray program should be started at the time of inoculation with the fungus or shortly thereafter. The type of sprayer for fungicide application depends on the size of the plot. Excellent results have been obtained using a 7.6-L (2-gal) CO_2-pressurized sprayer at 207 kPa (30 psi). For larger commercial-size plots, the sprayer should be tractor-mounted or self-propelled at constant speed. It is important to cover both leaf surfaces uniformly, and leaves should be sprayed until they are wet. The amount of water to mix with the fungicides and the volume used per hectare (acre) vary from 470 to 1,870 L (50–200 gal) depending on the size of the plants.

Because repeated applications of fungicides may be phytotoxic to foliage of strawberry plants, plants must be carefully examined during the course of the experiment. Spreader-stickers may be used in foliar fungicide applications, but care must be taken to be sure that the fungicide, and not the spreader-sticker or the combinations thereof, is causing injury to the plant.

Data Collection

Yield data should be taken wherever possible because fungicides may reduce yield without causing visible symptoms. Several methods can be used to determine which fungicides provide the best control of leaf spot. One method is to count the number of individual lesions on 10–20 leaves from each treatment. Number of lesions could be recorded for the entire plot, or number of lesions per leaf could be calculated. Another method is to rate each treatment on a disease scale of 0 to 9, with a 9 rating being severe foliar disease development and lesions completely covering the leaf surface. A rating of 1 should have at least one lesion per leaf.

GRAY MOLD

Gray mold is undoubtedly the most important fruit rot disease of strawberry in the United States. It originates at the stem end next to the calyx. *Botrytis* mycelium is present in senescent petals, stamens, and calyxes of marketable fruit (11). The disease, although regarded as a fruit rot, often causes a blossom blasting and rotting of green fruit during extended rainy periods or high humidity. The disease is easily identified on blossoms or fruit by the development of light gray masses of spores. These spores are windborne and carried to other blossoms and fruit. The direct infection of ripe fruit by airborne spores is believed to be of minor importance in the development of the disease. Berries resting on soil or touching another decayed berry or a dead leaf in dense foliage are most commonly attacked.

Test Plants

Strawberry cultivars vary in susceptibility to the gray mold fungus (8). These differences have been attributed to the growth habits of the cultivars. Cultivars with firm fruit, less dense leaf growth, frost-resistant flowers, or flowers protected by leaves from frost have been found to have less gray mold on the fruit. Plant or select a commercial strawberry cultivar that is known to be susceptible in the test area.

Environmental Conditions

Gray mold is a cool-weather disease. Humidity and rainfall are the most important environmental factors that influence the incidence and severity of the disease. Optimal conditions for the development of gray mold are temperatures of 15–20°C with over 90% relative humidity. Furthermore, it seems that these conditions must be maintained for more than 28 hr for an epidemic development of the disease (3). In areas with low relative humidity and rainfall, supplemental sprinkler irrigation is important to enhance development of a *Botrytis* epidemic. Our experience has shown that it is not necessary to add additional *Botrytis* inoculum over that normally present if the proper moisture conditions are present in the plot area.

Fungicide Application

Methods and equipment are similar to those used for the control of common leaf spot. Flowers, fruit, and foliage should be completely covered for adequate control of gray mold. Fungicide sprays should be applied approximately every 7–10 days. The number of sprays applied depends on the length of the flowering period.

Data Collection

It is imperative to take fruit yields and to determine the percentage of gray mold fruit rot in each treatment. Some researchers weigh the "healthy" fruit and compare that weight with the gray mold fruit weight from each treatment. Another method is to count the total number of fruit from each plot as well as compare the percentage of fruit rot in each treatment. Obviously, the goal of the fungicide application is to control gray mold fruit rot and produce healthy fruit of normal size.

As stated previously, strawberry plants can be very sensitive to repeated applications of some fungicides. Phytotoxic reactions can range from small, distorted fruit to a slight bronzing or complete blackening of the leaves. Plants should be observed for several days after any spray application to see if the fungicide treatments have had any negative effects.

LEATHER ROT

Symptomatology

An understanding of leather rot symptomatology is essential to conducting fungicide evaluations, especially in the field. Environmental conditions that favor the development of leather rot (primarily excessive moisture) also favor the development of other fruit rots such as gray mold. Both diseases often occur simultaneously in the same planting and even on the same fruit. Correct disease identification is essential to a successful test.

P. cactorum can infect berries at any stage of development. Infection of green fruit is common. On green berries, diseased areas may be dark brown or natural green outlined by a brown margin. As the rot spreads, the entire berry becomes brown, rough, and leathery. On fully mature berries, infection may result in little color change or discoloration ranging from brown to dark purple. When cross sections of diseased berries are taken, a marked darkening of the vascular system to each seed can be observed. In later stages of decay, mature fruit also become tough and leathery. A fine mycelium on the surface of infected fruit is not very common and usually occurs only when moisture is excessive over long periods, in contrast to gray mold, where a dusty coating of gray mycelia and spores is present on the surface of most infected fruit. Other distinctive characteristics of leather rot include an unpleasant phenolic odor and bitter taste. The odor is a useful tool in identification. We recommend that researchers smell individual berries when assessing disease incidence.

The Fungus

The isolate of *P. cactorum* used in all leather rot studies should be obtained from infected strawberry fruit. *P. cactorum* is easily isolated from strawberry fruit. Small sections (1 mm × 1 mm) are placed on pentachloronitrobenzene-benomyl-neomycin-chloramphenicol (PBNC) medium (12). After 48–72 hr at 24°C, *P. cactorum* can be observed at the edge of plated fruit sections. Mycelia can then be transferred to sterile PBNC plates. A technique that is useful in purifying cultures is to invert a section of medium bearing the fungus with a small scalpel or spatula, thus placing the mycelium at the bottom of the dish and forcing it to grow up through the selective medium, after which additional transfers can be made. Once cultures are purified, they are easily maintained on dilute V-8 juice agar (12) in screw-cap bottles or test tubes. To ensure pathogenicity, strawberry fruits should be inoculated with the test isolate and the fungus reisolated from diseased tissue before any test inoculations are made. We have found that the aggressiveness of the pathogen can be lost during long-term storage.

In Vitro Laboratory Procedures

Preliminary screening may help to eliminate compounds that show little or no activity against the fungus. In tests for inhibition of mycelial growth, the required amount of fungicide is incorporated into hot or warm lima bean agar (LBA) (13) after autoclaving. LBA is preferred in this test because *P. cactorum* grows rapidly on it. It is safest to allow the medium to cool to about 45–50°C before incorporating the fungicide, because some fungicides decompose at high temperatures. Fifteen milliliters of medium containing the desired concentration of active ingredient is placed in petri dishes of uniform size. There should be a minimum of 10 dishes per concentration tested. Medium without fungicide should serve as the control. LBA disks 5 mm in diameter colonized by the test fungus are taken from the margin of 5-day-old cultures. One disk is placed in the center of each dish. Dishes are incubated at 24°C under constant light, and linear growth of the fungus is measured daily for 7 days. Regression is the best form of statistical analysis for tests such as this that deal with fungal response to a range of fungicide concentrations.

Laboratory and Field Procedures

If a test site is selected where *P. cactorum* is endemic and leather rot occurs naturally, field inoculations may not be necessary. However, we have found that field inoculations generally ensure disease development providing climatic conditions are not hot and dry. When climatic conditions in the field are not conducive to disease development, inoculation of detached fruit may be more useful in obtaining a good evaluation.

Field Inoculation Procedures

A suspension of *P. cactorum* zoospores in water is used for field inoculations. Zoospores are produced by placing three colonized agar disks 0.5 mm in diameter in petri plates containing 15 ml of lima bean broth. Plates are incubated in light for 3 days at 24°C. To induce sporangium production, broth cultures are washed with Chen-Zentmyer salt solution (1). The lima bean broth is replaced with salt solution four times at 1-hr intervals. The fourth salt solution is left in the plates overnight, and abundant sporangia are produced. To induce zoospore production, cultures with sporangia are placed at 4°C for 30 min, then returned to room temperature (24°C). Zoospore discharge begins within 25 min. Inoculum can be adjusted to various concentrations by gently adding sterile distilled water and counting zoospores with a hemacytometer. Inoculations are performed by spraying zoospore suspensions onto desired plots with a low-pressure sprayer. The choice of pressure used to apply inoculum is subject to experimentation; we have found that pressures above 103 kPa (15 psi) cause the zoospores to lyse. Inoculations should be made in late evening or at night to avoid the hot sun and rapid drying conditions.

Detached Fruit Technique

Immature (green) strawberry fruits are collected from fungicide treatments and controls in field tests. At least 20 fruits per replication and treatment should be selected about 24 hr after fungicide application. Fruit should not be washed or surface-sterilized. Zoospore suspensions are prepared as previously described. Inoculations are performed in large plastic petri dishes or other suitable containers. Fifty milliliters of a suspension containing 10,000 zoospores per milliliter is placed in each dish, and fruits from each replication and treatment are placed in separate dishes arranged in a completely randomized design. Fruits are left in the suspension for 2 hr and then are transferred to moisture chambers and incubated at 24°C. After 96 hr, fruits are visually inspected for leather rot. To ensure the presence of *P. cactorum*, isolations can be made from randomly selected fruit on PBNC agar as previously described.

LITERATURE CITED

1. Chen, S. W., and Zentmyer, G. A. 1970. Production of sporangia by *Phytophthora cinnamomi* in axenic culture. Mycologia 62:397-402.
2. Dale, J. L., and Fulton, J. P. 1957. Severe loss from strawberry leaf spot in Arkansas in 1957. Plant Dis. Rep. 41:681-682.
3. Devaux, A. 1978. Etude épidémiologique de la moisissure grise des fraises et essais de lutte. Phytoprotection 59:19-27.
4. Ellis. M. A., and Grove, G. G. 1983. Leather rot in Ohio strawberries. Plant Dis. 67:549.
5. Fall, J. 1951. Studies on fungus parasites of strawberry leaves in Ontario. Can. J. Bot. 29:299-315.
6. Fulton, R. H. 1958. Studies on strawberry leaf spot in Michigan. Mich. Q. Bull. 40:581-588.
7. Horn, N. L. 1961. Control of Botrytis rot of strawberries. Plant Dis. Rep. 45:818-822.
8. Maas, J. L. 1978. Screening for resistance to fruit rot in strawberries and red raspberries: A review. Hortic. Sci. 13:423-426.
9. Nemec, S. 1969. Determination of leaf spot races in southern Illinois strawberry plantings. Plant Dis. Rep. 53:94-97.
10. Paulus, A. O., Voth, V., Bringhurst, R. S., and Welch, N. 1973. Comparison of fungicides and methods of application for the control of Ramularia leafspot of strawberry. Pages 471–476 in: Proc. 7th Br. Insectic. Fungic. Conf. British Crop Protection Council.
11. Powelson, R. L. 1960. Initiation of strawberry fruit rot caused by *Botrytis cinerea*. Phytopathology 50:491-494.
12. Schmitthenner, A. F. 1973. Isolation and identification methods for *Phytophthora* and *Pythium*. Proc. Woody Ornamental Dis. Workshop, Univ. Mo., Columbia. 128 pp.
13. Tuite, J. 1969. Plant Pathological Methods, Fungi and Bacteria. Burgess Publishing Company, Minneapolis, MN. 329 pp.

Evaluating Fungicides for Pecan Disease Control

P. F. BERTRAND, The University of Georgia Cooperative Extension Service, Tifton 31793, and T. R. GOTTWALD, USDA, ARS, Southeastern Fruit and Tree Nut Research Laboratory, Byron, GA 31008

Over 121,000 ha of commercial pecans are grown in the Coastal Plain region of the southeastern United States. Control of fruit and foliar diseases is essential for continued commercial pecan production in this region. Pecan scab caused by *Cladosporium caryigenum* (Ell. et Lang.) Gottwald (5) is by far the most important disease. There are numerous other pecan diseases, but they are minor compared to pecan scab (9–11). Accurate methods of disease assessment are essential in evaluating fungicide performance.

METHODS

The easiest way to evaluate pecan diseases is with a visual rating system based on the amount of infection. Two such systems are in use in pecan disease research. Phillips et al (11) illustrated five categories of pecan scab infection on pecan fruit in 1952. Hunter and Roberts (8) developed a pecan scab rating system based on this illustration (Fig. 1). The other system is a modification of the Horsfall-Barratt concept (Fig. 2) (6,7).

Both systems involve dividing diseased tissues into various categories based on the amount of infection present. An overall disease rating can be determined from these data as follows. Multiply the number of plant parts in each disease category ($N_1, N_2, \ldots, N_t$) by the numerical infection value for that category ($1, 2, \ldots, t$); add these products, and divide by the total number of plant parts evaluated:

$$\text{Disease rating} = \frac{[(N_1 \times 1) + (N_2 \times 2) + \ldots + (N_t \times t)]}{(N_1 + N_2 + \ldots + N_t)}$$

These systems can be used to detect rather small differences in the incidence of pecan scab (1,4,6).

DISCUSSION

Both methods outlined above have been used primarily for pecan scab. Either method could, however, be used to evaluate incidence of any pecan disease on

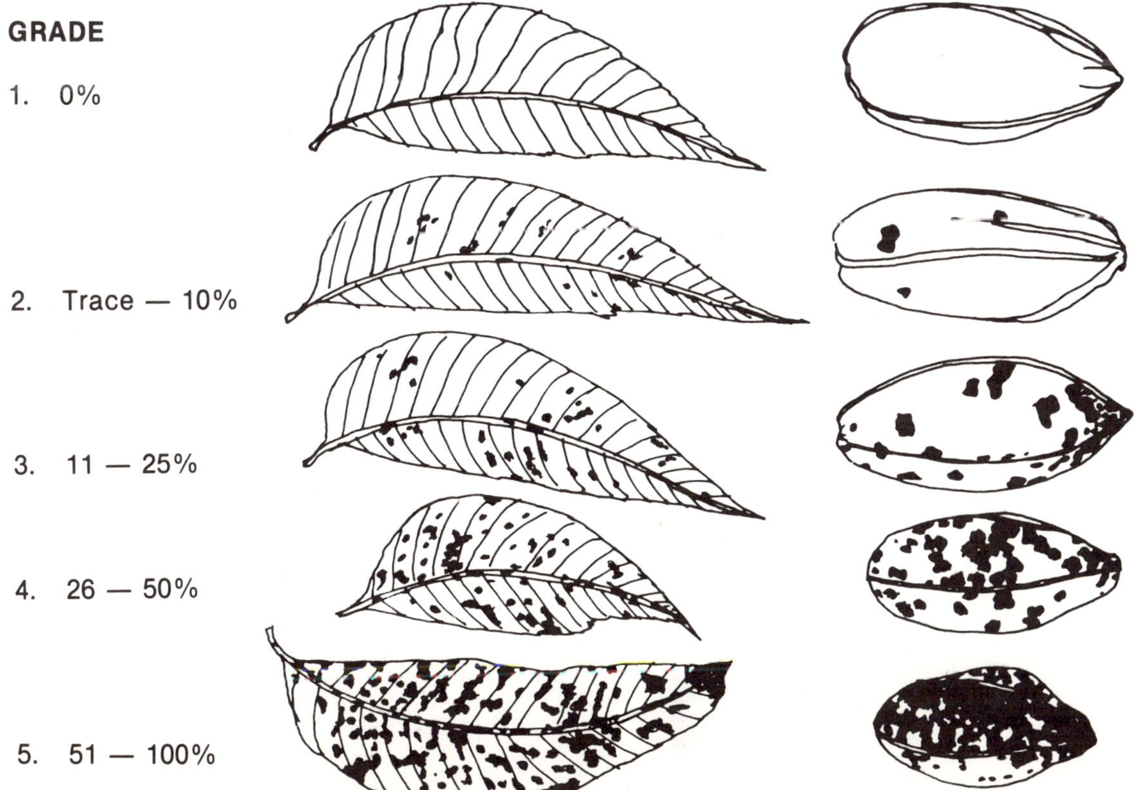

Fig. 1. The Hunter-Roberts (8) system for evaluating pecan diseases.

nuts, individual leaflets, or whole compound leaves (10,11). Entire trees could be rated for incidence of pecan bunch disease (3) using either method.

Researchers must keep in mind four important points when using either of these disease evaluation methods. First, pecan cultivars vary greatly in disease susceptibility. For any single study, all data must be gathered from a single cultivar. Second, overall disease ratings have value only in making comparisons. They can be used to compare two or more fungicide treatments or the disease susceptibility of two or more cultivars. Any single disease rating by itself is meaningless.

Third, and much more important, differences in overall disease ratings calculated from either method do not always reflect differences in actual disease losses (1,4,6). The disease ratings have no direct relationship to disease losses. In general, higher overall disease ratings mean higher levels of disease in an orchard, and higher disease levels increase the chances of disease loss. But both rating methods are based only on the amount of infection. They make no allowance for the time of infection, which, in the case of pecan scab, has been shown to have critical bearing on disease losses (6). In the case of foliar diseases, the situation is very complex.

The foliar diseases cause leaf drop (10,11). Premature leaf drop has been shown to result in depleted carbohydrate reserves and reduced crop quantity and quality the next season (13,14). Thus, disease loss from foliar infection may not be measurable until the following season. To further complicate matters, numerous insect and mite pests also cause leaf drop in pecans (10,11). Losses from foliar diseases have never been measured for pecans. The effect of pecan bunch disease on yields was studied (2) using a system similar to that of Hunter and Roberts (8).

Actual disease losses can be measured in terms of quality or quantity. Measuring total yield would obviously be valuable but is difficult. Pecan trees are very large when mature; in general, there are only 22–37 trees per hectare. The wood is brittle, and limb breakage is often considerable in heavy crop years, so trees of the same age may vary greatly in bearing potential. Pecans are also very subject to alternate-year bearing. All of these factors make large numbers of trees necessary for accurate replicated yield records. A fairly accurate method to quickly estimate yield has been developed (15). It reduces the time allotted to each tree but not the need for a large number of trees.

Other parameters can also be used to measure disease loss. The amount of premature nut drop from tagged clusters can give a measure of yield loss. Quality factors can also be considered. The pecan industry uses nut size as measured by the number of nuts in a standard unit of weight and the percentage of shell-out as measures of quality. Samples of 25–50 nuts are counted and weighed, then shelled and the kernels weighed to give these data. This should be done at about 3–4% kernel moisture, which is about the standard shelling moisture used by the industry.

Fourth, although disease rating systems provide fast and easy estimation of disease and allow simple statistical comparisons between fungicide treatments, they pose special problems when used in detailed epidemiologic studies. Data based on disease rating systems do not easily lend themselves to the basic statistical approaches used by epidemiologists to differentiate between treatments based on rates of disease increase over time. Such epidemiologic approaches give more accurate comparisons between treatments by allowing identification of infection periods, the points at which treatments begin to diverge from one another, and changes in host susceptibility over time. The main problem stems from disease rating categories that do not reflect equivalent groupings, e.g., 0%, trace–6%, 6–25%, 25–50%, etc. Data generated using these rating scales are not easily transformed back to accurate percent disease data. Use of nonuniform categories requires more complex nonparametric statistical analysis and does not allow direct logarithmic transformations such as those traditionally used to estimate rates of disease increase over time, which may be the best way to compare fungicide efficacy (12). This shortcoming can be overcome by generating a conversion equation by linear or curvilinear regression analysis of disease ratings of a small sample (about 500) of diseased plant parts (i.e., leaves, nut shucks, etc.) against actual percent disease estimations of these same plant parts. This conversion

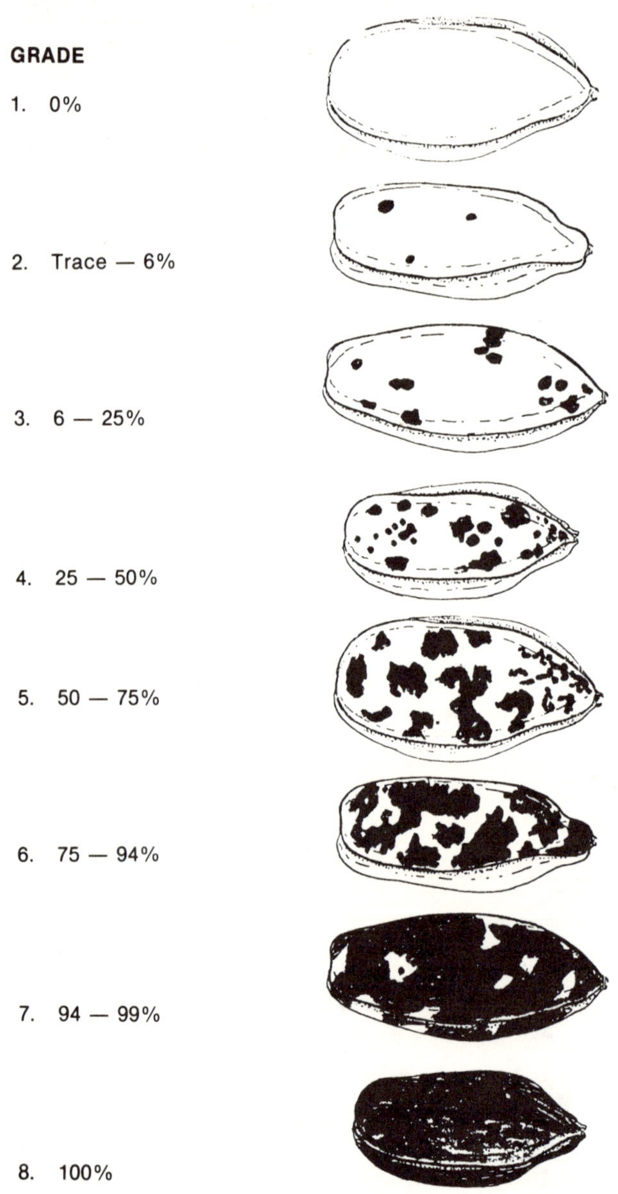

Fig. 2. A modified Horsfall-Barratt system for evaluating pecan diseases.

equation allows a fairly direct transformation of disease ratings to corresponding percent disease data for epidemiologic analysis.

Once the data have been thus transformed, curvilinear regression analysis and graphing the resulting regression lines are often helpful in visualizing disease increase trends over time. Another helpful statistical technique is multiple comparisons of statistical means via Duncan's multiple range test, least significant difference, etc. This allows the researcher to compare treatments on every sampling date and thus determine if the treatments diverge early, middle, or late in the season, indicating efficacy differences related to changes in disease pressure.

The statistical procedures outlined above require at least five or six sampling dates to enable the researcher to accurately track the progress of epidemics under the influence of various fungicide treatments.

The methods outlined here for evaluating pecan diseases are very useful. They do not, however, reflect actual disease losses and cannot be used to predict losses. Other measurements of crop quality and/or quantity are necessary to see if differences detected by the ratings reflect differences in actual disease losses. Special statistical considerations are necessary to use the data for intensive epidemiologic studies.

LITERATURE CITED

1. Bertrand, P. F., and Gottwald, T. R. 1981. Comparison of flowable and wettable powder formulations of triphenyltin hydroxide. Pecan South 8(2):12-15.
2. Bertrand, P. F., and Gottwald, T. R. 1984. The effect of bunch disease on yield and quality of pecans. (Abstr.) Phytopathology 74:809.
3. Cole, J. R. 1937. Bunch disease of pecans. Phytopathology 27:604-612.
4. Diener, U. L. 1962. Control of pecan scab in Alabama in 1961 and its relation to nut quality. Plant Dis. Rep. 46:232-235.
5. Gottwald, T. R. 1982. Taxonomy of the pecan scab fungus *Cladosporium caryigenum*. Mycologia 74:382-390.
6. Gottwald, T. R., and Bertrand, P. F. 1983. Effect of time of inoculation with *Cladosporium caryigenum* on pecan scab development and nut quality. Phytopathology 73:714-718.
7. Horsfall, J. G., and Barratt, R. W. 1945. An improved grading system for measuring plant diseases. (Abstr.) Phytopathology 35:655.
8. Hunter, R. E., and Roberts, D. D. 1978. A disease grading system for pecan scab. Pecan Q. 12(3):3-6.
9. Latham, A. J., Diener, U. L., and Garrett, F. E. 1972. Pecan disease research in Alabama. Auburn Univ. Agric. Exp. Stn. Circ. 199. 23 pp.
10. Payne, J. A., Malstrom, H. L., and KenKnight, G. E. 1979. Insect pests and diseases of the pecan. U.S. Dep. Agric. Sci. Ed. Admin. Agric. Rev. Man. AMR-S-5. 43 pp.
11. Phillips, A. M., Cole, J. R., and Large, J. R. 1952. Insects and diseases of the pecan in Florida. Univ. Fla. Agric. Exp. Stn. Bull. 499. 76 pp.
12. Vanderplank, J. E. 1963. Plant Diseases: Epidemics and Control. Academic Press, New York. 349 pp.
13. Worley, R. E. 1979. Pecan yield, quality, nutlet set, and spring growth as a response to time of fall defoliation. Proc. Am. Soc. Hortic. Sci. 104:192-194.
14. Worley, R. E. 1979. Fall defoliation data and seasonal carbohydrate concentration of pecan wood tissue. Proc. Am. Soc. Hortic. Sci. 104:195-199.
15. Worley, R. E., and Smith, M. 1984. A method of estimating pecan yield. HortScience 19:664.

Field Evaluation of Fungicides for Control of Celery Diseases

M. L. LACY, Department of Botany and Plant Pathology, Michigan State University, E. Lansing 48824-1312; J. O. STRANDBERG, University of Florida, Central Florida Research Center, Sanford 32771; and A. O. PAULUS, Department of Plant Pathology, University of California, Riverside 92521

Early blight of celery, caused by *Cercospora apii* Fres. (1,3), and late blight of celery, caused by *Septoria apiicola* Speg. (3,10), are two foliar diseases of celery that are of great importance in most celery-producing areas of the United States (Florida, New York, Michigan, Ohio, and California). Their relative importance in each area depends on the climatic conditions in that area. Since a high level of resistance to either disease is not commercially available, foliar-applied fungicides are the primary method of control. Each disease is discussed separately below.

EARLY BLIGHT

Early blight (Cercospora leaf spot) is especially destructive in Florida and Ohio, but it occurs in other areas as well (3,5). Conditions favorable for spread are relative humidity near 100% with heavy dews and temperatures of 15–30°C (1). Significant spore release occurs when these conditions prevail for 10 or more hours daily. Rain during the middle of the day greatly reduces spore numbers in the air, but wind velocity has little effect on numbers of spores released (1). Great numbers of spores are released during field operations, especially harvesting. Early blight spreads faster in narrow between-row spacings of 60 cm than in wider rows of 80 cm (2), probably because of the alteration in microclimate caused by row spacing.

C. apii causes small, irregularly circular, brown lesions that rapidly enlarge on both sides of the leaves (3,10). As lesions enlarge, the affected tissue takes on an ashen gray to tan color, and larger lesions assume a dry, papery texture. Under humid conditions, the lesion may bear a gray to dark mold that is comprised of masses of long, colorless, septate spores on long, brown conidiophores that are produced on the lesion (10). This mold can only be seen with difficulty with the unaided eye. Lesions lack the rather conspicuous pycnidia of *Septoria* lesions; otherwise the appearance is similar.

Methods. A well-drained test site should be selected to promote good celery growth. Seeds of a highly susceptible cultivar such as Tall Utah 52-70 are planted in germination flats about 8–12 weeks before being transplanted into the field. Three-year-old seed with good germination (at least 80%) is used to minimize the risk of seedling infection by either the early or late blight fungus, both of which can be seedborne (3,10). Seeds are scattered over a firm, smoothed bed of soil and covered with a thin (2-mm) layer of washed white sand, and a piece of glass is placed over the germination flat, covering 80–90% of the surface to keep the seedbed from drying out. Plants are watered daily with a fine mist. When they have two true leaves, seedlings are transplanted into flats with a spacing of 4–5 cm (1.5–2 in.) between plants. When plants are about 12–15 cm (4.5–6 in.) tall, they are transplanted into the field.

Procedure. After the experimental design is decided and celery seedlings are well established, the trial can be initiated. Treatments should be applied to at least 5 linear meters (15 linear feet) of row. Preferably, a guard row separates each treatment row to prevent fungicide drift from one plot to another; under conditions particularly favorable for the disease, however, there may be some risk of the guard rows generating excessive amounts of inoculum that could reduce apparent efficacy. Even if this should occur, there is always some point during the experiment when differences can be clearly seen if the plots are observed frequently.

In some areas it is possible to rely on natural infestation of plots with early blight for disease initiation. If natural infestations are unreliable, or if there is a need to control the timing of infections, *C. apii* conidia can be produced on celery leaf and muck soil medium, on carrot leaf decoction agar, or on carrot root disks and used to inoculate unsprayed guard rows or entire plot areas (11). Good infection levels can be obtained by spraying plants with an aqueous spore suspension (5–10 × 10^3 conidia per milliliter) during a period of favorable weather. Best results are usually obtained if inoculations are performed in late afternoon. The first spray is often applied just before inoculation so that fungicide is present when the spores arrive.

The most satisfactory method for application of fungicides is a conventional tractor-mounted sprayer with a spray boom rigged so that two cone nozzles are placed on either side of the sprayed row at a 45° angle or with a nozzle over the center and a drop nozzle on either side (9). The boom height must be adjustable so that young as well as maturing plants can be properly sprayed. This arrangement gives excellent coverage of the foliage and closely simulates spray application by growers. Pressures of 700–1,380 kPa (100–200 psi) and ground speeds of 3.2–4.8 km/hr (2–3 mph), with a delivery rate of 470–935 L/ha (50–100 gal/A), are common. Careful calibration of the sprayer is very important. Although other sprayers may be effective, this seems to be a good standard for comparison of materials.

If hand-held sprayers are used, they should preferably be the compressed air or CO_2 type with a pressure regulator so that a constant pressure and flow rate is assured. The disadvantages of these sprayers are imprecision in ground speed and low volume per hectare (acre)—234–327 L (25–35 gal)—which may not give adequate coverage under all conditions. A large percentage of the celery crop receives fungicide treatments applied by air. We know of no single, reliable way to simulate aircraft applications of celery fungicides in small plots. Innovation and experimentation are badly needed to solve this problem.

Sprays are usually applied at 7- to 10-day intervals for protectant fungicides and 7- to 14-day intervals for the systemics (9,12,13). Sprays are initiated as soon as plants are well established or just before inoculation, and they are terminated either when nontreated controls are 50% or more diseased or when the celery is ready for harvest.

Plot designs for simple efficacy testing or screening are usually of the randomized block type with a minimum of three, and preferably four, replications. Split-plot or factorial designs can also be used. Other designs have been used for specialized purposes.

Disease assessment. Visual disease ratings are often used to estimate the amount of disease damage. This can be done by directly estimating percentage of disease with the aid of the Horsfall-Barratt rating scale (8) or by using some simplified form of the Horsfall-Barratt scale. Lacy (9) devised a rating scale of 1–5, where 1 = 0% disease, 2 = 1–10%, 3 = 11–25%, 4 = 26–50%, and 5 = greater than 50% of foliage diseased. Using this scale, any fungicide receiving an average disease rating higher than 2.5 is considered unacceptable for commercial use.

Disease is rated two or three times at weekly intervals before the termination of the experiment. Any evidence of phytotoxicity is recorded. Herbicides and some insecticides can cause injury when applied under certain conditions, so these materials should be applied at the proper times and at the proper rates to minimize the chance of damage to the plants from unwanted pests or from the pesticides themselves. A careful record is kept of all pesticide applications and of rainfall and irrigations.

Yield data are always useful in assessing fungicide efficacy, and they should be taken if possible. Yields often reveal stunting or yield reduction resulting from disease damage not evident through visual inspection, and they can reveal unexpectedly high yields as well. At least 3 linear meters (10 linear feet) of row are harvested, trimmed, and weighed as marketable celery. It is often helpful to assess the level of trimming and removal of disease-damaged leaves needed to obtain a commercially acceptable product, and yields expressed in amount of graded commercial celery are helpful in determining expected economic benefits of fungicide treatment. Appropriate statistical tests are performed to reveal significant differences between treatments.

LATE BLIGHT

Celery late blight (Septoria leaf spot) is much more important economically in Michigan and New York than is early blight (10), but it causes severe damage in other production areas in some years. Originally, Cochran (4) thought that late blight was caused by two separate species of *Septoria*, but it was later shown that only one species, *S. apiicola*, is involved (6,14). Conditions favoring maximum infection are 36 or more hours of continuous high humidity (≥95%) and cool temperatures (15–20° C) with long periods of leaf wetness (3,15). The fungus produces its conidia inside pycnidia that erupt from the infected portions of leaves. The conidia are forcibly expelled from mature pycnidia in long, gelatinous masses when wet by free water on the leaves (3,7). These spores are spread by splashing rain or can be carried on people's clothing, on animals, or on farm implements. *Septoria* does not disseminate as far from centers of infection as does *Cercospora* because *Cercospora* spores are windborne, whereas *Septoria* spores apparently are not (3,10). As a consequence, *Septoria* damage tends to be patchy or clumped in its distribution within a field.

Septoria may also be carried as pycnidia on coats of seeds from *Septoria*-infected plants and will infect seedlings emerging from infested seeds (3,7). Because *Septoria* does not survive on infested seeds for more than about 2 years, 3-year-old seeds are usually free of the fungus (3,10). Infested seed can also be treated with hot water (3) to eradicate the pathogen. Hewett (7) devised a test method where *Septoria* can be detected on as little as one infested seed per thousand.

Methods. Because late blight spreads from plant to plant slowly, a field inoculation is often used to ensure uniform distribution of disease throughout the plots (9). Infected leaves from nonsprayed areas are collected, air-dried, and stored in plastic bags at 2° C until used (up to 12 months). When environmental conditions appear favorable (a cool, rainy period is predicted), a spore suspension is made by immersing about 500 g of dried leaves in about 10 L of water. After the leaves have soaked for 15 min, the plant material is removed by straining the suspension through four layers of cheesecloth. The spore suspension is adjusted to 1×10^5 conidia per milliliter and is sprayed evenly over the plots with a hand sprayer at the rate of about 15–20 L/930 m^2 (10,000 ft^2) of plot area. Inoculations are repeated once to ensure good infection.

Other methods are similar to those used for early blight.

Procedure. Planting, spraying, and data collection are similar to those discussed for early blight above. Late blight lesions require 14–21 days following inoculation to develop (9), whereas early blight lesions usually appear within 7–10 days (11).

In the past, fungicides effective on late blight have almost always been effective for early blight, and vice versa (9,12,13), but some newer classes of fungicides show differential responses to these two pathogens.

LITERATURE CITED

1. Berger, R. D. 1973. Early blight of celery: Analysis of disease spread in Florida. Phytopathology 63:1161-1165.
2. Berger, R. D. 1975. Disease incidence and infection rates of *Cercospora apii* in plant spacing plots. Phytopathology 65:485-487.
3. Chupp, C., and Sherf, A. G. 1960. Vegetable Diseases and Their Control. Ronald Press, New York. 693 pp.
4. Cochran, L. C. 1932. A study of two Septoria leaf spots of celery. Phytopathology 22:791-812.
5. Duggar, B. M. 1897. Early blight of celery. N.Y. Agric. Exp. Stn. Bull. 132:201-206.
6. Gabrielson, R. L., and Grogan, R. G. 1964. The celery late

blight organism *Septoria apiicola*. Phytopathology 54:1251-1257.

7. Hewett, P. D. 1965. Viable *Septoria* spp. in celery seed samples. Ann. Appl. Biol. 61:89-98.
8. Horsfall, J. S., and Barratt, J. W. 1945. An improved grading system for measuring plant diseases. Phytopathology 35:655.
9. Lacy, M. L. 1973. Control of Septoria leafspot of celery with systemic and nonsystemic fungicides. Plant Dis. Rep. 57:425-428.
10. Lacy, M. L., and Grafius, E. J. 1980. Disease and insect pests of celery. Mich. Agric. Exp. Stn. Bull. E-1427. 8 pp.
11. Murakishi, H. H., Honma, S., and Knutson, R. 1960. Inoculum production and seedling evaluation of celery for resistance to *Cercospora apii*. Phytopathology 50:605-607.
12. Paulus, A. O., Shibuya, F., Holland, A. H., and Nelson, J. 1970. Timing interval for control of Septoria leaf spot of celery. Plant Dis. Rep. 54:531-535.
13. Paulus, A. O., Otto, H., and Nelson, J. 1981. Combination, single, and alternate applications of fungicides for control of *Septoria apiicola* on celery. (Abstr.) Phytopathology 71:107.
14. Sheridan, J. E. 1968. The causal organism of celery leaf spot, *Septoria apiicola*. Trans. Br. Mycol. Soc. 51:207-213.
15. Sheridan, J. E. 1968. Conditions for infection of celery by *Septoria apiicola*. Plant Dis. Rep. 52:142-145.

Field Evaluation of Fungicides for the Control of Clubroot of Crucifers

T. K. KROLL, Mobay Chemical Corporation, Milford, NH 03055, and ROBERT C. LAMBE, Department of Plant Pathology, Physiology and Weed Science, Virginia Polytechnic Institute and State University, Blacksburg 24061

Clubroot, caused by the fungus *Plasmodiophora brassicae* Wor., is a serious disease of crucifers that has been responsible for the disappearance of crucifer production in certain areas of the United States. The disease is favored by moderate temperatures (20–25°C), acidic soils, and a period of moisture that allows *P. brassicae* zoospores to swim to the root (3). There are several races of *P. brassicae*. Because there is no resistance in many commercially desirable cultivars and cultural practices provide only limited control (3), the use of fungicides is an important control strategy.

METHODS

Soil preparation. There are two major methods of preparing clubroot-infested fields for research. The first method is to use a naturally infested field. Before chemicals are applied, the field plot is disked and cross-disked several times so that *P. brassicae* spores are thoroughly dispersed in the soil. Sufficient time should be allowed for diseased tissue to decompose completely, which may take a full year to occur. This increases the probability that spores will be exposed to test chemicals. Evaluations, of course, should be made with the consideration that growers would probably be applying chemicals when plant debris is still present. Because of this, it may be advisable to include a fallow period in clubroot control recommendations.

The second method of preparing plot land is to infest a plot artificially by transplanting diseased plants into the field. Transplants are spaced 60 cm (3 ft) between rows and 23–46 cm (9–18 in.) between plants (5) to obtain thorough infestation of the soil. After galls have formed on the roots, plants and roots are disked completely into the soil. Disking can be performed within 6 weeks of transplanting and tests conducted after an additional 2 weeks; however, it is best to wait for tissues to decompose.

Inoculum source, preparation, and storage. For artificially infested fields, the inoculum source should represent the pathogenic races prevalent in the area. Four distinct pathotypes have been found in North America (7). They may be differentiated in the greenhouse or growth chamber by sowing differentially sensitive crucifer seeds into infested soil or growth medium and rating them for disease 35 days later. For more specific information, the reader may consult Buczacki et al (1) or Williams (7).

Spore suspensions can be prepared by grinding fresh or frozen *P. brassicae*-colonized roots with distilled water in a blender at high speed for 3 min and straining the resulting suspension through eight layers of cheesecloth to remove any debris. Further clarification of the suspension may be made by centrifuging the suspension at 1,000 *g* for 7 min and resuspending the resulting spore pellet in distilled water, then repeating this procedure two times (6).

Transplants used for infesting fields artificially may be infected by soaking roots in a preparation of this suspension containing 1×10^5 spores per milliliter. Spores are clear and round and measure 4 mm in diameter. Plants are incubated in the dark at 25°C (77°F) for 24–36 hr (4). An inoculum source may be maintained for several years by storing diseased roots at −4°C (6).

Test plants. Most commercial crucifer cultivars are susceptible to clubroot (3) and can be used as test indicators. The authors have generally used cabbage as a test plant because it takes at least 90 days for the crop to mature, which allows enough time for an accurate evaluation of chemical persistence in the soil (5). These test plants may either be directly seeded or transplanted when they are 6 weeks old.

Plot size and layout. Cross-contamination is avoided by using a minimum of three rows per plot and taking data from only the center row. For artificially infested fields, it is wise to leave a few large areas for noninfested controls. From experience, we suggest that these plots have a minimum of 6×6 m (20×20 ft). Microplots may also be an effective method for maintaining such pathogen-free areas. A randomized block design of five replications per treatment and 20 observations per plot allows for satisfactory statistical separation of data (5).

Chemical tests are often conducted using a high inoculum density. In greenhouse tests, Kroll (4) found that high populations ($\geq 10^6$ *P. brassicae* spores per gram of medium) may decrease the effectiveness of certain chemicals. These results suggest that some control strategies be directed at preventing population increases while the population is still low. Finding plots with a reported low disease incidence, infesting fields with fewer diseased transplants, or maintaining known populations using microplots are some methods for testing chemicals at low population levels.

Chemical application. The method of chemical application may influence effectiveness. Since *P.*

brassicae spores are quite small and soil populations may be quite high, it is important that soil-applied chemicals be thoroughly distributed to obtain optimum control (2). Chemicals should be applied so that the roots do not grow out of the chemically treated zone during the growing season. This can be accomplished by 1) rototilling chemicals into the soil to a depth of 10–15 cm (4–6 in.), 2) applying chemicals with sufficient water to disperse the chemical adequately, 3) applying chemicals with the transplant water, 4) applying chemicals that are systemic and will protect developing roots, or 5) applying a gaseous compound. Pentachloronitrobenzene (Terrachlor) and hydrated lime are often applied as standard treatments (4). Pentachloronitrobenzene requires constant agitation when used with the transplant water because it rapidly settles out.

Disease and phytotoxicity ratings. Disease may be rated as early as 6 weeks after transplants are set in a field or 6 weeks after seedlings have emerged. However, since this does not provide for an accurate evaluation of the persistence of chemical treatments or the effectiveness of chemicals in preventing secondary infection, it is probably best to allow plants to mature fully. Symptoms of clubroot are expressed as galls on the taproot and lateral roots and, in advanced stages, as a complete root decay caused by secondary pathogens (3). Treatments are rated for disease incidence and yield. Disease intensity ratings that are similar to the following scale are often used: 1 = no disease, 2 = slight clubbing of lateral roots with no root necrosis, 3 = slight clubbing of lateral roots and taproot with some root necrosis, 4 = severe clubbing of lateral roots and taproot with more than 50% root necrosis, and 5 = almost complete root necrosis and near death or death of the plant (5). Club area (club length × diameter) has also been used as a measurer of disease intensity (4). Phytotoxicity should be rated on the basis of stunting, chlorosis, and necrosis.

LITERATURE CITED

1. Buczacki, S. T., Toxopeus, H., Mattusch, P., Johnston, T. D., Dixon, G. R., and Hobolth, L. A. 1975. Study of physiologic specialization in *Plasmodiophora brassicae*: Proposals for attempted rationalization through an international approach. Trans. Br. Mycol. Soc. 65:295-303.
2. Dobson, R. L., Gabrielson, R. L., Baker, A. S., and Bennett, L. 1983. Effects of lime particle size and distribution and fertilizer formulation on clubroot disease caused by *Plasmodiophora brassicae*. Plant Dis. 67:50-52.
3. Karling, J. S. 1968. The Plasmodiophorales. 2d ed. Hafner Publishing, New York. 256 pp.
4. Kroll, T. K. 1983. Clubroot: Affects of resistance, root-colonizing bacteria and chemicals on pathogenesis. Ph.D. thesis, Virginia Polytechnic Institute and State University, Blacksburg. 163 pp.
5. Kroll, T. K., Lambe, R. C., Lacy, G. H., Moore, L. D., and Reilly, T. G. 1981. Comparison of fungicides for the control of clubroot of cabbage in Virginia, 1980. Fungic. Nematic. Tests 36:56.
6. MacFarlane, I. 1958. A solution culture technique for obtaining root-hair or primary infection by *Plasmodiophora brassicae*. J. Gen. Microbiol. 18:720-732.
7. Williams, P. H. 1966. A system for the determination of races of *Plasmodiophora brassicae* that infect cabbage and rutabaga. Phytopathology 56:624-626.

Field Evaluation of Fungicides for Control of Lettuce Drop

S. A. JOHNSTON, New Jersey Agricultural Experiment Station, Rutgers University, Rutgers Research and Development Center, Bridgeton 08302, and J. W. LORBEER, Department of Plant Pathology, New York State College of Agriculture and Life Sciences, Cornell University, Ithaca 14853

Lettuce drop can be caused by either *Sclerotinia minor* (small sclerotia-producing species) or *S. sclerotiorum* (large sclerotia-producing species), and infection by both species may occur in the same field. *S. minor* is the predominant species on lettuce, and it is frequently consistently present with each crop. Infection of lettuce with *S. minor* can occur either at the soil line through senescent lower leaves or below ground through stem tissues. The majority of infections originate from sclerotia within the top 2 cm of soil (1,4,7). Sclerotia either undergo eruptive mycelial germination (a mass of hyphae that emerges through the sclerotial rind) or noneruptive mycelial germination (13) and then are capable of directly infecting lettuce. The optimum levels of eruptive sclerotial germination occur at −1/3 bar soil moisture tension and at 18° C (8). Therefore, lettuce drop is most prevalent during cool, moist environmental conditions. Apothecia have rarely been observed for *S. minor*, and ascospores are not considered important in the epidemiology of lettuce drop (1,5).

In contrast, the occurrence of *S. sclerotiorum* is generally more sporadic; it may be inactive for several years and then produce a widespread epidemic (1). In New Jersey, *S. sclerotiorum* occurs primarily during the spring lettuce crop and is seldom observed during the fall crop. Infection of lettuce with *S. sclerotiorum* predominantly occurs at the ground level, originating from ascosporic infection of lower leaves. Sclerotia present in the top 2–3 cm of soil germinate carpogenically to produce apothecia that release ascospores. In the eastern United States, ascosporic inoculum can originate within or outside of lettuce fields (1). Mycelial germination from sclerotia of *S. sclerotiorum* seldom occurs in the eastern United States (1), whereas in Arizona this form of germination is common (12).

Since the primary means of infection of lettuce by *Sclerotinia* spp. is by colonization of senescent leaves, a fungicide must be applied to prevent colonization of the leaves. Thus, soil surface coverage near the plant and timing of fungicide applications are the most important factors in obtaining control. The primary means of chemical control of lettuce drop in western areas of the United States, such as in California, has been a single postthinning application of dicloran directed toward the base of plants and adjacent soil at 4.5 kg/ha (4.0 lb a.i./A) (10). In contrast, under eastern conditions in New York and New Jersey, dicloran has not been effective. Generally, effective control with experimental fungicides—e.g., benomyl—requires up to three post-thinning applications (6,11). More recently, iprodione and vinclozolin have received federal registration for control of drop on head lettuce and have effectively controlled this disease in New Jersey and New York in commercial lettuce fields. Iprodione or vinclozolin is applied every 10–14 days after thinning for a total of three applications at 0.6–0.8 kg/ha (0.5–0.75 lb a.i./A) (9). Applications are directed toward the base of plants and adjacent soil.

METHODS

Test plot. The field used for experimental plots should have a history of losses due to lettuce drop. It should be level to prevent interplot interference caused by sclerotia being washed from one plot to another during heavy rains. This goal can also be achieved by growing lettuce on raised beds.

Lettuce should be grown according to state extension service recommendations. Soil tests should be collected far in advance of planting to ensure ample time to achieve proper pH and fertility levels. In New Jersey, lettuce is produced on rows that are 0.6 m (2 ft) apart, and plants are spaced 30.5–38 cm (12–15 in.) apart in the rows. Under New Jersey conditions, two field tests can be conducted in a field during one growing season. A spring test can be conducted using 8-week-old transplants grown in a peat-vermiculite-sand (1:1:1) mixture in flats containing cells 2.5 cm × 2.5 cm × 6.4 cm (1 in. × 1 in. × 2.5 in.) deep. Transplants are planted during late March, and fungicide applications begin about 1–2 weeks after transplanting. For a fall test, lettuce is direct-seeded during early August and fungicide applications begin shortly after the rows of lettuce are thinned to final stands (about 2–3 weeks after seeding).

In New Jersey, experimental plots used for fungicide trials on lettuce have consisted of two rows 7.6 m (25 ft) long on a bed 1.5 m (5 ft) wide (total of 50 plants per plot). Plots are separated from each other by a border bed containing two rows of lettuce to avoid spray drift from one plot to another. Treatments are replicated four times in a randomized complete block design. In New York, experimental plots on organic soils have consisted of single-row treatments (no border rows), each 7.7 m (25 ft 2 in.) long and replicated four times in a randomized complete block design with 17–25 plants per plot (13). Such designs have permitted statistical evaluation of replications as well as of treatments. This is useful in separating the differences among replications that result from field conditions. Replicates should be arranged in a field according to potential differences in drainage, soil type, etc., whenever possible (see the chapter on statistical design by L. A. Nelson, *this volume*).

In all field tests, an untreated control and standard recommended fungicide treatments are included. These treatments enable the investigator to determine whether an experimental treatment reduced disease incidence and whether it was better than the standard treatment.

Records of soil type, topography, field history, rainfall, amount and timing of supplemental irrigation, and details of all pesticides used in the field should be maintained. This information is valuable for interpretation of the data.

Test-plot inoculum level. Soil samples from lettuce fields with a history of continuous lettuce production generally have 0–82 sclerotia of *S. minor* per 100 g of soil in mineral soils (3,7) and 100–300 sclerotia of *S. minor* per 100 g of soil in organic soils (2). Fields should be sampled to determine the population level of *S. minor* far in advance of planting to ensure the presence of a satisfactory pathogen population. If the population level is low, inoculum will have to be added to the field to increase the incidence of lettuce drop in the test area.

Fields are sampled for the population level of *S. minor* in the mineral soils in New Jersey by use of the following procedure (3). Soil samples are collected with a 2-cm-diameter soil sampling tube from the top 20–25 cm of soil at random locations over the sampling area (about 20 cores per hectare). Soil samples are air-dried and screened through a 10-mesh sieve. Four 25-g samples are assayed from the soil sample, and each is washed on a 45-mesh sieve with running tap water. The residue is rinsed into a 250-ml plastic beaker and swirled several times to suspend the sclerotia in water. The suspension is poured onto the 45-mesh assay sieve. The process of suspending the sclerotia in the water in the beaker and pouring the suspension onto the assay sieve is repeated several times until only sand particles remain in the beaker. The assay sieve is examined under a dissecting microscope, and the sclerotia are picked out with a fine-tipped forceps.

For organic soils, such as in New York, a different procedure is followed to remove sclerotia of *S. minor* from the sample (2). Soil samples are air-dried, sifted on a 9-mesh sieve, blended in tap water in a Waring Blendor, and wet-sieved through 9-mesh and 48-mesh sieves. Residue on the 9-mesh sieve is discarded, and the residue on the 48-mesh sieve is transferred to a centrifuge tube containing 70% glycerol. After centrifugation for 3 min at 3,000 rpm, the sclerotia in the overlying liquid are removed and counted by using a stereoscopic microscope.

Test-plot infestation. If it is necessary to infest a field with *S. minor*, the following procedure is followed. Oat seed and water are added to polypropylene sterilizing pans (2.0 kg of seed + 2.5 L of water per pan) and wrapped with aluminum foil. After 16 hr, the pans are autoclaved for 1 hr on two successive days. Inoculum of *S. minor* is added to the pans and incubated at room temperature for 1 month. At least 1 month before planting, the inoculum is broadcast uniformly over the test area (500 g of inoculum per 9.3 m^2) and incorporated to a depth of 15 cm. Although this technique should result in a high incidence of lettuce drop, in some infested fields it may take two or three continuous crops of lettuce before uniform, severe disease incidence occurs.

Infestation of a field plot with *S. sclerotiorum* has been accomplished in a slightly different manner (12). Inoculum is prepared by boiling 5.5 kg of barley in 15 L of tap water containing 7 g of potassium chloride per liter for 1.5 hr. After boiling, the grain is sieved and added to 1,000-ml flasks. Twenty-four hours later, the flasks are autoclaved for 1 hr, cooked for 24 hr, and autoclaved again. After the final cooling, the grain is seeded with a single sclerotium and incubated at room temperature for 3 months under continuous light. The inoculum is dried and then distributed in lettuce plots after thinning at the rate of 400 cc of inoculum per 93 m^2.

Fungicide application. Equipment used for applying fungicides to lettuce for control of drop should be able to deposit fungicide uniformly around the base of the plant and the surrounding soil surface. In New Jersey, fungicides have been applied with a tractor-drawn boom sprayer that delivers 374 L/ha (40 gal/A) at 276 kPa (40 psi). The boom is equipped with nozzles containing TeeJet 8004 flat spray tips. Three nozzles, spaced 51 cm (20 in.) apart on the boom, are used for each bed, and the boom is positioned 51 cm above the top of the bed. Fungicide application begins after transplants recover from transplanting in the spring test and shortly after thinning in the fall tests. Additional applications are made at 10 and 20 days after the first application. In some cases, additional applications are made at 10-day intervals for some fungicide trials. In some areas, drop nozzles are used to ensure coverage of the undersurface of the leaves adjacent to the soil. Since the lettuce plants are small at the time of fungicide application, sufficient coverage has been obtained without the use of drop nozzles in experiments conducted in New Jersey. In New York, fungicide sprays have been applied with a hand-held CO_2-pressurized sprayer at 103 kPa (15 psi). The spray boom is equipped with two nozzles (TeeJet 8002, delivering a fan-shaped pattern) mounted so that the spray is directed at the plants from the sides or at a 45° angle from above. Fungicides are applied at the rate of 935 L of water per hectare (200 gal/A), and 0.25 L (0.25 qt) of formulated material is applied to each treatment replicate.

Data collection. Data are collected for the following parameters: phytotoxicity, disease control, and yield. When phytotoxicity is observed, it is measured by a rating system. Phytotoxicity has not been a problem with lettuce fungicides, except occasionally with early applications of dicloran (10). Disease control is evaluated by counting the number of plants with drop symptoms. The primary symptom of drop is the complete wilting of the plant. Evaluations conducted in the early morning are advantageous because wilting resulting from moisture and heat stress is not a problem at this time. Drop will occur at any time during the course of the test, but incidence is generally not high enough to separate differences among treatments until closer to harvest. Therefore, disease incidence determinations usually are not taken until 1–2 weeks before harvest. Each infected plant should be examined at the soil line to determine whether the infection is caused by *S. minor* or *S. sclerotiorum*. Infected plants produce an abundant amount of sclerotia, and the size difference between species is great enough to distinguish them easily. If both *Sclerotinia* species are present, species identification is important in lettuce drop fungicide evaluations because fungicide control can be different for each species. Benzimidazole fungicides are extremely effective against *S. sclerotiorum*. In New Jersey, in plots where both *Sclerotinia* species are present, *S. minor* is the predominant species in infected plants in the benzimidazole treatment; in other treatments, a more equal distribution of *S. minor* and *S. sclerotiorum* occurs (9). In New York, *S. minor* has been the most important in recent years, and infections by *S. sclerotiorum* have been infrequent. Yield may be calculated as the weight of healthy marketable plants or percentage of marketable heads. All data are subjected to an analysis of variance and to Duncan's multiple range test.

RESULTS

Data collected from tests of this nature are used for several purposes. In many cases, data are used in the formulation of a recommendation for lettuce drop control in specific states where the disease occurs. Results from these tests are usually submitted to respective agrichemical companies to support registration for the use of a particular fungicide on lettuce.

Residue samples may need to be collected from the test area. This need is accommodated either by expanding plots or establishing a separate plot solely for the purpose of residue collection. Residues are collected, placed in plastic-lined bags, labeled, and frozen until analyzed for residue content.

LITERATURE CITED

1. Abawi, G. S., and Grogan, R. G. 1979. Epidemiology of diseases caused by *Sclerotinia* species. Phytopathology 69:899-904.
2. Abd-Elrazik, A. A., and Lorbeer, J. W. 1980. Rapid separation of *Sclerotinia minor* sclerotia from artificially and naturally infested organic soil. Phytopathology 70:892-894.
3. Adams, P. B. 1979. A rapid method for quantitative isolation of sclerotia of *Sclerotinia minor* and *Sclerotium cepivorum* from soil. Plant Dis. Rep. 63:349-351.
4. Adams, P. B., and Tate, C. J. 1975. Factors affecting lettuce drop caused by *Sclerotinia sclerotiorum*. Plant Dis. Rep. 59:140-143.
5. Adams, P. B., and Tate, C. J. 1976. Mycelial germination of sclerotia of *Sclerotinia sclerotiorum* on soil. Plant Dis. Rep. 60:515-518.
6. Hawthorne, B. T. 1979. Effectiveness of benomyl for control of *Sclerotinia minor* on lettuce. N.Z. J. Exp. Agric. 7:215-219.
7. Imolehin, E. D., and Grogan, R. G. 1980. Factors affecting survival of sclerotia, and effects of inoculum density, relative position, and distance of sclerotia from the host on infection of lettuce by *Sclerotinia minor*. Phytopathology 70:1162-1167.
8. Imolehin, E. D., Grogan, R. G., and Duniway, J. M. 1980. Effect of temperature and moisture tension on growth, sclerotial production, germination, and infection by *Sclerotinia minor*. Phytopathology 70:1153-1157.
9. Johnston, S. A. 1982. Control of lettuce drop with foliar sprays and transplant drenches: Spring, 1981. Fungic. Nematic. Tests 37:71.
10. Marcum, D. B., Grogan, R. G., and Greathead, A. S. 1977. Fungicide control of lettuce drop caused by *Sclerotinia sclerotiorum* 'minor'. Plant Dis. Rep. 61:555-559.
11. Springer, J. K. 1973. Control of lettuce drop with fungicides. Fungic. Nematic. Tests 28:76.
12. Troutman, J. L., and Matejka, J. C. 1982. Establishing and maintaining a lettuce drop nursery. Plant Dis. 66:415.
13. Wymore, L. A. 1984. The epidemiology and control of *Sclerotinia minor* causing drop of lettuce grown on organic soils in New York. Ph.D. thesis, New York State College of Agriculture and Life Sciences, Cornell University, Ithaca. 305 pp.

Field Test Procedures for Evaluating Fungicidal Control of Botrytis Leaf Blight of Yellow Globe Onions Grown on Organic Soils in Northeastern North America

JAMES W. LORBEER, Department of Plant Pathology, Cornell University, Ithaca, NY 14853

Botrytis leaf blight of onion (*Allium cepa* L.), a fungal disease caused by *Botrytis squamosa* Walker, annually threatens onion production in Michigan, Wisconsin, New York, Ontario, Quebec, and other areas in northeastern North America where yellow globe onions are grown on organic soils (5,8,10,12,23). The disease is manifested first by leaf spots (lesions) on the leaves 24–48 hr after inoculation and then by maceration of leaf tissues by pectolytic enzymes during the next several days. About 5–7 days after inoculation, the leaves become partially to completely blighted. The leaf spots are white in color, range from 1 to 5 mm in length, and usually are surrounded by greenish white, water-soaked halos when first formed. The centers of the lesions typically are sunken and straw colored; they extend to the lacuna of the leaf and frequently develop a characteristic slit that is oriented lengthwise in the lesion. Older leaves are more susceptible to infection and blighting than are younger ones. An onion field in which the plants are severely affected by the disease will become yellow white in color as a result of heavy leaf blighting on most plants. Depending on the severity of the disease and the timing of recurring disease cycles during the growing season, onion bulb size may be substantially smaller, and yields may be reduced by as much as 15–40% or more under heavy disease pressure.

Because onion varieties resistant to leaf blight are not presently available (4), the disease must be controlled by protective fungicides. Thus, new fungicides and fungicide combinations must be tested regularly to develop knowledge on effective control of the disease by chemical treatments (11,13,15,16,18). In evaluating foliar fungicide sprays for control of leaf blight under field conditions, it is important that the disease and its symptoms not be confused with those of other leaf diseases or disorders, such as Botrytis leaf fleck caused by *B. cinerea* Fr., which is typified by small, shallow flecks on the leaves; ozone injury characterized by uneven patches of tiny necrotic stipples occurring irregularly on the leaves; leaf spots and leaf blighting caused by *Alternaria porri*; leaf-tip dieback resulting from water stress usually induced by hot or dry weather; rain and hail injury expressed as pale green to whitish silvery spots irregular in size and shape; and herbicide injury typified by irregular spots and other symptoms.

METHODS

Test site. Although several states and provinces in northeastern North America operate field stations for horticultural crops grown on organic (muckland) soils, New York, for example, has never developed such a facility; thus, Botrytis leaf blight control experiments performed in New York by experiment station personnel traditionally are conducted in commercial onion fields operated by cooperating growers. Therefore, when onion plantings on experiment station land are unavailable or inappropriate as the test site for leaf blight control experiments in New York or other areas, the location chosen for the field trial should be a commercial onion field that historically has been subject to natural outbreaks of leaf blight.

The site in the field for the experiment should be chosen about 2 months after the seed is planted, ideally in early to mid-June when uniformity of stand can be determined and when the first light outbreaks of leaf blight can occur. A level area of the field with an even plant stand located near a roadway to aid in the transport of spray equipment and chemicals should be selected as the site. Fields sheltered from the wind by headland configurations, woods or forested areas, and natural or artificial windbreaks frequently provide the best locations. If an experiment station planting is available, the same guidelines should be followed and the use of misting systems and artificial inoculum of the pathogen could be incorporated into the experimental design if desired. The techniques and parameters for these approaches have not yet been established and need to be developed and tested before application in fungicide field tests.

Test plants. Most, if not all, commercially available yellow globe onion varieties—including hybrids as well as those produced from homegrown seed of the yellow globe type—are susceptible to Botrytis leaf blight (4). Varieties that include germ plasm from sweet spanish parentage should be avoided because of their sensitivity to chlorothalonil, presently the standard fungicide for leaf blight control. Late-growing varieties (maximum number of days to maturity) are preferred to ensure the greatest opportunity for the occurrence of leaf blight and subsequent effect of the disease on plant growth and bulb size. Other than these considerations, the variety planted by the grower should suffice.

Plot design. In tests done in cooperation with growers, the plot design that fits well in commercial onion fields is the randomized complete block design. If the individual plots are large enough, interplot interference is generally minimized and data are reliable. Experiments in New York utilize onion plants grown in four-row beds averaging 38.1-cm (15-in.) row spacings and five replications per treatment. The plots are arranged with five contiguous four-row beds used for the five replications, allowing each treatment to be replicated in each bed. The individual treatment replicates are 6.63 m (21 ft 9 in.) long and 1.52 m (5 ft) wide, equaling 0.001 ha (0.0025 A). This design allows the incorporation of any number of treatments in an experiment and permits evaluations to be made on outer or inner rows. After the trial area is determined and the spray schedule is begun, the grower conducts all cultural procedures except the application of fungicide and insecticide sprays. Brightly colored flags mounted on flexible wire stakes placed at the corners and ends of the trial area remind the grower not to overspray the plants in the experiment with fungicides and insecticides.

Spray application. For full evaluation of the comparative abilities of different fungicides or combinations of fungicides to protect onion leaves from Botrytis leaf blight throughout the growing season, investigators apply weekly sprays from mid-June to late August or early September in New York to assure that the fungicides being tested are present at desired levels on the leaves when environmental conditions conducive to the disease occur (19). The fungicides are applied until 1–2 weeks before harvest. Sprays are necessary during the last several weeks of bulb growth before harvest to provide maximum protection from leaf blight during a period when the disease can be serious, few if any new leaves are being formed, and bulb growth is dependent on maximum leaf health.

A four-row, CO_2-pressured sprayer that is either hand-held or mounted on a one-wheel pushcart is used in New York to distribute the fungicide spray treatments to all plants in a uniform pattern by a single pass through each plot. With a little experience, the operator develops a constant walking speed so that the sprays are applied in an even pattern. Tractor-mounted or self-propelled sprayers also could be used and should be designed to provide a constant speed that can be adjusted easily. A compressed-air or CO_2-pressured sprayer fitted with a regulator provides a constant pressure and assures an even flow rate. The maximum speed for spray application does not exceed 6.44 km/hr (4 mph), and applications are not made when winds are strong enough to cause excessive drift or prevent good plant coverage.

Experimental sprays are applied in 378.5 L of water per hectare (100 gal/A) at 1.05–2.1 kg/cm^2 (15–30 psi) utilizing Teejet 8002 nozzles delivering fan-shaped spray patterns from a four-nozzle boom passing directly over the plants at a height that provides even coverage of all plants (19). The height of the boom is adjustable so that it can be raised as the plants grow taller. Although the water carrier may be reduced in volume, a 378.5-L (100-gal) amount assures greater accuracy in fungicide application than lower rates. Experimental accuracy makes the larger volume more desirable despite the fact that commercial onion growers frequently use lower rates (19). Many onion growers use aircraft to apply fungicides for leaf blight control, but no reliable procedures are presently available to simulate this form of application in small plots (22).

All sprays include an insecticide for onion maggot and thrips control to avoid or minimize insect injury to the plants in the trial. Diazinon 4EC at 2.3 L/ha (1 qt/A) is used effectively for this purpose in New York. In addition to the fungicide treatments being tested, each trial includes a control treatment (insecticide only) and a standard fungicide treatment. When wettable powder formulations of fungicides are used, Triton B-1956 spreader-sticker is included in each such treatment at the rate of 442.5 ml/946.3 L of water carrier per hectare (6 oz/100 gal per acre). When flowable formulations of fungicides that incorporate a spreader-sticker are utilized, Triton B-1956 is not included.

Data collection. After the first medium to heavy outbreaks of the disease, usually during late July or early August, the outer two rows of each treatment replicate are sampled for lesions by removing five typical plants selected at random. The number of lesions caused by *B. squamosa* are then counted on the three outermost leaves that have not yet blighted (when blight is severe, these may be the three youngest leaves on each plant), and the average number of lesions per leaf for the 15 leaves is calculated. This is best accomplished by cutting the leaves lengthwise and counting the lesions that extend to the inner side of the leaf. Because of time constraints for lesion counting, usually only one lesion count is made if the disease level is high. If the disease level is low and only a few lesions are present on each leaf, two or more lesion counts are made during the remainder of the growing period if time permits. The number of lesions per square centimeter of leaf surface for each treatment is determined by measuring the length and width of each leaf that was cut lengthwise, computing the area in square centimeters, and calculating the mean number of lesions per square centimeter.

Leaf greenness (a reciprocal measurement of disease incidence) is determined (0–10) by making greenness ratings (0 = no leaf greenness, 5 = half of leaves green, 10 = full leaf greenness) for each treatment replicate at two different dates late in the growing season and determining the average greenness rating for each treatment for each of the two dates. Objectivity is increased by having at least two individuals make separate ratings on each set of data, and the data are then combined.

Fresh weight yields of onion bulbs grown in the middle 5.5 m (18 ft) of the inner two rows for each treatment replicate are determined, and the mean yield in pounds per treatment is calculated. A final plant stand can be made just before harvest and a bulb count can be made immediately after harvest for each replicate to standardize treatments for the number of plants. If onion bulb quality and size determinations are to be made, the bulbs are field-stored until early to mid-October when decayed and waste onions are removed, the salable onions are graded for size, and the total weight for each size category is determined.

Phytotoxicity. Symptoms of phytotoxicity on onion plants are usually not similar for any two fungicides; they can include necrotic spotting, chlorosis, browning, tip burning, scalding and general necrosis of leaves, reduction in plant growth, and reduction in bulb size, as well as leaf and bulb deformity. Any phytotoxic effect of a fungicide used in the experiment should be described, particularly with respect to dosage, the number of sprays applied, stage of plant growth when noted, and general climatic conditions preceding the appearance of the symptoms.

Data analysis. After lesion counts, leaf greenness

ratings, and fresh weight yields have been determined, an analysis of variance is performed and Duncan's multiple range test is applied for significance at the 5% level (6). The average effectiveness index (AEI) also can be determined for each treatment based upon the relative effectiveness of the treatment (first, second, third, etc.) for each function (fewest lesions per leaf, greatest leaf greenness, and greatest fresh weight yield). The comparative effectiveness (CE) for each treatment is then determined by ranking the treatments in the trial on the basis of their AEI. The treatment with the lowest AEI would have a CE of 1 and the treatment with the greatest AEI would have a CE corresponding to the total number of treatments in the trial.

Reports. A detailed report on each fungicide trial conducted for Botrytis leaf blight control not only provides the information from the trial for agrichemical representatives, research workers, and growers, but also furnishes the needed background information and efficacy data for the Environmental Protection Agency in the product label registration process for those fungicides that are demonstrated to be effective in controlling the disease and needed for commercial application. Information needed from each trial for industry and Environmental Protection Agency use includes identity of the disease, host variety, and pathogen; description of the plot size; number of replicates; application dates for all sprays; dosage rates; name of fungicide, formulation, and dosage rate tested in each treatment; method of application of the sprays; efficacy data for each experimental treatment in the trial, as well as the control and standard treatment if the latter is included; yield data; phytotoxic responses; and environmental conditions at the time spray applications were made.

Authoritative sources of information regarding the efficacy of fungicides and documentation of their commercial need are critically needed by the Environmental Protection Agency to facilitate the label registration process. Publication of research results with candidate fungicides in journals such as *Plant Disease*, *Fungicide and Nematicide Tests*, and *Phytopathology* is becoming more important to aid the process by establishing peer review of the research and documentation in a respected journal.

Residue analysis. The task of collecting, transporting, and storing onion bulb samples for residue analysis is cumbersome and time-consuming for researchers, but it is necessary to assist in the registration process of those efficacious fungicides that are needed for commercial use. Samples of onion bulbs from treatments that are promising in controlling leaf blight should be taken at harvest, placed in separate bags, and stored in frozen or cold storage until analyzed for residue.

DISCUSSION

The procedures listed in this chapter generally are satisfactory for testing and evaluating experimental fungicides and fungicide mixtures for Botrytis leaf blight control. Several fungicides can be tested each year under careful statistical control at a reasonable cost following the procedures described. However, some limitations exist, particularly when the trials must be conducted in commercial onion fields. Natural inoculum levels, rainfall, and the duration of leaf wetness periods are not always adequate to provide conditions for ideal levels of leaf blight. This, at times, has resulted in performance results between experimental fungicides, standard fungicides, and controls that are not greatly different. Since conditions in grower's fields are not always satisfactory, particularly with regard to uniform plant density and growth in different beds (replicates), apparent differences between treatments are not always statistically significant at the desired level of analysis. This problem has sometimes resulted in judgment evaluations related to trend differences that, however useful and accurate, are not necessarily scientifically sound. Another problem is that the trial may be inadvertently oversprayed by unanticipated aircraft application of fungicides or may receive fungicide drift from commercial ground or aircraft sprays made outside the trial area. Also, since the individual plots of 0.001 ha (0.0025 A) are of appreciable size, it takes considerable time to harvest the onions manually in the trial, especially if there are a large number of treatments; if the grower or other interested growers can furnish three to five field workers for the harvest, this problem is overcome.

For many years, leaf blight was controlled by weekly applications of dithiocarbamate fungicides. During the early 1970s in New York, control by this group of fungicides failed. These failures have continued to the present, and chlorothalonil has become the standard fungicide for leaf blight control (14). Because *B. squamosa* produces a sexual stage in or near onion fields (7), and new genotypes of the fungus are potential threats as resistant forms to specific fungicides, it is becoming apparent that fungicide combinations may offer the best control of the disease for the future. Since the early 1970s, the disease has been controlled commercially by tank mixtures of chlorothalonil and a dithiocarbamate. Experimentally, the disease has been controlled by mixtures of benomyl and chlorothalonil; benomyl and a dithiocarbamate; and benomyl, chlorothalonil, and a dithiocarbamate.

In recent years, the dicarboximide fungicides at moderate to high rates have given control of the disease in experimental trials. Good control of the disease is also achieved with dithiocarbamate and dicarboximide mixtures. It is particularly noteworthy that when a standard rate of the dithiocarbamate and a low rate of the dicarboximide are used, leaf blight control is satisfactory. Since the cost to growers for new fungicides continues to escalate, it is becoming more important to develop fungicide combinations that not only control leaf blight but also are cost-effective. Combinations of new fungicides, such as the dicarboximides at low dosage rates, with the traditional fungicides, such as the dithiocarbamates and chlorothalonil at low to moderate rates or higher, offer this possibility. These combinations should also help to prevent the development of forms of *B. squamosa* resistant to the new fungicides.

Knowledge of the relationships of weather variables to the epidemiology of Botrytis leaf blight has increased in recent years (17,21,25,27). Application of integrated pest management (IPM) programs to commercial onion production practices requires that future field testing programs consider fungicide use for leaf blight control within the IPM framework, particularly as related to weather variation. Questions about timing of fungicide sprays and dosage rates will probably have to be combined with leaf blight predictive models (9,20,26), considerations of cost and yield responses (including yield quality), and environmental protection concerns in such experiments. Several experimental and commercial onion IPM programs have been or presently are being studied and operated in northeastern North America (1–3). Although field test procedures for fungicidal control of Botrytis leaf blight may need to conform to the IPM approach, continued testing following the procedures listed in this chapter may also result in information concerning the usage and performance of candidate fungicides that would modify or change the measures for leaf blight control within an IPM program. Also, as instrumentation for measuring, predicting, and recording weather and other environmental factors becomes more accurate and available at a reasonable cost (24), knowledge of the epidemiology of Botrytis leaf blight will advance and changes in the approach and procedures for experimental chemical control studies can be expected.

LITERATURE CITED

1. Andaloro, J. T., and Lorbeer, J. W. 1982. Implementation of an integrated pest management pilot program for onions in New York. (Abstr.) Phytopathology 72:257.
2. Andaloro, J. T., Rose, K. B., and Lorbeer, J. W. 1983. Application of a lesion counting procedure in determining the seasonal dynamics of Botrytis leaf blight in a commercial onion disease management program. (Abstr.) Phytopathology 73:361.
3. Andaloro, J. T., Rose, K. B., and Lorbeer, J. W. 1983. Assessment of disease control strategies on commercial onions in the New York pest management program. (Abstr.) Phytopathology 73:361.
4. Bergquist, R. R., and Lorbeer, J. W. 1971. Reaction of *Allium* spp. and *Allium cepa* to *Botryotinia* (*Botrytis*) *squamosa*. Plant Dis. Rep. 55:394-398.
5. Crête, R., Tartier, L., and Devaux, A. 1981. Diseases of onions in Canada. Agric. Can. Publ. 1716E. 37 pp.
6. Duncan, D. B. 1955. Multiple range and multiple *F* tests. Biometrics 11:1-42.
7. Ellerbrock, L. A., and Lorbeer, J. W. 1977. Sources of primary inoculum of *Botrytis squamosa*. Phytopathology 67:363-372.
8. Hancock, J. G., and Lorbeer, J. W. 1963. Pathogenesis of *Botrytis cinerea, B. squamosa,* and *B. allii* on onion leaves. Phytopathology 53:669-673.
9. Lacy, M. L., and Pontius, G. A. 1983. Prediction of weather-mediated release of conidia of *Botrytis squamosa* from onion leaves in the field. Phytopathology 73:670-676.
10. Lorbeer, J. W. 1962. Botrytis leaf diseases of onion. N.Y. State Veg. Grow. News 19(4):4.
11. Lorbeer, J. W. 1972. Control of Botrytis leaf blight of onion by protective and systemic fungicides and their combinations. (Abstr.) Phytopathology 62:773-774.
12. Lorbeer, J. W., and Andaloro, J. T. 1983. Botrytis leaf blight. Vegetable Crops, Cooperative Extension, New York State. Cornell Univ. Publ. 737.10. 2 pp.
13. Lorbeer, J. W., and Dumont, K. P. 1965. Botrytis leaf blight of onion. Fungic. Nematic. Tests 20:65-66.
14. Lorbeer, J. W., and Ellerbrock, L. A. 1976. Failure of ethylene bisdithiocarbamates to control Botrytis leaf blight of onion. Proc. Am. Phytopathol. Soc. 3:75-84.
15. Lorbeer, J. W., and Kawamoto, S. O. 1963. Botrytis leaf blight of onion. Fungic. Nematic. Tests 18:59.
16. Lorbeer, J. W., Kawamoto, S. O., Dumont, K. P., and Smith, H. A. 1964. Botrytis leaf blight of onion. Fungic. Nematic. Tests 19:76-77.
17. McDonald, M. R. 1981. Effects of environment and host factors on Botrytis leaf blight of onion. M.Sc. thesis, University of Guelph, Ontario. 108 pp.
18. Shoemaker, P. B., and Lorbeer, J. W. 1968. Onion leaf blight. Fungic. Nematic. Tests 23:76.
19. Shoemaker, P. B., and Lorbeer, J. W. 1971. Spray volume, interval, and fungicide rate for control of Botrytis leaf blight of onion. Plant Dis. Rep. 55:565-569.
20. Shoemaker, P. B., and Lorbeer, J. W. 1977. Timing initial fungicide application to control Botrytis leaf blight epidemics on onions. Phytopathology 67:409-414.
21. Shoemaker, P. B., and Lorbeer, J. W. 1977. The role of dew and temperature in the epidemiology of Botrytis leaf blight of onion. Phytopathology 67:1267-1272.
22. Shoemaker, P. B., Lorbeer, J. W., Muka, A. A., Steiner, P. W., and Brann, J. L. 1968. Control of Botrytis leaf blight of onion by aircraft application of a protective fungicide. Plant Dis. Rep. 42:469-472.
23. Small, L. W. 1970. The epidemiology of leaf blight disease of onions caused by *Botrytis squamosa*. M.Sc. thesis, McGill University, Montreal, Quebec. 249 pp.
24. Sutton, J. C., Gillespie, T. J., and Hildebrand, P. D. 1984. Monitoring weather factors in relation to plant disease. Plant Dis. 68:78-84.
25. Sutton, J. C., James, T. D. W., and Rowell, P. M. 1983. Relation of weather and host factors to an epidemic of Botrytis leaf blight in onions. Can. J. Plant Pathol. 5:256-265.
26. Sutton, J. C., James, T. D. W., and Rowell, P. M. 1984. Botcast—a forecaster for timing fungicides to control Botrytis leaf blight of onion. Department of Environmental Biology, University of Guelph. Circ. EB 07 84. 15 pp.
27. Sutton, J. C., Swanton, C. J., and Gillespie, T. J. 1978. Relation of weather variables and host factors to incidence of airborne spores of *Botrytis squamosa*. Can. J. Bot. 56:2460-2469.

Field Evaluation of Fungicides for Control of Allium White Rot

S. A. JOHNSTON, New Jersey Agricultural Experiment Station, Rutgers University, Rutgers Research and Development Center, Bridgeton 08302, and D. HALL, Plant Pathology Department, University of California, Davis 95616

White rot, caused by *Sclerotium cepivorum* Berk., is a serious disease that can occur on all commercially grown *Allium* spp. In the United States, it is an important disease on onion (bulb and bunching), garlic, leek, and shallot. The optimum temperature for disease development is 15–20° C (3,5). The disease is particularly destructive in areas of California, Nevada, New Jersey, Oregon, and Washington when the crops are planted in the fall and harvested in the spring or early summer. White rot can be avoided by growing *Allium* spp. during the summer months when the soil temperature is too high for disease development to take place. Within the genus *Allium*, there is little indication of resistance to *S. cepivorum*; however, a low level of resistance has been reported among onion cultivars (11). Therefore, disease control for overwintered *Allium* crops is dependent upon the use of effective fungicides.

Primary inoculum is from sclerotia present in the soil. In the case of direct-seeded onions, primary infection results in root infection with probable seedling death. In *Allium* crops started from transplant sets or cloves, initial infection occurs on the roots and the fungus progresses to the stem plate. Some plants previously infected will die during winter, resulting in reduced stands. During spring, as the soil temperature approaches 15–20° C, new infection from sclerotia in soil occurs and secondary infection takes place by plant-to-plant spread from external mycelia originating from primary infection sites on roots. At harvest time, infected plants are generally clustered as a result of secondary spread (4,7).

Control of white rot is based on seed treatment, proper placement of fungicides at seeding or planting, and proper placement of fungicides to the soil in the spring. The standard fungicide treatment for control of white rot is dicloran. More recently, iprodione has been shown to be effective (8,10). Iprodione has received commercial registration for control of Allium white rot on garlic in the United States and has been used on a temporary basis in New Jersey on onions and leeks through a Section 18 exemption of the Federal Insecticide, Fungicide, and Rodenticide Act.

METHODS

Design of experimental field trials, plot size, and planting configuration is variable. The methods given below have given satisfactory results, but modifications may be desirable. When working in a farmer's field, the planting configuration, timing, and other methods should conform to those of the grower to avoid interfering with normal cultural practices.

Test site. The experimental plots should be located in exact areas where white rot was a recent problem. The site should be level to prevent interplot interference caused by sclerotia being washed from one plot to another during heavy rains or irrigation. Growing the crop on raised beds will also achieve this goal.

Soil tests should be taken in advance of planting to permit proper pH and fertility levels for the crop. In New Jersey, bunching onions are seeded 1.3–3.8 cm (0.5–1.5 in.) apart and 1.3–1.9 cm (0.5–0.75 in.) deep in early September. Five rows, spaced 30.5 cm (12 in.) apart, are planted on beds 1.8 m (6 ft) wide. Different configurations are used in other areas. For example, in the Tulelake area of California, onions are planted in two-row beds on 91-cm (36-in.) centers and five to six rows per bed with 102-cm (40-in.) centers in the Salinas Valley.

A randomized complete block design with treatments replicated four or more times is recommended. Replicates are arranged, whenever possible, to avoid variables such as changes in soil type, drainage problems, and so forth. Of course, an untreated check and the standard control recommendation are included as treatments. These enable one to determine the relative efficacy of experimental treatments. The test site is described as to soil type, topography, moisture, cropping history, and disease levels.

Inoculum density. The distribution and population of sclerotia of *S. cepivorum* in infested fields vary. Population numbers as high as 1,288 sclerotia per 100 g of soil have been recorded (1). Sclerotia are uniformly distributed vertically within the top 20–25 cm of soil, whereas sclerotia are nonuniformly distributed horizontally within a field (1). Even heavily infested fields have nonuniform distribution of sclerotia within the field (1,6). Knowledge of the inoculum density within the test area to be used is important in interpreting results and should be included in the test report. It has been shown that as the inoculum level increased, the control with fungicides decreased (9).

The population of *S. cepivorum* may be estimated by collecting soil from the test area and recovering sclerotia by a wet-sieving technique. The procedure used in New Jersey is as follows (1,2). Samples are collected by taking 16 cores with a 2-cm-diameter soil sampling tube from the top 20–25 cm of soil at each 0.2-ha portion of the test area. The samples composited are air-dried and then screened through a 10-mesh sieve. Five 10-g subsamples are assayed from each composite sample. Each subsample is thoroughly washed on an 80-mesh sieve with running tap water. The residue is then backwashed into a 50-ml beaker and swirled several times to suspend the sclerotia. The suspension is immediately decanted onto the 80-mesh sieve. The last step is repeated several times until only sand particles remain in the beaker. The residue on the sieve may be examined directly with a microscope for the typical spherical, black sclerotia of the pathogen, or residue can be transferred to petri dishes for counting. Viability of the recovered sclerotia may be determined by transferring the sclerotia from the residue, surface-sterilizing them for 2 min in 0.5% NaOCl, crushing each sclerotium with fine-tipped forceps, and culturing on water agar. Viable sclerotia will produce relatively sparse mycelium in the agar, and "spermatia" will form in clumps within the agar in about 8 days at room temperature.

Test plot infestation. If it is necessary to infest a field with *S. cepivorum* to conduct an experiment, the following procedure is used. Oat seed and water are added to polypropylene sterilizing pans (2.0 kg of seed + 2.5 L of water per pan) and wrapped with aluminum foil. After 16 hr of incubation at room temperature, the pans are autoclaved for 1 hr on two successive days. Inoculum of *S. cepivorum* is added to the pans and incubated at room temperature for 1 month. At least 1 month before planting, the inoculum is broadcast uniformly over the test area (500 g of inoculum per 9.3 m^2) and incorporated to a depth of 15 cm. (Newly formed sclerotia have a constitutive dormancy and require conditioning in soil for 30–40 days before germination will occur.)

More precise inoculum densities in the test can be established by separating sclerotia produced in culture by wet-sieving and calculating numbers of sclerotia to be added to each plot.

Fungicide application. *Bunching onion.* Control of early infection of white rot can be obtained with fungicides applied as a seed treatment or as a preplant incorporation or an in-furrow spray to place a fungicide barrier between the fungus and the base of the stem of the onion. For seed treatment, a specified quantity of fungicide is added to seed in a capped 500-ml jar and gently rotated until thorough coverage is obtained. When using preplant-incorporated treatments, fungicides are applied with a tractor-drawn boom sprayer that delivers 374 L/ha (40 gal/A) at 2.76 kPa (40 psi). The boom consists of three nozzles containing TeeJet 8004 flat spray tips spaced 51 cm (20 in.) apart and positioned 51 cm above the ground. Fungicides are incorporated to a depth of 3.8 cm (1.5 in.) with a tractor-drawn roterra equipped with sides to prevent spread to adjacent plots. Although in-furrow sprays are desirable, application is difficult with bunching or other direct-seeded onions because of the shallow depth of planting. Applications of compounds into the furrow blow the seed away and often result in poor stands. In place of in-furrow applications, a preemergence fungicide application is made using the same equipment as used for applying the preplant-incorporated treatment. This technique is believed to be successful because seeds are placed near enough to the soil surface to allow subsequent sprinkler irrigations to move the fungicide physically to an effective level in the soil.

Secondary spread of white rot is controlled with a fungicide applied to the soil in the early spring (mid-March). Fungicide treatments may be applied with the same equipment as that used for the preplant and preemergence fungicide treatments.

Garlic. Garlic is planted commercially by specially designed planters that are not readily usable for small-plot work. For convenience, experimental plots are done by hand. Garlic is planted in late fall. Preformed beds with two furrows per bed are made with tractor-drawn equipment. Planting furrows are 5 cm × 10 cm (2 in. × 4 in.) wide. Fungicides are applied by a constant-pressure, hand-held sprayer to the bottom and sides of the furrows of each appropriate plot. Cloves are dropped into each furrow at the rate of seven to eight cloves per 0.3 m (linear foot) of row. Plot length may vary, but plots 4.6–6.1 m (15–20 ft) long are suggested. Cloves are then covered by garden rake or hoe.

Data collection. Data are collected for the following parameters: plant stand, phytotoxicity, disease level, and yield. Plant stands are generally taken 30 days after seeding, late winter, and at harvest. The number of diseased plants is counted and percentage of stand reduction and degree of infection are calculated. A specified portion of each plot—for example, a 4.6-m (15-ft) section in the middle of the plot—is harvested, and all of the plants are evaluated for symptoms of white rot. Symptoms at harvest include lesions covered with black sclerotia and white mycelium present at the base of the bulb. Yield is calculated as the weight of healthy plants. All data are subjected to an analysis of variance and to Duncan's multiple range test. A record of the weather (rain, irrigation, and temperature), cultural practices, and the presence of other diseases and pests is maintained.

RESULTS

Data collected from tests of this nature are used to support registration (either state or federal) for effective products for onion white rot control. Additionally, residue samples can be collected from portions of the plot not used for data collection. Residues are collected, placed in plastic-lined bags, labeled, and frozen until analyzed for residue content.

LITERATURE CITED

1. Adams, P. B. 1979. A rapid method for quantitative isolation of sclerotia of *Sclerotinia minor* and *Sclerotium cepivorum* from soil. Plant Dis. Rep. 3:349-351.
2. Adams, P. B. 1981. Forecasting onion white rot. Phytopathology 71:1178-1181.
3. Adams, P. B., and Papavizas, G. C. 1971. Effect of inoculum density of *Sclerotium cepivorum* and some soil environmental factors on disease severity. Phytopathology 61:1253-1256.
4. Adams, P. B., and Springer, J. K. 1977. Time of infection of fall-planted onions by *Sclerotium cepivorum*. Plant Dis. Rep. 61:722-724.
5. Crowe, F. J., and Hall, D. H. 1980. Soil temperature and moisture effects on sclerotium germination and infection of onion seedlings by *Sclerotium cepivorum*. Phytopathology 70:74-78.
6. Crowe, F. J., Hall, D. H., Greathead, A. S., and Baghott, K. G. 1980. Inoculum density of *Sclerotium cepivorum* and the incidence of white rot of onion and garlic. Phytopathology 70:64-69.
7. Entwistle, A. R., and Munasinghe, H. L. 1978. Plant Disease Epidemiology. P. R. Scott and A. Bainbridge, Publishers. Pp. 187-191.

8. Entwistle, A. R., and Munasinghe, H. L. 1981. The effect of seed and stem base spray treatment with iprodione on white rot disease (*Sclerotium cepivorum*) in autumn sown salad onions. Ann. Appl. Biol. 97:269-276.
9. Hall, D. H., and Somerville, P. A. 1983. Effect of inoculum potential on control of white rot by fungicides. Pages 113-114 in: Proc. Int. Workshop Allium White Rot. U.S. Dep. Agric., Beltsville, MD.
10. Johnston, S. A. 1982. Fungicide control of white rot of onion, 1980–1981. Fungic. Nematic. Tests 37:72.
11. Utkhede, R. S., and Rahe, J. E. 1980. Stability of cultivar resistance to onion white rot. Can. J. Plant Pathol. 2:19-22.

Field Testing Foliar Fungicides for Potato Late Blight Control

F. E. MANZER, Department of Botany and Plant Pathology, University of Maine, Orono 04469; W. E. FRY, Department of Plant Pathology, Cornell University, Ithaca, NY 14853; D. P. WEINGARTNER, University of Florida, Agricultural Center, Hastings 32045; R. C. ROWE, Department of Plant Pathology, Ohio State University, Ohio Agricultural Research and Development Center, Wooster 44691; and ROSEMARY LORIA, Department of Plant Pathology, Cornell University, Long Island Horticultural Research Laboratory, Riverhead, NY 11901

Late blight, caused by *Phytophthora infestans* (Mont.) De Bary, is a major disease of potato (*Solanum tuberosum* L.). Present in most potato production areas of the world (2), it first appeared in Europe and the United States about 1830 and was the cause of the 1845–46 potato famine in Ireland.

Many studies have been conducted to determine the environmental conditions that are favorable for *P. infestans* and the disease that it incites. In general, the conditions favorable for late blight are 90–100% relative humidity (RH) for 10 hr or more and air temperatures of 7.2–29.4°C (45–95°F). Optimum conditions for the various stages in the life cycle of *P. infestans* are, for sporangiospore formation, 18–22°C (64.4–71.6°F) and 100% RH; direct germination of sporangiospores, 24°C (75.2°F) and free moisture; zoospore formation (indirect germination of sporangiospores), 9–15°C (48.2–59.0°F) and free moisture; zoospore germination, 12–15°C (53.6–59.0°F) and free moisture; infection, 15–21°C (59.0–69.8°F) and free moisture; and incubation, 18–23°C (64.4–73.4°F). Under optimum conditions, sporangiospores form in 8–14 hr, indirect germination and infection occur in 2.5–7 hr, and new lesions appear in 3–5 days (2,3). Under field conditions, the incubation period is usually 5–10 days.

Although many control measures have developed since 1846, spraying plants with an effective fungicide has remained the chief control measure since about 1900 when Bordeaux mixture became widely accepted in Europe and the United States. The control of late blight constitutes one of the major uses of fungicides. Consequently, many candidate fungicides are screened for control of late blight before being tested for control of other vegetable diseases.

Most agricultural chemical companies have developed their own routine laboratory and greenhouse methods for determining the potential activity of candidate fungicides. The methods and procedures described in this chapter are intended as a guide for conducting field tests to determine fungicidal efficacy. These procedures can also be used in field tests to determine proper dosage, application interval, and other factors necessary to obtain satisfactory control of late blight with fungicides.

METHODS

Test site. Tests are conducted in an area adaptable for potato culture and where climatic conditions favor development of late blight. The soil type and fertility should be as uniform as possible. Irrigation should be available to supplement rainfall during periods of drought. Overhead sprinkler systems are preferred. If possible, the test site should be isolated from commercial or other experimental potato or tomato plantings.

Test plant. Certified or foundation quality seed potatoes of a susceptible cultivar adapted to the area are used. The cultivar should have little or no known horizontal or vertical resistance to late blight. Acceptable cultivars include Chippewa, Green Mountain, Hudson, Katahdin, Norgold Russet, Red LaSoda, and Russet Burbank. Careful cultivar selection is advisable where confounding due to the presence of early blight (*Alternaria solani* Sorauer) may be a problem. All plots within an experiment are planted with seed of the same cultivar from the same source. Cut seed pieces can be treated with a recommended fungicidal dust (e.g., 8% mancozeb or 7% captan) or dip treatment to prevent seed piece decay. Seed pieces average between 42.5 and 56.7 g (1.5 and 2.0 oz) and are spaced 20.3–30.5 cm (8–12 in.) apart in rows 0.76–1.02 m (30–40 in.) apart. Plants are maintained in good growing condition by use of fertilizer; weed, insect, and nematode control measures; irrigation; and other cultural practices recommended for the area. If nonregistered pesticides or those labeled exclusively for experimental purposes are used, the tubers or foliage must not be used for human consumption or animal feed.

Plot design. The experimental design depends on the number of compounds to be tested and on the land and type of equipment that are available. Acceptable designs include paired plots, randomized complete blocks, Latin squares, and randomized split plots (14). Treatments are randomly allocated by using such tables as Snedecor's random digits (14) or other appropriate means of randomization. Each treatment is replicated at least three times. Plot size depends on objectives of the test. For screening and ranking fungicides, small plots one to three rows wide by 3.0–7.6 m (10–25 ft) long are adequate. For determining fungicidal rates, application intervals, or specific materials for recommendation, larger plots 3–10 rows wide and 7.6–15.2 m (25–50 ft) long may be used. Ends of plots are separated by buffer zones of at least 1.5 m (5 ft) to prevent drift or cross-contamination. Additional buffering may be necessary when fungicides that are effective at low dosages, such as those with systemic activity, are to be used. Buffer zones can be planted to potatoes or left fallow. Plots should be arranged so that test rows are not damaged by application equipment.

Recently, interplot interference has been discussed as a confounding factor in field experiments. Inclusion of a no-fungicide or ineffective fungicidal treatment may result in underestimating the efficacy of more effective treatments. Conversely, placing a highly effective fungicidal treatment next to an inferior one can result in overestimating the efficacy of the inferior fungicide. Errors contributed by interplot interference can be limited by increasing plot size, increasing space between plots, and grouping treatments of similar efficacy (10). For example, when disease intensity is expected to vary tenfold among treatments, the interplot interference error can be limited to 10% of the observed value by locating small plots (3 m × 3 m) about 5 m apart. When disease intensity is expected to vary only twofold, the same small plots need be only 3 m apart (10). Many researchers prefer to separate nonfungicide control plots downwind from the experimental area by seven or more rows that are sprayed with an effective fungicide on a 7- to 10-day schedule. Uniformly high inoculum or disease pressures can be achieved by having one or more no-fungicide guard (buffer) rows adjacent to each treatment plot. This is desirable when screening and ranking fungicides for effectiveness, but may be undesirable when comparing application intervals.

A standard fungicide is included in each experiment. The standard selected is a fungicide commonly used by commercial growers in the area, e.g., mancozeb or maneb. These are usually applied at the rate of 1.34–1.79 kg a.i./ha (1.2–1.6 lb a.i./A) per application. Other acceptable standards include chlorothalonil, captafol, and metiram. These are used at locally recommended dosages.

Fungicide application. The sprayer is tractor-mounted or self-propelled and provides a constant speed that can be adjusted readily. The sprayer is calibrated to apply between 468 and 936 L/ha (50 and 100 gal/A) at sufficient pressure for good coverage. High-volume, high-pressure boom sprayers usually are operated at 1,724–2,758 kPa (250–400 psi). The maximum speed should not exceed 6.4 km/hr (4 mph), and applications should not be made when winds cause excessive drift or interfere with good plant coverage. The sprayer has either a brush or modified-brush boom or booms, with at least three dripless, hollow-cone nozzles per row. The nozzles are angled forward about 0.26 rad (15°) from the vertical. The sprayer must be equipped with either paddles or an overflow agitator or agitators that keep wettable powders in suspension.

The first spray is applied when plants are 20.3–25.4 cm (8–10 in.) tall. Subsequent sprays are applied at weekly intervals for routine evaluations. When application intervals are being studied, the standard is compared at the same intervals as candidate fungicides.

Organism and maintenance of cultures. Race 0 (field race) of *P. infestans* is preferred. Cultures of the fungus have been maintained by serial transfer on frozen lima bean agar (13), rye agar and liquid medium (4), potato tuber slices, foliage of susceptible cultivars of potato and tomato, and in potato tubers (9). Cultures grown on agar are incubated at 4–20°C (39.2–68.0°F). The cultures are transferred at weekly to biweekly intervals when incubated at 20°C (68°F) and monthly when at 4°C (39.2°F). Stock cultures under mineral oil have remained viable for at least 2 years (4). Isolates maintained on agar frequently become attenuated and should be tested for pathogenicity on plants of susceptible potato cultivars before being used to produce inoculum for field plots. Some workers have found autoclaving the rye agar for 30 min at 116°C (240.8°F) necessary to eliminate bacterial contaminants from the rye seeds.

Inoculum for field plots. Sporangiospores of *P. infestans* may be produced on any one of the media indicated above. They may also be produced by planting infected tubers in a special "blight garden" where environmental conditions favorable for late blight development are maintained.

To obtain good zoospores, cultures should not be more than 7–10 days old, the younger the better. Sporangiospores are washed off the culture medium (lima bean agar, potato tuber slices, etc.) with precooled (5°C) distilled water or with nonchlorinated and nonfluorinated tap water. Tap water, especially if copper pipe is involved, should be tested for its effect on spore germination before being used to prepare spore suspensions. If it seriously inhibits sporangiospore or zoospore germination, it should not be used. The suspension is filtered through four or more layers of cheesecloth to remove debris and adjusted to a concentration of at least 5–10 sporangiospores per 100× microscope field (5,000–10,000/cm^3) or to an equivalent measure of zoospores.

Inoculation of field plots. One or more foliar fungicidal sprays are applied before the plots are inoculated when protective foliar fungicides are being evaluated. At least 12 hr should elapse between application of the fungicide and inoculation of the plants. Plants should be wet with dew, rain, or irrigation water when inoculations are made and foliage should remain wet for 10 hr or more following inoculation. This requirement is often met when the plants are inoculated in the evening. If the natural environment does not provide these conditions, overhead sprinkler irrigation at 0.25 cm/hr (0.10 in./hr) can be used to provide moisture for spore germination and infection.

Plots may be inoculated by spraying the spore suspension onto plants in the no-fungicide guard or buffer rows adjacent to each plot, onto all plants in the test (middle) row or rows of each plot, or onto plants in a swath about 0.91 m (3 ft) wide across all rows in the center of each plot. Many researchers prefer the first method, allowing natural spread of the disease into the test plots, because uniform inoculation within and among plots is difficult to achieve.

Inoculum can be applied with a small hand-sprayer or with a one-nozzle-per-row, compressed-air or power sprayer. The sprayer must be free of all chemicals and preferably one used solely for this purpose. Repeated inoculations may be required or desired to establish an epiphytotic of sufficient severity to differentiate treatment efficacies relative to the standard.

Environmental conditions favorable for development and spread of late blight must be maintained following completion of the primary cycle. If not provided naturally, supplemental overhead sprinkler irrigation [about 0.25 cm/hr (0.10 in./hr)] can be used, in the morning or evening, to extend natural dew periods.

DATA

Foliar blight, phytotoxicity, and yield data are collected from most routine efficacy tests. Some researchers examine the tubers for late blight on a routine basis. When candidate fungicides reach a given stage in their development, manufacturers often request tuber samples for residue analyses. Occasionally, information may be desired concerning effects of the fungicidal treatments on specific gravity, taste, cooking, or processing qualities of the tubers.

Foliar blight. The Horsfall-Barratt grading system for measuring plant disease (5,11) is commonly used to estimate disease severity visually. Some researchers, especially in Canada and Europe, prefer the British Mycological Society key (1,2). Disease severity is determined at least twice following inoculation. One reading is made, if possible, before the plants in the no-fungicide controls are 50% defoliated, and the second is made when they are 90–100% defoliated. The final reading, however, should be made immediately before the plants are killed by frost or by chemicals in preparation for harvest. If the plants are allowed to die naturally, the final reading is made before natural senescence prevents proper estimation of disease severity. If disease progress curves are desired, visual estimates of disease severity are made at intervals of 5–7 days beginning with completion of the primary cycle. If eradicative properties of a fungicide are of interest, observations on lesion development, sporulation, and spore viability are taken.

Disease severity is commonly expressed as percentage of defoliation. Some researchers, however, prefer to express their data as percentage of control (PDC), using the formula $PDC = [(DIC - DIT) \div DIC] \times 100$, where DIC is disease incidence in control and DIT is disease incidence in treatment. When percentage of disease control is reported, it should be in addition to percentage of defoliation since defoliation data provide information on the severity and reliability of the test.

Special techniques may be necessary for evaluation of systemic fungicides. For example, Rowe (12) described a bioassay for monitoring the degree of systemic protection of foliage over time as a result of various fungicide treatments. Fully expanded, healthy compound leaves were removed from treated potato plants and immersed for 30–40 min in a tap water suspension of 2,000–4,000 *P. infestans* zoospores per milliliter at ca. 12°C. After soaking, leaf petioles were inserted into moist vermiculite in 10-cm-diameter plastic pots. Pots were randomly arranged in a controlled environment chamber equipped with a fog generator and kept at 20°C in continuous fog under a 12-hr photoperiod for 7–10 days. At that time, disease development on each leaf was evaluated on the following scale: 0 = no visible symptoms, 1 = pinpoint lesions without sporulation, 2 = a few expanded lesions with light sporulation, and 3 = large lesions with heavy sporulation.

Phytotoxicity. Symptoms of phytotoxicity seldom are the same for any two compounds. Some symptoms that may be observed are necrotic flecking or spotting of leaflets, marginal chlorosis or necrosis, stunting resulting from shortening of the internodes, smaller or darker green leaflets or both, thicker and crisper leaflets, upward rolling of margins of the leaflets, and deformed leaflets. In addition to describing phytotoxic effects, it is important to indicate their severity in relation to dosage, number of sprays, and stage of plant growth.

Yield. In the absence of late blight and other foliar diseases, nonphytotoxic foliar fungicides seldom have a significant effect on yield of tubers. In the presence of late blight or other foliar diseases, the difference in yield between a fungicidal treatment and the no-fungicide control depends on when the epidemic begins in relationship to the stage of tuber development, rate at which the epidemic develops, the severity of the epidemic, and the effectiveness of the fungicidal treatment (6,8). Consequently, many researchers think that yield data are relatively unimportant in evaluating protective foliar fungicides. Yield data, at least from some tests at most locations, are required to demonstrate that the candidate fungicide has no adverse effect on yield. Furthermore, yield data are required by governmental agencies responsible for registering and labeling pesticides.

Yield data are obtained by harvesting tubers from the center row or rows of each plot. They are usually reported as total and U.S. No. 1 size A (4.8 cm or 1.875 in. minimum diameter) and are expressed as either kilograms or pounds per plot, quintals or metric tons per hectare, or hundredweight per acre.

Tuber rot. The amount of late blight tuber rot that develops depends upon the amount of fungicidal residue on the soil surface or in the tuber, persistence of the residue, soil type, depth at which the tubers develop, amount of rainfall and supplemental irrigation applied during periods of sporulation of the fungus, and amount of fungicide used. Consequently, the amount of tuber rot that develops often is not what might be expected based on the amount of foliar blight. When these data are taken, all size A tubers are washed and examined for tuber rot within 2 weeks of harvest. A second reading taken 1–2 months later would be beneficial. These data can be expressed as percentage of rot based on either weight or number of tubers.

Residues. Residues of candidate fungicides in the tubers must be determined to comply with the regulations of governmental agencies responsible for registering and labeling pesticides. This normally is the responsibility of the manufacturer, who usually prescribes how and when tuber samples should be taken. The samples normally consist of a composite of 2.3–4.5 kg (5–10 lb) of tubers from the desired treatments. An equal number of tubers of similar size are selected randomly from each of three or more replications of each treatment. To prevent contamination, the no-fungicide control sample or samples should be collected first. The tubers normally are not washed and are placed in heavy (1.5–2 mil) plastic bags or in special sample bags provided by the manufacturer. Unless otherwise indicated, the samples are frozen at −17.8°C (0°F) immediately after harvest and packed with dry ice in insulated boxes when shipped to the residue laboratory for assay.

Specific gravity. Specific gravity is a measure of the internal quality of tubers and is correlated with total solids. Nonphytotoxic protective fungicides usually have no significant effect on specific gravity (8). If desired, specific gravity can be determined by using a potato hydrometer with a 3.6-kg (8-lb) capacity (available from Potato Chip/Snack Food Assn., 1711 King St.,

Alexandria, VA 22314) or by the weight-in-air versus weight-in-water method. These readings are based on a random sample of size A tubers from each plot.

Off-flavor. Taste tests normally are not considered in evaluating protective foliar fungicides. As we move into the use of systemic compounds, however, this may become a more important aspect of fungicidal evaluation programs. An effective compound that imparts an off-flavor to tubers would be restricted to use only on seed potatoes. Therefore, sufficient taste tests should be conducted to demonstrate that the candidate fungicide does not adversely affect flavor. Samples for taste tests are collected in a manner similar to those described for residue analyses. Preliminary taste tests can be conducted in the researcher's own laboratory. Peeled tubers are boiled until tender, and a panel of three or more people tastes the cooked potatoes. More refined tests often can be conducted in cooperation with home economics schools or colleges in many universities.

Cooking and processing. Tests to determine the effect of foliar fungicides on cooking and processing qualities of tubers normally are not part of routine valuations of fungicides for late blight control. However, if such information is desired, tuber samples can be collected from routine fungicidal experiments by a method similar to those described for residue analyses.

Additional tests. The manufacturer or cooperating agency may conduct additional tests that are beyond the scope and intent of this publication. Examples of such tests include the following: (i) efficacy of candidate fungicides when applied by ultralow- or low-volume ground sprayers, row-crop airblast sprayers, aircraft, overhead sprinkler irrigation, or any other method that may be developed; (ii) effect of various adjuvants on efficacy of the candidate, including studies to determine the minimum and maximum amount of the adjuvant required to improve efficacy or prevent drift or both; and (iii) compatibility of the candidate fungicides with other pesticides that commercial growers may include in the spray mixture.

Statistical analyses. Some measure of experimental error is required for proper interpretation of results. Appropriate statistical procedures for data reduction and tests of significance are discussed in detail elsewhere (see L. A. Nelson, *this volume*). However, consultation with a competent statistician before conducting experiments is strongly recommended to ensure validity of the methods used.

REPORTS

Each investigator has the responsibility of reporting results to the manufacturer of the candidate fungicide and, where applicable, to the public. In reporting to the public, the format that the publishing agency recommends (e.g., *Fungicide and Nematicide Tests*) should be followed. Reports to manufacturers are used to make decisions concerning future development of the candidate fungicides. These reports often are used in preparing applications for registration and labeling of the candidates with governmental agencies for commercial growers' use. They should be as brief but complete as possible. The following is a checklist of information that most manufacturers require:

1. Product name or code number, including trade name, chemical name, or common name, or all three if established and not confidential.
2. Formulation and percentage of each active ingredient unless confidential.
3. Formulation of spray mixture. This includes the amount of water and formulated product put into the spray tank.
4. Rate of application. This includes the amount of spray mixture and the amount of pesticide, expressed as amount of active ingredient or formulated product, applied per hectare or acre.
5. Actual dates of application, not the intended interval between applications.
6. Method of application, including a brief description of the application equipment.
7. Experimental design. This includes such specifications as the number of replications, distance between rows, spacing of plants within the row, number of treated rows per plot, length of treated plots, number of nontreated rows between plots, length of buffer areas between blocks or ranges of plots, and location of no-fungicide control plots.
8. Cultivar used, when planted, soil type, fertilization, and other pertinent cultural practices.
9. Date or dates and method or methods of inoculation.
10. Number and dates of irrigation, including the approximate amount of water in centimeters or inches applied per irrigation and the time required to apply the water for each irrigation.
11. Brief summary of the environmental conditions between planting and termination of the test. This should include the amount of rainfall in centimeters or inches per month, number of rainy periods each month, the monthly average maximum and minimum temperatures, and the dates and severity of late blight infection periods as indicated by either Blitecast (7), the Wallin system (15), or other methods of forecasting late blight that may be developed.
12. Other pesticides used, including rates, dates of application, and whether they were applied separately or in combination with the fungicides.
13. Percentage of defoliation on each date that data were taken.
14. Percentage of disease control if calculated.
15. Type and severity of phytotoxicity.
16. All other data taken from experimental plots.
17. A brief written evaluation and interpretation of the data to indicate the confidence that the investigator has in the results.

LITERATURE CITED

1. British Mycological Society. 1947. The measurement of potato blight. Trans. Br. Mycol. Soc. 31:140-141.
2. Cox, A. E., and Large, E. C. 1960. Potato blight epidemics throughout the world. U.S. Dep. Agric. Agric. Handb. 174. U.S. Government Printing Office, Washington, DC. 230 pp.
3. Crosier, W. 1934. Studies in the biology of *Phytophthora infestans* (Mont.) DeBary. N.Y. Agric. Exp. Stn. (Ithaca) Mem. 155. 40 pp.
4. Hodgson, W. A., and Grainger, P. N. 1964. Culture of *Phytophthora infestans* on artificial media prepared from rye seeds. Can. J. Plant Sci. 44:583.
5. Horsfall, J. G., and Barratt, R. W. 1945. An improved

grading system for measuring plant diseases. Phytopathology 35:655.
6. James, W. C., Shih, C. S., Hodgson, W. A., and Callbeck, L. C. 1972. The quantitative relationship between late blight of potato and loss in tuber yield. Phytopathology 62:92-96.
7. Krause, R. A., Massie, L. B., and Hyre, R. A. 1975. Blitecast: A computerized forecast of potato late blight. Plant Dis. Rep. 59:95-98.
8. Manzer, F. E., Cetas, R. C., Partyka, R. C., Leach, S. S., and Merriam, D. 1965. Influences of late blight and foliar fungicides on yield and specific gravity of potatoes. Am. Potato J. 42:247-252.
9. Mills, W. R., and Stevens, R. B. 1967. Zoospore production by the late blight fungus and techniques in inoculation. Pages 28-29 in: Sourcebook of Laboratory Exercises in Plant Pathology. A. Kelman, ed. W. H. Freeman, San Francisco. 378 pp.
10. Paysour, R. E., and Fry, W. E. 1983. Interplot interference: A model for planning field experiments with aerially disseminated pathogens. Phytopathology 73:1014-1020.
11. Redman, C. E., King, E. P., and Brown, I. F., Jr. 1962. Tables for converting Barratt and Horsfall rating scores to estimated mean percentage. Eli Lilly and Company, Indianapolis.
12. Rowe, R. C. 1982. Translocation of metalaxyl and RE26745 in potato and comparison of foliar and soil application for control of *Phytophthora infestans*. Plant Dis. 66:989-993.
13. Smoot, J. J., Gough, F. J., Lamey, H. A., Eichenmuller, J. J., and Gallegly, M. E. 1958. Production and germination of oospores of *Phytophthora infestans*. Phytopathology 48:165-171.
14. Snedecor, G. W., and Cochran, W. G. 1967. Statistical Methods. 6th ed. Iowa State University Press, Ames. 593 pp.
15. Wallin, J. R. 1962. Summary of recent progress in predicting late blight epidemics in United States and Canada. Am. Potato J. 39:306-312.

Field Evaluation of Fungicides for Control of Early Blight, Septoria Leaf Spot, Anthracnose, and Other Ripe Rots of Tomato in Northern Growing Regions

CHRISTINE T. STEPHENS, Department of Botany and Plant Pathology, Michigan State University, East Lansing 48824-1312; ALAN A. MacNAB, Department of Plant Pathology, Pennsylvania State University, University Park 16802; and THOMAS C. STEBBINS, Department of Botany and Plant Pathology, Michigan State University, East Lansing 48824-1312

Major diseases of fresh market and processing tomatoes (*Lycopersicon esculentum* Mill.) include early blight caused by *Alternaria solani* Auct., Septoria leaf spot caused by *Septoria lycopersici* Speg., anthracnose caused by *Colletotrichum coccodes* (Wallr.) Hughes, and other ripe fruit rots caused by *A. solani* and other fungi (3,8). Leaf spot diseases lead to premature leaf death and defoliation. Extensive defoliation reduces the size, quality, and quantity of fruit. Fruit exposure resulting from defoliation can result in sunscald. Fruit rot reduces both yield and fruit quality. Fruit rots can also result in high mold counts in tomato products during processing.

These three pathogens are easy to identify in the field. Anthracnose occurs only on fruit. The early symptoms include the appearance of small, slightly depressed, water-soaked, circular spots. The lesions increase in size, become sunken, and often contain a series of concentric rings. The lesions gradually darken, and black acervuli appear in the center. In smaller lesions, the center may turn dark because of the numerous fruiting structures just below the epidermis of the tomato fruit. As the fungus spreads within the fruit, a semisoft decay is evident. Lesions often coalesce, forming large rotted areas on the fruit. In moist, warm weather, masses of slimy tan or salmon-colored spores are exuded from the acervuli.

The *Septoria* organism only causes leaf and stem spotting. Small, water-soaked, circular lesions first appear on the underside of older leaves. As the lesions mature, they enlarge to about 0.5 cm in diameter. These lesions are generally gray or tan with a dark brown margin. In the center of these lesions, dark brown acervuli appear. Heavily infected leaves turn yellow, dry up, and drop.

Early blight causes both fruit rot, leaf spot, and stem canker. The symptoms on the leaves are dark brown to black lesions ranging in size from 0.8 to 2.4 cm. Within the lesion, a target spot pattern appears as the result of a growth flush of the fungus. First symptoms appear on older leaves. Lesions often coalesce to form large blighted areas on the leaf. When the leaf is covered with many lesions, the leaf yellows and drops off. Fruit infection generally occurs at the point of attachment to the stem, but it can occur through growth cracks and wounds on the fruit. The lesions gradually become discolored, sunken, and circular with concentric rings. In humid weather, infected areas on the fruit become covered with black, velvety spores.

These pathogens can overwinter in plant debris in the soil. Disease pressure is most intense in humid, mild (21.1–26.7° C; 70–80° F) weather with abundant rainfall.

This paper is intended as a summary of considerations that we consider important, and of some procedures currently used, in the evaluation of fungicides for control of early blight, Septoria leaf spot, anthracnose, and other ripe fruit rots of tomatoes. We hope that information presented will be of practical value to those with limited experience in tomato fungicide evaluation. Also, we hope that experienced fungicide evaluators will share additional considerations for future revisions of this paper.

PROCEDURES

Test site. It is preferable to select a site in which there is a high level of natural inoculum. Generally, a field that has been planted in tomatoes for several years contains the overwintering stage of these pathogens. If such a site is not available, a field can be artificially infested. This can be done in several ways. In the fall of the year, plant debris from tomatoes heavily infected by these pathogens can be placed in the test site. Of the three pathogens, the anthracnose fungus is the most difficult to establish in a chosen site. A technique developed by Stevenson and Braddock (7) is helpful in establishing the population of *Colletotrichum* in a site not previously planted to tomatoes. In this method, treatment rows are infested by spreading colonized, killed sorghum seed in the row. The chosen site should be favorable for tomato cultivation as well as environmentally suited for development of leaf spot and fruit rot. A well-drained site consisting of a sandy loam soil is preferable. Soil type and fertility should be uniform. Irrigation should be available to supplement natural rainfall, particularly at the time of transplanting. Overhead irrigation promotes the spread of these fungal pathogens as their spores are dispersed by water splashing.

Test plant. Test varieties must be susceptible to the pathogens, and results may be more useful by choosing varieties that are preferred locally. Most currently used commercial varieties are at least moderately susceptible to early blight, Septoria leaf spot, and anthracnose.

Either fresh market or processing varieties can be used; results with commercial varieties used locally may be most interesting to local producers.

Test plantings can be direct-seeded or established with transplants. Transplants usually facilitate establishment of an even stand. They should be obtained from a single source and be of uniform quality. Results may have greater value for local growers by using plant and row spacing similar to that used in local production fields.

Cultural procedures. Fertility levels should be adjusted on the basis of soil analysis results. Good weed control can be provided by using appropriate herbicides specific for broadleaf, grass, and sedge weeds. Insect control, usually for Colorado potato beetle, aphids, and fruit worms, should be provided as needed throughout the season.

Experimental design. Basic factors considered when choosing an experimental design include number of treatments and expected variation associated with sampling and treatments (6). A preferred experimental design may be modified to adjust to limitations in size, shape, and topography of test sites and to facilitate use of equipment available for maintenance and application of fungicide treatments. When desirable, experiments can be designed to produce results representative of losses in commercial fields.

Several designs are available for setting up field experiments (4). If no gradients exist in the test field, a completely randomized design allows the most statistical precision. However, a completely uniform test site is rarely available. A randomized complete block design probably is most frequently used. Statistical precision is increased when blocks run perpendicular to such field gradients as slope, soil texture, soil moisture, and soil type. A split-plot design may be preferable to the randomized complete block design when all levels of all factors cannot be assigned at random to treatment locations. Paired-plot and Latin square designs can also be used. Treatment locations must be randomly assigned.

Replication. Number and size of treatment replications depend on the extent of variation that occurs and the degree of precision needed in evaluation of the treatments. Usually, treatments are replicated at least three times and each treatment replication contains from 5 to 20 plants for each harvest time. At times of low disease incidence, more replications may be needed.

Border rows. Border rows, from which data is not collected, should be included between treatment rows. Such border rows act as a buffer to spray drift or other effects from adjacent treatments. Sometimes buffer rows are used as drive rows for equipment. A decision whether to spray border rows should be based on the level of secondary inoculum desired in the experiment; secondary inoculum from unsprayed border rows can be expected to be higher than from sprayed border rows. Unsprayed border rows result in high disease pressure, which may be desirable; however, they may create an artificially high level of disease pressure.

Standard treatments. An unsprayed check treatment is usually included in fungicide evaluation experiments and is useful in indicating disease severity. A disadvantage of including an unsprayed check is that high levels of inoculum associated with the treatment may result in excessive interplot interference. A commonly recommended fungicide treatment is included as a standard for comparison.

Fungicide application. Fungicide application techniques vary considerably. A wide variety of nozzle types and sizes, sprayer pressure levels, and spray application volumes are used. A mechanically propelled boom sprayer is generally preferred because it allows maintenance of uniform speed and nozzle height. Complete coverage is important. One or two hollow-cone nozzles mounted over the plants and on each side on drop pipes is a good arrangement. Many sprayer calibrations are available, including from 468 to 935 L/ha (50 to 100 gal/A) at pressures of 207–1,724 kPa (30–250 psi). Spray pressures of 345–552 kPa (50–80 psi) ensure good penetration of dense tomato foliage and help minimize drift. Accurate sprayer calibration is essential.

EVALUATION OF DATA

Data collection. Defoliation, fruit rot, marketable yield, healthy yield, usable yield, and potential yield data may be useful. Evidence of phytotoxicity should be recorded.

Foliar evaluation. Disease severity can be estimated by use of a visual rating system; it is usually expressed as percentage of defoliation. Defoliation can be estimated by means of the Horsfall-Barratt scale or modifications of it (2). Determining defoliation at periodic (e.g., weekly) intervals throughout the season allows the calculation and comparison of disease development rates.

Evaluation of tomato fruits grown for multiple harvests. Tomatoes grown for multiple harvests by hand (currently most fresh market and some processing tomatoes) can be harvested and evaluated as they ripen. The interval between harvests can vary; it may be influenced by factors such as the normal interval that is followed during commercial production of the variety and the incidence of fruit rot during the particular season. Because of the frequent harvests, little anthracnose develops on tomatoes harvested by hand. In seasons with low incidence of fruit rot, the interval between harvests may be extended. On harvest days, all ripe fruit should be harvested, rinsed if necessary, graded and sized according to standard commercial guidelines, and counted and weighed.

Evaluation of tomato fruits grown for a single destructive harvest. In tests that simulate single harvests by machine (currently, most processing tomatoes and some fresh market tomatoes), effect of treatments on incidence of fruit rot is confounded by effect of treatments on time and rate of ripening. The optimum time to harvest a no-fungicide control treatment might be 1 week earlier than the optimum time to harvest a superior fungicide treatment. When it is not important to determine the effect of treatments on yield, or the amount of fruit rot on the optimum-yield harvest day, a sample of fruit that ripened over a specific period (e.g., 1 week) can be harvested and observed for rot development over a 7- to 10-day period or can be observed in the field after sufficient fruit rot appears (1). When it is important to determine effect of treatments on yield and on the amount of fruit rot on the optimum-harvest day, weekly destructive harvests of subplot segments are necessary. Equations and curves can thus be developed that describe the incidence or proportion of ripe and rotted fruit throughout the harvest period, allowing an unbiased projection of the day when maximum usable and healthy yield is available, the rot incidence or proportion on this day, and the time period during which a specific percentage of the maximum yield would be available (5).

For destructive harvests, fruit can be removed by hand or by being uprooted and shaken over a sheet or container. Fruit can be assessed by rinsing when necessary and by examining fruits individually for specific rots such as early blight, anthracnose, and blossom-end rot. State or federal grades should be used. Fruit in each category and grade should be counted and weighed to allow presentation of data in terms of percentage and weight of fruit in each category.

LITERATURE CITED

1. Barksdale, T. H., and Koch, E. J. 1969. Methods of testing tomatoes for anthracnose resistance. Phytopathology 59:1373-1376.
2. Horsfall, J. S., and Barratt, R. W. 1945. An improved grading system for measuring plant disease. Phytopathology 35:655.
3. MacNab, A. A., Sherf, A. F., and Springer, J. K. 1953. Identifying Diseases of Vegetables. Pennsylvania State University Press, University Park. 62 pp.
4. Madden, L. V. 1983. Measuring and modeling crop losses at the field level. Phytopathology 73:1591-1596.
5. Pennypacker, S. P., Knoble, H. D., Antle, C. E., and Madden, L. V. 1980. A flexible model for studying plant disease progression. Phytopathology 70:232-235.
6. Steel, R. G. D., and Torrie, J. H. 1980. Principles and Procedures of Statistics. McGraw-Hill, New York. 633 pp.
7. Stevenson, W. R., and Braddock, R. L. 1979. Evaluation of fungicide treatment for tomato disease control. Fungic. Nematic. Tests 34:87.
8. Walker, J. C. 1952. Diseases of Vegetable Crops. McGraw-Hill, New York. 529 pp.

Evaluation of Chemical Control of Bacterial Diseases of Tomato

R. D. GITAITIS, Department of Plant Pathology, University of Georgia, Coastal Plain Experiment Station, Tifton 31793; J. B. JONES, Department of Plant Pathology, University of Florida, Agricultural Research and Education Center, Bradenton 33508; and S. M. McCARTER, Department of Plant Pathology, University of Georgia, Athens 30602

Foliar diseases of tomato (*Lycopersicon esculentum* Mill.) caused by phytopathogenic bacteria can be severe in most tomato-growing areas and are particularly devastating in the southern United States. Bacterial spot incited by *Xanthomonas campestris* pv. *vesicatoria* (Doidge) Dye is probably the most important bacterial disease of tomato and pepper. Favorable environmental conditions—high humidity, extensive rainfall, and high temperatures—make Florida and Georgia ideal habitats for bacterial spot epiphytotics. The bacterium attacks all aboveground tissue, causing stem lesions, severe blighting and defoliation of leaves, blossom blight, and fruit spotting. Yield losses, which can be severe, may be attributed to reduced leaf area and fewer fruits as well as a deterioration in fruit quality. Initial inoculum may be from contaminated seed, alternate hosts, volunteer tomato plants, and infested soil debris. The consistency of the distribution of different races of *X. campestris* pv. *vesicatoria* in Florida over a 19-year period (4) would incriminate local survival as a major source of inoculum.

Another (and until recently an insignificant) disease of tomato, bacterial speck, has occurred in widespread areas of the world. The causal organism, *Pseudomonas syringae* pv. *tomato* (Okabe) Young et al., was reported in Florida in the winter of 1977–78 (13). Except in 1981 when environmental conditions were extremely hot and dry, *P. syringae* pv. *tomato* has been found in Georgia on tomato transplants every year from 1978 through 1983. Tomato is the only known natural symptom host of *P. syringae* pv. *tomato*, although a diverse group of weed species supports epiphytic populations (11,14), and these may serve as one source of primary inoculum (11). As with *X. campestris* pv. *vesicatoria*, the exact epidemiologic role and importance of seed dissemination with *P. syringae* pv. *tomato* are unclear (1,11,14). The bacterium may or may not overseason in soil and host debris, depending on the soil environment (2,11). Temperature and soil microbial activity are apparently major factors influencing survival of *P. syringae* pv. *tomato*; it survives for up to 12 weeks at low temperatures, but populations decline within 2 weeks in soil or buried host debris under high temperatures (11) or in the presence of unfavorable microflora (2). Once present in a field, *P. syringae* pv. *tomato* spreads rapidly under conditions optimum for disease development, which are extended periods of high moisture; unlike *X. campestris* pv. *vesicatoria*, however, the pathogen has a low optimum temperature of below 25°C.

When conditions are adverse for disease development, *P. syringae* pv. *tomato* and *X. campestris* pv. *vesicatoria* can survive epiphytically on tomato foliage for a period of several weeks. The impact of this phenomenon is felt not only by the local fresh-market tomato industry but also by the tomato processing industry in the northern United States and Canada, since southern Georgia and Florida are major areas for production of field-grown tomato transplants destined for northern areas. The ability of phytopathogenic bacteria to colonize tomato foliage as an epiphyte without symptom expression complicates plant certification procedures that depend exclusively on visual inspection. Control measures include protective bactericides applied at 5- to 10-day intervals, but bacterial diseases are extremely difficult to contain once an epiphytotic has begun. This is especially true in the transplant industry, where cultural conditions such as high fertility, dense spacing, overhead irrigation, clipping of plants, use of highly susceptible cultivars, and continuous cropping all enhance bacterial disease development.

The unpredictable nature of many bacterial diseases has created difficulties in experiments dealing with evaluation of chemical control. When relying on natural inoculum pressure, disease is sometimes negligible or so aggregated in distribution that an evaluation becomes impossible. Another factor that has considerable influence on the experiment is the duration of the latent period for the disease under study under the environmental conditions being experienced. Also, knowledge of the variation of sensitivity to the test chemicals among strains of the bacterium is vital to the analysis of the results of an experiment. Strains of *X. campestris* pv. *vesicatoria* are known to be resistant to streptomycin (15), and in Florida it was found that certain copper-tolerant strains of this pathogen were sensitive only to a copper-mancozeb mixture, whereas other strains were equally sensitive to copper alone (9).

Obviously, the identity of the causal organism is important. Initial symptoms of several diseases of tomato, including early blight incited by *Alternaria solani*, could be confused with bacterial speck and bacterial spot. In particular, a disease in Florida and Georgia of relatively minor importance caused by *P. syringae* pv. *syringae* van Hall induces symptoms that can be mistaken for bacterial speck (7). Apparently, this pathogen is weak and more readily controlled by bactericidal sprays. Current policy allows the shipment of plants with syringae leaf spot as certified. However, 24–48 hr are needed to distinguish between syringae leaf spot and bacterial speck. Procedures required include the culturing of bacteria onto various media for a tentative identification. Bacteria then have to be reisolated and obtained in a pure state for accurate identification. *X. campestris* pv. *vesicatoria* can be separated from most bacteria by testing any yellow bacteria that grow aerobically; are monotrichous, oxidase negative, and nitrate reduction negative; and can hydrolyze gelatin. Syringae leaf spot can be separated from speck by the ability of *P. syringae* pv. *syringae* to produce acid in erythritol, serve as an ice nucleus, and be virulent on California Blackeye No. 3 cowpea. Consequently, confirmation of the identity of the causal organism is required for accurate analysis of any experiments with chemical control of bacterial speck, even if a known strain of *P. syringae* pv. *tomato* is used to initiate disease development.

The diversity found in the culture of tomato, as well as in the different bacterial diseases, is reflected by the various research techniques used. The methods presented here have been successfully used by the authors in studies with bacterial speck and bacterial spot on tomato grown to maturity and on tomato transplants that can be harvested 40–45 days after planting. Although there are differences in the culture of the plants, the cultivars used, methods for disease evaluation, and chemical application, it is agreed that inoculations should be used and that inoculum preparation and its deployment are critical factors in the success of initiating uniform disease pressure.

BACTERIAL SPECK ON TOMATO GROWN TO MATURITY

Since cool, moist weather is essential for bacterial speck development, tests are conducted either during the early part of the growing season or in the fall months in areas that have high midseason temperatures. In Georgia, where midseason weather in June is adverse to bacterial speck development, plants for test plots are produced in the greenhouse during March so that transplanting can be done as soon after the frost-free date as possible, usually about 15–20 April. These early tests allow at least a 6-week evaluation period before the onset of high-temperature or low-moisture conditions.

The site selected should be suitable for tomato production. In most areas it is risky to depend on natural rainfall, so the site should be located where irrigation is available. Sprinkler irrigation is preferred because foliar wetting enhances disease development.

A randomized complete block design, with a minimum of four replications, is preferred. Plot size depends on the type and design of the spray equipment being used. In our tests, where motorized, back-mounted mist sprayers are used, two- or three-row plots about 10 m long are convenient because the applicator can spray for good coverage without walking within the plots. Other plot arrangements may be more suitable where tractor-mounted sprayers are used. In our tests, rows are spaced 0.97 m apart, and plants within the rows are as close together as recommended for the cultivar being used. A continuous plant canopy promotes spread of the bacterium and provides a good environment for disease development. Cultivars with a determinate growth habit, as with the Chico III that is used in many of our tests, are spaced 45 cm apart; more indeterminate types are spaced 60–90 cm apart. The problem of spray drift must be considered in plot arrangement. Plots must be separated sufficiently either by bare alleys or border rows to prevent spray drift. Mist-type sprayers, such as we use, produce small droplets that tend to drift for considerable distances. Plots sprayed with such sprayers should usually be separated by at least 2 m to prevent drift problems. The researcher should consider the equipment available when spacing plots to minimize chemical drift problems.

The cultivar selected for the test is important because commercial cultivars differ in susceptibility (5). We have used the Chico III cultivar because it is moderately susceptible and represents an "average" disease reaction. More susceptible cultivars include FM 6203, Libby 8990-A, Heinz 1630, and Campbell 38, whereas Campbell 28, Ohio 7663, and Hunt 62 have considerable resistance (5). Plants for transplanting into the field are produced on a greenhouse bench. Seeds are planted in flats of sterile vermiculite 6–7 weeks before the projected field transplanting date. Ten days after seeding, the emerged seedlings are transplanted into 266-ml paper cups filled with a soil mix (soil, sand, and vermiculite, 3:1:1 by volume) previously treated with methyl bromide. At transplanting, each plant (20–25 cm tall) is given approximately 500 ml of a water-soluble fertilizer solution (Peters 20-20-20, 2.5 g/L).

Uniform disease development is ensured by producing diseased plants in the greenhouse and transplanting them among healthy plants in field plots. The plants used for inoculation are from the same lot used for the remainder of the plot. Diseased plants are produced by inoculating healthy plants 10 days before transplanting.

P. syringae pv. *tomato* is grown at 25°C and maintained between studies at 6°C on slants of nutrient yeast dextrose agar (NYDA; 23 g of nutrient agar, 5 g of yeast extract, 10 g of glucose, and 1 L of distilled water, adjusted to pH 6.8). For extended storage, it is also maintained at room temperature in tubes of sterile distilled water or in a dried state on glass beads or in a lyophilized state. Loss of virulence of this pathogen either during culture or in storage has not been a problem. For inoculum production, it is grown on plates of King's medium B (20 g of proteose peptone No. 3, 10 ml of glycerol, 1.5 g of K_2HPO_4, 1.5 g of $MgSO_47H_2O$, and 15 g of Bacto agar, adjusted to pH 7.2) (8). Inoculum is prepared by suspending cells from 24- to 48-hr-old cultures in sterile distilled water to obtain a concentration of 10^8 colony-forming units per milliliter (adjusted to 80% light transmission on a Bausch & Lomb Spectronic 20). Inoculum is applied to runoff on all plant surfaces with a paint sprayer (Burgess Model 862, Vibrocrafters, Inc., Grayslake, IL 60030). Inoculated plants are placed at high humidity (enclosed in polyethylene bags) for 36 hr in a growth chamber (alternate 12 hr of dark and 12 hr of light) at 20°C and are then held in the same chamber or on a greenhouse bench at 18–25°C to allow disease development. Plants usually develop 250–500 lesions within 8 days. Diseased plants are transplanted (usually about 25%) among healthy plants at preselected sites to ensure even distribution of inoculum within the plots. Healthy and diseased plants are marked for identification so that disease spread can be observed.

Routine land preparation and cultural practices are used, and plots are fertilized according to soil test results. Sprinkler irrigation is applied as needed either to promote normal vegetative and reproductive growth or to promote normal disease development. Fungal diseases, especially early blight caused by *A. solani*, must be controlled because young early blight lesions can be confused with bacterial speck, and extensive early blight damage can completely destroy the study. Chlorothalonil (Bravo 500, 3.4 L/ha) should be applied along with the bactericide being evaluated unless some other fungicide is being evaluated in combination or the bactericide alone also has fungicidal properties. Chlorothalonil is the best choice for routine use in bacterial speck evaluations because it provides excellent control of most fungal diseases and has no toxic effect on *P. syringae* pv. *tomato* (3). Insecticides are applied only as needed. Carbaryl (Sevin 80%, 2.2 kg/ha) is used for all insects except fruitworms, which are controlled with methomyl (Lannate 90%, 0.59 kg/ha).

The rate of disease spread varies greatly from year to year depending on weather conditions. Initially, disease spread is slow. However, once a few lesions appear on most plants, rapid increases occur if weather conditions are conducive. We follow the progress of disease by taking visual ratings at 7- to 10-day intervals. First, a diagram of each plot is prepared. Each plant is represented by a circle, and the originally diseased plants (inoculum sources) are marked. Each plant, except the diseased plants established as inoculum sources, is assigned a disease severity rating based on a 0–10 scale in which 0 = no bacterial speck lesions, 1 = a few isolated lesions, and 10 = several to many lesions on most leaves. When possible, we have two persons rate the same plants independently, and their readings are averaged. We have found the 0–10 scale useful in our tests. However, other scales are suitable as long as they show accurately the relative differences in disease severity among treatments. Bacterial speck lesions also occur on fruit, especially on fruit clusters formed early in the season. Percentages of fruit infected and disease severity (number of lesions per fruit) can be determined as a measure of chemical efficacy. Yield data should also be taken, as bacterial speck can reduce both yield and quality. We make five to seven hand harvests beginning when the first fruits are completely ripe. Diseased and healthy fruits are separated, counted, and weighed. Percentages of fruits with bacterial speck and yields from plots receiving the various treatments are calculated. Generally, we have not counted lesions or assigned severity ratings to individual fruits, but this would add another dimension to the disease severity ratings if time and labor are available.

BACTERIAL SPOT ON TOMATO GROWN TO MATURITY

A test area should be chosen that is conducive to tomato production and where excessive precipitation can be expected or where overhead irrigation is used to supplement rainfall and enhance bacterial spot development.

Tomato cultivar Sunny or any other cultivar that is highly susceptible to bacterial spot but resistant to gray leaf spot will prove useful. The plants are grown to transplant size and then set into raised beds covered with black plastic mulch. Plants are watered by means of subsurface seep irrigation in southwest Florida. Insecticides and cultural practices are as recommended. A recommended fungicide is applied on a regular basis for control of phytopathogenic fungi.

A randomized complete block design may be used. It should be composed of at least four replications with each plot consisting of a single row of 10–12 plants. Plant spacing may range from 30 to 90 cm. Plots are spaced approximately 1.5 m apart in all directions.

In Florida, the compounds that have given the most efficacious results through the years are copper compounds. To determine if the bacterial populations present in a particular area are resistant to copper, test strains and known copper-resistant strains are streaked on nutrient agar medium containing copper at 0, 10, 50, 100, and 200 μg/ml. Unlike the sensitive strains, copper-resistant strains will grow at the higher copper concentrations. In cases where copper strains are predominant, copper compounds are applied in combination with an ethylenebisdithiocarbamate compound.

The first bactericide treatments are begun 7 days after setting the plants in the field. Subsequently, sprays are applied weekly for the entire season. During conditions of heavy precipitation, two applications per week would be more effective. Bactericides are applied by means of a CO_2-powered hand sprayer pressurized to 244 kPa (40 psi). Plants are sprayed to runoff.

Pepper or tomato strains of *X. campestris* pv. *vesicatoria* are used for inoculating tomato, whereas only pepper strains are used for studies with pepper. The cultures are readily stored in tubes of sterile tap water for years without losing virulence. The cultures may also be stored on slants of NYDA at 4°C for several months. For inoculum production, the bacterium is grown on petri dishes of NYDA for 24–48 hr at 25–28°C. Inoculum is prepared by suspending the bacterium in a 0.01 M $MgSO_4$ solution. The concentration is adjusted to 10^8 colony-forming units per milliliter. This is achieved by adjusting the percentage of transmission of the suspension to 80% at an absorbance of 590 nm.

The bactericidal compounds can be applied one or more times before bacterial inoculation. Inoculum should be applied when the plants are wet. Leaf moisture may be achieved by rain, dew, or overhead irrigation. The plants should remain wet for several hours after inoculation. The inoculum may be applied with a small hand-powered backpack sprayer.

Disease ratings can be obtained by a number of different methods. One consists of collecting individual leaflets from each plot. Total lesions are determined on each leaflet and the average number of lesions per leaflet is determined. If the disease is too intense on individual leaflets, estimates of the percentage of leaf area affected are determined using the Horsfall-Barratt rating system (6), or just the percentage of leaf area affected is used.

A second method that is quite useful is to make a field rating by estimating the percentage of defoliation. This measurement is based on the dead portions of the plant that are still attached to the plant and those that have actually defoliated.

Fruit spotting may also be of major concern, primarily with fresh market tomatoes (12). Upon harvesting fruit, both incidence and severity of fruit spotting can be determined. Incidence is determined by counting the number of diseased fruit and determining the percentage of total fruit diseased. Severity can be determined by separating fruit into three categories: not diseased (no spotting of the fruit), diseased but marketable (few lesions present), and unmarketable (many lesions present on the fruit).

Symptom expression varies based on the type of compound tested. Some compounds may cause aberrant growth (malformation of growing points). Copper compounds can cause stunted plants and smaller leaves. Yield on plants sprayed with certain copper compounds has been shown to be reduced as a result of fewer fruits and spotting of fruit.

The importance of collecting yield data is twofold. First, bactericides may be phytotoxic and their application may affect yield by reducing fruit number and marketable yield. Second, yield data are important in determining the efficacy of bactericides in reducing bacterial spot severity and thus increasing yield.

Fruit is harvested from all plants in each plot. Harvestable fruit for fresh market consists of mature greens, pinks, and ripes. Unmarketable fruit is the result of excessive spotting of the fruit by bacterial spot. Generally, fruit with several spots is considered marketable. Yield is reported as kilograms per plot or as metric tons per hectare.

BACTERIAL DISEASES OF TOMATO TRANSPLANTS

Tomato transplants are produced in southern Georgia from the end of February to the beginning of June. The preparation and planting date of the field plots are dictated by the pathogens that are of interest. An early planting at the end of February or beginning of March is suggested if *P. syringae* pv. *syringae* is being studied. However, if *P. syringae* pv. *tomato* is of interest, a planting in early April would be best, whereas a later planting in April or May would be sufficient if *X. campestris* pv. *vesicatoria* is the pathogen under study.

A typical experimental design such as randomized complete block with four to six replicates is recommended. Tomatoes are direct-seeded with a Stanhay precision planter in raised beds (1.52 m wide) with four rows (35 cm apart) per bed. Plant beds should be of sufficient length to contain at least 5,000 plants (13.72 m). Plant spacing is approximately 1 cm within the row. NPK fertilizer (6-12-6) is applied at seeding at 2.5 cm below the seed (896 kg/ha) and again after 4 weeks as a side dressing. A herbicide, such as napropamide (Devrinol) at 1.68 kg a.i./ha, is applied at planting for weed control. Fungicides and insecticides are applied weekly for control of fungal and insect pests. Transplant fields are irrigated daily by overhead sprinklers until an acceptable stand is achieved. This procedure breaks the soil crust and permits the emergence of small-seeded vegetables such as tomato.

Treatments are begun after plant emergence and development of the first true leaf. Under normal circumstances, a spray schedule of 7- to 9-day intervals is adhered to, but it is shortened to 5 days when there is abundant rain. A standard bactericide treatment such as cupric hydroxide (2.24 kg/ha) is used for a comparison. It is best to use a pressurized boom sprayer with fan nozzles and a delivery of 233.72 L/ha to simulate a commercial operation. Success has also been achieved using hand-held, pressurized garden sprayers (Acme Burgess, Grayslake, IL 60030) calibrated for a similar delivery. After 25–30 days, plants should be clipped at least once per week and at shorter intervals as harvest approaches. Clipping helps to harden off plants for transplanting, maintains plant size for packaging in standard vegetable crates, and deblossoms plants. However, clipping disseminates plant pathogens mechanically (10) and should be immediately followed by a spray of the chemical being evaluated.

Inoculum is prepared and standardized in a similar manner as previously described for the bacterial speck and bacterial spot studies. Because transplants are harvested only 40–45 days after planting (depending upon environmental conditions), it is considered best to inoculate the entire plot rather than to wait for spread of secondary inocula from a point source or border row. This is achieved by circulating through the field while allowing a mist of inoculum to drift onto the plants until all plots have been exposed. A fine mist can be generated with a battery-run, spinning-disk, hand-held, ultra-low-volume applicator. Inoculum should be applied during a rainstorm, irrigation, or in a heavy dew. One or two applications of test materials are usually applied before inoculation.

Disease is extremely difficult to evaluate on transplants because of the dense plant canopy created by the close spacing of plants. The problem is compounded by the fact that nearly all of the symptoms occur on the lower foliage. We sample each bed at five different sites and uproot a sample of 50 plants per site for a total of 250 plants per treatment. Most disease rating scales are applicable, and we use a 0–10 scale with 10 representing maximum disease. When evaluating chemical treatments of transplants, one should be especially aware of the latent period (defined here as that time between initial infection and symptom expression). Because transplants are intended to continue their growth in northern areas, the amount of infection in the latent period at harvest is critical. Transplants could easily be shipped in good faith as being disease free but eventually show signs of disease upon arrival at their final destination. Likewise in research plots, evaluations at harvest of transplants may not be true indicators of the efficacy of chemical control. Consequently, we make ratings 1 week before harvest, at harvest, and 5–7 days after harvest to evaluate fully the efficacy of a compound for control of bacterial diseases on transplants.

LITERATURE CITED

1. Bashan, Y., Okon, Y., and Henis, Y. 1982. Long-term survival of *Pseudomonas syringae* pv. *tomato* and *Xanthomonas campestris* pv. *vesicatoria* in tomato and pepper seeds. Phytopathology 72:1143-1144.
2. Bosshard-Heer, E., and Vogelsanger, J. 1977. Uberlebensfahigkeit von *Pseudomonas tomato* (Okabe) Alstatt in verschiedenen Boden. Phytopathol. Z. 90:193-202.
3. Conlin, K. C., and McCarter, S. M. 1983. Effectiveness of selected chemicals in inhibiting *Pseudomonas syringae* pv. *tomato* in vitro and in controlling bacterial speck. Plant Dis. 67:639-644.
4. Cook, A. A., and Stall, R. E. 1982. Distribution of races of *Xanthomonas vesicatoria* pathogenic on pepper. Plant Dis. 66:388-389.
5. Gitaitis, R. D., Phatak, S. C., Jaworski, C. A., and Smith, M. W. 1982. Resistance in tomato transplants to bacterial

speck. Plant Dis. 66:210-211.

6. Horsfall, J. G., and Barratt, R. W. 1945. An improved grading system for measuring plant diseases. Phytopathology 35:655.
7. Jones, J. B., McCarter, S. M., and Gitaitis, R. D. 1981. Association of *Pseudomonas syringae* pv. *syringae* with a leaf spot disease of tomato transplants in southern Georgia. Phytopathology 71:1281-1285.
8. King, E. O., Ward, M. K., and Raney, D. E. 1954. Two simple media for the demonstration of pyocyanin and fluorescin. J. Lab. Clin. Med. 44:301-307.
9. Marco, G. M., and Stall, R. E. 1983. Control of bacterial spot of pepper initiated by strains of *Xanthomonas campestris* pv. *vesicatoria* that differ in sensitivity to copper. Plant Dis. 67:779-781.
10. McCarter, S. M., and Jaworski, C. A. 1969. Field studies on spread of *Pseudomonas solanacearum* and tobacco mosaic virus in tomato plants by clipping. Plant Dis. Rep. 53:942-946.
11. McCarter, S. M., Jones, J. B., Gitaitis, R. D., and Smitley, D. R. 1983. Survival of *Pseudomonas syringae* pv. *tomato* in association with tomato seed, soil, host tissue, and epiphytic weed hosts in Georgia. Phytopathology 73:1393-1398.
12. Pohronezny, K., and Volin, R. B. 1983. The effect of bacterial spot on yield and quality of fresh market tomatoes. HortScience 18:69-70.
13. Pohronezny, K., Volin, R. B., and Stall, R. E. 1979. An outbreak of bacterial speck on fresh-market tomatoes in south Florida. Plant Dis. Rep. 63:13-17.
14. Schneider, R. W., and Grogan, R. G. 1977. Bacterial speck of tomato: Sources of inoculum and establishment of a resident population. Phytopathology 67:388-394.
15. Stall, R. E., and Thayer, P. L. 1962. Streptomycin resistance of the bacterial spot pathogen and control with streptomycin. Plant Dis. Rep. 46:389-392.

Methods for Field Evaluation of Fungicides for Control of Foliar Diseases of Alfalfa

DONALD L. STUTEVILLE, Department of Plant Pathology, Kansas State University, Manhattan 66506

Alfalfa (*Medicago sativa* L.) is subject to many diseases (3) that prevent it from producing to its full potential. Foliar diseases reduce yields of forage and alter its quality. Resistant cultivars offer the most practical means of controlling alfalfa diseases, but cultivars with high levels of genetic resistance to all diseases are unavailable. Thus, the possibility of a role for fungicides as a complement to genetic resistance in disease control needs to be investigated.

Foliar fungicides have reduced alfalfa disease (1) and have substantially increased forage yields, e.g., an average increase of 44% per year for two years in Iowa (7) and increases of 15–25% for each of four years in Kansas (9–11,13) and 15% for the first cutting in California (12). Such information is useful in assessing losses from pests, but the large amount of chemicals and water diluent used in these treatments and the frequency of application (usually performed weekly) are not economical.

Few reports have been made of tests that evaluate foliar fungicides applied at economical rates on alfalfa, and thus little basis exists for establishing guidelines for such research. This chapter provides some general information based mostly on limited fungicide testing in Kansas; this testing evolved from research initiated to measure losses in alfalfa forage yield and quality caused by diseases (13).

MATERIALS AND METHODS

Test plots. An open site was chosen on a level, deep, sandy loam soil near Manhattan, Kansas. Recommended high fertility levels were maintained, and supplemental sprinkler irrigation was provided as needed, usually once immediately after each cutting.

Plots were 0.9 m wide (containing five rows, 15 cm apart) and 4.3 m long and were 0.3 m apart, arranged in a randomized complete block design with four replications per treatment. A seeded border of Cody alfalfa, 0.9 m wide, separated the blocks, and one at least 2.7 m wide surrounded the trial.

The treatments were applied to the cultivars Cody and Lahontan. Cody is recommended and consistently highly productive in Kansas, largely because of its comparatively high level of resistance to diseases and insects that are common there. Because of these characteristics, it was chosen as a test cultivar to provide a measure of the economics of fungicide use. Lahontan has high yield potential in Kansas but is very susceptible to foliar pathogens, particularly *Cercospora medicaginis* Ell. & Ev., and therefore was a logical choice for measuring fungicide efficacy.

Inoculum. No artificial inoculations were performed, and no attempts were made to increase disease severity by management practices.

In these tests, endemic diseases that caused serious losses under environmental conditions favorable for their development were spring black stem and leaf spot (caused by *Phoma medicaginis* Malbr. & Roum.) on the first cutting; Leptosphaerulina leaf spot (caused by *Leptosphaerulina briosiana* (Poll.) Graham & Luttrell) on the second cutting; summer black stem and leaf spot (caused by *C. medicaginis*) on the second, third, and fourth cuttings; bacterial leaf spot (caused by *Xanthomonas campestris* pv. *alfalfae* (Riker et al) Dye) on the third and fourth cuttings; and rust (caused by *Uromyces striatus* Schroet.) on the fourth cutting.

Diseases that occurred less often in these tests but caused substantial losses in some years were downy mildew (caused by *Peronospora trifoliorum* d By.) on the first and second cuttings of the first harvest year; yellow leaf blotch (caused by *Leptotrochila medicaginis* (Fckl.) Schüepp) and common leaf spot (caused by *Pseudopeziza medicaginis* (Lib.) Sacc.) on the first and second cuttings; and anthracnose (caused by *Colletotrichum trifolii* Bain) on the second and third cuttings.

Inoculum levels can often be increased by planting highly susceptible cultivars (2) in the borders or by frequent overhead irrigation, which allows free water to stand on the plants and provides high relative humidity for long periods. Altering or delaying the time of harvesting of a portion of the border or nearby alfalfa can also raise inoculum levels.

Controls. Control plots that received no fungicides were always included. Treated controls often were also included (9–11); these plots were sprayed weekly with a mixture of 908 g Dithane M-45 80W and 177 ml Triton B-1956 per 378 L applied at a rate of 1,870–3,740 L/ha to provide maximum disease control. Differences between the yields of treated and untreated controls were used to assess losses from disease and the efficacy of test fungicides.

To control insects, methoxychlor was routinely applied once to all plots about 1 week after each cutting, as insect damage can complicate the assessment of damage from diseases. For example, Norton (7) concluded that a 44% increase in alfalfa forage yields from weekly applications of maneb to control stem and leaf spot diseases was mainly due to its effect on leafhoppers.

Fungicide application. Fungicides were applied with a 7.5-L hand sprayer. About 935 L of water per hectare was required for adequate coverage. Plots were sprayed in the late afternoon or evening after the wind had subsided. A piece of rigid corrugated cardboard was held between plots when it was needed to prevent drifting during spraying.

In commercial operations, the fungicides would have been applied with conventional spray equipment, which requires less water diluent for coverage than do hand sprayers, but such equipment was unavailable.

Test fungicides were applied once per cutting, for economic reasons. The first application was 3–4 weeks before the first cut. Other applications were made about 2 weeks after each harvest, when regrowth had formed a canopy that prevented drying of the foliage and thus favored the foliar pathogens. This application schedule was arbitrarily chosen to control *C. medicaginis*, which usually is a principal pathogen on three cuttings. The schedule appeared to be satisfactory for the control of most pathogens. However, when extended periods of wet weather occurred at or a few days after the first harvest, *L. briosiana* often caused severe losses before the plots were sprayed.

Applying fungicides to the stubble before regrowth to reduce inoculum levels has the advantage that, at that stage, the wheel tracks of the spray rig cause less damage to the regrowth than they do later, although such damage can still be a factor (6). Also, less diluent should be required at the stubble stage.

Data collection. Just before each harvest, and when obvious differences existed among treatments, the individual diseases were rated on a scale of one (no disease) to nine (severe damage) to measure fungicide efficacy and provide a record of the diseases and their relative importance.

Forage was harvested at the 1/10 bloom stage (4) with a Carter flail-type plot harvester, 0.9 m wide, which harvested and deposited the green chop from each plot into a burlap bag. The green weight of the plot was recorded, and a representative sample of about 500 g was immediately removed, placed in a paper bag, and weighed. The samples were oven-dried, and the dry-weight yield was calculated for each plot. Dry weight is important because green weight is biased in favor of plots with the least disease, since healthy green leaves contain more moisture than do diseased ones.

Data on dry-weight yields and disease ratings were subjected to an analysis of variance. The least significant difference procedure was usually used to compare means.

The method of harvesting can alter the results of testing. For example, Summers and McClellan (12) noted that common leaf spot caused no significant defoliation before harvesting but that diseased leaves abscised during drying and were lost during the raking and turning of windrows. Therefore, the method of harvesting used in testing should be comparable to those used in commercial operations. The flail-type harvester does not simulate commercial hay-harvesting practices. However, in Kansas, most alfalfa grown for hay is cut, conditioned, and windrowed in one operation, and usually the windrows are not turned.

Besides increasing forage yield, disease control may also increase forage value by improving its grade, quality components, or nutritive value; some or all of these characteristics may thus be improved. Evaluation of these characteristics is costlier and more difficult than measuring forage yield. Carotene content was considerably reduced by disease (13). However, protein content was not changed (5,12,13), even with severe defoliation (13), although leaves have a higher concentration of protein than do stems (4,8). The effects of disease on the nutritive value of forage are impossible to measure without animal performance trials. Alfalfa forage with low levels of certain diseases, which caused no defoliation and altered none of several tested quality factors, affected the intake and gains of test animals (5). Little is known about the effects on animals of toxins and increased levels of coumestrol associated with alfalfa diseases (4).

DISCUSSION

Economic feasibility is the ultimate determinant of fungicide use on alfalfa. Therefore, fungicides should be evaluated as a complement to recommended management practices and not as a substitute for them. The use of a recommended high-yielding cultivar and proper timing of harvests are particularly important, since they have little influence on production costs but can greatly affect the value of the forage produced. One permissible deviation from recommended management practices is the use of sprinkler irrigation in dry periods to produce high relative humidity and free water (which occur naturally during wet periods) and thus favor disease development.

Little interest has been shown in evaluating foliar fungicides on alfalfa despite increasing evidence that their use is economically warranted. The use of insecticides on alfalfa, however, is widespread. Perhaps cooperative work with entomologists is needed to determine the effects of applying fungicides with insecticides to reduce application costs.

LITERATURE CITED

1. Banttari, E. E., Sundheim, L., and Wilcoxson, R. D. 1963. Chemical control of spring black stem in alfalfa. Plant Dis. Rep. 47:682-685.
2. Elgin, J. H., Jr., ed. 1984. Standard tests to characterize pest resistance in alfalfa cultivars. U.S. Dep. Agric. Agric. Res. Serv. Misc. Pub. 1434. 38 pp.
3. Graham, J. H., Frosheiser, F. I., Stuteville, D. L., and Erwin, D. C. 1979. A Compendium of Alfalfa Diseases. American Phytopathological Society, St. Paul, MN. 65 pp.
4. Hanson, C. H., ed. 1972. Alfalfa Science and Technology. American Society of Agronomy, Madison, WI. 812 pp.
5. Leath, K. T., Shenk, J. S., and Barnes, R. F. 1974. Relation of foliar disease to quality of alfalfa forage. Agron. J. 66:675-677.
6. Massee, T. W., Romanko, R. R., and Osgood, C. E. 1984. Evaluation of Bravo for disease control in alfalfa cultivars grown for hay, 1983. Fungic. Nematic. Tests 39:102.
7. Norton, D. C. 1965. *Xiphinema americanum* populations and alfalfa yields as affected by soil treatment, spraying, and cutting. Phytopathology 55:615-619.
8. Smith, D. 1970. Yield and chemical composition of leaves and stems of alfalfa at intervals up the shoots. J. Agric. Food Chem. 18:652-656.
9. Stuteville, D. L., and Sorensen, E. L. 1972. Alfalfa: Cercospora leaf spot (*Cercospora zebrina*), Leptosphaerulina leaf spot (*Leptosphaerulina briosiana*), spring black stem (*Phoma medicaginis*). Fungic. Nematic. Tests 27:106.
10. Stuteville, D. L., and Sorensen, E. L. 1973. Alfalfa: Cercospora leaf spot (*Cercospora zebrina*), common leaf spot (*Pseudopeziza medicaginis*), downy mildew (*Peronospora trifoliorum*), Leptosphaerulina leaf spot (*Leptosphaerulina briosiana*), rust (*Uromyces striatus*). Fungic. Nematic. Tests 28:95.
11. Stuteville, D. L., and Sorensen, E. L. 1974. Alfalfa: Anthracnose (*Colletotrichum trifolii*), bacterial leaf spot (*Xanthomonas alfalfae*), Cercospora leaf spot (*Cercospora zebrina*), Leptosphaerulina leaf spot (*Leptosphaerulina briosiana*), spring black stem (*Phoma medicaginis*). Fungic. Nematic. Tests 29:87.
12. Summers, C. G., and McClellan, W. D. 1975. Effect of common leafspot on yield and quality of alfalfa in the San Joaquin Valley of California. Plant Dis. Rep. 59:504-506.
13. Willis, W. G., Stuteville, D. L., and Sorensen, E. L. 1969. Effects of leaf and stem diseases on yield and quality of alfalfa forage. Crop Sci. 9:637-640.

Field Evaluation of Fungicides for Management of Peanut Foliar Diseases

F. M. SHOKES, University of Florida, North Florida Research and Education Center, Quincy 32351; D. H. SMITH, Texas A&M University, Texas Agricultural Experiment Station, Plant Disease Research Station, Yoakum 77995; R. H. LITTRELL, University of Georgia, Coastal Plain Experiment Station, Tifton 31794

Foliar diseases of peanut can cause yield losses of 50% without multiple applications of fungicidal sprays (18). Early and late leaf spot, caused by *Cercospora arachidicola* Hori and *Cercosporidium personatum* (Berk. & Curt.) Deighton, respectively, are the most widely distributed and destructive peanut diseases (8). Peanut rust, caused by *Puccinia arachidis* Speg., is also widely distributed, occurring annually in some areas of the United States, but is less important economically than the leaf spots. It is commonly a late-season disease of limited economic importance (18).

Primary inoculum of the leaf spot fungi survives in soil and on crop residues (8). Where peanuts are grown in monoculture, the leaf spot diseases usually appear early in the growing season. Secondary inoculum consists of conidia, which are produced on leaf lesions when humidity is high and temperatures are 22–35°C. The conidia are dispersed by wind, rain, insects, and farm implements. If the leaf spot diseases or rust develops early in the growing season and causes extensive defoliation, significant pod yield losses are common.

Although disease-resistant cultivars may eventually become available, multiple applications of fungicidal sprays are now necessary to manage peanut foliar diseases. The incubation period of the leaf spot pathogens varies from four to 22 days (9). In some areas, foliar sprays are applied from five to nine times at intervals of seven to 14 days. However, in areas where disease onset occurs later and disease severity is reduced, fungicide is applied according to a disease forecasting method based on temperature and relative humidity during the growing season (15).

In southern Texas, peanut rust reaches epidemic proportions in some years (4,5) and is of major importance in some counties where peanuts are cultivated. Researchers in the United States do not usually evaluate fungicides especially for efficacy against rust, but peanut rust severity is assessed when the disease occurs in field plot tests.

This chapter describes methods for evaluating the efficacy of compounds in controlling the principal foliar diseases of peanut. These methods can be used to study rates, application intervals, degrees of resistance in peanut genotypes, and disease management strategies with reduced dosages or extended application intervals.

METHODS

Test site. For compounds used to control peanut foliar diseases, fungicide tests should be conducted where epidemics of early or late leaf spot are highly likely. A fungicide that is being evaluated primarily for its effectiveness against leaf spot should be tested at a site where peanut rust and other diseases are not likely to occur. Selecting a field planted with peanuts in the previous season or seasons can increase the probability of having sufficient inoculum and adequate disease pressure. County extension agents and producers are often helpful in locating problem fields with a history of disease.

If data on the efficacy of fungicides against rust are required, the test site should be in an area with a high probability of annual epidemics of the disease. Parts of Texas and Florida are acceptable locations for peanut rust experiments. Field inoculation with urediospores of *P. arachidis* is effective but not prudent where peanuts are an economically important crop.

A test site should have uniform soil. Access to sprinkler irrigation is advantageous for test plots. Supplemental irrigation may be important in obtaining good plant growth and providing the moisture on leaves that is required for epidemics of foliar disease. Since pod yield data are usually collected in fungicide tests, a well-drained sandy loam or sandy clay loam with pH 6.0–7.0 is preferred in order to obtain good plant growth and favorable soil texture for maximum pod recovery.

On new land or fields where peanuts have not been planted in recent years, a *Rhizobium* inoculant should be applied as a seed treatment or an in-furrow treatment. Since calcium is required for peanut production, lime should be applied to the soil in some production areas three to four months before planting, and gypsum may be applied as a band application at flowering and pegging. Application of gypsum is essential for satisfactory culture of Virginia market cultivars.

The soil should be tested for fertility and nematodes. Appropriate fertility practices for peanut production were reported by Cox et al (3). A nematicide should be used if necessary.

Weed control is essential in fungicide application tests, because weeds can reduce spray deposition and coverage and diminish pod yields. Control of weeds in peanut fields is attainable with a combination of preplant and postemergence herbicides and limited cultivation. In cultivation for weed control, plant injury can be avoided by adjusting the implement so as to prevent throwing soil onto the plants. This practice of nondirting cultivation reduces the probability of predisposing plants to invasion by *Sclerotium rolfsii* Sacc. and other soilborne plant-pathogenic fungi. In some instances, pulling weeds or removing them with hoes may be necessary, but care should be taken to avoid crop injury and subsequent crop losses associated with this method of weed control.

Scouting of fields to afford early detection of insects or other pests is also essential. Insecticides and acaricides should be applied only when needed to prevent yield loss. Severe foliar damage or defoliation caused by insects or mites makes evaluation of foliar diseases difficult, if not impossible. If problems from southern stem rot or other soilborne diseases are expected in a test area, an appropriate fungicide (such as PCNB) should be used to minimize damage from these diseases.

Cultivar selection and planting. High-quality seed of a susceptible cultivar with a germination rate of 80% or higher should be planted. Peanut cultivars currently planted in the United States are susceptible to early leaf spot, late leaf spot, and rust. Runner market types, such as Florunner, Sunrunner, and Tifrun, can be used where these cultivars are adapted. Spanish market types, including Tamnut 74, Starr, and Pronto, are useful in fungicide tests. Virginia market types include Florigiant, Early Bunch, and Virginia 81 Bunch. The traits and yield potential of these cultivars were described by Henning et al (6,7).

Seed should be from one cultivar grown at a single location. It should be treated with a fungicide or fungicidal mixture with a broad spectrum of activity. Seed should be planted when soil moisture is adequate and soil temperature is 21°C or higher at 7:00 A.M. for three consecutive days. It should be planted on a flat or slightly raised bed at a depth of 3.75–5.0 cm. On coarse-textured soils, the planting depth may be increased up to 8.75 cm (6). For runner and Spanish market types, four to six seeds should be planted per 30 cm of row; for Virginia market types, the preferred planting is three or four seeds per 30 cm of row. Row spacing varies from 15 to 90 cm, depending on the preference in a given area.

Field plot design. The experimental design depends on the number of treatments, plot size, application equipment, and harvesting equipment. Usually, a randomized complete block design is used, because each treatment appears in each block, so that variability within blocks is minimized (11). Blocks fall within the ranges of the plot in some tests and cut across the ranges in others. A completely randomized design can be used in small tests with few entries if the test site is uniform.

The number of treatments in a test should be limited to those required to meet the test objectives. Yields from adjacent plots are usually more similar to one another than are those from widely separated plots. Treatments should be assigned by a random number table (11,19) or a random number list. Each treatment should be replicated at least four times.

Individual plots should contain at least two rows if they are to be lifted with a conventional two-row digger-shaker-inverter. Four rows per plot are preferred, to reduce the interplot interference associated with spray drift and inoculum production. In four-row plots, data are collected from the two inner rows, and the outer two serve as buffers between adjacent plots. Two unsprayed rows, one on each side of the two sprayed rows, may be included to increase inoculum quantity and the uniformity of distribution. The unsprayed rows are especially desirable if inoculum levels are low and the onset of epidemics is delayed.

If high inoculum pressure is not desired, the buffer rows should be treated, and the unsprayed check plot should be separated from the treated plots by buffer rows sprayed with a highly effective fungicide. This practice reduces disease pressure on adjacent plots. In this case, the unsprayed check plot is used to monitor disease levels, and data from this plot are not included in the statistical analysis. Exclusion of these data is valid because the objective of fungicide tests is to compare the efficacy of individual treatments.

Unplanted alleys should be left between ranges to prevent interrange spray drift. If plots are harvested mechanically, the alleys should be at least 2 m wide. If a tractor-propelled sprayer is employed, the alleys should be wide enough (6.1–9.1 m wide, depending on the tractor) to permit the tractor to move beyond a plot, turn, and proceed to the next one.

Unnecessary trips through plots with machinery should be avoided, because the machinery can cause plant injury and soil compaction, which predispose plants to infection by soilborne pathogens. Control plots may be sprayed with water, and even if not sprayed, they should be traversed with the tractor to impose the same conditions of soil compaction and plant injury on all test plots.

Fungicide application. Fungicides are applied with tractor-propelled sprayers, fixed-wing aircraft, and sprinkler irrigation systems. In tests evaluating numerous fungicides, a tractor-propelled sprayer (two-, four-, or six-row) or a CO_2-powered backpack sprayer is useful. Application with sprinkler irrigation systems or fixed-wing aircraft is impractical for routine fungicide tests, because excessive amounts of plot land and costly equipment are required and because uniform dosage and deposition of fungicides are difficult to achieve by these methods.

Tractor-propelled sprayers are often equipped with three hollow-cone nozzles per row, one centered over the row and the other two, which are adjustable swivel-type nozzles, equidistant from the center. Hollow-cone spray tips provide satisfactory delivery and are used by many commercial producers. Stainless steel tips are recommended for application of abrasive wettable powder and flowable fungicide formulations.

Propelled sprayers must be designed to move at a constant, easily adjustable speed. A sprayer speed of 3.2–6.4 km/hr is adequate to deliver the proper spray volume for most tests with conventional boom sprayers. Volumes ranging from 93.5 to 467.0 L/ha may be used to provide good coverage. Typically, fungicides are applied to peanut foliage at 93.5–187.0 L/ha at 207–483 kPa (10).

Tractor-mounted tanks should be equipped with some means of agitating the spray mixture. Uniform pressure can be supplied by CO_2, air pressure from the compressor of an automobile air conditioner, or a power takeoff (PTO) circulating pump. CO_2 or regulated air pressure provides more uniform pressure than a PTO pump. If pressure is supplied by an air compressor, a separate steel supply tank with an adjustable regulator may be used to maintain the pressure.

In small-plot tests, sprays may be applied with a CO_2-regulated backpack sprayer equipped with a lightweight aluminum boom having three hollow-cone nozzles per row. Spray delivery is controlled by a hand-operated valve in the boom handle. Small (1.0–2.0-L) plastic beverage bottles or 7.5–11.3-L stainless steel cans may be used as containers for spray mixtures. Agitators are not necessary with portable tanks, provided that the container is shaken before and during application.

Controlled droplet application (CDA) of fungicides may also be used. In a CDA system, electric or hydraulic spinning serrated disks produce spray droplets of uniform size. A low spray pressure (about 138 kPa) is supplied with a PTO pump or with CO_2. CDA units are usually placed 102 cm apart on a spray boom and oriented at an angle of 15° away from the tractor. Since they produce a circular spray pattern approximately 1.8 m in diameter, they should be mounted at least 1.5 m behind the tractor wheels to prevent disruption of the pattern. Spray volumes as low as 10.3 L/ha can be applied with CDA units. A greater volume of 28 L/ha is preferred for some products, because a larger orifice may be necessary to avoid restriction of the flow. All lines and tanks on CDA sprayers must be cleaned thoroughly. Particulate matter from dried spray residues can plug the small orifices and contaminate new spray mixtures.

The timing of the initial application of fungicides to control leaf spot depends on the test objective. If the main objective is to determine efficacy, the first spray might be applied 30–35 days after planting. In some cases, it is applied at the first appearance of symptoms or in accordance with local recommendations. In other cases, the timing of the initial application is determined by meteorological conditions, i.e., temperature and relative humidity. Properly applied, effective fungicides provide control of early and late leaf spot on a 14-day spray schedule with the final application two weeks before the anticipated harvest. Since the major purpose of fungicide tests is to rank the efficacy of fungicides, the spray interval should remain constant for all treatments. The influence of spray intervals of different lengths should be studied in separate tests.

If early leaf spot, late leaf spot, and rust occur at the same location, the efficacy of fungicides for rust control can be evaluated by the method developed by Harrison (5), who used benomyl to control the leaf spot diseases in test plots and then sprayed those plots with fungicides being evaluated for rust control. Some fungicides (such as benomyl, in areas where benomyl-tolerant populations of *Cercospora arachidicola* and *Cercosporidium personatum* are not present) control the leaf spot diseases but not rust. Therefore, the separate and combined effects of these diseases on defoliation and crop loss can be studied, and fungicidal efficacy in controlling them separately and in combination can be tested.

An effective fungicide should always be included in tests. Chlorothalonil applied at 2.48 L/ha in a 500 g/L flowable formulation is an acceptable treatment in leaf spot tests. Benomyl at 0.56 kg/ha plus maneb at 1.34 kg/ha or flowable sulfur at 3.36 kg a.i. per hectare are acceptable where benomyl-tolerant strains of *Cercospora arachidicola* and *Cercosporidium personatum* are absent. Other reference standard fungicides may be useful. Chlorothalonil applied at 2.48 L/ha on a seven-day schedule is an acceptable standard for rust control.

Data collection. Since the leaf spot diseases and rust can result in chlorosis, necrosis, defoliation, and leaf withering, these symptoms should be considered in evaluating fungicidal efficacy. Several assessment methods have been reported (1,14,16), but the method of Subrahmanyam et al (20) is practical and rapid. It consists of a scale from one to nine, with one representing no disease and nine representing severe disease (50–100% defoliation of plants with early or late leaf spot and withering of 50–100% of leaves of plants with rust). Estimates of the percentage of leaf area covered by lesions can be made with diagrams of leaves having symptoms on known percentages of their surfaces (13).

Disease assessment should be performed at least twice during the growing season. The first assessment should be made when disease is at its maximum severity. For runner market types in the southeastern United States, the optimum time of this rating is generally 107–127 days after planting (12). The first rating is good for separation of treatments on the basis of efficacy. The second rating should be made within two weeks of the anticipated harvest and usually gives a better correlation with treatment effect on yield.

If *Cercospora arachidicola* and *Cercosporidium personatum* are present in the area of testing, the relative numbers of each should be noted at the time of the initial fungicide application. Noting the amount of disease caused by each fungus may also be worthwhile in other evaluations. Fungicides differ in their effectiveness against the two organisms. If rust appears late in the season, the final rating of the leaf spot diseases may be less reliable because of the confounding effects of rust.

A separate estimate of the percentage of defoliation should be made (1). This percentage may be estimated visually or determined by counting the total number of leaflets on the main stem and the number of abscised leaflets and computing as follows:

$$d = 100(L - l)/L$$

where L is the total number of leaflets on the main stem, l is the number of abscised leaflets, and d is the percentage of defoliation.

Peanuts should be lifted (dug) with a digger-shaker-inverter and dried in windrows for three days or longer, depending on the weather. Pods can then be removed with a large combine or with a thresher designed for small-plot tests. If a combine is used, some peanuts from outside the test plots should be harvested before the test plots. Some plot-to-plot carry-over of pods occurs with large combines. To completely prevent plot-to-plot mixtures of pods, a small-plot thresher is necessary.

After threshing, pod samples should be dried in a forced-air drier until the kernel moisture is 8–10%. Before the samples are weighed, foreign material (e.g., stems, leaves, stones, soil clods) should be removed. If data on market quality and monetary value are required, a sample should be retained and inspected in accordance with the standards established by the Federal State Inspection Service of the U.S. Department of Agriculture (21).

Notes on the incidence of insects and pests other than the target pests should be made periodically, especially when disease assessments are made. Other diseases, such as southern stem rot (caused by *S. rolfsii*) and Cylindrocladium black rot (caused by *Cylindrocladium crotalariae* (Loos) Bell & Sobers), should also be noted.

Data on the number of disease loci for a given length of row may be needed if disease incidence is high enough to affect yield. Other factors that may have adverse effects on yield should be noted.

If analyses of chemical residue are required, samples can be collected and dried or frozen, depending on the test protocol. Sampling methods and sample sizes should be determined before the field test is begun.

Data analysis. Data should be subjected to an analysis of variance and an appropriate test for ranking treatment means. Percentage data should be transformed before analysis if the range of percentages is greater than 40 points (11). If the F value is significant, a multiple comparison test, such as the Waller-Duncan K ratio t test (22), should be performed. This test is conservative for low values of F and is indicative of homogeneous or nearly homogeneous treatments. This characteristic of the test increases the difference required for concluding that two treatment means are significantly different and thereby decreases the probability of a Type I error. At higher values of F, indicative of heterogeneous treatments, it is as powerful or more powerful than Fisher's protected least significant difference test and Duncan's multiple range test (17). The Waller-Duncan test is available as a statistical package from Statistical Analysis System, SAS Institute, Inc., Box 8000, Cary, North Carolina 27511.

A multiple comparison test is not appropriate for comparing structured quantitative treatments such as different dosages of a fungicide (2). For comparing such treatments, a regression analysis should be performed, response curves drawn, and the equation of best fit determined. If other diseases, such as southern stem rot, are a problem in the test area, a covariant analysis of the data on disease loci may be necessary to eliminate interference from those diseases.

REPORTING RESULTS

Data from efficacy tests and a concise interpretation of results should be reported to the manufacturer of the compound and scientific peers and, if the results are of public concern, to the public. These reports help manufacturers to make decisions about product development and to compile data required by government agencies for product registration. Reports may also be helpful to extension services in preparing recommendations for disease control.

Reports should include the following information: 1) the product identity—code number, trade name, or common name, or all of these, depending on the audience; 2) the product formulation—the type of formulation and the amount of the active ingredient in a unit of the formulated product; 3) the composition of the spray mixture, including fungicide, water, spray adjuvants, and other components; 4) the application rate—the amount of spray mixture per hectare and the amount of formulated product or active ingredient per hectare; 5) application dates for treatments and other products used in the management of plots; 6) dates and amounts of rainfall and supplemental irrigation; 7) wind speed, especially at the time of application; 8) a description of the application method and equipment; 9) information about crop management—e.g., the name of the cultivar, the planting date, harvest dates, fertilization, weed control, insect control, growth regulation, tillage practices; 10) details of the experimental design—the number of replications, the field plot design, row length and spacing, the number of unsprayed rows between plots, and the location of unsprayed plots; 11) data on disease assessment, yield, market quality, and monetary value, usually presented in a table with information on treatments and application rates, disease ratings, and data on defoliation, yield, and market quality, together with pertinent statistical information; and 12) observations of phytotoxicity and other diseases. Any other pertinent observations should be included.

LITERATURE CITED

1. Backman, P. A., Rodríguez-Kábana, R., Hammond, J. M., Clark, E. M., Lyle, J. A., Ivey, H. W., II, and Starling, J. G. 1977. Peanut leafspot research in Alabama. Ala. Agric. Exp. Stn. Res. Bull. 489.
2. Chew, V. 1977. Comparisons among treatment means in an analysis of variance. U.S. Dep. Agric. Bull. ARS/H/6.
3. Cox, F. R., Adams, F., and Tucker, B. B. 1982. Liming, fertilization, and mineral nutrition. Pages 139-163 in: Peanut Science and Technology. H. E. Pattee and C. T. Young, eds. American Peanut Research Education Society, Yoakum, TX.
4. Harrison, A. L. 1967. Some observations on peanut leaf rust and Cercospora leaf spots in Texas. Plant Dis. Rep. 51:687-689.
5. Harrison, A. L. 1973. Control of peanut leaf rust alone or in combination with Cercospora leaf spot. Phytopathology 63:668-673.
6. Henning, R. J., Allison, A. H., and Tripp, L. D. 1982. Cultural practices. Pages 123-138 in: Peanut Science and Technology. H. E. Pattee and C. T. Young, eds. American Peanut Research Education Society, Yoakum, TX.
7. Henning, R. J., McGill, J. F., Samples, L. E., Swann, C. W., Thompson, S. S., and Womack, H. 1979. Growing peanuts in Georgia. A package approach. Univ. Ga. Bull. 640. 46 pp.
8. Jackson, C. R., and Bell, D. K. 1969. Diseases of peanut (groundnut) caused by fungi. Ga. Agric. Exp. Stn. Res. Bull. 56. 137 pp.
9. Jackson, L. F. 1981. Distribution and severity of peanut leafspot in Florida. Phytopathology 71:324-328.
10. Kucharek, T. A., Riabob, J., and Cromwell, R. 1981. The effect of spray volume and spray pressure on peanut leafspot control. Pages 64-74 in: Proc. Fla. Conf. Pestic. Appl. Technol. Institute of Food and Agricultural Sciences, University of Florida, Gainesville.
11. Little, R. M., and Hills, F. J. 1978. Agricultural Experimentation—Design and Analysis. John Wiley and Sons, New York. 350 pp.
12. Monasterios, T. de la Torre. 1980. Genetic resistance to Cercospora leafspot disease in peanut (*Arachis hypogaea* L.). Ph.D. dissertation, University of Florida, Gainesville. 102 pp.
13. Neundorfer, R. 1983. Evaluation of *Arachis hypogaea* genotypes for tolerance to *Cercosporidium personatum* under varying disease management strategies. M.S. thesis, University of Georgia, Athens. 38 pp.
14. Pedrosa, A. M. 1975. Evaluation of several methods of taking disease severity readings for Cercospora leafspot of peanuts. Ph.D. dissertation, Oklahoma State University of Agriculture and Applied Science, Stillwater. 103 pp.
15. Phipps, P. M., and Powell, N. L. 1983. Criteria for effective utilization of peanut leafspot advisories in Virginia. (Abstr.) Am. Peanut Res. Educ. Soc. Proc. 15:99.
16. Plaut, J. L., and Berger, R. D. 1980. Development of *Cercosporidium personatum* in three peanut canopy layers. Peanut Sci. 7:46-49.
17. Smith, C. W. 1977. Bayes least significant difference: A review and comparison. Agron. J. 70:123-127.

18. Smith, D. H., and Littrell, R. H. 1980. Management of peanut foliar diseases with fungicides. Plant Dis. 64:356-361.
19. Snedecor, G. W., and Cochran, W. G. 1980. Statistical Methods, 7th ed. Iowa State University Press, Ames. 507 pp.
20. Subrahmanyam, P., McDonald, D., Gibbons, R. W., Nigam, S. N., and Nevill, D. J. 1982. Resistance to rust and late leafspot diseases in some genotypes of *Arachis hypogaea*. Peanut Sci. 9:6-10.
21. USDA. 1963. Inspection Instructions for Farmers Stock Peanuts, rev. ed. U.S. Dep. Agric. Agric. Mark. Serv., Washington, DC.
22. Waller, R. A., and Duncan, D. B. 1969. A Bayes rule for the symmetric multiple comparison problem. J. Am. Stat. Assoc. 64:1484-1503.

Evaluating Chemicals to Control Soilborne Diseases of Peanut

P. M. PHIPPS, Tidewater Research Center, Virginia Polytechnic Institute and State University, Suffolk 23437; and D. M. PORTER, Agricultural Research Service, U.S. Department of Agriculture, Tidewater Research Center, Suffolk, VA 23437

Soilborne diseases of peanut (*Arachis hypogaea* L.) can damage all parts of the plant and reduce the quality as well as the quantity of pods, seed, and forage. The subterranean nature of peanut pod and seed development after flowering and soil penetration by pegs makes reproductive as well as vegetative tissues highly vulnerable to attack by soilborne pathogens. Thorough understanding of the pathogen and factors triggering disease epidemics is essential to experimentation in chemical control. This information, coupled with knowledge of specific properties of test chemicals, facilitates the design of critical evaluations in disease control. Valuable information on chemicals includes the spectrum of activity and mode of action against microorganisms, previously tested rates and methods of application, formulations available, chemical and physical properties, phytotoxicity or other effects on plants, and mammalian toxicity as well as hazards to people and the environment.

Soilborne pathogens of peanut may be grouped on the basis of location of primary infection courts, either near the soil surface (e.g., *Sclerotinia minor* and *Sclerotium rolfsii*) or below the soil surface, in the rooting as well as fruiting zone of peanuts (*Cylindrocladium crotalariae, Pythium myriotylum,* and *Verticillium albo-atrum*). This chapter focuses on field strategies for evaluating chemicals to control pathogens representative of each group. Sclerotinia blight, caused by *S. minor* Jagger, and Cylindrocladium black rot, caused by *C. crotalariae* (Loos) Bell & Sobers, were selected for detailed discussion.

GENERAL PREPARATIONS

Site and Cultivar Selection

Field sites with a history of the specific disease problem should be selected. Aerial infrared imagery of previous crops can be extremely useful in determining the location and proper orientation of plots in heavily infested areas (14,16). Preplant soil assays to assess inoculum density and distribution (9,15) are recommended. Soil type, fertility conditions, and level of nematode infestation should be determined. Both an agronomist and a nematologist should be consulted on inputs necessary to prepare land for good crop management.

A commercially accepted cultivar should be selected and planted according to conventional practices. If space and resources permit, planting a resistant as well as a susceptible cultivar in a split-plot design is desirable. Data from such evaluations often provide valuable insight into the mechanism of disease control and allow detection of nontarget effects of chemicals.

Experimental Design

Field plots commonly consist of two treated rows 7.6–12.2 m long and spaced 0.9 m apart. A minimum of four replications in a randomized complete block design is recommended. Adjacent plots should be separated by two untreated buffer rows and ends of plots by alleyways of fallow soil at least 1.5 m wide to maintain separation of plots during harvest. Larger alleyways are necessary if farm tractors and sprayers are used to apply treatments. In Virginia, investigators commonly use 7.6-m fallow alleyways to separate commercial peanuts or other crops from the ends of tests. These alleyways provide adequate space for turning tractors and other implements without risk of damage to crops.

Plot Maintenance

Peanuts require intensive management to produce the yields normally attained by commercial producers. Agronomists and extension specialists with expertise in weed, insect, and disease control should be consulted on standard production practices for plot maintenance. Plots require routine scouting at weekly intervals for early detection and treatment of unexpected outbreaks of weed, insect, or disease problems that are not a subject of the study. All applications of materials for crop management should be recorded and should be made in a manner simulating commercial practices. When tractors are used in crop management, the degree of plant injury from tires or other implement parts should always be the same in each plot.

After digging, peanuts should remain in windrows at least 3–5 days to dry the vines and reduce seed moisture below 30% (w/w). Commercial peanut combines equipped with a platform for collecting and bagging peanuts from individual plots allow a rapid, efficient harvest. Immediately following harvest, peanuts should be placed in dryers to lower seed moisture to levels near 8 or 10%. Peanuts at this moisture level should be weighed to determine yield, and 1-kg samples from each plot should be saved for grading. Moisture should be determined on samples from several plots and the data used to express yield (kg/ha) on the basis of 8% moisture. The quality and value of peanuts from each plot should be determined in accordance with current federal and state inspection service methods.

CYLINDROCLADIUM BLACK ROT

Pathogen

C. crotalariae, the causal organism of Cylindrocladium black rot of peanut (1), overwinters in soil as microsclerotia. Microsclerotia in naturally infested fields are distributed neither randomly nor uniformly but are clumped or clustered. Clumping of inoculum accounts for the occurrence of groups of diseased plants in foci within rot-infested fields.

The primary infection court for *C. crotalariae* on peanut is believed to be near the region of the root tips, but all below-ground parts of plants may become colonized and severely decayed. Soil temperatures near 25°C and moisture near field capacity are most conducive for infection and rot of peanut roots. In dry seasons, irrigation may enhance disease development, particularly in well-drained soils with sandy texture.

The longevity of microsclerotia in soil largely depends on soil temperatures during winter months. Cold winters resulting in freezing of soil water in the plow layer or sustained periods of cold at or below 5°C markedly reduce microsclerotia populations (8). Two- or even 3-year cycles of nonhost crops (corn, cotton, tobacco, small grains) and tillage practices have little impact on the survival of microsclerotia and disease severity in subsequent peanut crops. Well-defined procedures for determining microsclerotia populations in soil have been developed. In Virginia, soil samples are processed by elutriation (9) and sieve residues are assayed on a sucrose-QT medium (4).

Perithecia of *Calonectria crotalariae* (the sexual stage of *C. crotalariae*) commonly develop on diseased plant tissues at or near the soil surface. Because ascospores are sensitive to desiccation, their epidemiologic importance is limited to short-distance spread of the disease within the field. The characteristic shape and orange-to-red coloration of perithecia make these fruiting structures a useful diagnostic sign for Cylindrocladium black rot.

Soil Treatment

Fumigant-type chemicals containing chloropicrin, metam-sodium, and methyl bromide have shown the greatest promise for control of Cylindrocladium black rot (6). The need for more testing is sustained by the release of new peanut cultivars as well as the potential for development of improved fumigants. Tractor-mounted coulter applicators (Ferguson Mfg., Suffolk, VA) or conventional chisel shanks are required to apply chemicals at depths of 15–20 cm below the soil surface. The flow of liquid fumigants to applicators can be controlled by commercial gravity-flow regulators or by a commercial pump and transfer system (Pearman Eng., Chula, GA). Row treatments may afford a reduced fumigant rate but require matching of applicator and planter spacing. A disk-bedder or other implement to produce raised beds of soil directly over centers of chemical deposition facilitates subsequent planting of seed directly above the treated area.

Phytotoxic effects of fumigants can generally be avoided in light to medium texture soils by planting at least 14 days after application of soil treatments. A longer waiting period may be necessary if soil is heavy in texture, contains high levels of organic matter, or remains excessively cold (below 15°C) and/or wet after treatment. All tillage operations as well as preplant incorporation of herbicides should be completed before soil treatment. Cultural practices that mix treated soil in beds with untreated soil should be avoided.

In addition to a nonfumigated check treatment, including some nematicide treatments and/or combinations of nematicides and fumigants may be useful. Northern root-knot nematode (*Meloidogyne hapla*) and ring nematode (*Macroposthonia ornata*) have been shown to increase the severity of Cylindrocladium black rot in soils with moderate to heavy infestations (3). Preplant, midseason, and preharvest soil samples from individual plots should be assayed to assess the significance of populations of plant-parasitic nematodes in the control of Cylindrocladium black rot.

Data Collection

Stand counts and/or visible symptoms of phytotoxicity should be recorded during the first 4 weeks after planting. Numbers of symptomatic and/or dead plants in plots should be determined monthly after planting, and a final count should be made just before digging plants in preparation for harvest. Records of the incidence and severity of other diseases in plots should also be maintained. Root rot severity indexes and pod rot indexes (6) are made most easily immediately after inverting peanut vines with a commercial peanut digger-inverter. Infection of roots by *C. crotalariae* can be determined by collecting 25 taproot samples at random from each plot, treating a biopsy sample of each root in 0.5% NaOCl for 1 min, and placing the sample on a selective isolation medium (8).

Yield and grade data are necessary to determine the value of peanuts harvested from each plot. Where plots 12.2 m long are used, yield differences of 1 kg or more are usually required to show significant differences in treatments by Duncan's multiple range test. The germinability of seed may be assessed to evaluate treatment effects on seed quality.

SCLEROTINIA BLIGHT

Pathogen

S. minor, the causal organism of Sclerotinia blight of peanut, was first found in the United States in 1971 (12). This disease is very severe during some growing seasons. Yield losses exceeding 1,500 kg/ha are not uncommon in some heavily infested fields (14).

The sudden wilting of a lateral branch of the peanut plant is usually the first symptom of disease. Foliage of infested branches becomes chlorotic, turns brown, withers, and dies. These symptoms result in the blight characteristic of this disease. Once the infection process begins, white, fluffy mycelium develops on the diseased tissues. Shortly thereafter, numerous black, irregularly shaped sclerotia (0.5–3 mm in diameter) are produced on all infested plant parts. Cool temperatures (21°C), moist soils, and high relative humidities (95–100%) generally favor infection.

Sclerotia of *S. minor* overwinter in the soil and serve as the primary source of inoculum. Populations of sclerotia in fields with severe symptoms of Sclerotinia blight often exceed 25 sclerotia per 100 g of soil (15). Although sclerotia in soil decrease in viability over time, enough remain viable to cause severe disease in peanuts after 2- or 3-year rotations with nonhost crops. In fact, sclerotia are often recovered from soil not planted to a susceptible crop for 5 years. Sclerotia germinate myceliogenically.

All plant tissues including branches, pegs, pods, and leaflets near or in direct contact with the soil are invaded rapidly. Senescing tissues and mechanically injured tissues are especially prone to colonization (13). Infrequently, infections have been observed in the foliage canopy. Such infections, thought to be initiated by ascospores, occur only in fields where *S. sclerotiorum* is present.

Production practices often enhance the severity of Sclerotinia blight. Growers are cautioned against the use of overhead irrigation in fields with a history of Sclerotinia blight and are urged to minimize foliage injury that might predispose plants to invasion (13). Fungicides such as chlorothalonil and captafol, normally used in peanut production for foliar disease control, enhance disease severity and losses from Sclerotinia blight (10).

Application of Treatments

The timing, frequency, and mode of chemical application should be selected after review of information on the mode of action and phytotoxicity of chemicals. Applications of soil fumigants (e.g., chloropicrin, metam-sodium) in the manner described for control of Cylindrocladium black rot have not provided control of Sclerotinia blight. Metam-sodium at 187 L/ha has suppressed Sclerotinia blight when applied in 2.5 cm of overhead irrigation. Metam-sodium must be applied at least 12 days before planting because of its phytotoxic properties. Plowing and other tillage practices that may alter the vertical distribution of sclerotial inoculum in soil should be done before the application of soil fumigants.

An overhead plot irrigator with a chemical injection pump has been developed and used in Virginia (17) to evaluate the utility of irrigation water as a mode for applying chemicals in crop management. This irrigator can be used to evaluate preplant as well as mid- to late-season applications of chemicals. Check valves to prevent contamination of water supplies are required at the point of chemical injection into the irrigation line and at a point between the irrigation pump and the source of water. Safety features for emergency cutoff of the injector pump should also be used in accordance with specifications of the manufacturer.

Fungicides are commonly formulated for application in water with ground sprayers. Applications of sprays by a commercial tractor-mount sprayer can be simulated with a CO_2 pressure-regulated backpack sprayer. These sprayers are available from commercial vendors or may be assembled by the researcher. Initial evaluations of a fungicide should be made by application with nozzles (8008LP, 8010LP, TK7.5) that deliver a high volume (≥327 L/ha) in large droplets that fall through the canopy of leaves and reach the soil surface. A single nozzle should be centered over each row, and height should be adjusted to completely cover rows with some overlap. Nozzles that deliver a fine mist and low spray volume (112–140 L/ha) should be used to evaluate an alternative strategy for applying sprays, especially when systemic activity by a fungicide is suspected. Three nozzles per row equipped with D2-13 or D3-23 (disk-core combination) tips are used routinely for low-volume applications of fungicides to control leaf spot diseases of peanuts. Fungicides with proven efficacy for control of Sclerotinia blight should also be evaluated in tank mixtures with fungicides commonly used for leaf spot control. Results of these trials may enable detection of compatibility problems and/or synergism as well as antagonism.

The rate and timing of fungicide applications are equally critical to achieving disease control. Laboratory tests using soil plates provide a rapid, efficient way to screen fungicides for activity against *S. minor* (5). Rates for evaluation in field trials can be selected by determining the ED_{50} value of a fungicide and making comparisons to the published values for reference standard fungicides, such as vinclozolin and dicloran (2). Preventive applications should be initiated before any evidence of disease, or generally about the time vines in adjacent rows are within 5 cm of overlapping (about July 15 in Virginia). A total of three applications on a 28-day schedule are commonly used, with vinclozolin applied at 0.84 kg/ha as a reference standard (7). Iprodione at 1.12 kg/ha should also be included as a reference standard treatment. Both fungicides have been used in commercial peanut fields in Virginia and North Carolina for control of Sclerotinia blight of peanut. Additional evaluations of materials should include application at the time of initial detection of active mycelium and infection of vines by *S. minor*. Proper timing of these treatments requires weekly scouting for disease in plots but may result in good disease control with only two applications of fungicide in some years.

Granular formulations of fungicides may or may not provide disease control that is superior to that provided by wettable or flowable formulations. Terraclor 10G has provided good suppression of Sclerotinia blight, whereas sprays of Terraclor 75W were ineffective (7). Granular formulations of dicloran and iprodione have performed about as well as wettable formulations. Fungicide granules should be applied in a 30- to 36-cm band over the row as a preventive treatment, followed by another application 4–6 weeks later. Commercial applicators and row banders for making these applications are available in small hand-tow units and tractor-mounted units. Unlike spray applications, granular formulations of fungicides can be delivered through the foliage canopy and onto the soil surface with a minimum of interference from the foliar canopy. Activation and dispersion of fungicide in granules, however, may require rainfall or irrigation. For this reason, granular formulations of fungicides should be evaluated primarily as preventive treatments, whereas wettable formulations may prove efficacious applied either as preventive or prophylactic treatments.

Data Collection

Stand counts and/or phytotoxicity, disease incidence and severity, yield, and quality should be recorded as described for Cylindrocladium black rot. Disease incidence should be assessed by counting infection centers (7) at least monthly. An infection center is a point of active growth by *S. minor*, characterized by symptoms and signs of disease. To reduce the time required for assessing disease incidence as well as confusion caused by overlap of infection centers, each count should encompass 15 cm of row on either side of an infection center. Disease severity indexes can be developed by rating plants in 40 selected sites on a scale from 1 (no disease) to 5 (death of plant) as described by Porter (11).

Pod rot and pod mycoflora can be determined by selecting plants at random from each plot and removing the pods by hand. The percentage of rotted pods can be determined following washing. Pieces of shell (about 1 cm^2) and seed should be surface-sterilized with 0.5% NaOCl and plated on potato-dextrose agar for mycofloral assay (11). Fungi associated with the shell and seed are identified after a 10-day incubation at 25° C. Soil samples for sclerotial assay should be taken from the center of each row and to a depth of 2.5 cm.

LITERATURE CITED

1. Bell, D. K., and Sobers, E. K. 1966. A peg, pod, and root necrosis of peanuts caused by a species of *Calonectria*. Phytopathology 56:1361-1364.
2. Brenneman, T. B., Phipps, P. M., and Stipes, R. J. 1983. Sensitivity of *Sclerotinia minor* from peanut to dicloran, iprodione and vinclozolin. (Abstr.) Phytopathology 73:964.
3. Diomande, M., and Beute, M. K. 1981. Effects of *Meloidogyne hapla* and *Macroposthonia ornata* on Cylindrocladium black rot of peanut. Phytopathology 71:491-496.
4. Griffin, G. J. 1977. Improved selective medium for isolating *Cylindrocladium crotalariae* microsclerotia from naturally infested soils. Can. J. Microbiol. 23:680-683.
5. Phipps, P. M. 1980. Soil plate and field evaluation of fungicides for control of Sclerotinia blight of peanut. (Abstr.) Phytopathology 70:692.
6. Phipps, P. M. 1982. Efficacy of soil fumigants in control of Cylindrocladium black rot (CBR) of peanut in Virginia. Fungic. Nematic. Tests 37:96.
7. Phipps, P. M. 1982. Control of Sclerotinia blight of peanut in Southampton County and the City of Suffolk, Virginia. Fungic. Nematic. Tests 37:97-99.
8. Phipps, P. M., and Beute, M. K. 1979. Population dynamics of *Cylindrocladium crotalariae* microsclerotia in naturally-infested soil. Phytopathology 69:240-243.
9. Phipps, P. M., Beute, M. K., and Barker, K. R. 1976. An elutriation method for quantitative isolation of *Cylindrocladium crotalariae* microsclerotia from peanut field soil. Phytopathology 66:1255-1259.
10. Porter, D. M. 1980. Increased severity of Sclerotinia blight in peanuts treated with captafol and chlorothalonil. Plant Dis. 64:394-395.
11. Porter, D. M. 1980. Control of Sclerotinia blight of peanut with procymidone. Plant Dis. 64:865-867.
12. Porter, D. M., and Beute, M. K. 1974. Sclerotinia blight of peanuts. Phytopathology 64:263-264.
13. Porter, D. M., and Powell, N. L. 1978. Sclerotinia blight development in peanut vines injured by tractor tires. Peanut Sci. 5:87-90.
14. Porter, D. M., Powell, N. L., and Cobb, P. R. 1977. Severity of Sclerotinia blight of peanuts as detected by infrared aerial photography. Peanut Sci. 4:75-77.
15. Porter, D. M., and Steele, J. L. 1983. Quantitative assay by elutriation of peanut field soil for sclerotia of *Sclerotinia minor*. Phytopathology 73:636-640.
16. Powell, N. L., Garren, K. H., Griffin, G. J., and Porter, D. M. 1976. Estimating Cylindrocladium black rot losses in peanut fields from aerial infrared imagery. Plant Dis. Rep. 60:1003-1007.
17. Wright, F. S. 1985. Irrigation equipment for plot research. Trans. Am. Soc. Agric. Eng. 28:1741-1743.

Evaluating Foliar Fungicides Applied Through Irrigation Systems for Control of Peanut Diseases

PAUL A. BACKMAN, Department of Plant Pathology, Alabama Agricultural Experiment Station, Auburn University, Auburn 36849

Because irrigation equipment represents a large capital investment for a farmer, the more services that it can perform, the more quickly it will return a profit. One of the major areas of research to expand irrigation system utility has been chemigation, defined as the application of chemicals (including both fertilizers and pesticides) through an irrigation system.

The application of fungicides through an irrigation system (fungigation) has been evaluated extensively on vegetables (5,8) and some field and row crops (2,7); positive results have been demonstrated for both foliage and soilborne pathogens. The techniques reported here were primarily developed in work with peanuts (*Arachis hypogaea* L.) for control of leaf spot diseases caused by *Cercospora arachidicola* and *Cercosporidium personatum* and southern stem rot caused by *Sclerotium rolfsii*, but they could be used for foliar or soilborne diseases of many crops.

Peanuts traditionally have been grown in the southern United States as a high-value crop. Successful managers typically use seven to nine fungicide applications spaced regularly throughout the growing season to control the peanut leaf spot diseases. To stabilize yields by preventing drought, many farm operators have installed irrigation systems costing many thousands of dollars. The ability to use these systems to apply fungicides would save seven to nine trips over the field with ground sprayers, thereby reducing soil compaction, plant damage, and per-acre application costs. The need for efficient leaf spot control is especially important for farmers who irrigate, because irrigation increases the number and length of infection periods, resulting in increased disease pressure (6). In addition, farmers who have irrigation systems do not rotate their crops as consistently as other farmers, preferring to continue planting peanuts on their most productive (irrigated) land, further increasing disease pressure.

METHODS

Test Site

Tests should be conducted in an area with a history of peanut production, preferably where peanuts were grown the previous crop year or the year before. The land should be uniformly productive, and the area should be large enough to contain the necessary treatments. Soil should be sampled for nematodes and fertilization requirements and treated accordingly. The field should be equipped with an irrigation system capable of applying water at a uniform rate over the entire test area. The uniformity of the system may be tested simply by placing open cans along the length of the system and measuring the water in each can after the irrigation unit has passed over the entire area. Areas of nonuniformity should be corrected or omitted from the test area. Nonuniformities that are not corrected inevitably increase variation within plots and between replications and can compromise the usability of test results.

In the Southeast, the most common irrigation systems are the pivot system and the traveling gun. However, the latter typically is not uniform enough for test purposes, particularly because it requires a very large area to start up and stop and requires overlapping passes to deliver water uniformly. Many research workers have used pivot systems with success for applying pesticides, and many peanut farmers with pivot systems have adopted fungigation. In the peanut production areas of the Southwest, side-roll, lateral-move, and hand-move (portable) systems have also been used successfully, both in research and in commercial production.

When a uniform field and irrigation unit have been located, the appropriate safety devices should be installed and the appropriate metering equipment purchased. Check valves to prevent the backflow of pesticides into the water source must be installed, or the risk of contaminating either the underground aquifer or the farm pond will be extremely high (3). Several states have laws defining the safety precautions necessary before irrigation systems can be used to apply pesticides.

Equipment Calibration

Metering devices for delivering a controlled quantity of fungicide into the irrigation system should be purchased with an understanding of the irrigation system to which they will be coupled (4). Constantly moving systems require a metering pump that constantly applies fungicides at a uniform rate. These pumps can be either a positive-displacement (piston) type or a diaphragm type. Both types must be calibrated against the head pressure at the point of injection into the system, using the actual solution to be injected, because both pressure and viscosity can affect flow rate. The pump can be calibrated by using a standard pressure relief valve adjusted to duplicate the pressure at the injection point. Flow rate per unit time can then be accurately determined for each pump setting.

For hand-move or wheel-roll systems, a simpler passive injection system can be used. Enough fungicide to treat the "set" is placed in a holding tank, and during the latter portion of the application period, the tank is slowly siphoned by the use of a venturi or by the application of pressure. The tank is refilled for each set, and the entire contents of the tank are delivered at each location. For both passive and mechanical injection systems, the mixing in the line is sufficient, assuming that there are several turns in the pipeline before the first sprinkler head. The researcher should know how much time is required after injection of the pesticide begins before the pesticide reaches the final sprinkler head at the end of the system. This can be determined by injecting a dye (such as rhodamine) and timing the color change.

Final calibration for moving systems requires determining the area traversed per unit time. Typically, the system should be moving as fast as possible (2.2–4.5 mm/ha of applied water [0.1–0.2 in./acre]) to maximize retention of fungicide on the leaf surface. If a soil fungicide is to be applied, a higher water rate may improve performance.

Once the area covered per minute and the dilution level of the fungicide in the holding tank are known, it is a simple task to adjust the flow rate from the metering pump to deliver a known amount of active ingredient per unit area. Recent research by entomologists (9) has indicated that vegetable or petroleum oils used as adjuvants in the injection mixture improve the retention of insecticides on leaves. Although this has not been reported for fungicides, the methodology deserves further evaluation. Care should be taken to provide sufficient agitation in the holding tank to prevent pesticide settling.

Plot Establishment

When a constantly moving irrigation system is used for research on fungigation, the investigator should be aware that each change in product or rate will be accompanied by an area of nonuniformity equivalent to the wetting diameter (i.e., the area wetted at one moment in time) of the system being used. In addition, the pattern will be displaced commensurate with the direction and velocity of the wind. After allowances are made for wind displacement and overlap, an area of treatment uniformity should be defined and used for all experimental evaluations (Fig. 1). For a pivot system, this uniform treatment area is typically a pie-shaped area that disappears well before the pivot point is reached. For a lateral-move system, the area of uniformity of treatment is rectangular. Uniform treatment areas are more easily defined for stationary systems, because all that must be done is to stay away from the areas of wind displacement, partial treatment, and overlap.

Appropriate Controls

Typically, the appropriate controls for fungicide evaluations on peanuts, or any other crop grown under irrigation, are as follows: 1) a ground-sprayed treatment, applied several hours before the fungigation and over which the irrigation unit applies the same amount of water as to plots receiving a fungigation treatment; or 2) an aerial application of fungicide to an irrigated plot area, applied as in 1) or soon after irrigation; 3) an untreated control of irrigated peanuts; and occasionally, 4) applications of the same product delivered either by air or from the ground but to nonirrigated plots, at the same time as the fungigation.

Replications

Movable irrigation systems use large areas of land before uniformity is obtained. Typically, only six to eight treatments can be located under a 16.2-ha (40-acre) half-pivot (Fig. 1). Consequently, the number of

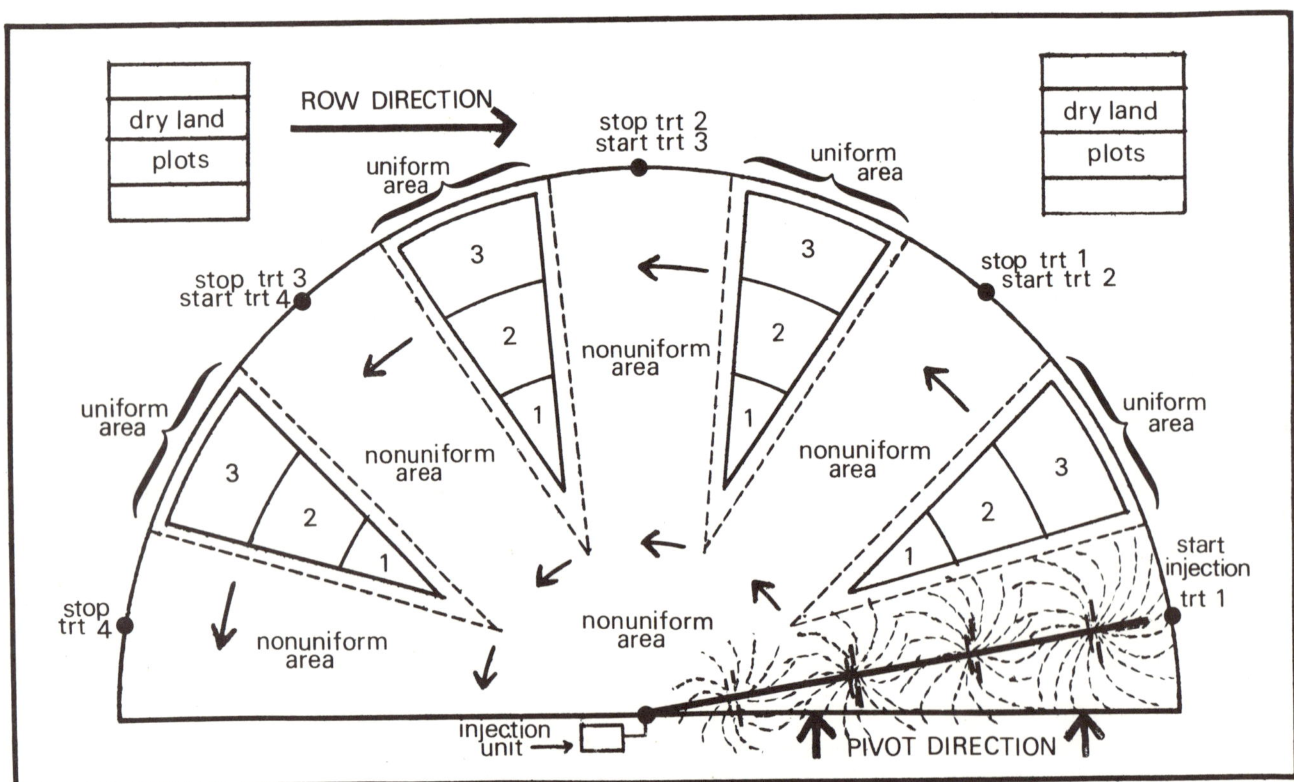

Fig. 1. Areas of treatment uniformity and nonuniformity in a research test evaluating the performance of fungicides applied through a pivot-type sprinkler irrigation system to control foliar diseases on peanuts. Areas labeled 1, 2, and 3 are for sampling and yield determinations.

treatments is limited to only two or three, or replications must be developed by using multiple locations sometimes covering a period of more than 1 year.

Scheduling of Applications

When comparing different fungicides applied through the irrigation system or different delivery systems for control of foliar diseases, the researcher should be sure that all applications are made on the same day and preferably close together in time. This precaution minimizes differences in environmental conditions and/or timing of infection periods that could otherwise confound the study. Typically, fungicides are applied to peanuts at approximately 14-day intervals regardless of the delivery system. However, timing should be adjusted for local environment and disease pressure.

Sampling

After control practices have been implemented for a month or more and disease symptoms are visible, terminal stems can be removed for determinations of leaf spot severity. Stems selected should be apically dominant and nonbearing, and they should be cut from the plant as close to the ground as possible. For my studies, triplicate samples were removed, each consisting of 10 stems removed in a zigzag pattern throughout a defined area of treatment uniformity. Each sample was removed from a different region of the same plot (Fig. 1), which allowed me to determine variability of control efficiency within subsections of these rather large plots. Ground-sprayed or aerially treated plots were sampled in the same way, as were replications when available.

Evaluation

Aerial infrared photography can easily detect and give a visual record on film of differences in peanut plant health caused by foliar or soilborne diseases. More precise results can be obtained by determining percentage of defoliation and/or percentage of infection on the samples removed by the previous procedures (1).

Yield and product quality are the variables that affect the producer. Yield should be determined at two or three locations in each plot if the test has not been replicated, and pod samples should be removed for grading by federal and state inspection service standards. Data should be reported as net return per hectare (acre).

DATA INTERPRETATION

The objective of fungigation is to adequately control a foliage disease, under irrigated conditions, while reducing overall production costs and increasing profits. The procedure described here is really an ultrahigh-volume sprayer compared to a standard low-volume ground or aerial sprayer. Typically, only 10% of the applied fungicide may stay on the plant parts of fungigated peanuts, whereas 90% may stay on the leaves of plants sprayed by more conventional means. Clearly, improvements in disease control and yield will not be achieved through deposition and must come through other means. These other means may include improved coverage (distribution) from drenching the plant in the fungicide solution; control of primary inoculum contained in plant debris, typically sheltered by the growing plant; effects on other diseases or pests not usually affected by other application methods, such as soilborne pathogens that are susceptible to the 90% of the fungicide that ends up in the soil; and improved plant growth because of reduced traffic from spray operations, resulting in less soil compaction and less damage to lateral stems.

Data obtained on peanuts with the fungicides chlorothalonil and fentin hydroxide reveal two interesting points. First, if a fungicide is effective in controlling disease when applied by conventional systems, it may or may not be effective when applied through an irrigation system. Second, yields of fungigated peanuts typically exceed yields of conventionally treated peanuts with similar levels of peanut leaf spot control. These points indicate that although fungicides applied through the irrigation system can often be effective, information on each active ingredient and formulation must be obtained before the practice is recommended to farmers.

LITERATURE CITED

1. Backman, P. A., Munger, G. D., and Marks, A. F. 1976. The effects of particle size and distribution on performance of the fungicide chlorothalonil. Phytopathology 66:1242–1245.
2. Backman, P. A., Rochester, E. W., and Crawford, M. A. 1981. Application of fungicides to peanuts through the irrigation system. Highlights Agric. Res. Auburn Univ. Ala. Agric. Exp. Stn. 28(2):8.
3. Fischbach, P. E. 1973. Antipollution devices for applying chemicals through the irrigation system. G73-43 Coop. Ext. Serv. Univ. Nebr., Lincoln.
4. Harrison, D. S. 1982. Selection, operation, calibration and maintenance of chemigation equipment. Proc. Natl. Symp. Chemigation Univ. Ga. 2:80–87.
5. Potter, H. S. 1981. Fungigation on vegetables. Proc. Natl. Symp. Chemigation Univ. Ga. 1:74–81.
6. Rochester, E. W., Backman, P. A., Curtis, L. M., Starling, J. G., and Ivey, H. 1984. Irrigation schedules for peanut production. Auburn Univ. Ala. Agric. Exp. Stn. Bull. 556. 35 pp.
7. Sumner, D. R. 1981. Application of foliar and soil fungicides and a soil fumigant through overhead irrigation water. Proc. Natl. Symp. Chemigation Univ. Ga. 1:82–88.
8. Sumner, D. R., Phatak, S. C., Smittle, D., Johnson, A. W., and Glaze, N. C. 1981. Control of cucumber foliar diseases, fruit rot, and nematodes by chemicals applied through overhead sprinkler irrigation. Plant Dis. 65:401–404.
9. Young, J. R. 1981. Chemigation: Insecticides applied in the irrigation water for control of the corn earworm and fall armyworm in sweet and field corn. Proc. Natl. Symp. Chemigation Univ. Ga. 1:56–64.

Field Evaluation of Fungicides for Control of Foliar Diseases on Small Grains

JAMES A. FRANK, Agricultural Research Service, U.S. Department of Agriculture, and HERBERT COLE, JR., Department of Plant Pathology, The Pennsylvania State University, University Park 16802

Foliar fungicides currently are receiving attention as a feasible means of disease control on cereals in the United States, in part because of the development of systemic foliar fungicides that provide significant disease protection for several weeks with only one application. The interest in high-technology wheat production systems, which were first developed in Europe, has also prompted the use of fungicides in the United States. Fungicides have been tested on small grains in field trials in Pennsylvania since 1978. The purpose of these trials was to select fungicides that could be used as tools in crop loss studies and also in integrated pest management systems.

TEST PLANTS AND PLOT DESIGNS

The cultivar and management practices used have a significant influence on the success of the test. The cultivar should be sufficiently susceptible to make control, or untreated, plots visually obvious when compared to fungicide-protected plots. Plots that are adequately protected with a fungicide generally have a greater proportion of green leaf area in later stages of growth, which contributes significantly to increased kernel weights. This is essential if treatments are to be evaluated for possible yield increase as well as disease reduction.

The cultivar should also be selected in relation to its reaction to a range of foliar diseases. For example, if the purpose of the experiment is to evaluate fungicides for their effectiveness in controlling powdery mildew, then a cultivar that is resistant to other diseases but susceptible to powdery mildew would be ideal. In this case, the differences in yield between treated and untreated plots generally could be attributed to powdery mildew. However, if the purpose of the experiment is to evaluate the overall effectiveness of the fungicide against various foliar pathogens, then a cultivar that is susceptible to two or more diseases would be preferable.

In our experiments, we have used two different plot sizes. Our small plots, which are used to evaluate chemicals in their preliminary developmental stages, measure 0.9 × 3.6 m, with five rows spaced 17.8 cm apart. These plots are planted with a cone-type planter mounted on a crawler tractor, at a seeding rate of 167 kg/ha. Every other drill strip is planted with a small grain different from the test grain, to provide a buffer zone between plots. The plots are separated within the drill strips by alleyways. These alleyways can be mowed or sprayed with glyphosate. Large plots measure 2.4 × 9.2 m, with rows spaced 17.8 cm apart, and are planted with a conventional grain drill. As in small plots, every other row is planted with a nontest grain species and alleyways are established. The nontest strips must be wide enough for a tractor to travel through during spray applications.

The spacing between the plots is generally adequate to minimize interplot interference with regard to Septoria leaf and glume blotch (caused by *Leptosphaeria nodorum*). However, based on published data, this is not true for powdery mildew of wheat or barley (caused by *Erysiphe graminis* f. sp. *tritici* and *E. graminis* f. sp. *hordei*, respectively) or leaf rust of wheat or barley (caused by *Puccinia recondita* f. sp. *tritici* and *P. hordei*, respectively). In general, any interplot interference in the experimental design described here would tend to reduce the effectiveness of the chemicals, because the pathogens would be present in the control plots and could reinfect the treated plots. This interference would make the interpretation of the results more conservative. However, because most of the new fungicides for control of powdery mildew and leaf rust are systemic in nature, they provide extended periods of disease protection, and the effect of interplot interference should be less significant than when nonsystemic pesticides are evaluated.

In the case of fall-planted grains, plots are fertilized with 67.2 kg of 5-10-10 per hectare before planting. Plots are planted approximately 1 week earlier than the recommended planting date to ensure fall infection by obligate parasites, such as the powdery mildew and leaf rust pathogens. Plots are top-dressed with 67.2 kg of NH_4NO_3 per hectare as early in the spring as weather conditions permit.

Spring grains are planted as soon as the soil can be worked. The NH_4NO_3 fertilizer is applied as a topdressing shortly after plants reach the three-leaf stage, at a rate of 67.2 kg/ha for wheat and barley and 56 kg/ha for spring oats.

Fungicides generally are not applied before growth stage 5 (GS-5, Large scale) (2) or after GS-10. Small plots are sprayed with a hand-held CO_2-powered boom sprayer with a water dilution rate of 300 L/ha. The boom has two hollow-cone nozzles spaced 45.7 cm apart.

Large plots are sprayed with a tractor-mounted unit powered by CO_2 or N. The 2.4-m boom has hollow-cone

Contribution 1473, Department of Plant Pathology, The Pennsylvania State Agricultural Experiment Station. Authorized for publication as Journal Series Paper 7025.

nozzles spaced 45.7 cm apart. Fungicides are mixed for an 11.3-L volume, and this solution is placed in a fire extinguisher can. These cans are then placed individually into a cast-steel pressure housing and pressurized with the gas. The fungicide is applied with 2.1 kg/cm^2 pressure at a rate of 280 L/ha. The sprayer is calibrated by measuring the number of milliliters delivered from one nozzle per second based on the number of meters traveled per second. Ground speed is adjusted to obtain the 280 L/ha rate.

Disease symptoms are scouted on a weekly basis when plants reach GS-5. For example, the foliar diseases we monitor on wheat in Pennsylvania are powdery mildew, Septoria leaf and glume blotch, and leaf rust. Notes are kept on the first appearance of each disease. Disease evaluations are made at GS10.1 (ear emergence) or as close to this growth stage as possible. This rating period generally follows the last fungicide appications by about a week. Severity is assessed visually on individual leaves of 10 tillers, selected at random throughout each plot, as the percentage of the leaf area covered by the individual disease lesions using the standard area key devised by James (1). Severity values are then averaged over the 10 tillers for each leaf position. Means for each leaf position are then totaled to provide the severity values used in statistical analysis. Different diseases present on the same leaves are recorded and analyzed separately. The primary diseases evaluated at this early stage are powdery mildew and Septoria leaf blotch.

The second disease rating is made at GS-11.1 (milky kernels) and generally involves diseases present on the flag leaf and the heads. The disease severity values are combined and analyzed as previously described.

When the grain is mature, it is harvested with a plot combine. The large plots are harvested with an International Harvester combine that has been modified with a blower to provide complete seed cleanout between plots. Small plots are harvested with a Hege plot combine. Moisture is measured on the harvested grain, and yields are adjusted to 13% moisture. Seeds from each sample are counted and weighed to calculate the 1,000-kernel weight.

DATA ANALYSIS

The disease and yield data are analyzed using analysis of variance and a modified Bayesian LSD ($K = 100$). The response of each disease to the fungicide treatments is analyzed separately. The yield data are used to indicate the possible economic benefits of chemical control. To better estimate the effects of disease control on yield, the 1,000-kernel weights are also analyzed. Because disease pressure is greater in the later growth stages, the major component of yield that is affected is the 1,000-kernel weight. Correlation coefficients may be calculated to relate disease severity to 1,000-kernel weights.

LITERATURE CITED

1. James, W. C. 1971. An illustrated series of assessment keys for plant diseases, their preparation and usage. Can. Plant Dis. Surv. 51:39-65.
2. Large, E. C. 1954. Growth stages in cereals: Illustration of the Feekes scale. Plant Pathol. 3:128-129.

Evaluating Fungicides for Control of Foliar Diseases of Tobacco

WILLIAM C. NESMITH, Department of Plant Pathology, University of Kentucky, Lexington 40546

Tobacco (*Nicotiana tabacum* L.) is subject to many destructive foliar diseases in the field. The subject has been reviewed well by Lucas (6). Foliar diseases cause damage by reducing weight and leaf quality and can be of great concern because the leaf is the material used in the manufactured product. Some cigar wrapper grades have a zero tolerance for leaf lesions.

Historically, resistant cultivars and modified cultural practices have been the primary tools used to minimize tobacco diseases. Foliar fungicides are not widely used on tobacco outside of the tropics except for short-term use when a new problem appears suddenly. The tobacco industry is reluctant to use foliar fungicides because of concerns that leaf residues may affect leaf quality and/or raise additional health concerns for laborers and users.

Blue mold (caused by *Peronospora tabacina*), brown spot (caused by *Alternaria alternata*), and frogeye leaf spot (caused by *Cercospora nicotianae*) are the most destructive foliar diseases of tobacco in the United States. Fungicides have been extensively tested worldwide, primarily for blue mold control. Evaluations of fungicides tested for control of foliar diseases are available locally in the regions where testing was done, but the information is not easily obtained from the scientific literature. This chapter discusses mainly the techniques I and my U.S. colleagues have used to evaluate chemicals for blue mold control in the field. Brief comments appropriate for evaluating chemicals for control of other foliar diseases are included, but the techniques used for blue mold can be modified to fit other foliar diseases.

Researchers considering testing materials on tobacco should realize the significance of different tobacco types (burley, flue-cured, cigar wrapper, cigar filler, dark-fired, Virginia, etc.). Although all are tobaccos and many similarities exist, they differ strikingly in genetic makeup, cultivation, curing, and use. These differences should be carefully considered during the planning, conducting, and evaluating of fungicide trials. I have neither the space nor the experience to address here the ramifications of the procedures required for the various types of tobacco. Instead, I will describe only general procedures, which may serve as a guide to fungicide testing for foliar diseases of tobacco, especially blue mold, in the field.

METHODS

Preliminary Considerations

Blue mold is the downy mildew disease of tobacco. The pathogen is a strict, obligate parasite in the class Phycomycetes. Thus it is important to realize that host physiology and leaf moisture are critical to disease development. Precisely what set of field conditions is necessary for an epidemic to develop is not known, but they are complex and interacting (6). Many plant pathologists have become frustrated working with this disease in the laboratory because only slight changes in the host's status, brought on by changes in the environment, can result in temporary immunity during otherwise ideal conditions for disease development (6). Certain environmental factors, especially light intensity and leaf wetness, must fall within a narrow range for a major epidemic to develop. The literature should be thoroughly and carefully reviewed; before testing begins, the reviews by Lucas (6) and Shultz (15) should be consulted. Very careful planning is required for successful field testing with this disease. The planning is much greater than that required for brown spot and frogeye leaf spot, which are caused by facultative saprophytes, but is similar to that required for downy mildew diseases of other highly managed crops. The greatest attention to detail must be given to any factor that affects either juvenility of the host or leaf wetness, because both greatly affect disease development (2,4).

Researchers should realize that their actions can initiate an epidemic. Because it is disseminated by wind, the pathogen can spread rapidly over great distances (11), so extreme care must be exercised in locating test plots and handling inoculum. Tobacco production is geographically concentrated, and the proceeds from its production are often the major income source for the communities where it is produced. Tobacco growers are very aware of the devastation this disease can cause, and they realize how it is spread. They fear blue mold more than any other tobacco disease. The political consequences (for both administration and researcher) of being labeled a "blue mold spreader" must be carefully considered before blue mold field trials are undertaken. In the United States, the local blue mold coordinator should be kept advised of the status of blue mold research plots.

Site Selection

To improve the chances of a successful test, a site should be chosen where the disease will develop and where steps can be taken to enhance the chances of disease development. Sites where fog prevails are best. Natural shade from tree lines, especially where more than one side of the field is shaded, greatly enhances disease development but often creates a disease gradient in the field that must be accounted for in the experimental design. Overhead irrigation from a nonchlorinated water source should be available to supplement rainfall. The soil fertility should support excellent growth. Soils high in organic matter are preferred. My best field trials have been located on sites with a history of blue mold, because the environmental conditions from season to season favored disease development once the pathogen arrived. A variety of sites in several regions should be selected because of the erratic nature of blue mold development (6).

Test Plants

Disease development depends greatly on the growth stage and host physiology, because both are sensitive to changes in the environment (11). Young tissue is most susceptible; as tissue ages, it becomes more resistant (15). Systemic colonization by the pathogen often occurs in young plants.

In general, cultural steps taken to support succulent growth favor disease development. Nitrogen and potassium fertilization at the high end of the recommended rates have promoted disease development by inducing rapid growth and increased susceptibility in my plots of burley and dark tobaccos. Water supply should be adequate to maintain excellent growth. Rotem et al (13) reported that plants cultivated under low soil moisture regimes are highly susceptible once the leaf again becomes turgid. However, my experience during the prolonged droughts of 1980 and 1983 in Kentucky do not support this observation. Instead, except for the new growth, once the plants were drought-stressed for a few weeks they were resistant to infections, even after ideal weather for infection returned. Plants irrigated weekly to prevent stress remained susceptible, whereas those drought-stressed for several weeks before irrigation began were resistant.

Low light intensity within the planting favors disease development (6). Close spacing of the plants results in considerable plant-to-plant shading and also favors leaf wetness, which greatly favors blue mold development in Kentucky. However, in recent studies (*unpublished*) I have noticed that plants remain susceptible to blue mold longer when planted at lower populations. Some techniques I and my co-workers have used successfully to enhance disease for fungicide efficacy trials are to transplant the border rows about 3 weeks earlier than the evaluation rows, to increase the plant population in the border rows from 19,760 to 27,170 plants per hectare (from 8,000 to 11,000 plants per acre), and to place strips of corn between the plots. Besides shading the plots, these techniques greatly enhance disease incidence next to the plots, thereby increasing the availability of inoculum.

The cultivar selected should not have horizontal or vertical resistance unless resistance is a variable in the experiment. Instead, it should be highly susceptible to local strains of *P. tabacina*. However, selecting cultivars that are resistant to other diseases, especially soilborne diseases, increases the number and variety of sites where trials can be placed.

Plot Design

The experimental design used (paired comparison, randomized complete block, etc.) should be appropriate for the experiment (3; Nelson, *this volume*). Shoemaker (14) routinely uses a randomized complete block design with at least four replications for initial fungicide evaluations. Plots are single rows 7.62 m (25 ft) long, separated by unsprayed buffer rows. The check plots are located at random within the test site. We routinely apply test materials to four-row plots but collect data only from the center rows (10). Several rows of unsprayed plants are placed around each replication to act as buffers and to serve as a local site for inoculum production. Additional check plots, spaced at random, are also included to increase inoculum density within the plots.

Larger plots are used to evaluate spray intervals and reduce interplot interference. Shoemaker (14) uses plots seven rows wide by 7.62 m (25 ft) long. We routinely use plots eight rows wide by 61 m (200 ft) long, replicated three times, for these tests, because eight-row strips with a "drive through" are commonly used by Kentucky growers. Instructional markers can easily be placed in the driveways to allow the test to be conducted under actual farming conditions. This plot design has both provided valid research data and served as a demonstration site for the extension program.

Interplot interference has been a major problem in our test plots, especially with the new systemic fungicides like metalaxyl (Ridomil 2E and other trade names). These materials are effective at very low rates, so drift can destroy a test (12). The materials inhibit sporulation, which may greatly affect the uniformity of inoculum production (7). This is especially troublesome in evaluating inferior products in plots adjacent to those treated with excellent products. Shoemaker (14) has attempted to minimize these concerns by placing protective and systemic materials in separate tests. We have attempted to compare protective and systemic materials within the same test by using an untreated buffer and extra checks to maintain inoculum density, similar to techniques reported for other downy mildews (1). From a practical standpoint, the researcher needs to avoid overestimating the control potential of inferior treatments. Growers are seldom disappointed when control is better than expected but develop serious mistrust when the opposite occurs.

A standard fungicide should be included in each test. Mancozeb or maneb, at the locally labeled rates, is used commonly for this purpose (10,14). Also, systemic plus protective fungicide treatments should be included to evaluate the level of resistance to the systemic material. Sprayed and unsprayed check plots should both be included where space permits. Most authors prefer an unsprayed check to a water-sprayed check when only one check is included. We have never observed a significant difference in blue mold incidence between sprayed and unsprayed checks in our field testing program.

Fungicide Application

Both systemic and protective fungicides are being evaluated for blue mold control (8–10,14). Protectants are being applied as foliar sprays, and systemics are being applied in a number of ways, including soil application before planting or at layby, application in transplant water, foliar spray application, and application in irrigation water. Foliar sprays are applied as they are on other row crops. Backpack mist-blower, CO_2-pressurized, and tractor-mounted sprayers are being used for foliar applications to tobacco. Shoemaker (14) uses a model 423 Solo mist-blower in fungicide screening studies, applying the material as a

3× concentrate, using 112 L/ha (12 gal/acre) early in the season, when plants are small, increasing to 309 L/ha (33 gal/acre) by the time of flowering. We (8–10) use a CO_2-pressurized backpack sprayer operating at 275.8 kPa (40 psi) to deliver 187–280 L/ha (20–30 gal/acre) when plants are small, 468 L/ha (50 gal/acre) in midseason, and 935–1,430 L/ha (100–150 gal/acre) at topping time. A tractor-mounted hydraulic boom sprayer is used for larger plots in both our and Shoemaker's tests. The sprayer is operated at 689.5–1,723.7 kPa (100–250 psi) and no more than 107 m/min (4 mph). Fungicide nozzles (Spraying Systems D_2 or D_3 orifice size) are placed over each row and on each side of the row, spaced 23–46 cm (9–18 in.) apart. The booms are made up to spray four-row plots with a driveway placed either between each plot or every other plot. High-clearance sprayers with similar nozzles are also used to spray large field plots. Row spacing should not be closer than 1.07 m (42 in.) if reasonable coverage is expected in mature tobacco. At present, adequate spray equipment for applying protectant-type foliar fungicides to mature burley tobacco growing under current cultural practices is not available.

Preplant and cultivation applications are applied as a broadcast spray using 187–280 L/ha (20–30 gal/acre) followed by incorporation to a depth of 5–10 cm (2–4 in.) by disking or cultivation (8). Layby applications are applied the same way, except that they are made at or just before the final cultivation (9). Transplant water applications (8) are made by adding the fungicide to transplant water solutions and then transplanting either mechanically or by hand. The amount of solution varies from 468 to 4,675 L/ha (50–500 gal/acre). A mechanical transplanter can be modified with separate containers for each treatment to facilitate this approach. The water flow is more accurately controlled using hand rather than mechanical transplanting. However, soil compaction, which greatly affects growth around the plant, varies considerably between hand and mechanical transplanting, prohibiting direct comparison of the two methods in the same test. Mechanical methods are preferred, because essentially all commercial tobacco is transplanted in this manner.

Inoculum for Field Plots

Because *P. tabacina* is an obligate parasite, inoculum production is through a host plant. Field inoculum can be prepared by any of several laboratory means (15). However, in the United States plant pathologists do not generally place laboratory-produced inoculum in the field for fear of repercussions associated with being accused of starting an epidemic. Instead, we rely on natural infections and inoculum for field plots, but we do take steps to encourage early disease development and enhance the inoculum density once the disease has appeared naturally.

One way to encourage early disease development is to place an untreated plant bed next to the plot area. Abandoned plant bed sites serve as a major inoculum source for development of epidemics in the United States (6,11). These sites provide an excellent environment for blue mold; that is, many highly susceptible plants in a crowded environment trapping moisture for extended periods of time. Once the disease develops, infected plants can be transplanted from the plant bed to the field plots in a variety of patterns to increase inoculum density and provide a uniform distribution of inoculum in the plots. One plant per 7.6 m (25 ft) placed in each guard row is usually adequate.

Inoculum can also be collected from sporulating lesions found in the field or old bed site by brushing the sporangiospores from the leaf into cold distilled water. The sporangiospore suspension (50,000 sporangiospores per milliliter) is then sprayed onto the border rows with a small hand-held sprayer. The sprayer must be free of chemicals and must be used solely for that purpose. Inoculum is applied in darkness on nights with heavy dew or fog. Irrigating with about 1.3 cm (0.5 in.) of water before inoculations is helpful.

Primary lesions are usually evident within 5–7 days. Environmental conditions favorable for sporulation and infection (wet leaf surface in the dark) are needed for further development of the epiphytotic (11). If such conditions do not occur naturally, supplemental overhead irrigation is required. The epiphytotic develops slowly but becomes explosive with multiple cycles of the pathogen (11).

DATA

Disease incidence, phytotoxicity, and yield are minimum data requirements for most tests. However, effects on leaf quality and residues in the smoke should also be ascertained before the fungicide is labeled for use.

Disease Incidence

Lucas (6) and Shultz (15) state that blue mold can affect plants at any stage throughout the growing season. Symptoms vary greatly depending on genotype, age of plant, and environment. Chlorotic spots, which become necrotic with age, are common. The spots may remain distinct or coalesce to form large, unusable areas of the leaf. Systemic infection, evident by the vascular discoloration of stems, results in stunting and considerable distortion of the plant. Systemic infections confined to the leaf veins result in puckered and distorted leaves. Sporulation—sporangiophores with sporangiospores emerging from the leaf stomata—usually occurs within a week after infection. Sporulation usually starts before significant chlorosis is observed and continues until chlorosis is advanced.

Rating systems should take into account the range of symptom expression and differences in sporulation. The Horsfall-Barratt (5) grading system is commonly used for scoring the foliar phases of the disease, especially when disease incidence is significant. Usually each rating is expressed as percentage of leaf damage. Number of affected leaves per plant is also valuable when disease pressure is great. When disease pressure is low, lesion counts per plant or leaf are used. The number of systemically infected plants per plot should be ascertained if this symptom is present.

Sporulation should be evaluated in the early part of the day while the leaf is still wet. A four-tiered rating system (no sporulation, low-moderate, medium, heavy) appears useful if exact spore counts per leaf area cannot be made.

Disease severity and sporulation should be measured at least weekly once symptoms develop and more often when conditions are favorable for infection and sporulation. Coresta (Centre de Cooperation pour les Recherches Scientifiques Relative au Tabac) (*personal communication*) rates germ plasm by multiplying lesion severity by sporulation and adding systemic lesions, where each category is rated on a scale of 1 to 5, with 5 as the highest incidence. The minimum score is 2, and the maximum is 30.

Phytotoxicity must be carefully monitored when dealing with diseases caused by obligate parasites, because slight changes in host physiology may greatly affect disease development (6). A wide range of phytotoxicity symptoms may be observed. The more common symptoms are chlorosis, necrotic flecking, bronzing, increased greening, puckering of leaves, stunting, and time of flowering. All symptoms of phytotoxicity, regardless of how minor, should be noted because phytotoxicity can greatly affect disease development and test results with this obligate parasite. Suspicions of phytotoxicity can usually be clarified by careful analysis of the various rates of a product in the test. Excessive rates of products should be included in trials to obtain phytotoxicity data, because grower errors often result in similar use rates.

Yield

Marketable yield data should be obtained from field tests, even in the absence of the desired disease. Yield data are not only important to obtain efficacy data but are also needed by growers and manufacturers. Financial data can be obtained by marketing the yield from plots. Most major U.S. tobacco companies will cooperate in such research and have trained personnel to assist. The plots should be harvested and cured by standard techniques for the type of tobacco. For example, in air-cured tobaccos, this can be accomplished by attaching paper tags (nursery tags) to the plants at harvesttime. Wire tags work best because they do not rot and are not eaten by insects. Water-resistant markers should be used to label the tags. After the lengthy curing period (2–6 months), the leaves are removed (stripped) into the various categories for the type and tied into bundles while still "in case" (tissue can be handled without crumbling). The nursery tags can be used to mark the lot while securing it. The tobacco is taken to a warehouse and graded by federal graders. The Federal Grading Service has been most cooperative in our studies. Crop value can be determined through mock sales. A more accurate sale occurs with larger plots. The experimental tobacco can be displayed on the warehouse floor next to farmers' tobacco offered for sale. The owners (researchers) reserve the right to reject any sale by tearing the marketing tag in half, thus ensuring that tobacco treated with unlabeled chemicals does not reach the consumer, yet a true market value is obtained.

Evaluating Smoke Flavor

Flavor is extremely important in tobacco products, so all chemically treated tobacco should be test-smoked to ensure that flavor changes are evaluated. Tobacco manufacturers have "smoking test panels" for maintaining quality control. These panels should evaluate tobacco before fungicide labels are granted. These evaluations are coordinated in the United States through the Tobacco Pesticide Committee of the Tobacco Workers Conference. The following guidelines are used at present. Once the maximum planned rate is determined, the tobacco is treated at 0, 1×, and 2× the rate with the candidate chemical, in the same manner as the planned usage. The tobacco is grown and cured under standard conditions for the type(s) to be labeled, to include all routine cultural practices, including use of other pest control chemicals. After curing, 22.7 kg (50 lb) of whole leaf is obtained per treatment, composited equally from various parts of the plant. The tobacco is placed in polyethylene bags, labeled, and shipped as directed by the Committee. The Committee then arranges for use products (cigarettes, etc.) to be made and smoked by the test panels.

Residues

The Environmental Protection Agency (EPA) requires data on residues of candidate fungicides before registration and labeling. In addition, some states and universities require residue information before recommending new products, because EPA requirements may not be appropriate for the type of tobacco produced locally.

The chemical manufacturer is responsible for outlining the protocol necessary for residue collection. Researchers should participate carefully in these tests, ensuring that rates and time of application are correct. Green and cured material should be obtained by collecting the desired samples from at least three replications. Samples should be stored as directed by the chemical manufacturer, which usually requires quick freezing to −18°C and shipment in insulated boxes with dry ice to a designated residue laboratory. Cured material must also be assayed. Residue collection in common rotational and cover crops is also appropriate.

Statistical Analysis

All data should be analyzed statistically with the appropriate test for the experiment (3,16; L. A. Nelson, *this volume*).

REPORTS

Investigators should submit a complete report of their results to the manufacturer, and the results of public investigations should be submitted to the public unless secrecy agreements prohibit this. Preliminary reports are often necessary because companies need to make management decisions before tests are complete. The basic guidelines for reporting are discussed well by Cetas et al (1). Many manufacturers have special computerized forms that require similar information.

BROWN SPOT AND FROGEYE LEAF SPOT

Little information is available in the literature concerning fungicide testing for brown spot and frogeye leaf spot control in tobacco. Lucas (6) has adequately reviewed the diseases. Fungicide trials are conducted essentially as described here for blue mold control evaluations, with a few modifications. These diseases are caused by facultative saprophytes, so steps taken to weaken the plant generally favor disease development. Sites should be selected with a history of the diseases where leaf wetness is encouraged. Both diseases are favored by potassium deficiency and overmaturity of the plant. Brown spot is favored by high nitrogen levels late in the season, whereas frogeye leaf spot is favored by low nitrogen levels late in the season. Both diseases are encouraged by not topping the plants.

LITERATURE CITED

1. Cetas, R. C., Manzer, F. E., Fry, W. E., Schultz, O. E., Wade, E. K., Weingartner, D. P., Dooley, H. L., Leach, S. S., Heidrick, L. E., and Richards, B. L., Jr. 1978. Procedures for Field Testing Foliar Fungicides for Potato Late Blight Control. Pages 63–68 in: Methods for Evaluating Plant Fungicides, Nematicides, and Bactericides. E. I. Zehr, ed. Am. Phytopathol. Soc., St. Paul, MN.
2. Clayton, E. E., and Gaines, J. G. 1945. Temperature in relation to development and control of blue mold of tobacco. J. Agric. Res. 71:171-182.
3. Cochran, W. G., and Cov, G. M. 1962. Experimental Design. 2nd ed. John Wiley and Sons, New York.
4. Hill, A. V., Paddick, R. G., and Green, S. 1967. Epidemiology of blue mold of tobacco in the Ovens Valley District of Victoria. Aust. J. Agric. Res. 18:575-600.
5. Horsfall, J. G., and Barratt, R. W. 1945. An improved grading system for measuring plant diseases. (Abstr.) Phytopathology 35:655.
6. Lucas, G. B. 1975. Diseases of Tobacco. Biological Consulting Association, Raleigh, NC. Pages 235-266.
7. Morton, H. V. 1983. Blue mold control strategies, industry viewpoint. In: Blue Mold Symposium III, presented at the 30th Tobacco Workers Conference, Jan. 10-13, 1983, Williamsburg, VA.
8. Nesmith, W. C. 1982. Evaluations of some preplant uses of two systemic acrylalanine fungicides against blue mold of tobacco. Fungic. Nematic. Tests 38:157.
9. Nesmith, W. C. 1982. Evaluations of Ridomil 2E for blue mold control in burley tobacco. Fungic. Nematic. Tests 38:160.
10. Nesmith, W. C. 1983. Evaluation of foliar fungicides for blue mold control in the field. Fungic. Nematic. Tests 39:263.
11. Populer, C. 1981. Epidemiology of downy mildews. Pages 57–105 in: The Downy Mildews. D. M. Spencer, ed. Academic Press, New York.
12. Reuveni, M., and Siegel, M. R. 1984. A bioassay for the sensitivity of *Peronospora tabacina* to systemic fungicides. (Abstr.) Phytopathology 74:631.
13. Rotem, J., Cohen, Y., and Spiegel, S. 1968. Effect of soil moisture on the predisposition of tobacco to *Peronospora tabacina*. Plant Dis. Rep. 52:310-313.
14. Shoemaker, P. B. 1983. Fungicide evaluation for blue mold on burley tobacco. In: Blue Mold Symposium III, presented at the 30th Tobacco Workers Conference, Jan. 10-13, 1983, Williamsburg, VA.
15. Shultz, P. 1981. Downy mildew of tobacco. Pages 577-599 in: The Downy Mildews. D. M. Spencer, ed. Academic Press, New York.
16. Steel, R. G., and Torrie, J. H. 1960. Principles and Procedures of Statistics. McGraw-Hill, New York.

Evaluating Chemicals For Control of Soilborne Pathogens on Tobacco

A. S. CSINOS, Department of Plant Pathology, University of Georgia, Coastal Plain Experiment Station, Tifton 31793; B. A. FORTNUM, Department of Plant Pathology, Pee Dee Experiment Station, Florence, SC 29501; S. K. GAYED, Agriculture Canada, Delhi Tobacco Experiment Station, Delhi, Ontario, Canada N4B 2R1; J. J. REILLY, Virginia Polytechnic Institute and State University, Southern Piedmont Research and Continuing Education Center, Blackstone 23824; and H. D. SHEW, Department of Plant Pathology, North Carolina State University, Raleigh 27650

Endemic soilborne diseases are responsible for major yield reductions in tobacco (*Nicotiana tabacum* L.) throughout the United States. The three major tobacco diseases caused by soilborne bacteria or fungi are black shank (caused by *Phytophthora parasitica* Dast. var. *nicotianae* (B. de Haan) Tucker), black root rot (caused by *Thielaviopsis basicola* (Berk. & Br.) Ferr.), and Granville wilt (caused by *Pseudomonas solanacearum* (Smith) Smith). These diseases have remained difficult to control despite the development of resistant tobacco cultivars. Crop rotation is diminishing in popularity as a means of disease control because of limitations in available land (12). However, a combination of control practices, including the use of crop rotation, resistant varieties, and chemicals, is necessary for the long-term management of soilborne tobacco diseases.

BLACK SHANK

Disease and Organism

Black shank is one of the most serious soilborne diseases of tobacco. Direct losses from the disease vary from $16 million to $49 million annually in the United States. Because the fungus is persistent in soil, cultural practices such as use of rotations and resistant cultivars are of limited value. Thus the use of chemicals for black shank control is an important consideration.

Symptoms include stunting, wilting, root decay, black stem lesions extending upward from the soil line, and plant death. Early development of black shank in the seedling stage results in brown to black lesions near the soil line with typical damping-off symptoms. In established plants, initial root colonization often results in wilting and yellowing of the leaves. If the plant is stressed for moisture at this point, it may lose turgidity and succumb to the disease with relatively small portions of the root system blackened and discolored. If moisture is adequate, the disease progresses by colonization of the crown and pith. Affected roots, crown, and stem are typically black. If the stem is split, disking of the pith is frequently observed (9). Highly resistant cultivars may have only a small area of the base of the stalk infected but under stress may still be killed.

P. parasitica var. *nicotianae* is a heterothallic fungus, and the sexual stage is rarely observed in culture or nature. The role of oospores in pathogen overwintering is unknown. Chlamydospores are frequently observed in diseased tissue and soil and are believed to be the primary source of inoculum. Sporangia are produced over a wide temperature range; however, production is most pronounced at 24–28° C. Under conditions of warm temperature and adequate moisture, zoospores may be a significant source of secondary infection within a field and in pond water.

The severity of black shank varies from year to year, depending on soil temperature and soil moisture. The conditions most favorable for rapid growth of the tobacco plant are also conducive to rapid propagule generation and pathogen dissemination.

Candidate chemicals for control of black shank may be evaluated in the laboratory, greenhouse, or field. Laboratory and greenhouse evaluations may be conducted almost any time of the year and provide preliminary data on phytotoxicity, satisfactory methods of application, and efficacy. However, testing under field conditions provides the most informative and conclusive data on the performance of a chemical.

Inoculum Production, Identification, and Isolation

Inoculum used for making cork borer disks can be grown on V-8 juice agar, cornmeal agar, or potato-dextrose agar (PDA). Chlamydospores or sporangia can be produced by growing the pathogen in 10% V-8 juice broth at 28° C. Chlamydospore formation is induced by transferring mycelial mats to sterile, deionized or distilled water for about a week. Chlamydospores are separated from mycelium by blending and sieving through four layers of cheesecloth. Sporangia are produced by incubating chlamydospores in sterile water at 25° C for 2–3 days (15). Zoospore release is triggered by chilling sporangia to 4°C for 30 min, then returning them to room temperature (24–28°C). Zoospores can then be decanted or pipetted from the surface layer of water.

Infected tissue washed in running tap water and rinsed in distilled or deionized water may be placed in sterile water for 24–48 hr. If the tissue is fresh and *P. parasitica* var. *nicotianae* is active, the characteristic papillate, ovoid sporangia will appear. Because no other pathogen of tobacco causes symptoms characteristic of black shank or has the fungal morphology described above, the symptoms, signs, and presence of the described sporangia identify the organism and disease. Lucas (9) provides additional information on taxonomy and identification.

The pathogen may be isolated from fresh tissue on acidified PDA or PDA amended with 250 ppm ampicillin. Specific media such as those developed by Tsao and Ocana (18) and Flowers and Hendrix (3) may also be used to isolate the pathogen from tissue and soil. However, the PAR medium developed by Kannwischer and Mitchell (7) has given the best performance of the three media.

Laboratory Methods

The simplest of all tests is inhibition of radial mycelial growth in petri plates. A rich medium such as V-8 juice agar, amended with the candidate fungicide at various concentrations, can be used to study radial growth and inhibition to compute an ED_{50} for the compound. Because most chemicals are heat-labile, the test materials should be added to the agar after it has been cooled to 40–45° C, stirred, and then dispensed into petri plates. Inoculum may consist of No. 2 cork borer disks of mycelium obtained from an actively growing culture of the fungus. The inoculum plug should be inverted on the fungicide test plate so that the fungus comes into direct contact with the medium. Ten or more replications of each treatment should yield reliable results. The optimum temperature for growth of *P. parasitica* var. *nicotianae* is between 28 and 32°C. Light is not necessary for growth, but both light and relative humidity should be controlled for reproducible results. Radial growth can be measured easily with calipers or simply a ruler; the maximum and minimum diameters should be measured so that an average for each plate may be determined.

In a more detailed laboratory test, the roots of small (3-to 4-week-old) tobacco plants are inoculated with zoospores and the plants are placed in petri plates (15). Tobacco plants grown in organic medium for 3–4 weeks, then transferred to water in petri plates for 3–4 days have young, undamaged, actively growing root tips. Inoculum, either zoospores or chlamydospores, may then be added to the system amended with various concentrations of the candidate chemical. Disease or fungal development can be observed directly with a microscope or by examining samples transferred to microscope slides. Fungal development in root systems may be more easily observed by clearing and staining with trypan blue or cotton blue.

Greenhouse Methods

Greenhouse experimentation using soil systems is a reliable method for testing candidate fungicides. Transplant water, preplant incorporated, drench, and foliar applications may all be simulated in a greenhouse using potted plants.

Inoculum may consist of chlamydospores and mycelium mixed with soil or zoospores directly inoculated onto roots; or naturally infested field soil may be used. The researcher must be aware of other pest problems that may occur in natural soil, such as other pathogens, nematodes, insects, and weeds. In general, heat treatment or methyl bromide treatment of soil eliminates organisms that may complicate the evaluation of candidate fungicides. Then, inoculum may be added and thoroughly mixed with soil. Disease progresses more rapidly at temperatures optimum for tobacco and fungal growth (28–32° C), high humidity, and optimum moisture.

Fungicides may be evaluated by counting dead plants and/or washing roots free of soil at specified periods of time and examining and rating roots and stems for disease. Isolations from lesions may be necessary to confirm *P. parasitica* var. *nicotianae* in some instances.

Systemic fungicides may be tested using plants grown in a Hoagland's hydroponic solution. Tobacco plants seeded and grown to transplant size in Leach containers (1) have roots free of soil and can be easily treated with fungicides. Plants allowed to take up a known concentration of a candidate fungicide, placed in opaque containers for 24–96 hr, and then transferred back to Hoagland's solution can be challenge-inoculated with *P. parasitica* var. *nicotianae* zoospores and chlamydospores. Alternatively, stems of plants can be aseptically split open with a scalpel, a No. 2 cork borer disk of inoculum inserted, and the stem then sealed with masking tape to prevent dehydration. Stems should be 8–10 mm in diameter to ensure that plants will not be killed by the incision alone. Untreated, susceptible plants will be killed in 4–5 days. Plants protected by a fungicide may show symptoms that vary from stem discoloration and wilting to no symptoms, depending on the control afforded by the treatment (1).

Field Tests

Site selection. Selection of the test site may be the single most important factor in conducting evaluations of chemicals for control of tobacco black shank. For flue-cured tobacco, nurseries may be established that can be cropped to tobacco continuously; inoculum will reach a high level and remain there. However, in the burley tobacco-growing areas of Kentucky, this apparently does not occur. Levels on specific sites in Kentucky may be high for a few years but generally decrease with time. If a (flue-cured tobacco) nursery is established or has been established, the incidence of disease across the area should be uniform after a few years. Any areas not being used for testing during a specific year should be planted to a cultivar with low to moderate resistance (e.g., NC 2326) to maintain high populations of *P. parasitica* var. *nicotianae.* A highly susceptible cultivar such as Hicks Broadleaf will die too early in the season to maintain a high population of the pathogen. When possible, two or more cultivars should be used—one more susceptible (NC 2326) and one more resistant (McNair 944 or Coker 48) in flue-cured tobacco, and KY 14 (susceptible to both races 0 and 1), KY 14 × L8 (resistant to race 0, but susceptible to race 1), and KY 17 (generally resistant to races 0 and 1) in burley tobacco—to maximize the data obtained from the test.

A well-drained test site with soil pH ranging from 5 to 6 (6 to 7 for burley) should be selected. Naturally occurring inoculum may not be present in sufficient quantities to provide uniform disease, and artificial infestation may be required. Areas to be used as nurseries may be infested with infected tobacco stalks or sterilized inoculated oat or wheat seed spread over the area and tilled into the soil. An alternative method is to inoculate individual tobacco plants transplanted into the field. The inoculum may be oat cultures homogenized in a Waring Blendor, mixed with transplant water, and applied to roots of plants either by hand or through a commercial transplanter. To reach acceptable levels of infestation for testing may take 2–3 years of continuous tobacco culture. Caution is necessary in selecting and managing the test site, because *P. parasitica* var. *nicotianae* is easily disseminated by water and farm machinery. Because metalaxyl has activity against *P. parasitica* var. *nicotianae*, it should not be used for blue mold control in the plant beds or in the field where black shank fungicide candidates are being tested.

Experimental design and statistical analysis. The most common field design for evaluating fungicides for control of black shank is the randomized complete block. However, any other statistically sound design is acceptable (16). Experimental plots should be large enough and sufficiently replicated to produce reliable data (e.g., at least 20 plants per plot, replicated four times) and should be bordered with treated or untreated tobacco.

In most cases, multiple comparison procedures such as Duncan's multiple range test may be used. However, where rates of chemicals are being tested together, Johnson and Berger (6) suggest using regression analysis or curve fitting.

Application methods. Row treatments are used to apply multipurpose fumigants such as those containing chloropicrin. The fumigant is injected 35 cm below the top of a wide, high bed using a gravity flow or wheel-driven pump applicator. The fumigant must be applied a minimum of 21 days before transplanting to avoid phytotoxicity. Soil must be in good tilth and should not be too wet or too dry. Wet soil generally retards diffusion of the fumigant and may affect the exposure period. Aeration of the bed is not recommended after the exposure period, since recontamination might occur.

Fungicides may be applied preplant broadcast to the soil surface in enough water for adequate coverage and incorporated into the upper 15 cm of soil (2,14) by rototilling or disking the treated area. Immediately after the incorporation treatment, a high, wide bed should be formed. Soil-incorporated treatments should be applied immediately before transplanting.

Nonphytotoxic fungicides may be applied effectively in the transplant water solution. The volume of transplant water applied can vary from 1,870 to 3,740 L/ha, and this volume may have a bearing on the efficacy or phytotoxicity of the treatment.

Fungicides applied after transplanting can be applied in a directed spray toward the base of the plant and incorporated into the upper 5 cm of soil immediately afterward by cultivation. Various combinations of the four procedures may be used, depending on the fungicide.

Disease evaluation. To evaluate fungicides for reduction of disease under field conditions, measurements of several parameters may be needed. Observations of phytotoxicity may be complemented by a subjective vigor rating, phytotoxicity rating (if leaves are affected), or height measurement at specific times after transplanting. Disease may be evaluated at final harvest and percentage of disease for each plot calculated. A disease index may be calculated if the numbers of dead plants are counted several times during the season. The following formula for a disease index (DI) may be used (2):

$$DI = \frac{\sum_{i=1}^{n} X_i\,[100 - (i - 1)(100/n)]}{I},$$

where i is an ordinal evaluation number, n is the number of evaluations (excluding the initial stand count), X is the number of dead plants since the last count, and I is the initial number of plants in the plot. The formula is weighted so that plots that have plants that die early in the season have a higher DI than plots that have plants that die near the end of the season, even if the percentage of disease is the same for both. DI values may indicate relative fungicide residual activity and may differ for different formulations of a fungicide.

Yield data are important and should accompany disease data. Flue-cured tobacco should be harvested three to five times during the season as the crop ripens. Only leaves that the researcher considers would cure properly should be harvested. Leaves that have turned yellow and are permanently wilted from disease should not be harvested, since the resultant cured leaves would not be salable. Yield on a cured-weight basis may be calculated as follows:

$$G \times \frac{N_1}{N_0} \times 0.2 = C,$$

where G is the total green weight of the plot, N_1 is the number of plants per hectare, N_0 is the initial number of plants per plot, and C is the cured dry weight per hectare.

An alternative would be to cure and weigh tobacco from plots. Because variations in leaf quality may not be apparent in green leaf yield data, tobacco should be cured and the quality evaluated when possible.

Reports. All pertinent information concerning the testing of fungicides should be accurately reported, including product name and formulation; batch number; dates of application and rates used; application methods; volumes of material in spray, etc.; crop and cultivar; stage of crop; soil type and pH; soil and air temperature at application; other environmental information such as moisture, wind, cloud cover, etc.; crop history of area; experimental design, replications, and statistical analysis; inoculation information; rainfall and irrigation; fertilization; other pesticides used (rates and dates of application); and data on the evaluation criteria, including phytotoxicity, vigor ratings, percentage of disease, disease index, yield, quality, and value per hectare. An interpretation of the results is optional and in most instances should be based on at least 2 years of data.

BLACK ROOT ROT

Disease and Organism

Black root rot caused by the fungus *T. basicola* was first reported on tobacco in Ohio in 1899 and within a few years caused economic losses in the Connecticut Valley (5). The pathogen has a wide host range mainly in the Leguminosae, Solanaceae, and Cucurbitaceae (9). In addition, the fungus can persist indefinitely as a soil saprophyte (9).

In the field, black root rot often forms circular areas of

stunted, yellow plants among others that look normal in height and color. The stunted plants usually flower prematurely. The roots of field plants show varying degrees of root rot, not distinguishable from most other rots. In the plant bed, very young plants are killed and older plants turn pale. The roots, especially the tips, are blackened and pruned.

Resistance

There are cultivars tolerant to *T. basicola* in all tobacco types. The resistance is controlled by a single gene pair in *N. debneyi*. This type of resistance was transferred to Burley 49 in 1965 by Clayton and coworkers (9). Resistance in *N. tabacum* is controlled by several factors. Plants are quite susceptible when young and increase in resistance with age (9). Test plants should be uniform in size, because plants are rated for severity of black root rot by size after 6–8 weeks' growth in the field.

The following cultivars have little or no resistance: White Mammoth, NC 95, and Hicks (flue-cured); Burley 37 and Ky 16 (burley); Hastings and Lizardtail Orinoco (dark-fired); Little Sweet Orinoco (sun-cured); and Consolidated T and Connecticut 49 (shade). Resistant cultivars, to serve as standards in test plots, should include Va 115 or Va Gold (flue-cured); Ky 12 or Burley 49 (burley); Ky 170 or Va 312 (dark-fired); Va 407 or Va 409 (sun-cured); and Consolidated P or Consolidated L (shade). All Maryland cultivars have some resistance.

Inoculum

T. basicola grows and sporulates very quickly on carrot broth (200 g of carrots and 20 g of dextrose per liter) but even more rapidly on V-8 juice broth supplemented with 1 g of $CaCO_3$ per liter. Cultures in medicine bottles filled with 50 ml of broth, inoculated with a plug of mycelium from V-8 juice agar, and kept in the dark at 20°C first produce endoconidia (3–4 days), then a luxuriant lawn of black chlamydospores in about 8 days. Spores form most rapidly when the broth is shallow in relation to the surface area to be overgrown. If necessary, the endoconidia can be harvested before chlamydospores form.

Greenhouse Methods

Chemicals to control black root rot should first be tested in the greenhouse using a procedure similar to that of Troutman (17) for screening cultivars for resistance to *T. basicola*. Plants are germinated in a pan containing vermiculite until they are 2 cm in diameter. They are transplanted to 12-hole muffin pans with six holes 3 mm in diameter drilled in the bottom of each well to allow water absorption. The pans are filled with dry vermiculite, then dipped into a water suspension of endoconidia and/or chlamydospores (10^4–10^5/ml) until the vermiculite is saturated. The muffin pans are supported in loaf pans filled with the test chemical in aqueous solution. The seedlings are then transplanted to the muffin pans. Subsequent irrigation should be done with only water to avoid concentrating the chemical. Several concentrations are tested to determine phytotoxicity. The test should be replicated at least three times. Plants irrigated with water only serve as the controls. It takes approximately 10–14 days for growth differences to be obvious enough to detect. Air and water temperatures should be kept between 15 and 20°C for optimum disease development. Because seedlings are often more sensitive to chemicals under these conditions than under field conditions, field testing is recommended.

Studies may also be conducted in greenhouses using steam-sterilized muck (30 min at 82°C to a depth of 15 cm) as used for production of seedlings in Canada (4). Galvanized metal collars 45 cm in diameter and 20 cm deep should be surface-sterilized with 4% formalin and embedded in the sterilized muck so that the rim remains 2.5 cm above the soil level. Collars should be spaced enough apart to eliminate interaction between any volatile chemical treatments used in the tests. Collars can be infested with either endoconidia or chlamydospores (e.g., 2.0×10^7 endoconidia per collar).

Plot Design and Test Site

G. J. Griffin and C. Rittenhouse (*personal communication*) have shown that *T. basicola* is distributed in field soils according to a negative binomial distribution; that is, it occurs in clumps rather than being randomly or uniformly distributed. This may explain why normal plants appear next to stunted plants in infested fields. Plots should be separated by buffer rows of a resistant cultivar to avoid variability in growth caused by missing plants in adjacent rows increasing light and nutrition to the plants in the test plot. Variability makes rating difficult because the rating system is based on vigor. No fungicides other than fumigants are recommended for black root rot control, so fumigated plots should serve as the standard.

Black root rot is more severe in heavy soils with poor drainage and high organic matter. Because low soil temperatures (21°C or less) and pH of 6.0 or higher favor *T. basicola*, the test site should be in a cool place and the field should be limed periodically. The field should be planted continuously to tobacco. Cover crops of rye, barley, and timothy are known to increase the susceptibility of tobacco to *T. basicola* by releasing toxic compounds that stimulate tobacco roots to release stimulators of chlamydospore germination and reduce the resistance of tobacco by allowing the establishment of more infection sites (9). Rotation with corn and small grains reduces inoculum levels and should be avoided. Because legumes are also hosts for *T. basicola*, they can be used in a cropping sequence to enhance inoculum levels. Chemicals effective against phycomycetes, such as metalaxyl, should be incorporated before planting to avoid complications from interactions. In the plant bed, overfertilization (especially with nitrogen), high bed temperatures, and abundant water produce succulent plants that are very susceptible to damping-off from *T. basicola*.

Any plot design using randomly allocated treatments is acceptable with enough replications to minimize variability, as discussed in the preceding section on black shank.

Chemicals can be injected, preplant incorporated, broadcast before disking, banded before disking, or placed in the transplant water as discussed in the preceding section on black shank.

Detection and Isolation

Plants suspected of *T. basicola* infection can be eased out of the soil with a potato fork. The roots are rinsed and examined for dark lesions, especially on root tips. Often, mycelium and chlamydospores can be seen by making thin sections through the cortex and examining them with a light microscope. Occasionally, root tissue must be surface-sterilized for 2 min in 0.53% (w/v) NaOCl and plated on carrot agar. Sometimes even this procedure fails to detect the fungus, and small pieces of root and soil must be placed on raw carrot slices and incubated under cool, moist conditions in the manner described by Yardwood (19). Papavizas (13) describes a selective medium, VDYA-PCNB, for the detection of *T. basicola* that can also be used to quantify the pathogen to some degree. However, quantifying *T. basicola* from soil is difficult unless inoculum levels are high, and depending on the sensitivity of isolates and the presence of other microorganisms, the medium of Papavizas may give only partly quantitative results. Until better selective media are developed, the carrot slice technique of Yardwood, although laborious, can be used for quantitative work.

Disease Evaluation

Several systems have been used to evaluate black root rot in fungicide tests, including estimating the percentage of root rot and the size of root systems. However, it is not always possible to obtain the entire root system for size evaluation; furthermore, stunted cultivars may have only trace amounts of rot, whereas vigorous plants may have severe rot. This suggests differential sensitivity to a fungal toxin rather than diminished water and nutrient supplies caused by root decay.

In tobacco, because the leaf is of interest, a rating system based on plant vigor is suggested. Actual height measurement or any convenient subjective scale can be used. In the field, because inoculum is clumped, plants will not be uniformly stunted. The most vigorous and the most stunted examples in the field or greenhouse plot should be selected and made the extremes of the scale. Each plant in the row should be rated; totals for each replicate should be added, then the sum divided by the maximum score possible and multiplied by 100 to get the percentage of the maximum score possible for treatment comparisons. Depending on the environmental conditions, plants may begin to recover by 6–8 weeks after transplanting, so rating should be done then. Statistical analysis, plot design, and reporting of data are similar to those described in the preceding section on black shank.

GRANVILLE WILT

Organism and Disease

P. solanacearum, the incitant of Granville wilt or bacterial wilt, is a non-spore-forming, noncapsulate, aerobic, gram-negative, rod-shaped bacterium (9). The organism has a great deal of genetic diversity; its various races or strains differ in pathogenicity and in their chemical and serologic properties.

The disease can first be observed in tobacco by the drooping of one or several leaves during the heat of the day. The drooping is typically one-sided on the plant or leaf. As the disease progresses, the entire plant may be involved. Marginal leaf dehydration and necrosis between the veins are sometimes observed.

The presence of the organism is easily confirmed by suspending an excised section of stem tissue in water. Milky white strands of bacteria will be seen oozing from the vascular system and provide an easy diagnostic tool for the identification of the bacterial wilt pathogen.

Early symptoms of bacterial wilt in root tissue are inconspicuous, with few roots showing signs of decay. As the disease progresses, the entire root system becomes dark brown to black. Root wounds are generally considered necessary for the bacterium to enter the plant in the field. Transplanting wounds and damage caused by root-knot nematodes are two examples of root wounds that increase the incidence of Granville wilt. However, Kelman and Sequeira (8) reported the spread of Granville wilt from plant to adjacent plant without root wounding. They suggested the points of emergence of secondary roots as possible points of entry for the bacterium.

Inoculum and Host

Inoculum can be prepared by using a PDA or triphenyltetrazolium chloride (TTC) agar petri plate culture of *P. solanacearum* incubated for 3 days at 35°C and washed with distilled water (10). Virulent isolates are white and irregularly round (fluidal), whereas avirulent isolates are uniformly round, butyrous, and deep red on TTC (9).

The inoculum may be applied at a rate of 8×10^6 virulent bacterial cells per milliliter in transplant water, with 50 ml of inoculum applied to each transplant (11). Inoculum also can be applied to established plantings. About 60 days after transplanting, individual plants should be cut across the stem to remove the top portion of the plant. A virulent culture of *P. solanacearum* (10^8 cells per milliliter) should be sprayed on the freshly cut stem (10).

Cultivars without resistance to Granville wilt should be chosen to maximize disease development. If acceptable levels of disease are observed, the test site can be used the following year. Because of the longevity of *P. solanacearum* in soil and its wide host range, test sites should be carefully chosen, since future crop production can be affected.

At least two cultivars should be used, one with low-to-moderate resistance and one with high resistance to Granville wilt. All test plants should be obtained from plant beds previously fumigated with methyl bromide to ensure uniform, healthy transplants.

Lucas (9) provides further information on media, race determinations, and host-parasite relations.

Test Site and Plot Design

The test site should be well-drained and of a pH and soil type suitable for tobacco production. Field tests should be conducted in areas with a history of severe bacterial wilt or an established disease nursery. Areas with low disease incidence should be infested with the bacterium as described above. Other pest problems such as nematodes should be controlled to minimize interactions (9).

Plots should be large, usually four rows wide by 13 m long. In each plot, two rows may be planted to a susceptible cultivar and two rows to a resistant cultivar. Because incidence of bacterial wilt is related to soil type, soil temperature, and soil moisture, tests should be conducted at several locations and over several years. In all tests, untreated controls are included to determine efficacy of the chemicals tested. Chemicals tested are primarily fumigants and are preplant incorporated in the row as described previously in the section on black shank. Plot design and statistical analysis are similar to those described previously in the section on black shank.

Data Collection

A record of environmental conditions during the application of fungicides provides valuable information in the event of phytotoxicity or inadequate disease control. Data such as soil and air temperature, humidity, wind (direction and speed), soil moisture, soil type, and light intensity should be recorded at the time fungicides are applied. Fertilizer, herbicide, and nematicide treatments should be described in all reports.

The principal test of efficacy in a disease control trial is yield. Harvesting should proceed using normal harvesting techniques, and if possible the tobacco should be cured. Variations in leaf quality that may not be apparent in green leaf yield data but might have an impact on crop value may show up in cured leaf samples. Percentage of disease and a disease index may be recorded as described previously in the section on black shank. Reporting of the data is similar to that described previously in the section on black shank.

LITERATURE CITED

1. Csinos, A. S. 1979. *Phytophthora parasitica* var. *nicotianae* stem-inoculated tobacco treated with experimental systemic fungicides. Tob. Sci. 23:123-125.
2. Csinos, A. S., and Minton, N. A. 1983. Control of tobacco black shank with combinations of systemic fungicides and nematicides or fumigants. Plant Dis. 67:204-207.
3. Flowers, R. A., and Hendrix, J. W. 1969. Gallic acid in a procedure for isolation of *Phytophthora parasitica* var. *nicotianae* and *Pythium* spp. from soil. Phytopathology 59:725-731.
4. Gayed, S. K. 1976. The effect of steam compared with fumigants and benomyl on black root rot of tobacco in tobacco seedbeds. Phytoprotection 57:109-115.
5. Gilbert, W. W. 1909. The Root-Rot of Tobacco Caused by *Thielavia basicola*. U.S. Dep. Agric. Bur. Plant Ind. Bull. 158. 55 pp.
6. Johnson, S. B., and Berger, R. D. 1982. On the status of statistics in *Phytopathology*. (Letter to the editor) Phytopathology 72:1014-1015.
7. Kannwischer, M. E., and Mitchell, D. J. 1978. The influence of a fungicide on the epidemiology of black shank of tobacco. Phytopathology 68:1760-1765.
8. Kelman, A., and Sequeira, L. 1965. Root-to-root spread of *Pseudomonas solanacearum*. Phytopathology 55:304-309.
9. Lucas, G. B. 1975. Diseases of Tobacco. Harold E. Parker and Sons, Fuquay-Varina, NC. 621 pp.
10. McCarter, S. M. 1973. A procedure for infesting field soils with *Pseudomonas solanacearum*. Phytopathology 63:799-800.
11. Moore, E. L., Kelman, A., Powell, N. T., and Bunn, B. H. 1963. Inoculation procedures for detecting resistance of tobacco to *Pseudomonas solanacearum* in the field. Tob. Sci. 7:17-20.
12. Nesmith, W. C. 1983. Current fungicide testing in tobacco. Plant Dis. 67:707-708.
13. Papavizas, G. C. 1964. New medium for the isolation of *Thielaviopsis basicola* on dilution plates from soil and rhizosphere. Phytopathology 54:1475-1481.
14. Reilly, J. J. 1980. Chemical control of black shank of tobacco. Plant Dis. 64:274-277.
15. Staub, T. H., and Young, T. R. 1980. Fungitoxicity of metalaxyl against *Phytophthora parasitica* var. *nicotianae*. Phytopathology 70:797-801.
16. Steel, R. G. D., and Torrie, J. H. 1960. Principles and Procedures of Statistics. McGraw-Hill Book Co., New York.
17. Troutman, J. L. 1964. Indexing tobacco for black root rot resistance. Tob. Sci. 8:21-23.
18. Tsao, P. H., and Ocana, G. 1969. Selective isolation of species of *Phytophthora* from natural soils on an improved antibiotic medium. Nature (London) 223:636-638.
19. Yardwood, C. E. 1946. Isolation of *Thielaviopsis basicola* from soil by means of carrot discs. Mycologia 38:346-348.

Evaluation of Fungicides for Control of Ascochyta Blight of Field-Grown Chrysanthemums

ARTHUR W. ENGELHARD, Institute of Food and Agricultural Sciences, University of Florida, Gulf Coast Research and Education Center, Bradenton 34203

Outdoor chrysanthemum culture started in Florida in the late 1940s. For flower production, cuttings are set in the soil starting around the first part of August, with individual plantings continuing at intervals into March. Flowers are harvested 14–16 weeks after planting. This practice varies among growers, as a few produce chrysanthemum flowers during the summer months. Growers producing cuttings grow stock plants throughout the year.

Ascochyta blight, also called ray blight, is the major fungus disease of the flowers, leaves, and stems. It occurs throughout the year but more frequently and seriously during December, January, and February. The disease can cause extensive damage and loss during rainy periods accompanied by moderate temperatures. It attacks chrysanthemums at any stage of growth in the field, in the rooting bench, and in transit.

Ascochyta blight of chrysanthemum is incited by *Didymella ligulicola* (Baker, Dimock & Davis) von Arx (formerly *Mycosphaerella ligulicola* Baker, Dimock, & Davis), the perfect stage of *Ascochyta chrysanthemi* Stev. (5). Pycnidia and conidia are produced year-round on flowers, stems, and sometimes leaves. Perithecia and ascospores are found less frequently in Florida. The temperature range for ascospore development is 14–24°C (57–75°F). Conidia form between 18 and 32°C (64 and 90°F), with the optimum at 26°C (79°F) (4). The conidia are spread by water, blowing rain, infected plant material, or infested hands or tools. Ascospores are spread in a similar manner, and they also are windborne.

A method is described for evaluating foliar fungicides for control of Ascochyta blight of chrysanthemum (*Chrysanthemum morifolium* Ramat.). The system evolved over a period of 15 years, during which time many experiments were conducted successfully on crops grown during the fall (September to December) and spring (February to June) (1,2).

CULTURE OF CHRYSANTHEMUMS

The culture of chrysanthemum plants varies with location. A successful procedure followed in Florida uses well-prepared, well-drained sandy soil that is fumigated to control fungi, nematodes, and weeds. Since good drainage is necessary for chrysanthemum growth and control of soilborne diseases, plants are usually grown on raised beds that are about 10–15 cm (4–6 in.) high and 0.9 m (3 ft) wide. This is especially important if the area is subject to sudden heavy rains. Bedding wire with cells measuring 15 × 20 cm (6 × 8 in.) or 20 × 20 cm (8 × 8 in.) is placed on the beds, and one or two plants are planted in each cell. The wire is raised and hooked onto strategically placed posts near the outer edges of the bed to support the plants as they grow. The crop is grown vegetatively under long days (more than 14.5 hr of light) by lighting about 4 hr at night for 2–4 weeks; thereafter, the chrysanthemums are induced to flower by exposure to short days of less than 13 hr of light. Light is controlled during natural long days by covering the plants with black cloth. The plants flower, depending on the cultivar, 9–10 weeks after the initiation of short days. A nutritional program that allows for vigorous plant growth throughout the test period should be maintained.

PLOT DESIGN

The bedding wire mentioned above provides a good guide for designing the experiment. An excellent plot size is six cells across the bed and three parallel with the bed, for a total of 18 cells. Splitting such a plot into two subplots, each with 3 × 3 cells, permits the planting of two cultivars at one location across the bed. Each treatment can then be sprayed and evaluated on two cultivars at one time (split plot). Each cell may be planted with one or two plants, allowing 9 or 18 plants per subplot. Generally, if plants are pinched, a single plant will suffice; but if plants are grown with a single stem, two plants will increase foliage density and create a better environment for disease development. Each experimental treatment, standard fungicide, and water control is replicated four times. The treatments may be placed in a completely randomized block design.

Spraying and disease development are made easier by leaving an unplanted row adjacent to a test plot and by planting a plot (border plot) two cells wide next to the unplanted row. The unplanted row permits the placing of a thin spray board, 0.6 m × 0.9 m × 6.5 mm (2 ft × 3 ft × 1/4 in.), on each side of a test plot during spraying to reduce drift. A hole measuring 5 × 15 cm (2 × 6 in.) is cut near the top of each board to help in carrying. A suitable stake about 1 m long on which to lean the boards is placed on each side of the plot. A string, with an attached weight (7 cm of pipe), tied to the handle and placed around the stake, secures the board in case of wind. The unplanted row also permits better observation of the test plot.

The border plot is inoculated but not sprayed. Thus, there is an unsprayed but diseased plot on both sides of each experimental plot. These unsprayed but diseased plots are strategically and uniformly located throughout the entire test area.

INOCULATION OF PLANTS

Preparation of inocula. Diseased stems and flowers are harvested from a previous season's unsprayed plants. They are dried at room temperature in the laboratory and stored in paper bags in a dry place until needed. Pycnidia are produced in abundance on diseased stems and flowers and, occasionally, on leaves. The pathogen can be stored for at least 1 year at room temperature. When a spore suspension is needed, the dried, diseased material is placed in deionized or distilled water for 15 min to allow the conidia to eject from the pycnidia. Wetting agents are not needed. The suspension is filtered through a double layer of cheesecloth. Several gallons of spore suspension can be prepared within 30 min with this method. The concentration of spores in the suspension is determined using a Levy Corpuscle Counting Chamber (hemacytometer).

A second method of preparing inoculum is to grow the pathogen on potato-dextrose agar. Most isolates produce pycnidia and conidia when grown under continuous cool-white fluorescent lights in an incubator at 24–27°C (75–80°F). Both spores and mycelia cause infection. After conidia are produced, a suspension can be prepared for inoculation by adding either distilled or deionized water to cultures in petri plates for 15 min and filtering them through cheesecloth, or the entire contents of the petri plate can be comminuted in a Waring Blendor and filtered through cheesecloth. The inoculum should be sufficiently free of particles to be applied with a backpack sprayer. This method is more laborious than the one mentioned previously. Hadley and Blakeman (3) have discussed in detail factors affecting conidiospore production in culture.

Inoculation procedure. Plant inoculation has always been necessary to ensure infection at our location in Bradenton. Test plants should be inoculated during a rain, or at least during a rainy period. Inoculation of plants in hot, dry weather is usually not successful. A spore suspension containing 1.5×10^5 spores per milliliter yields good infection under rainy conditions. Ninety linear meters (300 linear ft) of bed 0.9 m (3 ft) wide can be inoculated with 11.5 L (3 gal) of spore suspension applied with a backpack sprayer with two nozzles on the spray wand. The plants are sprayed uniformly, once from each side of the bed and once directly down over the bed. Uniform application of the inoculum is extremely important for establishing disease uniformly throughout the test area. One replication should be completed before starting the second. A wet period of 24 hr is adequate for obtaining good field infection when the temperature is in the 21–27°C (70–80°F) range. Under these conditions, symptoms become apparent within 2–3 days on leaves and 30 hr on flowers.

Test plants may be inoculated after the first fungicide application, in which case the fungicides would be evaluated as protectants; or the plants may be inoculated 24, 48, or 72 hr before the first fungicide application to test the eradicant activity of the fungicides. Plants may be inoculated at any age, but preferably within 2–3 weeks after transplanting in the field. Inoculating early in the season allows the disease to spread naturally in the test area and border rows for the rest of the crop's duration (12–14 weeks). In the absence of adequate natural moisture, disease development and spread can be facilitated with overhead irrigation applied late in the day to keep plants wet overnight.

FUNGICIDE APPLICATION

Procedures. The first fungicide is applied either before the plants are inoculated (protectant test) or 24–72 hr after the plants are inoculated (eradicant test). Subsequent sprays are applied weekly for the entire season. Variations can be made to suit individual needs. Fungicides are conveniently applied with a CO_2-pressured hand sprayer at 380 kPa (55 psi) of pressure. Extreme care should be taken to apply sprays to the lower and upper surfaces of the leaves (infection occurs through both), especially those near the soil where environmental conditions are best for disease development and spread.

Cultivars suitable for evaluating fungicides are Stingray, May Shoesmith (or other Shoesmith cultivars), Mrs. Roy, or any of the Albatross or Indianapolis cultivars. These are standard types (large blooms) of chrysanthemums that are susceptible to Ascochyta blight. Mrs. Roy is so susceptible that under ideal conditions for disease development, plants may be killed. The other cultivars mentioned are generally less susceptible, with fewer plants being killed under ideal conditions for disease development. Stem and leaf lesions are common with these cultivars. The foliage and stems of Iceberg cultivars are too tolerant to be used satisfactorily, but the flowers are susceptible.

Standard treatments. A standard treatment is benomyl 50W at 300 mg/L (0.25 lb/100 gal) of water tank-mixed with either mancozeb 80W, captan 50W, or chlorothalonil 75W at 900 mg/L (0.75 lb/100 gal). Mixtures containing benomyl control Ascochyta blight when applied as late as 72 hr after inoculation. Tolerance to benomyl is not known to occur when combinations are used consistently, but it could occur, especially if there were a history of using benomyl or thiophanate methyl or other substituted benzimidazole compounds. Other standard treatments that may be used are captan 50W, chlorothalonil 75W, or mancozeb 80W (or any equivalent) at 1,800 mg/L (1.5 lb/100 gal) of water.

COLLECTION OF DATA

Rating system. Plants are rated at harvest by cutting the plants in the plots and counting individual lesions. The percentage of disease control (PDC) may be obtaining using the following formula:

$$\text{PDC} = \frac{\text{total lesions in control} - \text{total lesions in treatment}}{\text{total lesions in control}} \times 100 .$$

The weights of the plants or the numbers of diseased plants or both may be used as further indicators of the effect of disease or phytotoxicity on plant production.

The data should be subjected to adequate statistical procedures.

Phytotoxicity. The crop should be carefully inspected before and after each fungicide application for any indication of chemical phytotoxicity. Some systemic chemicals may not show phytotoxicity symptoms until 8 days after application. Common symptoms include chlorotic or necrotic leaf margins, white or necrotic areas on leaves, retarded growth, crinkled leaves, and necrotic spots and tipburn on petals. Fungicide-treated plants should always be compared with the control plants, as some of the symptoms mentioned above occur on plants growing under high light intensity (normal summer light). The symptoms induced by excess light are especially noticeable after lateral buds are removed.

LITERATURE CITED

1. Engelhard, A. W. 1981. Control of Ascochyta blight of chrysanthemum with foliage fungicides. Fungic. Nematic. Tests 36:124-125.
2. Engelhard, A. W., and Schuster, D. J. 1977. Combinations of fungicides and insecticides for a disease-insect control program for chrysanthemums. Fungic. Nematic. Tests 32:135-136.
3. Hadley, G., and Blakeman, J. P. 1968. Asexual sporulation of *Mycosphaerella ligulicola* in relation to nutrition. Trans. Br. Mycol. Soc. 51:653-662.
4. McCoy, R. E., Horst, R. K., and Dimock, A. W. 1972. Environmental factors regulating sexual and asexual reproduction by *Mycosphaerella ligulicola*. Phytopathology 62:1188-1195.
5. Walker, J., and Baker, K. F. 1983. The correct binomial for the chrysanthemum ray blight pathogen in relation to its geographical distribution. Trans. Br. Mycol. Soc. 80:31-38.

Evaluating Fungicides for Control of Foliar Diseases of Foliage Plants

A. R. CHASE, University of Florida, Institute of Food and Agricultural Sciences, Agricultural Research and Education Center, Apopka 32703, and D. D. BRUNK, Plant Disease Diagnostics, Inc., Apopka

Greenhouse, shadehouse, and full-sun production of tropical foliage plants occurs throughout Florida. Fungal foliar diseases often interfere with the growth, quality, and salability of the crop during its production. Because Florida's temperature and humidity favor disease development, cultural control is often difficult and fungicides must be used (10). They are applied to plant foliage routinely as eradicative and preventive treatments for these diseases. The methods used in Florida to evaluate chemical control of fungal foliar diseases are reviewed in this chapter.

The most common fungal foliar diseases are caused by species of *Alternaria, Bipolaris, Colletotrichum, Corynespora, Exserohilum, Fusarium, Myrothecium,* and *Rhizoctonia* (1). Important genera of foliage plants are infected by these fungal pathogens. Table 1 lists the pathogen-suscept (host) combinations that are commonly used to test fungicides for foliar disease control.

Tests are performed on specific pathogen-suscept combinations to provide disease control recommendations for growers on that crop and for pesticide manufacturers for possible label expansion or registration of experimental compounds. The most important information generated from fungicide evaluations on foliage plants concerns efficacy and phytotoxicity. Since foliage plants are produced for their appearance, both of these factors must be considered in a fungicide evaluation trial. Elements of most trials include host plant preparation; pathogen production and application; environmental factors; fungicide rates, intervals, and application methods; test evaluation; and experimental design and statistical analysis.

Table 1. Pathogen-suscept (host plant) combinations used to test fungicide activities

Pathogen[a]	Suscept (host plant)
Alternaria alternata	*Calathea bella*
A. panax	*Brassaia actinophylla*
Bipolaris setariae	*Chrysalidocarpus lutescens*
	Maranta leuconeura
Colletotrichum gloeosporioides	*Euonymus japonica*
Coniothyrium concentricum	*Yucca elephantipes*
Corynespora cassiicola	*Aphelandra squarrosa*
	Ficus benjamina
Exserohilum rostratum	*Chrysalidocarpus lutescens*
Fusarium moniliforme	*Dracaena marginata*
F. solani	*Dieffenbachia maculata*
Myrothecium roridum	*Aglaonema commutatum*
	Aphelandra squarrosa
	Dieffenbachia maculata
Rhizoctonia solani	*Hedera helix*
	Nephrolepis exaltata

[a] Representatives of the major foliar fungal pathogen groups: Dematiaceae, Melanconiaceae, Moniliaceae, Mycelia Sterilia, and Sphaeropsidaceae. Perfect stages are not included because they are generally not significant etiologic agents on foliage plants.

HOST PLANT PREPARATION

For the most part, plant material should be produced by the researcher under controlled conditions since only then can it be determined that the plants are free from both pathogens and pesticides. Plants obtained from commercial producers frequently have unknown problems, such as pesticide residues (visible or otherwise) and latent or active diseases. Several diseases can be latent in foliage plants and appear at inopportune times, thereby rendering data questionable. For example, Erwinia blight and dasheen mosaic virus commonly exist as latent infections in *Dieffenbachia* spp. (dumb cane). Plants selected for foliar disease control tests should have roots free of pathogens to preclude plant death from a root disease during a trial. For these reasons, the vast majority of test plants must be produced from pathogen-free seeds or cuttings grown in a steam-treated potting medium (78–82°C) for 2 hr.

Plants are grown for 1–6 months before experimentation under glasshouse or shadehouse conditions, depending upon the trial to be performed. In addition to pest considerations, the nutritional condition of the host plant must be controlled to ensure a nonstressed plant. Several studies have demonstrated the need for adequate nutritional control, since plants that are fertilized too little or too much do not respond to pesticides in the same manner as those that are fertilized at the manufacturer's recommended rate (5,6). The recommended rate and application interval of a slow-release fertilizer are used for each crop (8) in all of our fungicide evaluation trials.

Just before use, plants are examined carefully for foliage defects and defective plants are excluded. If plants are extremely variable in size (height and width), they are grouped from smallest to largest and each size placed together in blocks to minimize variation. Optimum light, temperature, irrigation, and fertilizer conditions are maintained throughout the trial.

PATHOGEN PRODUCTION AND APPLICATION

The source of the pathogen can be natural or artificial. Since uniform inoculation of experimental plots is dependent upon irrigation method, size of plot, and an adequate source of inoculum (among other factors), only trials that are performed outdoors in a shaded or full-sun area can use inoculum from a natural source. When the source is natural, tissue isolations from the host should be performed periodically during the trial to confirm identity of the pathogen. Of the pathogen-suscept combinations listed in Table 1, only leaf spots of *Chrysalidocarpus lutescens* are tested routinely at this research center using a natural inoculum source. A relatively large culture collection must be maintained and continually revitalized with fresh isolates to ensure a virulent inoculum source for a wide variety of trials. Initial selection of isolates, as well as subsequent transfers, must be of conidial types and not mycelial types (9). The fungal pathogens listed in Table 2 grow well on either V-8 agar medium or potato-dextrose agar medium (4). Reliable conidial production has been difficult to achieve for isolates of *A. panax, B. setariae,* and *E. rostratum*. Cultures of these pathogens kept over 6 months often fail to produce conidia under standard conditions. Wounding a fully grown culture by drawing a sterilized dissecting needle across its surface sometimes stimulates conidial production. Inoculation of the original host and reisolation sometimes revitalize an isolate; however, fresh cultures must often be isolated from naturally infected host tissue.

Culture plates are prepared using a fungal stock maintained on a V-8 juice agar or potato-dextrose agar slant at 24–26°C. They are grown at the same temperature under approximately 25 $\mu E/m^2$ per second (fluorescent cool-white light, from 0800 to 2000 hr daily) for 1–3 weeks, depending upon the fungus. Conidia are removed from culture plates using sterilized, deionized water and a rubber policeman (spatula). Concentrations are adjusted to the desired level using a hemacytometer (Table 2). Inoculum is applied to plant foliage using a pump-action hand sprayer to runoff. Plants are immediately placed in polyethylene bags for 2–5 days to maintain high moisture for conidial germination and infection.

Of those fungi listed, only *Rhizoctonia* cannot be handled in this manner. Culture dishes of this fungus are blended in a Waring Blendor for 15 sec with 200 ml of sterilized, deionized water per plate. The mycelial slurry is added to the surface of the potting medium and watered in lightly. The rate of addition ranges from 10 to 50 ml per pot, depending upon the size of the pot and the plant used.

ENVIRONMENTAL FACTORS

Overhead irrigation is the single most important environmental factor for a successful foliar disease control trial. Trials that are performed outdoors in a shade or full-sun area are exposed to overhead irrigation and rainfall, creating the high-moisture conditions needed for conidial germination, infection, lesion expansion, and secondary infection. Little, if any, foliar disease develops if the plant foliage remains dry for long periods of time (2). In the greenhouse, high-moisture conditions can be obtained with a misting system (15 sec/30 min from 0800 to 2000 hr daily) or hand watering to simulate overhead irrigation (two or three times daily). In addition, water applied to the area under benches and to walkways creates high relative humidities, which also favor leaf spot development.

Temperature requirements for disease development in foliage crops are similar to those for the same diseases on other crops. Tests must be scheduled during times of the year when optimal temperatures are achieved easily. Myrothecium leaf spot occurs optimally between 21 and 27°C, with temperatures below 15°C or above 32°C greatly inhibiting disease development (7). For this reason, investigators should test for Myrothecium leaf spot in central Florida during the spring and fall months. In contrast, optimum disease development for Rhizoctonia aerial blight occurs between 40 and 44°C (A. R. Chase and C. A. Conover, Florida Agricultural Experiment Stations, *unpublished data*), and trials should be run during the summer months when air temperatures reach 32°C and above. Table 3 indicates the optimum time of year for specific tests in central Florida.

Test location should be based upon the conditions under which the crop is normally produced and the site designation of the products to be tested. Since many fungicides labeled for ornamental crops specify greenhouse or shadehouse use, subsequent use by the pesticide manufacturer of the data generated is reduced if tests are performed under inappropriate conditions. Some plants are best tested under shadehouse conditions, since they are rarely produced in greenhouses, whereas others are produced only in greenhouses. The choice of location is often based upon the optimum conditions for plant growth and not the optimum pathogen requirements (Table 3).

Table 2. Preparation and application of inoculum for artificial infection of foliage plants

Fungal pathogen	Culture medium	Inoculum concentration	Method of application
Alternaria spp.	V-8 juice agar	10^3–10^4/ml	Sprayed to runoff
Bipolaris sp.	V-8 juice agar	10^4/ml	Sprayed to runoff
Colletotrichum sp.	Potato-dextrose agar	10^6/ml	Sprayed to runoff
Corynespora sp.	V-8 juice agar	10^4/ml	Sprayed to runoff (+ or − wounding)[a]
Exserohilum sp.	V-8 juice agar	10^4/ml	Sprayed to runoff
Fusarium moniliforme	Potato-dextrose agar	10^6/ml	Added to center of plant (1 ml)
F. solani	Potato-dextrose agar	10^6/ml	Sprayed to runoff (+ or − wounding)
Myrothecium sp.	Potato-dextrose agar	10^6/ml	Sprayed to runoff (+ wounding)
Rhizoctonia sp.	Potato-dextrose agar	1 plate/200 ml of water, ground for 15 sec	Various amounts added to potting medium

[a]With or without wounding. Wounding is accomplished by puncturing the leaf with a sterilized dissecting needle at various sites just before application of inoculum.

APPLICATION OF FUNGICIDES

The fungicide rate frequently corresponds to the labeled rate or to the rate for which the compound may be labeled. Efficacy trials should include rates above and below the recommended level to ensure determination of an effective use rate for that pathogen-suscept combination. When phytotoxicity is especially important, rates of at least $4x$ (x = recommended rate) should be included. Generally, $1/2x$, $1x$, $2x$, and $4x$ rates provide data to answer most questions posed. In addition to experimental fungicide treatments, inoculated and uninoculated control plants and a fungicide standard treatment should be included.

Another type of trial involves testing the recommended rate of many fungicides to identify the most effective product for control of a specific disease. Tests like these are not usually as useful to the pesticide manufacturer as the other tests, but they can be very useful to researchers and commercial plant producers. Direct comparison among different fungicides at a single rate appears to be the best method to identify the most effective product, since all products are tested in a single trial under the same conditions.

Because of the small experimental plot size in trials, fungicides are usually applied with a stainless-steel hand sprayer (7.6 L in volume) at between 80 and 160 kPa (1 atm = 101.3 kPa) of pressure, with care taken to spray upper and lower leaf surfaces to runoff. All sprays are applied when air temperature is between 15 and 32°C to avoid possible interactions between temperature and phytotoxicity (6,11). All of the foliar diseases listed in Table 1 can be tested using foliar fungicide sprays. Since *Rhizoctonia* is soilborne, Rhizoctonia aerial blight can also be controlled by applying a drench of the fungicide to the potting medium. In this case, the fungicide is added to the pot using a water volume of 5.1–10.2 L/m^2 (surface area).

Fungicides should be applied weekly or semimonthly, depending upon the time of the year, rainfall, irrigation, and disease pressure. Most greenhouse tests include seven weekly applications, whereas those outdoors (shade or full sun) include two or four applications per month for 3–6 months. This time period is especially important when natural infection is involved, since some diseases require a long period to infect and cause symptoms. For example, Coniothyrium leaf spot of *Yucca* sp. and leaf spots of *Chrysalidocarpus* sp. all require a minimum of 3 months for symptom expression under natural infection conditions.

The first fungicide application can be made either before or following infection, depending upon whether the test fungicide has preventive or eradicative characteristics. At least 3 days should elapse between inoculation and the first fungicide application to allow conidial germination and penetration.

Table 3. Desirable conditions for foliar disease control trials on foliage plants in central Florida

Pathogen-suscept	Location[a]	Time of year
Alternaria-Calathea	Greenhouse	Year-round
Alternaria-Brassaia	Greenhouse	Year-round
	Shadehouse	Summer
	Full sun	Summer
Bipolaris-Chrysalidocarpus	Shadehouse	Summer
	Full sun	Summer
Colletotrichum-Euonymus	Shadehouse	Year-round
Coniothyrium-Yucca	Shadehouse	Year-round
Corynespora-Aphelandra	Greenhouse	Spring and fall
Corynespora-Ficus	Shadehouse	Spring and fall
Exserohilum-Chrysalidocarpus	Shadehouse	Summer
	Full sun	Summer
Fusarium-Dracaena	Greenhouse	Year-round
	Shadehouse	Year-round
Myrothecium-all hosts	Greenhouse	Spring and fall
Rhizoctonia-all hosts	Greenhouse	Summer

[a]Growth of many foliage plants in central Florida is limited to greenhouses and other heated structures during the winter months. Therefore, most shadehouse and full-sun testing is conducted during the summer months.

TEST EVALUATION

Test evaluations include measurements of phytotoxicity, fungicide residue on the leaves, and efficacy. Quantitative measurements of host plant response can include such factors as height, number of leaves, and fresh weight of tops. These factors help establish the safety of products used on the plant. Other factors such as foliage color and quality are also important in evaluating the response of the plants to pesticides, but they are more difficult to standardize. In these cases, plants can be rated on a scale from 1 (none or minimum) to 5 (severe or maximum) for each factor. Since these factors are qualitative judgments, photographs of plants exemplifying the rating scale are essential.

Many times, the quantitative rating does not reflect the phytotoxicity incurred and the qualitative rating must be included (3). For relatively short trials (4–6 weeks), ratings of plants before pesticide application and at the conclusion of the trial are sufficient. When tests are performed for more than 6 weeks, ratings on a monthly basis during the trial should also be included. More ratings are made than are necessary to evaluate phytotoxicity or residue potential since the most appropriate rating may not be obvious until the trial is completed.

Efficacy rating can be handled in much the same manner as that of phytotoxicity, since both quantitative and qualitative data are often needed for an adequate description of the level of disease present. Quantitative rating scales include lesion number, number of leaves infected, and lesion size, and they are most helpful on

Table 4. Disease rating methods developed for use with foliage plants

Pathogen-suscept combination	Rating	Description
Alternaria-Brassaia	1	No symptoms
Fusarium-Dracaena	2	1–10 lesions/plant
	3	11–25 lesions/plant
	4	26–50 lesions/plant
	5	51 or more lesions, with coalescence
Bipolaris-Chrysalidocarpus	1	No symptoms
Exserohilum-Chrysalidocarpus	2	1–10 lesions/plant
	3	11–50 lesions/plant
	4	51–100 lesions/plant
	5	101 or more, with frond shredding
Corynespora-Aphelandra *Myrothecium*-all hosts	Number of lesions/plant	
Rhizoctonia-all hosts	Percentage of the crown that is symptomatic	

plants in 15-cm pots or smaller. Because obtaining data for plants in larger pots is often tedious, qualitative rating scales have been adopted for many trials. Specific rating scales have been developed for use on each pathogen-suscept combination (Table 4). On long-term tests, the disease severity rating is made monthly, if possible, to gain information on disease development, eradicative or preventive control, and sometimes environmental effects on disease control. This can be especially important if different intervals or methods of application are being evaluated.

EXPERIMENTAL DESIGN AND STATISTICAL ANALYSIS

Whether in a greenhouse, shadehouse, or full-sun area, all trials should be arranged in a randomized complete block design to minimize environmental effects. Since greenhouse conditions may vary considerably along a bench, especially when fan and pad cooling are used, such a design is essential for development of dependable data (7). Plot size varies from 3 to 10 pots per replication and a minimum of four replications per treatment. When a single pot is used as the experimental unit (replication), pots are removed from each experimental unit, placed together and treated, and then replaced in the original pattern. In this case, 10–20 pots per treatment are used. Data obtained from these trials are analyzed using standard statistical methods, including analysis of variance with the F test, multiple range tests, and regression analysis, depending upon the treatments used. Multiple range tests are generally inappropriate when analyzing multiple rates of a single fungicide. In such cases, regression analysis is required.

CONCLUSIONS

Foliage plants are sold for their appearance and therefore have a zero tolerance level for disease, phytotoxicity, and fungicide residues. Since foliage plants are grown commercially in greenhouses, shadehouses, and full-sun areas, tests performed under similar conditions are easily applicable to those of commercial producers. The high degree of botanical diversity and relatively low acreage for even the most economically important crop decreases the interest of chemical manufacturers in these plants. On the other hand, the high cash value of the crop allows many costly fungicide applications for disease control, which increases the interest of the manufacturers. Nevertheless, establishing registrations for foliage plants is difficult and depends at least partially upon the willing participation of land-grant universities. The ultimate goal in most disease control trials is the generation of reliable information in a form useful to extension plant pathologists, commercial plant producers, chemical manufacturers, and other agricultural scientists.

LITERATURE CITED

1. Chase, A. R. 1981. Common fungal leaf spot diseases of foliage plants. Foliage Dig. 4(5):7-12.
2. Chase, A. R. 1982. Influence of irrigation method on severity of selected fungal leaf spots of foliage plants. Plant Dis. 66:673-674.
3. Chase, A. R. 1983. Phytotoxicity of some fungicides used on tropical foliage plants. Fla. Nurseryman 30(5):26, 27, 47.
4. Chase, A. R., and Osborne, L. S. 1983. Influence of an insecticidal soap on several foliar diseases of foliage plants. Plant Dis. 67:1021-1023.
5. Chase, A. R., and Poole, R. T. 1984. Influence of foliar applications of micronutrients and fungicides on foliar necrosis and leaf spot disease of *Chrysalidocarpus lutescens*. Plant Dis. 68:195-197.
6. Chase, A. R., and Poole, R. T. 1984. Severity of acephate phytotoxicity on *Spathiphyllum* Schott. cv. Clevelandii as influenced by host nutrition and temperature. J. Am. Soc. Hortic. Sci. 109(2):168-172.
7. Chase, A. R., and Poole, R. T. 1984. Development of Myrothecium leaf spot of *Dieffenbachia maculata* 'Perfection' at various temperatures. Plant Dis. 68:488-490.
8. Conover, C. A., and Poole, R. T. 1981. Guide for fertilizing tropical foliage plants. Fla. Foliage 7(3):53-60.
9. Hansen, H. N. 1938. The dual phenomenon in imperfect fungi. Mycologia 30:442-455.
10. Knauss, J. F. 1978. Test procedures for fungicides and bactericides used to control foliar and soilborne pathogens of ornamental tropical plants. Pages 27-30 in: Methods for Evaluating Plant Fungicides, Nematicides, and Bactericides. American Phytopathological Society, St. Paul, MN.
11. Lund-Hoie, K. 1983. The influence of temperature on phytotoxic effect of glyphosphate on Norway spruce (*Picea abies* L.). Crop Prot. 2(4):409-416.

In Vivo Fungicide Screening on Field-Grown Turfgrasses

P. SANDERS and H. COLE, JR., Department of Plant Pathology, Pennsylvania State University, University Park 16802

Turfgrasses are excellent host plants for in vivo evaluation of fungicides. The advantages of turfgrasses for such screening include the following: (i) large host populations are possible in relatively small areas (10,000–13,000 individual plants per square meter); (ii) grass plots are relatively easy to establish or reestablish; (iii) turfgrasses are perennial; (iv) representatives of major fungal pathogen genera cause diseases on turfgrasses (e.g., species of *Pythium, Fusarium, Rhizoctonia, Drechslera, Bipolaris, Sclerotinia*, and several of the Basidiomycetes); (v) fungicide registration is more easily obtained for use on turfgrasses than on food crops, allowing developers of commercial fungicides to market more quickly so that recovery of research and development costs can begin while collection of data for registration on food crops proceeds; (vi) data can be obtained both on fungicides and on chemicals that may elicit antifungal responses in host plants; and (vii) safety of chemicals to plants can be determined on turfgrasses, where phytotoxicity is variably expressed as chemical burn, growth changes, or color changes (2).

The experimental protocols described here are those followed in our fungicide screening tests, and they may be modified as circumstances dictate. Procedures used in individual disease trials are largely determined by the nature of the disease. Number of replications, individual plot size, and rating system used depend on disease expression and uniformity. Treatment intervals depend on efficacy duration of test fungicides. Chemicals that are formulated for spray application are applied with a small, wheel-mounted, CO_2-powered boom sprayer with 0.9 or 1.8 m (3 or 6 ft) booms, using TeeJet 8004 nozzles at 207–276 kPa (30–40 psi), and at dilution rates compatible with manufacturer directions. Granular formulations are hand-distributed on small plots (less than 5.6 m^2 [60 ft^2]) or applied with a drop spreader on larger plots.

Environmental modification and artificial inoculation are used when required by the etiology and development of particular diseases. Fungicides are screened against some diseases without inoculation or environmental modification. Leaf spot disease (*Drechslera* and *Bipolaris* spp.), brown patch (*R. solani*), Fusarium blight (*Fusarium, Leptosphaeria*, and *Phialophora* spp.), and red thread (*Laetisaria fuciformis*) are screened against in this manner. Artificial inoculation is used in tests against dollar spot (*S. homoeocarpa*), and environmental modification plus artificial inoculation are required in tests against Pythium blight (*Pythium aphanidermatum*) in the climate of central Pennsylvania.

Disease loss evaluations are conducted at regular intervals throughout trials, using rating methods that vary with disease expression. Data obtained are subjected to analysis of variance and Waller-Duncan *K*-ratio *t*-tests.

SCREENING WITHOUT ENVIRONMENTAL MODIFICATION OR ARTIFICIAL INOCULATION

Drechslera and Bipolaris diseases. Various species of *Drechslera* and *Bipolaris* cause serious foliar, crown, and root diseases on turfgrasses. Screening trials against Drechslera diseases should commence so that fungicide application is coincident with onset of favorable disease environment. The most important Drechslera disease on turf in the northern United States is melting-out of Kentucky bluegrasses (*D. poae*). It is a serious problem on Kentucky bluegrass home lawns and golf course fairways, and it is most damaging in the cool, wet weather of early spring and fall.

Trials against Drechslera melting-out are begun as soon as Kentucky bluegrass growth begins in spring. Screening is conducted on a susceptible cultivar such as Cougar, Delta, Glade, Kenblue, Newport, Park, or Vantage. High levels of soluble nitrogen are applied to increase disease severity. Individual plots approximately 0.9 × 6 m (3 × 20 ft) are arranged in a randomized block design with three replications. Both systemic and contact fungicides can be tested against melting-out. Applications are made on a 2-week schedule, beginning in mid-April and continuing through early June. Disease evaluations are made as soon as disease losses are apparent and continue at 2-week intervals through mid-June. Turf loss to melting-out is generally unpatterned, but it is uniform in a susceptible variety; the loss is rated using a 0–10 visual scale, which corresponds to percentage of reduction in stand density (1 = 10%, 2 = 20%, and 10 = 100%).

Fusarium blight. This disease is a serious and devastating midsummer disease on Kentucky bluegrass turf areas. Although it has been ascribed to species of *Fusarium*, Koch's postulates have not been fulfilled with these fungi. Recent research reports indicate that the pathogens may be species of *Leptosphaeria* and *Phialophora* (1). Fusarium blight appears to occur on Kentucky bluegrasses in particular environmental settings year after year with some regularity. Screening trials against this disease, therefore, are conducted in areas prone to Fusarium blight with a history of outbreaks. Susceptible Kentucky bluegrass cultivars include Bonnieblue, Fylking, Kenblue, Merion, Newport, Nugget, and Pennstar. The disease is most damaging during the heat and drought stress of midsummer. Symptoms first appear as small circles, rings, and crescents of wilted grass that quickly collapse and die, leaving in some cases the "frogeye" symptom that characterizes the disease.

Because infection centers of Fusarium blight are fairly large (15 cm to 0.6 m [6 in. to 2 ft]) and tend to be nonuniform in distribution, individual plot size should be relatively large (1.8 × 15 m [6 × 50 ft]) or the number of replications should be increased to minimize error resulting from variation in the experimental site. High nitrogen fertility is maintained in Fusarium blight test areas, and drought stress is often required for disease development. Fungicides should be applied before the onset of symptoms, with treatments commencing in early June. In years when hot, dry weather occurs in May, however, treatments must begin earlier so that applications are made preventively. Only systemic fungicides have been effective in control of Fusarium blight. Because of the longer efficacy duration of these chemicals, treatments are made monthly, with applications in early June, July, and August. Shorter treatment intervals may be required in particularly high-stress environments. Disease loss assessments are made monthly. Numerical rating of the percentage of loss or the number of infection centers per plot can be made to evaluate fungicide efficacy in control of Fusarium blight.

Red thread. Red thread, which is caused by the basidiomycete *Laetisaria fuciformis*, is a foliar disease affecting most turfgrasses in temperate climates. Perennial ryegrasses and fine fescues are particularly susceptible, making them the grasses of choice for fungicide testing against this disease. Most fine-leaf fescue cultivars are susceptible, and susceptible perennial ryegrass cultivars include Yorktown, Yorktown II, and Manhattan. Low nitrogen fertility is maintained in red thread test areas. The disease manifests itself under moderate temperatures of 16–21°C (60–70°F) and adequate moisture as irregular patches of pinkish or bleached white grass. The pink to red, threadlike or antlerlike hyphal aggregates that project from the tips of blighted leaves are routinely used to diagnose this disease.

Red thread is usually fairly uniform in distribution over the test area. Individual plots, approximately 0.9 × 7.6 m (3 × 25 ft), are arranged in a randomized block design with three replications. Since red thread is primarily a foliar disease, efficacy data can be obtained for both contact and systemic fungicides. Treatments are applied at 2-week intervals beginning in early May and continuing through mid-June. Disease loss assessments are made at 2-week intervals, using a visual rating scale that corresponds to percentage of plot area blighted.

Large brown patch. Brown patch of turfgrasses is caused by the important plant pathogen *R. solani*. The disease attacks all turfgrasses, but bentgrasses, ryegrasses, and tall fescues are most susceptible and are used for fungicide screening tests. Most bentgrass varieties are susceptible, and susceptible perennial ryegrass cultivars include Manhattan, Yorktown, Campus, and Linn. Brown patch appears as a foliar blight at daytime temperatures above 27°C (80°F) and under conditions of high humidity. Infection centers are irregularly circular patches of brown, thinned, or blighted turf that, under high moisture conditions, may be covered with mycelium.

High nitrogen fertility is maintained in brown patch test areas. Individual plots, approximately 0.9 × 6 m (3 × 20 ft), are arranged in a randomized block design. Since brown patch may be nonuniform in distribution, four replications are preferable. Both systemic and contact fungicides can be tested against this disease. Applications are made at 2-week intervals beginning in late June or early July. Biweekly fungicide applications are continued until a brown patch outbreak occurs, at which time disease loss assessments are made. Experience has shown that damage from a severe outbreak of brown patch is such that evaluation of subsequent outbreaks is difficult. Our procedure, therefore, is to discontinue the experiment after one outbreak on the experimental area. Disease loss assessments make use of a visual rating scale that estimates percentage of plot blighted.

SCREENING WITH ARTIFICIAL INOCULATION

Dollar spot. Dollar spot is a disease that affects most turfgrasses during warm, humid weather; it is a serious problem on bentgrass golf greens, where it destroys the putting quality of the turf. Fungicide screens against dollar spot are typically conducted on greens-managed creeping bentgrass. Maintenance of creeping bentgrass is labor-intensive, since it requires daily mowing. However, symptom expression of dollar spot on close-cut (0.6 cm; 0.25 in.) bentgrass is most striking, where it appears as 1.3–2.5 cm (0.5–1 in.) sunken patches of bleached dead grass. Testing may be carried out on other susceptible grasses if creeping bentgrass is not available. Low nitrogen fertility is maintained in dollar spot test areas.

Before fungicide application, the experimental area is inoculated by uniformly hand-scattering *S. homoeocarpa*-infected rye grains at an approximate density of 215/m^2 (20/ft^2). Inoculum is prepared by growing virulent isolates of *S. homoeocarpa* separately on autoclaved rye grain for about 2 weeks. Rye for inoculum is prepared by placing 225 g of rye grain, 4 g of $CaCO_3$, and 275 ml of water in 1,000-ml Erlenmeyer flasks, stopping the flasks with cotton plugs, and autoclaving for 45 min at 103 kPa (15 psi). The experimental area is thoroughly watered following inoculum distribution, and mowing is suspended for 2 days to allow fungal colonization of the grass adjacent to inoculum propagules. Because disease will be quite uniform in the test area, individual plots may be as small as 0.9 × 4.6 m (3 × 15 ft), arranged in a randomized block design with three replications. Since both contact and systemic fungicides can be tested against dollar spot, applications are made at 2-week intervals, beginning 2 days after inoculation in mid-June and continuing through mid-August. Disease loss assessments are usually made at 2-week intervals. Numerical ratings of the percentage of loss or the number of infection centers per square meter (or square foot) are made to evaluate fungicide efficacy in control of dollar spot.

SCREENING WITH ENVIRONMENTAL MODIFICATION AND ARTIFICIAL INOCULATION

Pythium blight. Pythium blight is a devastating disease on various turfgrass species under conditions of high temperature and moisture. Bentgrasses and ryegrasses are especially susceptible. Environmental conditions for disease development are critical, however, and relying on natural inoculum and the natural environment for disease occurrence in research plots has often failed. Therefore, we have devised a field-inoculation procedure that combines movable, plastic-covered humidity chambers with artificial inoculation, which allows us to induce Pythium blight at will in a test area. This procedure may be used anywhere that susceptible grass species are available. The only requirement is that ambient air temperatures be high enough to reach 29–32° C (85–90° F) in the humidity chambers during the incubation period or that a supplemental heating system (described below) be available.

Although Pythium blight is a problem on both bentgrasses and ryegrasses, ryegrasses are more easily established and maintained and the cultivars are uniformly susceptible. *Pythium aphanidermatum* is the primary Pythium blight pathogen in Pennsylvania, and we thus use this species in our fungicide evaluations. Isolates are maintained at room temperature on kernels of rye grain autoclaved in tubes of distilled water.

Our movable humidity chambers (Fig. 1) are rectangular and are constructed of 2.5-cm (1-in.) polyvinyl chloride pipe covered with transparent polyethylene sheeting. The polyethylene is attached to the frame with plastic tape. TeeJet spray nozzles with No. 45 dis-core type cone-spray tips are positioned at various angles in the pipe so that mist can be directed to cover the area under the chambers adequately (Fig. 1). Water supply to the chambers is controlled by an electric intermittent interval timer.

The procedure consists of fungicide application followed by successive weekly inoculations to evaluate long-term fungicidal suppression of disease. Each successive inoculation is made on a "new" uninoculated portion of the plot. The humidity chamber is moved to the new site each time successive inoculations are made. The width of individual treatment strips is 0.9 m (3 ft), and length depends on the number of weekly inoculations to be done. If the test fungicides are to be evaluated at weekly intervals over a 5-week period, the individual treatment strips must be 4.6 m (15 ft) long to allow for five successive weekly inoculations. Our humidity chambers are 0.9 × 6.4 m (3 × 21 ft). When the chamber is placed across the treatment strips, this size accommodates the simultaneous weekly inoculation and incubation of six adjacent treatments plus an unsprayed check (Fig. 1). Our procedure requires three such chambers to permit three replications of each weekly inoculation. If more than six fungicides are to be tested, more chambers are needed. The experimental design is a replicated randomized block.

The entire plot area is treated once with the test fungicides. Since the procedure imposes no restriction on application methods, chemicals may be applied as granules, foliar sprays, or soil drenches. Individual fungicides are applied in 0.9-m (3-ft) strips. The length of these treatment strips depends on the anticipated duration of the experiment. If long-residual fungicides are to be tested, an evaluation period of 4–5 weeks after treatment may be required.

Inoculation is done immediately after fungicide application and weekly thereafter until the residual effect of the test fungicide is no longer apparent. Inoculum is prepared by growing virulent isolates of *P. aphanidermatum* separately on autoclaved rye grain for approximately 1 week. Just before inoculation, rye grain is pooled and homogenized in a Waring Blendor with sufficient water to make a thick homogenate. Inoculation is carried out by uniformly spreading the *Pythium*-infested rye homogenate over a 0.9-m (3-ft) strip across all treatments. After inoculation, the inoculated areas are covered with the humidity chambers (Fig. 1).

During periods of bright sun, chamber ends must be open to prevent interior air temperatures from exceeding 38°C (100°F). During early evening, night, and early morning hours, the chamber ends are kept closed. The intermittent mist system in the chambers is on during daylight hours (two 1-min mists per hour) and off at

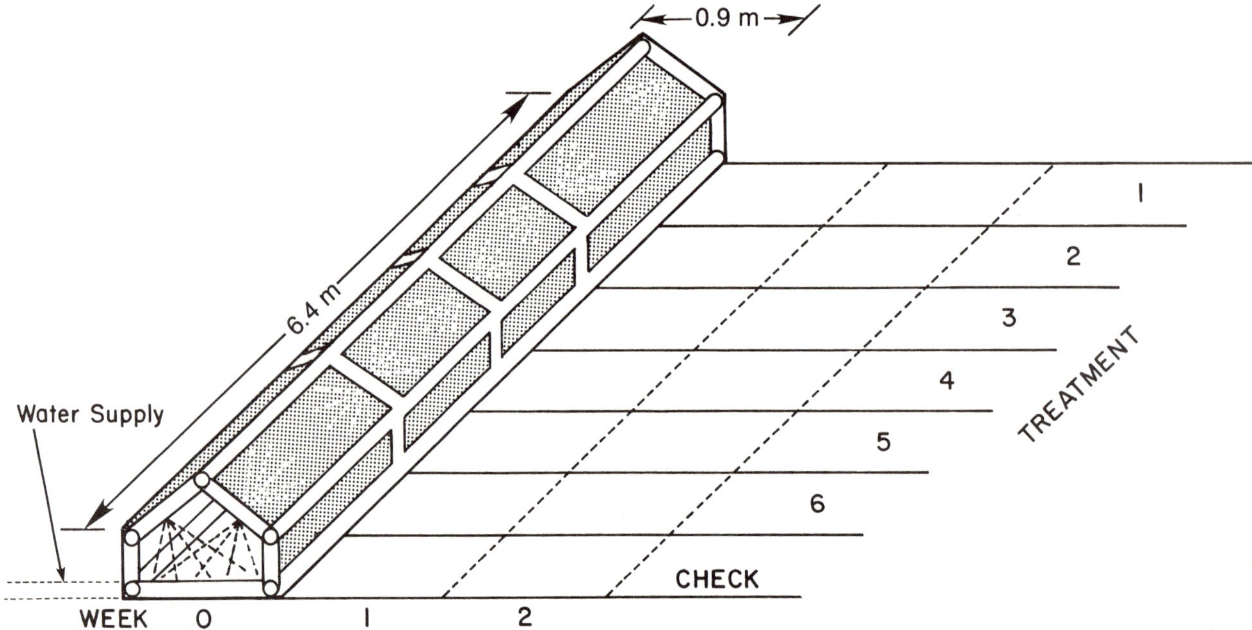

Fig. 1. Placement of humidity chamber to incubate inoculated strip across all fungicide-treatment strips. Chambers are moved to incubate successive weekly inoculations.

night. If maximum daytime air temperatures fall below 27°C (80°F) during the test period, heat can be supplied to the chambers via perforated metal heating ducts from small portable electric space heaters with fan blowers. These environmental modifications provide the high temperatures and humidities in the foliar microclimate that are essential for *Pythium* infection and disease development.

The humidity chambers are kept in place for about a week. Chambers are typically removed when the untreated check areas are completely blighted. Disease severity is evaluated visually by estimating the percentage of treatment area blighted.

This experimental procedure ensures infection pressure at regular intervals after fungicide treatment and helps to offset some of the difficulties of uncertain environmental conditions encountered in field evaluations of fungicides for control of Pythium blight.

LITERATURE CITED

1. Smiley, R. W., and Craven Fowler, M. 1984. *Leptosphaeria korrae* and *Phialophora graminicola* associated with Fusarium blight syndrome of *Poa pratensis* in New York. Plant Dis. 68:440-442.
2. Vargas, J. M., Jr. 1981. Management of Turfgrass Diseases. Burgess Publishing Co., Minneapolis. 204 pp.

Evaluating Seed-Treatment Fungicides

D. E. MATHRE, Montana State University, Bozeman 59717, and EARL D. HANSING, Kansas State University, Manhattan 66506

Treating seed with fungicides to control crop disease is an important facet of plant disease control (2). Many plant pathologists test fungicides in the field as seed treatments. Techniques for testing vary, and they need to be improved and standardized so that test results can be used more effectively.

Seedborne pathogens are of special concern to growers because of the risk of introducing them into a new area. Seedborne pathogens can become established early in the life of a plant when it is often most vulnerable. In addition, their presence throughout the growing season greatly increases the probability of an epidemic. Because seed treatments prevent or reduce losses from seed decay, seedling blights, damping-off, and other diseases caused by seed- and soil-inhabiting organisms, many acres of cultivated crops are planted each year with fungicide-treated seed (4). In the United States, more than 99% of corn (*Zea mays* L.), cotton (*Gossypium hirsutum* L.), and sorghum (*Sorghum vulgare* Pers.) seed and various percentages of seed from other agronomic, horticultural, and forest crops are treated before being planted. The percentage treated depends largely on the expected economic benefits from treatment.

Seedborne pathogens may infest or infect a seed. For example, pathogenic fungi on cereals may be on or in the lemma and palea, on or in the caryopsis seed coat, or in the meristematic tissue of the embryo. They may also be mixed with seed in the form of sclerotia, smut balls, or infested plant parts. Seeds are treated to reduce all such sources of infection. They are also treated to protect them from soilborne fungi such as *Fusarium, Pythium,* and *Rhizoctonia* spp., already present in soil, that cause seed decay and preemergence and postemergence seedling blights. Seed-treatment fungicides kill or inhibit pathogenic fungi in the soil near treated seeds. If the fungicide is systemic, the *zone of protection* may be larger than that provided by nonsystemic protective materials.

Many fungicides are available for seed treatment. They may be organic or inorganic, volatile or nonvolatile, systemic or nonsystemic. The treatments are formulated and applied to seed as powders, wettable powders, liquids, or flowables. Methods and techniques discussed here are restricted to fungicide tests with true seed.

An earlier version of this chapter was published in *Fungicide and Nematicide Tests* 30:1-3.

PROCEDURES

Selecting a cultivar. The cultivar selected should be so highly susceptible that nearly 100% of the plants from untreated seed become infected. For example, when Red Chief wheat (*Triticum aestivum* L.) seed infested with bunt (*Tilletia foetida* (Wallr.) Liro) is planted under favorable environmental conditions, 90% of the spikes subsequently produced may be bunted. The use of an effective seed-treatment fungicide reduces bunted spikes to 1% or less. This gives a researcher a range of 0 to 90% for determining the effectiveness of seed-treatment formulations at different rates on bunt. For most diseases, however, plant cultivars are less susceptible than in this example. In any case, the researcher must develop procedures to ensure adequate disease development.

Seed. Seed viability and vigor are highly important, especially in experiments involving seed decay and seedling blight. Experiments should be conducted with seed of average germination and vigor to approximate field conditions. Experiments with seed of high germination and strong vigor planted under highly favorable conditions may not give a true picture of differences between treatments.

Using uniform seed from the same source is important, particularly in tests involving seed decay, seedling blights, and yields of forage or grain. Percentage of germination, moisture content, and vigor of seed near the top of a container stored for some time may differ significantly from seed in the middle or bottom. Seed for experiments must be thoroughly mixed before being separated into different sets for treatment. Some seed, such as peanuts (*Arachis hypogaea* L.) and soybeans (*Glycine max* (L.) Merr.), must be handled carefully to avoid damage during mixing.

Naturally infested or infected seed is preferred, but the inoculum level must be considered carefully. A low level of infestation may produce sufficient infection for evaluating treatments, but high infestation may require impractical or injurious quantities of fungicide to effect control.

The volume or weight of seed to be treated should be as large as can reasonably be handled in the laboratory to minimize errors, since the rate of application of most seed-treatment materials is quite low. Normally, a minimum of at least 100 g of seed should be treated.

Fungicide rates for initial evaluation. One well-known standard treatment is included at the recommended rate in all seed-treatment tests. Also included are treatments at one-half and twice the normal rates. Rates for entries should differ by a factor of at least two.

These rates may give early information on phytotoxicity or limits of fungicidal activity. Although rates are based on active ingredients, appropriate formulation rates are preferred in reports. The formulation of the fungicide, the amount of product used, and the quantity of seed treated in tests should be stated clearly.

Application of fungicide to seed. The fungicide should be applied to seed as uniformly as possible, either with a mechanical mixer or by hand. Many workers mix by hand as follows. A clear glass jar is used so that the point at which most of the seeds are covered with fungicide is easy to determine. The volume of the jar should be about four times that of the seed—e.g., a 2-L jar for 0.5 L (2-qt jar for 1 pt) of seed. The jar is capped immediately after the seed and product are in the jar. With hands on the bottom and top of the jar, the jar is rotated rapidly at a 45° angle, first clockwise and then counterclockwise. The jar is inverted and rotation is continued until the glass is relatively clear. The treated seed may then be poured into a double, heavy-duty paper sack. Untreated seed may also be mixed and used as a control to assess mechanical injury to the seed.

Another, possibly better method of mixing seed and chemicals is by use of a 0.45-L (1-pt) metal can to which a 2.5-cm-wide baffle is soldered on one side. The capped can plus seed and chemical can be rotated by hand or by machine for several minutes to coat the seed with the fungicide.

For powder formulations, the seed is placed in the can or jar, the calculated amount of powder scattered over the top of the seed, and all mixed as described above, usually for 3–5 min.

Wettable powder formulations must not be applied directly to seeds because some will receive more fungicide than others. Wettable powders are mixed with the correct amount of water in a beaker and agitated with a magnetic stirrer to prevent settling. The jar is held at a 45° angle, the correct amount of suspension is added by pipette to one side of the jar, the jar is rotated 180°, and the seed is added while the jar is held at a 45° angle. Mixing is done as described above.

For liquid or flowable formulations, the correct volume of fungicide is added to a volumetric flask, and water (or other diluent, such as 95% ethanol) is added to volume. In cases with extremely viscous formulations, it is best to add the correct volume of diluent to the volumetric flask first and then add the "flowable" fungicide with a Pasteur pipette to bring the total volume to the volumetric mark. Normally, no more than 5 ml of total volume should be added to 100 g of seed. To simulate the method of applying these materials commercially, the 5-ml volume of fungicide plus diluent is atomized onto the seed in the can as it rotates. After the entire volume has been applied to the seed, the seed is tumbled for another 1–2 min to assure uniform coverage. The treated seed is then placed in an open paper sack for 1–2 days to dry.

Another technique for liquids or flowables is to add the correct amount of fungicide by pipette to one side of a jar held at a 45° angle. The jar is rotated 180° and is held at a 45° angle while the seed is added. Mixing is done as described above. When less than 28.3 g (1 oz) of product per 35.2 L (1 bu) is to be applied and the fungicide is miscible in water, uniform distribution may be accomplished by diluting one part fungicide to four parts water and applying five times the quantity to compensate for the dilution. In both cases, the jar must be rotated rapidly because the first seeds to come in contact with the liquid fungicide may absorb more fungicide than those contacted later.

The theoretical amount of formulation to be used must be increased by 10% to compensate for the fungicide that adheres to the container. Flowable fungicides and liquid formulations give better coverage of seeds than do wettable powders at low rates. They create essentially no chemical dust during application, and when dry they create less dust than do wettable powders. A flowable formulation should be mixed thoroughly before use.

Volatile formulations of fungicides are applied by the same methods as are nonvolatile fungicides. The treated seed, however, should be left in the airtight container 5–10 min and then aerated in a double, heavy-duty paper sack for 24–48 hr before being planted. Aeration is especially important when the lemma and palea remain attached to each seed (e.g., cultivars of barley [*Hordeum vulgare* L.], oats [*Avena sativa* L.], and other grasses).

Pelleting seed. Pelleting is another method used to apply fungicides to seed when the desired amount of chemical per seed exceeds the amount adhering to a given amount of seed. Pine (*Pinus* spp.) and other conifer seeds are pelleted primarily to protect them against soil fungi; pelleting has also been used to protect onion (*Allium cepa* L.) seed against the soilborne smut fungus *Urocystis cepulae* Frost (2).

Methyl cellulose, a standard pelleting agent, may be prepared as follows. A weighed quantity of methyl cellulose is placed in a flask and dried at 110°C for several hours. Then 40% of the final volume of water (at 80–90°C) is added during vigorous stirring. When the particles of methyl cellulose are wetted thoroughly, the suspension is brought to 50% of the final volume of water by adding water at room temperature. The mixture is vigorously stirred until it cools. The mixture is stored at 4–5°C until the methyl cellulose is completely dispersed (about 12 hr). Finally, more water is added to bring the final solution to the desired concentration.

The amount of sticker (e.g., methyl cellulose) required per 0.45 kg (1 lb) of seed depends on size of seed, sticker used, and amount of fungicide to be fixed to the seed. To pellet 0.45 kg (1 lb) of red pine (*P. resinosa* Ait.) seed, a jar is held at a 45° angle and 56.7 g (2 oz) of 4% methyl cellulose is added to one side of the jar. The jar is rotated 180° and seed is added and mixed as described above. The process is repeated using the desired amount of fungicide powder. The fungicide should be uniformly mixed with methyl cellulose as the seeds are coated.

Systemic fungicide. Systemic fungicides were first used as seed treatments to control loose smut of wheat and barley about 1965 (1). Carboxin, registered for use as a seed treatment for several crops, inhibits development of several systemic diseases. It not only controls loose smut of wheat and barley but is also an excellent replacement for volatile mercurial fungicides to control *Ustilago avenae* (Pers.) Rostr. and *U. kolleri* Wille, which cause loose smut and covered smut of oats, respectively, and *U. hordei* (Pers.) Lagerh., which causes covered smut of barley. More recently, other systemic fungicides, such as metalaxyl, have been approved for use as seed treatments.

Combinations of fungicides and of fungicides and insecticides. Organic nonmercurial fungicides developed during the last 35 years are more specific than were ethyl and methyl mercurial fungicides. Some nonmercurial fungicides are effective against the lower fungi and others against the higher fungi. Combinations of two or more selected nonmercurial fungicides are often active against a broad range of pathogens. Some combinations are registered for commercial use. They may be applied to seed by using methods similar to those used to apply one fungicide.

Insecticides as seed treatments should be evaluated in combination with fungicides because insecticides may predispose seedlings to seed decay and seedling blights or may react synergistically with the fungicides to cause phytotoxicity—for example, reduced germination or seedling vigor.

Phytotoxicity. A good seed treatment leaves no residues that are detrimental to seed germination, seedling development, plant growth, or final yield of foliage or seed. Phytotoxicity may be determined by observing the percentage germination of seed, the speed of emergence of seedlings, or the vigor of growth of emerged seedlings. In some cases, an abnormal seedling color may signify significant phytotoxic effect. It is best to obtain yield data to substantiate that the effects observed early in plant growth are deleterious to yield. Because some diseases also reduce the percentage of emerged seedlings, relatively uninfested seed and soil should be used to test fungicides for phytotoxicity.

Soils. Sites for seed-treatment tests should have uniform soil, topography, and cropping histories because these factors can influence the activity of soilborne pathogens and the performance of seed-treatment chemicals. In field tests with seed treatments for control of soilborne diseases, soils are usually not artificially infested because most seedling pathogens are ubiquitous in cultivated soils. To assure uniformity and severity of infection, however, inoculum can be added to the seed furrow at the time of seeding. One technique is to grow the pathogen on autoclaved plant material (e.g., oat kernels) and to add this inoculum to the furrow with the seed. This can be done by mixing the inoculum with the seed or adding it to the furrow separately by hand. Many procedures also enhance infection and disease, including minimum tillage, incorporation of organic matter, fertilization, irrigation, and flooding.

Planting. In general, crops used in seed-treatment experiments should be planted when environmental conditions favor the disease, not the crop. This procedure accentuates the value of seed treatments when compared with controls and permits better evaluation of treatments. The incidence of disease is higher when seed is planted slightly deeper than normal. For systemic diseases such as wheat bunt and barley stripe, deeper planting gives the fungus additional time to become established in the meristematic tissue of the seedlings. Infected seedlings must survive, however, so that experimental data may be taken as the crop matures. For these experiments, researchers should try for a balance between high infection and seedling survival.

In seed decay and seedling blight tests of sorghum, however, the seedlings need not survive. Most seedlings die from either seed decay or preemergent seedling blights. Data are collected when about 50% of the seedlings are in the two-leaf stage. Generally, planting slightly deeper than normal gives the fungi more time to infect and destroy the sorghum seedlings. One should attempt to obtain a 50% decrease in stand for the untreated control plot.

One planting date is usually adequate for seed treatment experiments, but, for certain diseases (e.g., wheat bunt, damping-off), soil temperature and moisture for the first few days after planting are critical in determining disease level. Therefore, two or more planting dates may ensure the collection of information.

Replications and plot design. The number of replications depends on the experiment conducted, susceptibility of cultivar, range in expected data, and soil variations. For example, when Red Chief wheat seed is infested with bunt teliospores and then planted, 90% of the spikes subsequently produced may be bunted. Treating the seed with an effective fungicide reduces the infected spikes from 90% to 1%. As few as three replications may be sufficient to separate mean differences. Experiments with seed decay and seedling blights of the same crop, however, require at least five replications. When seed is treated before planting, 75% of the emerged seedlings may be observed, compared with 60% when the seed is not treated. Replications should not be reduced to test more formulations or rates! Layout of plots (randomization of treatments, etc.) should be such that standard statistical procedures can be used to analyze the results (3).

Plot size depends on the crop, disease, and data desired—for example, number of emerged seedlings, infected seedlings, infected heads, and yield. Differences expected between the control and treated plots also affect plot size—the greater the difference, the smaller the plot needed. For control of bunt of wheat, the plot is a 3-m (10-ft) row with 0.3 m (1 ft) between rows; for covered kernel smut of sorghum (*Sphacelotheca sorghi* (Lk.) Clint.), it is a 12-m (40-ft) row with 0.6 m (2 ft) between rows. For yield of barley or wheat, it is four or six rows in width. The middle two or four rows are harvested to minimize border effects from adjacent plots. Size of plots should not be reduced to test more formulations or rates.

Publication of seed treatment data. In general, percentages of emerged seedlings, tillers, final stand of plants, infected seedlings, infected plants, and infected heads are expressed in whole numbers. Figures rounded off to one or two decimal places contribute little additional information when compared with whole numbers. Furthermore, comparing whole numbers is much easier for readers. Yields of grain or foliage may be expressed in either whole numbers or with one decimal, whether kilograms or metric tons per hectare are used.

Soil type and conditions are stated (e.g., organic matter content, pH, moisture, and temperature), as well as recent cropping history and weather conditions during the test. Seed-treatment data are analyzed statistically to show the magnitude of significant differences between treatment means. Newman-Keuls multiple comparison test and Duncan's multiple range and least significant difference tests are commonly accepted methods (3). Regression analysis may be more appropriate in some cases.

CONCLUSION

Extension and research and development personnel involved in seed-treatment testing may follow the above suggestions, but they will have to make modifications depending on the crop, cultivar, disease, and type of data desired. Each worker benefits from experience in developing and improving seed-treatment tests to meet individual needs and to satisfy professional standards and legal requirements.

CAUTION

As with all pesticides and chemicals, seed-treatment fungicides must be used with caution. All safety precautions and directions on the label of the container or accompanying fact sheet that the manufacturer provides should be followed. Direct contact with the chemical can be avoided by wearing impermeable gloves, working in a vented chemical hood or a well-ventilated room, and using a suction bulb with a pipette. All treated seed should be identified (preferably with a dye) and safely stored in a locked room to prevent accidental human or animal consumption.

LITERATURE CITED

1. Hansing, E. D. 1967. Systemic oxathiin fungicide for control of loose smut (*Ustilago tritici*) of winter wheat. (Abstr.) Phytopathology 57:814.
2. Hanson, E. W., Hansing, E. D., and Schroder, W. T. 1961. Seed treatments for control of disease. Pages 272-280 in: Seeds. The Yearbook of Agriculture. U.S. Dep. Agric., Washington, DC.
3. LeClerg, E. L., Leonard, W. H., and Clark, A. G. 1966. Field Plot Technique. 2d ed. Burgess Publishing Co., Minneapolis.
4. Mathre, D. E., Metz, S. G., and Johnston, R. H. 1982. Small grain cereal seed treatment in the postmercury era. Plant Dis. 66:526-531.

Evaluating Fungicides for Cottonseed

EARL B. MINTON, U.S. Department of Agriculture, Agricultural Research Service, Cotton Physiology and Genetics Research Unit, Stoneville, MS 38776; R. H. GARBER, U.S. Department of Agriculture, Agricultural Research Service, U.S. Cotton Research Station, Shafter, CA 93263; and J. E. DeVAY, Department of Plant Pathology, University of California, Davis 95616

Since the inception of the Cotton Disease Council in 1936, members of the council have convened almost every year to discuss and plan methods to control diseases of cotton (*Gossypium hirsutum* L.). Since 1953, the council has published an annual compilation of losses caused by each disease. Yield reductions caused by the seedling disease complex from 1953 through 1977 averaged 2.9% annually (21). Seedling diseases are also important factors in reducing plant stands and uniformity of plant development and in increasing losses from other diseases. The council, through its national cottonseed treatment program, coordinates advanced evaluations of experimental and registered fungicide seed dressings throughout the United States (Cotton Belt).

The effectiveness of the national cottonseed treatment program was apparent in 1971 when the use of alkyl mercury fungicides for coating of planting seeds was prohibited. Soon after notice of prohibition, other registered, effective fungicides, evaluated through this program, were available to replace the alkyl mercuries previously used for treating cottonseed (8,19). Also, other commodity groups used the data from the national cottonseed treatment program to select replacement fungicides for their crops. Industry uses the data from the national cottonseed treatment program to assess the performance of new fungicides and to aid in the registration of these fungicides with the U.S. Environmental Protection Agency.

A prerequisite of controlling seedling diseases in cotton is the planting of high-quality seeds. Planting seeds should be produced under conditions ensuring full development of the embryo and handled to maintain maximum quality. Seeds exposed to high moisture and high temperature, either before or after harvesting, may deteriorate rapidly. Seeds damaged mechanically or chemically during harvesting, ginning, or processing may be of inferior quality. Thus, high-quality seeds are not always available for planting. Utilization of high-quality seed does not always ensure uniform stands of vigorous seedlings, since most seeds and seedlings are very susceptible to soilborne microorganisms. A direct relationship occurs among inoculum levels of pathogens, disease incidence, and the effectiveness of fungicides to control seedling diseases (4,15).

The effects of fungicides on seed germination and seedling vigor should also be determined (18,20). Seed quality can affect germination, as well as plant development, and should not be misinterpreted as a fungicidal response. Some variations in seed quality are desirable when evaluating fungicides because seeds with a wide range in quality are planted by cotton growers.

The objectives of the cottonseed treatment program are to develop more effective techniques to use in evaluating fungicides applied to planting seeds and to identify more effective seed dressings to control the seedling disease complex. The active ingredient of fungicides on the seeds must protect seed and seedlings from both seedborne and soilborne pathogens.

EVALUATION PROCEDURES

Fungicides. In the early years of the national cottonseed treatment program, the same fungicide treatments were evaluated beltwide. Today, different treatments may be evaluated in the eastern (east of Texas) and in the western (Texas and westward) regions.

During the annual meetings of the Cotton Disease Council, chemical company representatives nominate fungicide treatments to be included in the eastern regional program. Many treatments nominated, especially combinations of fungicides, involve two or more chemical companies. All treatments nominated must have performed as well in industry field and greenhouse programs as the standard treatment in the national program. A minimum of three industry tests over a wide range of conditions is required before potentially effective treatments are accepted for evaluation.

The western regional program was begun in 1968. Only treatments that have increased stands of cotton significantly in the eastern regional or in special tests are evaluated in the western program.

Seeds. During the early years of the national cottonseed treatment program, cooperators treated their own seed lots. Thus, a wide range of lots, in both variety and seed quality, was used. It became apparent that a common seed lot from a single variety should be used for all evaluations and that a given treatment should be from the same batch of seed. Therefore, each year a single seed lot with 75–85% germination and naturally infested with seedborne pathogens was used for all evaluations. Since 1959, the seeds used in the eastern program and in most of the western program have been grown in Mississippi by the Stoneville Pedigreed Seed Company, Stoneville. In the western program, seeds for some years have been produced in California.

Originally, all of each seed lot was machine delinted; later, one-half of each machine-delinted lot was also acid delinted before the fungicides were applied. From 1972 through 1979, one-half of the seed lot was machine delinted and the other half was acid delinted only. Because the fungicides performed similarly with both types of delinted seed (5,13), only acid-delinted seeds have been evaluated since 1980. The seeds used in these programs are assumed to be representative in quality of a high percentage of the seeds planted by cotton growers.

Fungicide application. The chairman of the Seed Treatment Committee of the Cotton Disease Council supervises the application of the fungicide to 1.6- to 6.4-kg lots of seed in a rotating drum treater. This procedure requires smaller quantities of fungicide, less labor, and provides more uniform treatments than would be possible if cooperators treated their own seeds. Emulsifiable liquid concentrates of fungicides are diluted with water and sprayed with an atomizer at 2% by weight of the seeds as they tumble in the rotating drum. Oil formulations without dilution are applied to the seeds with an atomizer. Wettable powder formulations are mixed with water and applied in a volume equivalent to 2% by weight of the seeds. Wettable powder suspensions are applied to the seeds with a small hand sprayer, which disperses the suspension into small droplets. The water reduces the loss of fungicide while the seeds are being treated and ensures uniform distribution. The seeds are tumbled for 5 min or longer to assure uniform coverage and to aid in water evaporation. Treated seeds are stored in opened paper bags at room temperatures for 24 hr to allow all excess water to evaporate. Subsamples of each of the treated and untreated seed lots are then placed in plastic bags and mailed to each cooperator.

In both the eastern and western programs, untreated seeds are included for comparisons. Ethylmercury p-toluene sulfonamide (7.7%) (Ceresan M) was used as a standard from the 1950s until 1982. Since 1971, 22.8% pentachloronitrobenzene (PCNB) and 11.4% 5-ethoxy-3(trichloromethyl)-1,2,4-thiadiazole (ETMT) (Terra-Coat L-21) has been used as a second standard, and it became the sole standard in 1983. The standard treatment is used to compare the effectiveness of fungicides of known performance and to determine the progress being made in the development of more effective treatments.

Field test. Conventional cultural practices representative of those used across the Cotton Belt are used to plant the seeds. More precise or uniform planting techniques may allow for more accurate determination of differences among treatments (17), but the ones used are similar to those used commercially.

One hundred seeds per plot (treatment) are planted in a randomized complete block design with eight replications. Each plot consists of one row from 6 to 9 m long. The treated seeds are planted about 14 days earlier than the suggested planting date for each respective test site. The seed and seedlings are usually exposed to conditions more favorable for the development of diseases than would be desirable for commercial cotton. Because diseases are usually severe under these conditions, the more effective fungicide treatments can be readily identified based on the surviving and dead seedlings recorded at appropriate times after emergence.

Data analyses and report. Percentages of live and dead seedlings are calculated from the number of seeds planted, and statistical analyses of the data for individual and combined test sites are performed. Since seed from the same treated lots are planted at several locations, data from both individual and combined test sites are used to select effective treatments for further evaluations or for commercial use (9,10). Ineffective treatments are identified and eliminated after the first year. These intensive evaluations permit the identification and documentation of effective treatments within 3 years because a wide range of climatic conditions, soil types, seed qualities, and microorganisms are encountered.

The chairman of the Seed Treatment Committee assembles and summarizes the data and compiles a list of suggested treatments received annually from cooperators for commercial use in each state. The data are published annually in *The Cotton Gin and Oil Mill Press* and in the *Proceedings of the Beltwide Cotton Production Research Conferences.* The policy of the Cotton Disease Council requires that all treatments included in the suggested list for commercial use be evaluated in the regional program for a minimum of 3 years.

Pathogen identification. Another area that has received major emphasis is the identification of organisms that cause seedling diseases of cotton. More recently, soil samples from selected test sites have been examined to determine the inoculum levels of *Pythium* spp., *Rhizoctonia solani* Kühn, and *Thielaviopsis basicola* (Berk. & Br.) Ferr., important soilborne seedling disease pathogens. Several additional but less virulent seedborne and soilborne organisms may influence disease severity, especially when the host is under stress. The prevalence and virulence of primary disease-causing fungi differs from field to field and from eastern to western United States.

Fungicide specificity. In addition to the national tests discussed above, greenhouse and field evaluations conducted in California since 1958 have been instrumental in the identification of effective new fungicides. The greenhouse tests have effectively identified new fungicides and the rate of active ingredient needed to control the major fungi that cause seedling diseases. Data from field tests have supported greenhouse results.

In greenhouse tests, the potting mix (Yolo loam and sand, 1:1 ratio) is infested with either *P. ultimum* Trow., *R. solani*, or *T. basicola* alone or in combination. *P. ultimum* and *R. solani* are grown on separate potato-dextrose agar plates for 1 week, then 12 plates of each fungus are blended separately in 600 ml of water. One-sixth of the suspension of each fungus is mixed with the growth medium in six separate flats (45 × 60 × 15 cm). *T. basicola* grown in potato-dextrose broth is mixed with the growth medium at a rate to give an inoculum density of 100 conidia per 100 g of growth medium. These inoculum concentrations for *P. ultimum* and *R. solani* kill about 90% of the untreated germinating seeds and seedlings.

For seed treatment, fungicides are added to 5 ml of distilled water in a 1-L (clear) container. The fungicide suspension is uniformly distributed over the inside walls of the container by rotation. Then 100 g of Acala SJ2 cottonseed is added. The container is closed and shaken until the fungicide suspension is transferred from the walls of the container to the seed and all seeds are uniformly coated. The treated seeds are air-dried in a fume hood overnight.

Each greenhouse test consists of six randomized seed treatments per flat. Twenty seeds for each treatment are planted in each flat. An untreated check and a standard, fenaminosulf (Lesan; 3 oz cwt) + chloroneb (Demosan; 10 oz cwt), are included in each test. Several concentrations of new fungicides are evaluated to determine the most effective rate.

Emergence and postemergence damping-off of seedlings are recorded after 1 and 2 weeks. Seedlings in flats infested with *T. basicola* alone or in combination with the other pathogens are assigned a disease index (1–5) on the basis of root browning. A disease index of 1 indicates no symptom, whereas 5 indicates complete browning and rotting of lateral roots. If the initial test indicates that the fungicide controls one or more of the pathogens, it is used with other fungicides in subsequent greenhouse tests.

Effective treatments are then evaluated in a field test at the U.S. Department of Agriculture Cotton Research Station, Shafter. In each field test, treatments are replicated at least eight times. One hundred seeds are planted per plot. Each treatment is planted through each planter unit the same number of times to eliminate planter rate and depth of seed covering as variables. After 4 weeks, emergence and survival data are recorded and analyzed.

Additional tests are conducted in at least eight different fields in the San Joaquin Valley of California. Fields with a history of seedling pathogens are selected to test the fungicides that were effective in the test at Shafter. Treatments are replicated at least eight times in single rows 30 m long. Several stand counts are made and analyzed. Isolations are made from diseased seedlings from the field tests to determine the causal seedling disease pathogens. Soil samples are also assayed for the presence and quantity of *Pythium* spp., *R. solani*, and *T. basicola*.

RESULTS AND DISCUSSION

PCNB was one of the first fungicides reported to control *R. solani* (1,2,6). Since then, several chemicals, such as chloroneb and carboxin, have been identified and used commercially to control this pathogen (11). Furmecyclox (EPIC) is a promising new systemic material that controls *R. solani*. Fenaminosulf was identified in the mid-1950s as a control for *P. ultimum* (6). Later, ETMT and more recently metalaxyl (APRON), a systemic fungicide, have been shown to be very effective against *P. ultimum*. Several broad-spectrum fungicides, such as captan, thiram, TCMTB, captafol, and mancozeb (Dithane M-45), are somewhat effective against both *R. solani* and *P. ultimum*, but they are not as effective as the more selective fungicides mentioned above. The latter fungicides are usually combined with others to increase the overall control of diseases.

The fungicides mentioned above do not effectively control *T. basicola*, which delays the growth of seedlings but rarely kills them. In recent tests, seed treated with benomyl, imazalil, CGA 64250, and CGA 64251 produced seedlings with significantly less *T. basicola* (3).

The level of disease control with seed treatment has been increased by combining two or more seed protectants or by combining seed protectants with systemic fungicides. Protectant fungicides are applied to the seed to kill or inhibit the growth of seedborne pathogens and those that occur in the soil near the seeds. Systemic fungicides are absorbed by the roots of seedlings, and the chemical remains in the roots or is translocated to the hypocotyl of seedlings to provide protection against postemergence seedling diseases. Many fungicides selectively control specific pathogens. To provide maximum protection against the pathogens anticipated, two or more selective fungicides are used in combination. Plant stands have been similar when seeds from the same seed lot were coated with identical fungicide treatments by commercial seed processors and in small cement mixers similar to the one used for the regional tests (16).

SUMMARY

Under the auspices of the Cotton Disease Council, the agricultural chemical industry has cooperated with federal and state agricultural researchers in evaluating fungicides for cottonseed treatment. Through this program, many promising new seed protectant and systemic fungicides have been developed and evaluated. Seed protectant fungicides primarily control the organisms on or in the immediate vicinity of the seed. Systemic fungicides provide protection to both the seed and the seedlings. Significant progress has been made in the development of systemic fungicides for control of *R. solani* and *Pythium* spp. Many of these are registered for use on planting seed. During the last few years, the first fungicides evaluated that produced a significant level of control of *T. basicola* were identified. Coverage and adherence of the fungicides to the seed coats have been improved significantly by the development of flowable fungicides. Comparable levels of disease control are obtained with lower rates of active ingredients with flowable than with wettable powder fungicides.

More effective fungicide treatments have been identified by statistical analysis of the data from both individual and combined test sites (9,10,12,14). Treatments that control disease complexes over a wide range of environments, soil types, and organisms have been more readily selected from the combined analyses of the data across test sites than from individual locations. The list of fungicides suggested for coating cottonseed for each state is revised as warranted on the basis of performance data (7).

LITERATURE CITED

1. Arndt, C. H. 1953. Evaluation of fungicides as protectants of cotton seedlings from infections by *Rhizoctonia solani*. Plant Dis. Rep. 37:397-400.
2. Brinkerhoff, L. A., Oswalt, E. S., and Tomlinson, J. T. 1954. Field tests with chemicals for the control of *Rhizoctonia* and other pathogens of cotton seedlings. Plant Dis. Rep. 38:467-475.
3. DeVay, J. E., Garber, R. H., Weinhold, A. R., and Wakeman, R. J. 1981. Comparisons of selected fungicide seed treatments against *Rhizoctonia solani, Pythium* and *Thielaviopsis basicola*. Proc. Beltwide Cotton Prod. Res.

Conf. 1981. Pp. 24-28.
4. Garber, R. H., DeVay, J. E., Weinhold, A. R., and Matherson, D. 1979. Relationship of pathogen inoculum to cotton seedling disease control with fungicides. Plant Dis. Rep. 63:246-250.
5. Garber, R. H., and Hoover, M. 1973. A comparison of acid- and machine-delinted cotton seed planted at two fungicide seed treatment rates. Proc. Beltwide Cotton Prod. Res. Conf. P. 30.
6. Leach, L. D., Garber, R. H., and Lange, W. H. 1959. Cottonseed treatment trials in California, 1954–1958 with special references to specific fungicides. Plant Dis. Rep. (suppl.) 259:213-221.
7. Minton, E. B. 1983. Report of the Cottonseed Treatment Committee—1982. Proc. Beltwide Cotton Prod. Res. Conf. Pp. 11-16.
8. Minton, E. B. 1974. Status of non-mercurial seed treatments, 1967–73. Proc. Western Cotton Prod. Conf. 1974. Pp. 5-9.
9. Minton, E. B., and Fest, G. A. 1975. Seedling survival from cottonseed treatment experiments at several locations. Crop Sci. 15:509-513.
10. Minton, E. B., Fest, G. A., and Sciumbato, G. L. 1975. Effect of cottonseed treatment on stand. Texas Agric. Exp. Stn. MP-1210c.
11. Minton, E. B., and Garber, R. H. 1983. Controlling the seedling disease complex of cotton. Plant Dis. Rep. 67:115-118.
12. Minton, E. B., and Green, J. A. 1980. Germination and stand with cottonseed treatment fungicides: Formulations and rates. Crop Sci. 20:5-7.
13. Minton, E. B., and Quisenberry, J. E. 1980. Comparison of cottonseed delinting methods in evaluating seed-treatment fungicides. Agron. J. 72:573-575.
14. Minton, E. B., and Quisenberry, J. E. 1981. Beltwide evaluation of fungicides applied to acid-delinted cottonseed. Agron. J. 73:684-686.
15. Papavizas, G. C., Lewis, J. A., Minton, E. B., and O'Neill, N. R. 1980. New systemic fungicides for the control of cotton seedling disease. Phytopathology 70:113-118.
16. Paulus, A. O., Nelson, J. A., Shibuga, F., and Garber, R. H. 1976. Treating cotton seed cut losses to damping-off. Calif. Agric. 30 (10):15.
17. Pinckard, J. A., and Ivey, J. 1971. Chemical treatments for cottonseed. La. Agric. Exp. Stn. Bull. 655.
18. Pinckard, J. A., and Melville, D. R. 1975. Some new developments in cottonseed dressings. Plant Dis. Rep. 59:262-266.
19. Ranney, C. D. 1971. Effective substitutes for alkyl mercury seed treatments for cottonseed. Plant Dis. Rep. 55:285-288.
20. Ranney, C. D. 1972. Multiple cottonseed treatments: Effects on germination, seedling growth and survival. Crop Sci. 12:346-350.
21. Watkins, G. M. 1981. Cotton diseases—general concepts. Pages 2-10 in: Compendium of Cotton Diseases. American Phytopathological Society, St. Paul, MN. 87 pp.

Evaluating Fungicides as Soybean Seed Treatments

N. G. WHITNEY, Texas A & M University, Agricultural Research and Extension Center, Beaumont 77706

No seed treatment provides protection against all seedling diseases. Seedling pathogens are either seedborne or soilborne, and their interaction with different environmental conditions compounds the problem of disease control. Stress, both physical and chemical, alters the rate at which a seedling emerges. A slowly emerging seedling is more vulnerable to attack by both seedborne and soilborne pathogens.

Few soybean seeds are treated with fungicides before planting, partly because treated seed that is not sold for planting cannot then be sold for the soybean oil market: treated seed can only be sold for planting. Seedsmen and producers want a treatment that is inexpensive, safe to handle, and compatible with their planting operation. In the past, dust treatments were popular but were a health hazard. With the development of liquid formulations and machinery for their application, dust formulations are being phased out (5).

Soybean seed-treatment tests have been conducted at various locations in the United States since 1925 (4). In the southern states, the combination of poor seed quality and cool, moist environment after planting sometimes results in the loss of 20–25% of the potential stand if soybean seeds are not treated with fungicides. Under such conditions, researchers have demonstrated yield increases from seed treatments (3).

The method described below is intended for field evaluation of the efficacy of fungicides used to control seedling diseases of soybeans. It provides guidelines for establishing test plots, treating and planting the seed, obtaining the data, and interpreting the results.

METHODS

Test site. Soil must be suitable for growing soybeans and climatic conditions favorable for disease development. Fungi causing seedling disease should be endemic to the test area. These include *Pythium, Phytophthora, Rhizoctonia, Diaporthe* spp., and *Phomopsis* spp. Naturally occurring inoculum may not be present in sufficient quantities to assure uniform disease development throughout the experimental area. Consequently, selected pathogenic fungi may need to be cultured in the laboratory for incorporation into the soil of the research area. For example, *R. solani* can easily be increased by transferring a pure fungal culture to an autoclaved, moist, natural grain medium. Two to three weeks should be allowed for fungal growth. The inoculum is incorporated into the test site at the rate of 40 g/m^2 of soil. Tuite (6) suggests other methods for increasing *Pythium* and *Phytophthora*. Seedborne fungi such as *Phomopsis* or *Diaporthe* are usually found in planting seed (2,7). Cultural practices for the test site should approximate those of commercial producers.

Before field evaluation, seed lots are commonly evaluated for germinability using the "rag doll" technique. In some cases, accelerated aging tests are performed to predict seedling vigor.

Seed treatment. Any locally grown commercial variety is acceptable. Seed germination should range between 60 and 80% to reveal differences in treatments. It is advisable to run a dosage series for each fungicide to ascertain the effective rate for maximum disease control. The dosage series should range between 0.5 and 8.0 g of formulated product per kilogram of seed. Each fungicide, whether dry or flowable, is mixed with 10–15 ml of distilled water. The slurry is placed with 1 kg of seed in a jar mill and tumbled for about 5 min. Whatever time is chosen, that time is used for each treatment. If a jar mill is not available, a plastic bag can be used. The seed and chemical are introduced into the bag and the bag is filled with air. After the opening is twisted to hold the air, the bag is agitated until seeds are coated. Again, the time is noted, with each treatment receiving the same amount of agitation. Care should be taken when using a plastic bag, as some solvents and carriers dissolve plastics. The plastic can be tested by placing a small amount of fungicide on it and waiting a few minutes. Slurry-treated seed should be allowed to dry 24 hr, then counted and packaged for planting 40 seeds per meter of row for individual plots.

There are two types of hopper-box treatments—dry and flowable. For the dry formulations, the calculated amount of fungicide can be added to individual row packets and shaken before planting. For flowable treatments, the calculated amount is added to the seed in a beaker or cup and swirled until coated.

A standard control treatment, such as captan or thiram at the recommended rate, and untreated controls are included to determine disease severity and provide a basis for treatment comparisons. Precautions should be taken to avoid inhalation of or exposure to fungicide dusts during treatment, packaging, and planting of the seed. Commercial inoculant at 0.25 g per package of treated seed is added just before planting, with the following precautions: the fungicide and inoculant should not be combined before being added to the seed, either in a slurry mix or hopper-box treatment; and seed treated with both inoculant and fungicide should not be held for longer than 6 hr. If this amount of time passes, the seed should be reinoculated.

Procedure. The test design is a randomized complete block with at least four replications per treatment. More replications are used in areas where disease incidence is light. A precision-drop planter is used to plant the correct number of seeds for the desired length of row. Each plot is three rows wide and about 6 m long. Row width varies, depending on cultural practices of local producers. Under good soil conditions, seedlings should be ready to count in 20 days. The center row of each three-row plot is used for taking stand counts. Plants are then grown to maturity under standard management practices. At maturity, the center row of each plot is harvested and yields are calculated for each treatment. Yields are adjusted to a constant 13% moisture level.

Data determination. Emergence is determined by counting all healthy seedlings that are free of abnormalities in the center row of each plot. Abnormalities include shortened, thickened stems, which may have deep lesions in them; shriveled or weakened stems; damaged or absent primary leaves; missing or moldy cotyledons; or missing terminal buds.

After the stand counts are recorded for each replicated treatment, the data are analyzed and Duncan's multiple range test (1) or the least significant difference (LSD) test is applied for significance at the 5% level. Effective fungicides must have significantly higher stand counts than does the untreated control. Other methods that help to clarify the data include calculations of the percentage of stand for each treatment and of the percentage of change from the untreated control, determined as follows:

$$\text{Percentage of change} = \frac{\left(\begin{array}{c}\text{Number of plants}\\ \text{in treatment}\end{array}\right) - \left(\begin{array}{c}\text{Number of plants}\\ \text{in control}\end{array}\right)}{\text{Number of plants in control}} \times 100.$$

Plants should be observed for any type of phytotoxicity and the percentage of severity reported. Common types of phytotoxicity are leaf burning, chlorosis, or stunting. At maturity, yields are analyzed statistically to show significant differences from the untreated control.

The following information is included in reporting test results: (i) product name and U.S. Environmental Protection Agency registration number, (ii) percentage and chemical identity of active ingredient or ingredients, (iii) treatment rate, (iv) method of application, (v) number of seeds per treatment, (vi) number of replications, (vii) treatment date, (viii) planting date, (ix) date on which emergence count was made, (x) percentage of stand, (xi) percentage of change from untreated control, (xii) emergence significance as determined by Duncan's multiple range test or LSD test at the 5% level, (xiii) percentage and type of phytotoxicity, and (xiv) yield and significance as determined by Duncan's multiple range test or LSD test at the 5% level.

DISCUSSION

In an area where fungal inoculum concentrations are adequate, the test yields a high degree of precision. Results vary from year to year, however, because of environmental conditions. A dry year produces smaller differences than does a wet year because some seedling disease fungi require water for movement on or through the soil. Humidity also plays a large role in fungal effects on host plants. Where necessary, overhead irrigation is desirable to establish high humidity. If inoculum levels are not high enough, artificially grown inoculum incorporated into the soil increases the precision of the test.

Data obtained from this type of test are reliable. Treatments showing an increased response over controls can be recommended to producers, provided the chemical has been cleared for use on soybeans by the Environmental Protection Agency.

LITERATURE CITED

1. Duncan, D. B. 1955. Multiple range and multiple *F* tests. Biometrics 11:1-42.
2. Ellis, M. A., Ilyas, M. B., and Sinclair, J. B. 1975. Effect of three fungicides on internally seed-borne fungi and germination of soybean seeds. Phytopathology 65:553-556.
3. Johnson, H. W. 1951. Soybean seed treatment. Soybean Dig. 11 (7):17-20.
4. Johnson, H. W., and Koehler, B. 1943. Soybean diseases and their control. U.S. Dep. Agric. Farmers Bull. 1937.
5. Phipps, P. M. 1984. Soybean and peanut seed treatment: New developments and needs. Plant Dis. 68:76-77.
6. Tuite, J. 1969. Plant Pathological Methods, Fungi and Bacteria. Burgess Publishing Co., Minneapolis.
7. Wallen, V. R., and Cuddy, T. F. 1960. Relation of seed-borne *Diaporthe phaseolorum* to the germination of soybean. Proc. Assoc. Off. Seed Anal. 50:137-140.

Screening Fungicides for Seed and Seedling Disease Control in Plug-Mix and Fluid-Drilling Plantings

R. M. SONODA, University of Florida, Institute of Food and Agricultural Sciences, Agricultural Research and Education Center, Fort Pierce 33454, and S. C. PHATAK, University of Georgia, College of Agriculture, Coastal Plain Experiment Station, Tifton 31793

Plug-mix planting (3) and fluid-drilling (1) are two relatively new techniques for the direct seeding of small-seeded crops. The media used in these planting techniques provide environments that promote quicker germination, quicker plant growth, and more uniform plants with greater yields than when seeds are planted directly in soil. In plug-mix planting, a 1:1 mixture of predetermined amounts of horticultural vermiculite and shredded sphagnum peat moss containing seed and starter fertilizer is deposited at desired intervals by hand or tractor-drawn implement (3). In fluid-drilling, primed or germinated seeds are pumped in a gel carrier into small furrows in continuous streams or in clumps (2).

Fungicides can be added to the planting media in both planting systems to protect seeds and seedlings against plant pathogens. When effective fungicides are added to these planting systems, the presence of a fungicide-infused medium around seeds and seedlings provides a fungicide-protected, pathogen-free environment for seed and seedling development. The amount of fungicide required on a per-hectare basis depends upon the concentration of fungicide needed in the planting media for effective disease control and the amount of media required per hectare. Usually only a small amount of fungicide is needed. For example, about 40 g active ingredient (a.i.) of ethazol was required to control *Pythium aphanidermatum* (Edison) Fitzp. on tomatoes on a crop having 7,500 hills per hectare (5).

Since both techniques are relatively new and evolving, especially fluid-drilling, different carrier materials are continually being tested. The following procedures for fungicide screening apply to the two named techniques and may be applicable to similar techniques where carriers of seeds also serve as media for seedling development. The procedures described below can be used for any small-seeded crop. Most of the information is provided using tomato, *Lycopersicon esculentum* Mill, as the test crop and the two most important preemergence and postemergence pathogens of tomato in many parts of the world, *P. aphanidermatum* and *Rhizoctonia solani* Kühn, as test pathogens.

PROCEDURES

Seed. Untreated tomato seed obtained from a commercial supplier is used. Percentage of germination of seed is determined, and seed lots with poor germination are not used. Seed can be added ungerminated or germinated to plug-mix media. For fluid-drilling, tomato seeds are germinated at 20°C for 3 days in nylon or cheesecloth sacks in aerated water in a clean glass container (beakers, aquariums, etc.) and then added to the gel carrier as soon as possible.

Media. Because of the many different materials used as carriers and media in plug-mix planting and fluid-drilling, safe levels of fungicides for a particular media must be determined before efficacy trials, even if the fungicide is one that has been used on the host in other situations. The media used may affect phytotoxicity as well as efficacy of the compound. Peat and vermiculite mixes used in plug-mix planting generally differ in coarseness of material and in such additives as fertilizer, bark, perlite, and growth stimulants. To standardize the amount of fungicide in plug-mix planting, fungicides are added to peat-vermiculite on a per-dry-weight planting mix basis. Commercially available peat-vermiculite mixes contain different amounts of moisture, and the addition of fungicides to these mixes based on weight of mix without thorough drying will lead to gross errors.

Many different types of gels are being tested as carriers for the primed or germinated seed used in fluid-drilling. The interaction of the gels with different additives is being studied. Presently there are few data on the interaction of fungicides with these gels. In addition to possible effects of the gels on efficacy and phytotoxicity of fungicides, fungicides or their carriers may affect some properties of the gels.

Greenhouse phytotoxicity tests. In plug-mix planting, fungicides are suspended in distilled water and added to peat-vermiculite. The volume of water added is usually enough to bring the moisture level of the mix up to about 60% of its total weight. Calculations on the amount of fungicides to add to water and the amount of fungicide suspension to add to peat-vermiculite can be based on the amount of moisture already present in the particular peat-vermiculite used. Treatments that can be used as standards are 0.6 g a.i. of captan 50W or 0.5 g a.i. of ethazol per kilogram (dry weight) of peat-vermiculite (for comparisons on *P. aphanidermatum*) and 0.15 g a.i. benomyl and 0.15 g a.i. pentachloronitrobenzene per kilogram (dry weight) of peat-vermiculite (for comparisons on *R. solani*). In initial tests, the amount of fungicide to use may be estimated from standard use rates for the fungicide. Once a good estimate of a new fungicide is determined, amounts 2 to 3 times that should be screened for phytotoxic effects.

Fungicide and gel powder can be mixed with water at the same time, or, preferably, fungicide can be added after the gel is prepared. The concentration of fungicide to be used is based on the volume of gel suspension. A starting point for the amount of fungicide required for control of a seedling pathogen may be obtained from the report by Ohep et al (4).

Uninfested, limed and fertilized, pathogen-free soil (steamed soil, uncropped soil, etc.) is placed in containers—for example, in 180- to 360-ml (6- to 12-oz) polystyrene cups or in alternate 5 × 5 cm (top opening) cells of Speedling trays. Depressions are made on the soil surface with a No. 7 rubber stopper. For plug-mix planting, 15 cm^3 (1 tablespoon) of peat-vermiculite containing the fungicide is placed in the depression and tamped down. Five or 10 tomato seeds are placed on the top of the mix. Another tablespoon of mix containing the same amount of fungicide is added and tamped down. For fluid-drilling planting, enough pregerminated tomato seeds are added to the gel to give about five seeds for 5 cm^3. Then 5 cm^3 of mix is transferred to the depression on the soil surface, and the gel may be lightly covered with uninfested soil. Controls for phytotoxicity screening in both planting systems consist of medium without added fungicides. When polystyrene cups are used as containers, they can be distributed in a randomized pattern when incubated. When Speedling trays are used, the treatments can be randomized within the tray or over several trays. Four to five replicates of each fungicide concentration should be adequate for separating differences. The treatments can be incubated on the greenhouse bench, preferably at temperatures favorable for the development of the pathogens in question. For the tomato pathogens, *P. aphanidermatum* and *R. solani*, greenhouse temperatures ranging from 24 to 36°C are suitable. The treatments can be incubated in growth chambers set at about 30°C. The number of emerged seedlings is counted daily until most of the seedlings emerge. Phytotoxicity symptoms, such as reduced amount and rate of seed germination, reduced amount and rate of seedling growth, and abnormal growth or coloration of cotyledons, leaves, stems, or roots, should be recorded.

Compounds or rates of compounds that produce marked symptoms of phytotoxicity are usually eliminated from further tests. In a few cases, compounds drastically reduced rate of seedling emergence or rate of growth but plants appeared healthy otherwise. Some of these compounds, as well as those eliciting little or no phytotoxic effects, are selected for further testing.

Greenhouse efficacy tests. *P. aphanidermatum* and *R. solani* can be grown in several ways to provide inoculum for greenhouse fungicide efficacy tests. In tests by the first author, the fungi were grown on twice-autoclaved brown-top millet seed for 2 weeks. About 30 infested seeds were added per 75 cm^3 of steam-sterilized sandy soil. The infested soil was placed in polystyrene cups or cells in Speedling trays. Planting media, seed, and fungicide were placed in the soil in the same sequence as described in the greenhouse phytotoxicity tests. Plantings made in uninfested soil served as controls. The soils in the containers were kept moist by daily watering. Emerged seedlings were counted daily for 5–7 days after the emergence of first seedlings. Damped-off seedlings were counted daily up to 20 days after planting. Each damped-off seedling was removed and replaced with a toothpick to avoid confusion in counting, especially in cases where emergence of seedlings occurred over a prolonged period. Twenty days after planting, seedlings were lifted and roots rated for incidence and severity of lesions. A disease severity index was calculated based on a rating system of 1 = healthy to 5 = plants dead.

Field tests. Field test sites are selected on the basis of past history of disease. The population levels of the pathogen in question can in many cases be assayed with selective media. Generally a crop grown immediately before the field test of fungicide efficacy will increase the population of the desired pathogens. *P. aphanidermatum* and *R. solani* populations are generally increased with cultivation of tomato. The time of year for the test should coincide with periods especially conducive for the development of the disease or diseases. *P. aphanidermatum* and *R. solani* are problems on tomato seed and seedlings during the warmer periods of the year. Along the southeast coast of Florida where the experiments cited above were conducted, May through September are suitable months for good seedling disease development. However, if yield data are required, July through September are most suitable, as earlier plantings would be subjected to excessively high temperatures and heavy rainfall that might increase losses to other factors and confound results.

In plug-mix planting, weighed amounts of seed can be mixed with peat-vermiculite in a small, portable cement mixer to provide an average of about five seeds per hill. This usually results in most hills receiving from 2 to 10 seeds. The planting mix is then divided into smaller batches, and candidate fungicides are added to the planting mix. The fungicides are thoroughly mixed into the planting mix. One batch of planting mix is left free of fungicide to serve as control. For fluid-drilling, in field tests, the material can be prepared in the same manner as for the greenhouse phytotoxicity tests. About 60 cm^3 of peat-vermiculite mix or 5 cm^3 of gel is deposited in depressions made in soil at selected intervals, depending on the crop and the purpose of the field tests. A gel planter for small plot work has been developed (2). Plans for the planter, which is easy to build and use, are available from the second author. If tests require only seedling data, the distance between hills can be short. For tests requiring data on effect on yield or other mature plant factors, distance between hills should be the standard for the crop in the area. Four to six replications of each treatment are usually adequate. For precision in determining differences in fungicide efficacy, based on the work with *P. aphanidermatum*, each replicate should consist of at least 20 hills. In some cases, even this number may be inadequate if conditions are extremely unfavorable for disease development. The critical period for development of seeds and seedling diseases is extremely short. For *P. aphanidermatum* on tomato, the period is from planting until about 2 weeks after planting. If little or no rainfall occurs during this period, free moisture must be provided to aid disease development. Overhead sprinkler irrigation has been helpful in aiding the development of *P. aphanidermatum* on tomato. Standard foliar fungicides and insecticides and other cultural practices should be applied when yield data are desired. These practices generally do not interfere with seedling fungicide efficacy tests.

Emerged seedlings are counted daily. Damped-off

seedlings are counted each day up to the period when the crop hardens and is no longer susceptible to the disease. Each damped-off seedling is replaced with a toothpick to avoid confusion with seedlings that emerge or are killed later. After the period for losses due to damping-off is over, seedlings can be lifted and rated for root damage and a disease severity index can be calculated. Means of disease severity indices for different treatments can be compared by least significant difference, regression analysis, or other appropriate procedures. The number of hills with all seedlings killed is also recorded and analyzed. When yield data are desired, hills are thinned to the desired number of plants and marketable plant parts are harvested at maturity. Determinations are made on total yield, quality, etc., and comparisons are made between treatments. In situations where disease development is low or nonexistent, yield data may be used to determine whether phytotoxic yield losses are occurring for some of the treatments.

SUMMARY

The information obtained from the greenhouse phytotoxicity tests, greenhouse fungicide efficacy tests, and field tests must be summarized and published in some form to aid in the development of the use of this relatively simple, inexpensive, and effective method of protecting seed and young seedlings of small-seeded crops.

LITERATURE CITED

1. Currah, I. E., Gray, D., and Thomas, T. H. 1974. The sowing of germinating vegetable seeds using a fluid-drill. Ann. Appl. Biol. 76:311-318.
2. Ghate, S. R., Phatak, S. C., and Jaworski, C. A. 1982. A gel planter to sow multitreatments. HortScience 17:582-583.
3. Hayslip, N. C. 1973. "Plug-Mix" seedling developments in Florida. Proc. Fla. State Hortic. Soc. 86:179-185.
4. Ohep, J., McMillan, R. T., Bryan, H. H., and Cantliffe, D. J. 1984. Control of damping-off of tomatoes by incorporation of fungicides in direct-seeding gel. Plant Dis. 68:66-67.
5. Sonoda, R. M. 1976. Incorporating fungicides in planting mix to control soilborne seedling diseases of plug-mix seeded tomatoes. Plant Dis. Rep. 60:27-30.

Introducing and Evaluating Liquid Fungicides in Elm Trees for the Control of Dutch Elm Disease and Other Disorders

R. JAY STIPES, Virginia Polytechnic Institute and State University, Blacksburg 24061, and RICHARD J. CAMPANA, University of Maine, Orono 04473

The introduction of pesticides, nutrients, plant growth regulators, and other substances into the vascular system of woody plants, especially trees, has enormous potential and has intrigued both scientist and lay person for centuries. The introduction of these substances has been prompted by many basic or practical end points: investigating the movement of liquids in xylem and phloem; stimulating growth, flowering, or fruiting; imparting protection of various tree tissues against insect attack; preventing or curing an array of diseases, such as those caused by vascular, foliar, and other pathogens; controlling tree growth or killing unwanted trees; preserving wood from decay after felling; and numerous other, less frequently sought purposes (17,23).

Because of the perennial and longevous nature and high individual value of fruit and landscape trees such as the elm (24), diseases and pests that effect their decline, loss of aesthetic beauty, reduction in fruiting, and eventual death are of paramount importance when compared with those of annual crop plants. An integrated approach to disease (pest) management is crucial, and the latest innovation in managing wilt and other fungous pathogens, as well as arthropod pests, is the introduction of disease-preventive or -curative, systemic compounds into the vascular (xylem and phloem) components where they translocate ideally in a uniform manner so that the problem is either prevented or brought under satisfactory control.

Tree injection experimentation, unlike human medical research, has been unregulated; as a result, practitioners with few or no scientific credentials continue to exploit the field by marketing disease control systems that are not only marginally effective or ineffective but damaging to tree health. At the research level, there has been a lack of the coordinated, in-depth, and long-term experimentation, primarily because of limited and discontinuous financial support, needed to arrive at a dependable and effective system for any pest control strategy. Early interest in a compound that exhibits systemicity and antifungal activity in vitro, or in a method that shows promise in preliminary trials, is often not sustained. If the sponsoring institution (research unit, industry) terminates research and withdraws support, the project is dropped until a new compound and system are devised, and the cycle is repeated. Therefore, we believe that tree injection is a valid principle with great potential but that the current state of the science and art lacks guidelines, standards, and uniformity.

TREE PHYSIOLOGY AND DISTRIBUTION PHENOMENA IN TREES

Any compound introduced into the vascular system sensu lato (phloem as well as xylem) of a tree is transported in liquids (i.e., sap, photosynthate) of the xylem and phloem conduits. Upward (acropetal) movement of water and other compounds in the xylem, for example, including the "ascent of sap," accompanied by loss of water from leaves via transpiration—not to mention downward (basipetal) and bidirectional movement—has a long history of intensive investigation and is currently not clearly understood (29). Even though many definitive and intricate experiments have been executed on the topic, translocation phenomena are neither clearly understood nor fully researched. It was not until 1894 that Dixon and Jolly in Ireland elaborated a satisfactory theory to explain the rise of sap to the tops of tall trees. Their cohesion theory has been attacked and tested, but it is now generally accepted by most tree physiologists (16,28). The theory proposes that transpiration from leaves initiates the ascent of sap and that water moves through plants because of its continuity from the leaves through the stems to the roots and via the cohesive forces among water molecules. Obviously, water must be under negative pressures in stems of tall trees, since atmospheric pressure can force water to only 10 m. The cohesion theory states that water is pulled along gradients of decreasing pressures through the vessels and tracheids of the xylem. Gravitational force requires that the pressure gradient be -1 atmosphere for each 10 m of tree height. Under negative pressure, water columns would be under liquid tension and in a metastable condition. Once broken, these continuous water columns would be difficult to repair, and this fact probably explains conduction only in vessels in the outer sapwood, particularly in the large-diameter vessels of ring-porous species (29). The breaking of the water columns by tree treatment techniques, discussed later in this chapter, also might explain the difficulty encountered in achieving better uptake and distribution of liquids.

A better understanding of the phenomenon of movement of compounds in xylem and phloem would aid the plant pathologist in developing methods and tailoring molecules of systemic fungicides, for example, for tree injection for the control of the recalcitrant wilt diseases. At this time, only fragmentary and insufficient information based on empirical observation is available on the translocation phenomena of the various pesticidal compounds introduced into trees; their metabolic fate and residual nature; and their transport from tissues in which they were introduced to those being synthesized during periods of growth (3,4,6,18,21,23,25–27).

ADMINISTERING FUNGICIDES AND OTHER COMPOUNDS TO TREES

At an international conference on Dutch elm disease (DED), Stipes (23) listed the various methods either used or that might be used to introduce fungicidal compounds into elm trees for the control of DED (Table 1). These methods would also apply to certain other pesticides. The fungicide (or other compound) can be administered in dry form (powder, crystals); as a liquid concentrate, dilute solution, or suspension; or as a paint, depending on the method elected for use. The various desirable and undesirable features of these methods are indicated. In a later paper, Smalley (21) collated the various strategies used to introduce chemicals into elm trees via foliar and branch sprays and trunk and root injection. In this chapter, however, we will highlight only infusion and injection methods.

The history, principles, and various methods of tree injection were monographed by May (17), and it is not our purpose to reiterate this information. May referred to *injection* in a broad sense to include all methods of introducing chemicals into tissues of trees. However, we prefer, in the interest of etymological accuracy, to use the word (which means "to force in") as any forceful introduction of liquids into freshly punctured, incised or drilled wounds (treatment ports) through the bark into the outer sapwood via syringes or other pressurized units that use compressed air or nitrogen. *Infusion* ("a pouring in"), on the other hand, is their introduction from external reservoirs through tubing to these treatment ports where uptake occurs via (i) the force of the atmospheric pressure on the liquid surface in the reservoir and its gravitational pull into the treatment ports and (ii) the negative pressure in the xylem columns as a result of cohesion and transpiration phenomena mentioned previously (28,29). Few, if any, reports have issued from critically executed studies in which uptake and translocation phenomena were evaluated in comparative injection and infusion treatments. In most cases, however, more liquids per unit time can be introduced under pressure, and this results in less labor and hence reduced costs to arborists who treat trees; also, it is well known to those who introduce liquids into injection ports that the port will accept liquids for only a short period of time, perhaps for only several hours, before a plugging occurs that is thought by some to be tyloses. In one study, injection of dyes enhanced the radial and tangential (circumferential) translocation between injection ports when compared with infusion (Himelick, *unpublished data*); Himelick (9) also found no difference in translocation of the dye whether it was introduced by injection or infusion in roots. According to the scarce literature extant, the most uniform distribution of compounds results from lateral root infusion or injection (13–15); but as one injects farther up the tree (e.g., from flare roots to the lower trunk to the upper trunk to individual branches), translocation of most compounds becomes less uniform and predictable (25). Another stumbling block in understanding translocation phenomena is the mistake of suspecting that all compounds have similar or identical mobility; the chemistries of the many compounds administered are vastly different, and all are different from the dyes used as indicators in pilot studies of translocation phenomena.

Table 1. Evaluation of methods of chemical administration for control of Dutch elm disease[a]

	Tissue treated and method used						
			Root		Stem (bole)		
Methodology feature	**Leaves (spray)**	**Bark (topical appl.)**	**Soil injection**	**Root injection**	**GFR[b]**	**TID[c]**	**Pressure injection**
No stem wounding	+	+	+	+	0	0	0
No root wounding	+	+	+?	0	+	+	+
Accessible treatment area	+	+	+?	+?	+	+	+
General skill required	+	+	+	0?	+?	+	+?
Annual treatments not required (long-term protection)	0	+?	+	0?	0	0	0
Effective uptake	0?	+?	+	+	+	+?	+
Uniform distribution	+	+?	+	+	+?	+?	+?
Small dosages required	+?	+	0?	+	+?	+	+?
Prophylactic (preventive)	+	+	+	+	+	+	+
Therapeutic (curative)	0	0?	0?	+?	+?	+?	+?
Aerial environment not contaminated	0	+	+	+	+	+	+
Soil not contaminated	+	+	0	+	+	+	+
Relatively economical (labor/materials)	+	+	0?	0	+	+	+?
Low vandalism potential	+	0?	+	+	0	+	+
Low disease-transmission risk	+	+	+	+	0	0	0
Demonstrated efficacy in disease control	+	+?	+	+?	+?	+?	+
EPA[d] registered	+	0	0	0	+[e]	0	0

[a]Adapted from Stipes (23). + = item description generally true; 0 = item description generally false.
[b]Gravity-feed reservoir.
[c]Trunk-implantation device.
[d]U.S. Environmental Protection Agency.
[e]Mauget reservoir only.

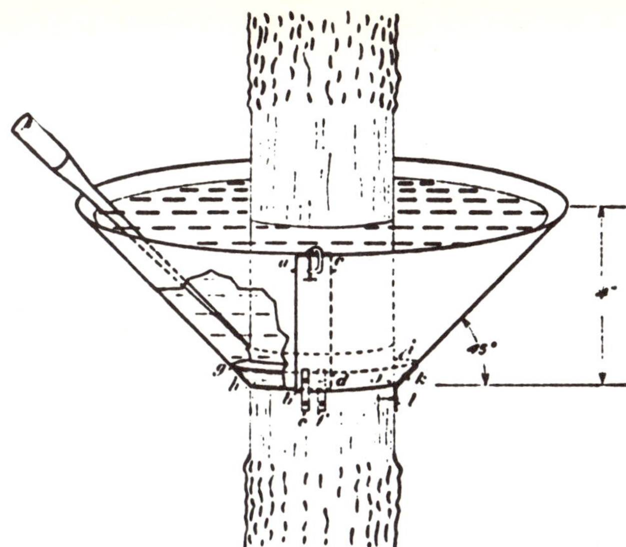

Fig. 1. Tree injection pan (45° cone frustum, 10 cm [4 in.] deep): *ab* and *cd* = free ends of pan; *e* and *f* = nails fastening ends of pan to stem; *ghij* = layer of grafting wax (or plastic clay) sealing lower edge of pan; *kl* = metal strap fastening pan to stem. Reprinted, by permission, from Banfield (1).

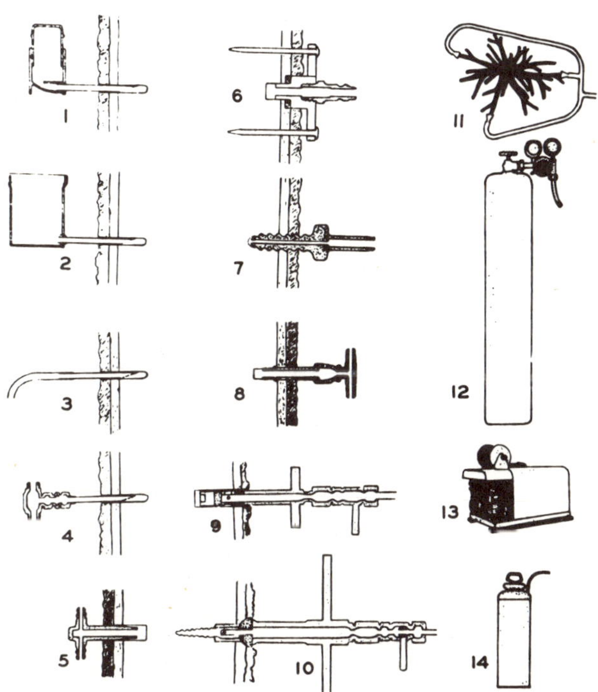

Fig. 2. Various injection heads and other devices for injection of elms with systemic fungicides. 1: Mauget injector (pressure type) with about 10-ml capacity. 2: Mauget injector (nonpressurized) with 65-ml cup. 3 and 4: Mauget injector, showing tube connected to pressure source. 5: Modified "maple tap" for low-pressure injection. 6: High-pressure injection head held in place with nails (after Gregory and Jones [7]). 7: Lag-bolt injector head connected to high-pressure system (after Himelick [9]). 8: Elm Research Institute pipe-nipple injector head for low pressures. 9 and 10: English-style injector lances (after a device marketed by Tree Saver Ltd.). 11: Direct root injection under low pressure (after Kondo [13]). 12: Nitrogen tank for high-pressure injection (>7 kg/cm, 100 psi). 13: Gasoline-powered hydraulic sprayer for high-pressure injection. 14: Garden sprayer tank for low-pressure injection (1.4–2.8 kg/cm, 20–40 psi). Reprinted, by permission, from Smalley (21).

Among the several infusion methods reviewed by May (17), probably the most commonly used in recent decades has been the cone-frustum tree injection pan used by Banfield (1), who introduced spore suspensions of the DED fungus—*Ceratocystis ulmi* (Buism.) C. Moreau—into elm trees to study their distribution (Fig. 1). Banfield's detailed description of this technique indicates that from 100 ml to as much as 89 L were taken up by individual trees in treatment periods of 20 min to 2 hr. Fungicides can be similarly introduced, and the small chisel wounds would appear to be preferable to the large, bored holes used currently so that bacterial and fungal infections that lead to discoloration and decay column development might be minimized. No one has attempted to develop either a better experimental system or a commercially available infusion method based on Banfield's design. Smalley (21) illustrated, in a quasi-evolutionary way, the various infusion and injection methods developed since that used by Banfield and others (Fig. 2). Depicted here are infusion/injection ports made by driving hollow delivery tubes into the outer xylem growth layers, drilling holes with a manual or electric drill bit, or the severing of lateral roots. Difficulties in liquid uptake have commonly been encountered by the Mauget method where the bark and xylem tissues have been compacted into the treatment port when the delivery tube is inserted; more recently, however, the Mauget port has been made by drilling. The variation to a drilled port injection system was developed by U.S. Department of Agriculture scientists (7,8,11) who removed a round plug of outer bark and phloem to the vascular cambium and, before pressure injecting, opened the xylem vessels with a clean chisel cut (Fig. 3). This appears to have the advantage, over a large, bored hole, of enhancing wound closure and

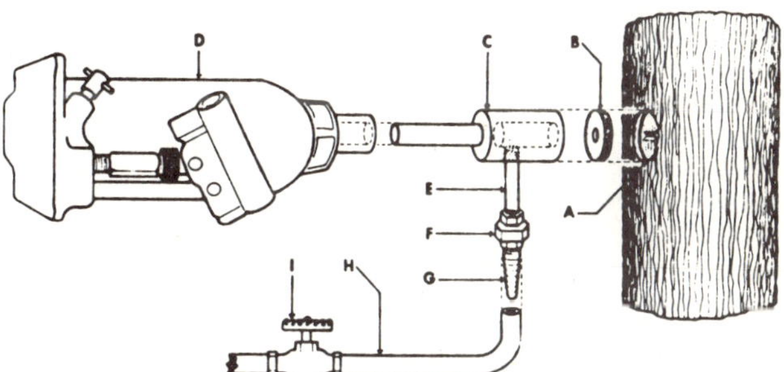

Fig. 3. Tree injection system developed by the USDA Forest Service. A: Prepared injection site on tree with hole 3.5 cm (1-3/8 in.) in diameter cut through bark and wedge-shaped hole two or three annual rings deep. B: Neoprene gasket, 3.2 cm (1-1/4 in.) in diameter by 0.3 cm (1/8 in.) thick with 0.45-cm (3/16-in.) hole in center. C: Injector head machined from steel rod measuring 2.9 cm (1-1/8 in.) by 13.9 cm (5-1/2 in.) long; half the length of the rod is turned down to fit into hydraulic jack; centered hole, 1.6 cm (5/8 in.) in diameter by 5.7 cm (2-1/4 in.) deep, is drilled lengthwise into rod from large end; second hole, 1.1 cm (7/16 in.) in diameter, is drilled radially in side of rod 4.4 cm (1-3/4 in.) from large end, deep enough to open into bottom of longitudinal hole, and threaded to accept nominal 0.6-cm (1/4-in.) galvanized iron pipe. D: Hydraulic jack, 1-1/2 tons, with threaded extension removed from ram. E: Galvanized iron pipe, nominal 0.6 cm (1/4 in.) by 12.7 cm (5 in.) long, threaded at both ends. F: Pipe union, 0.6 cm (1/4 in.). G: Tapered hose fitting. H: Fluid supply hose from reservoir to injector head, 0.9-cm (3/8-in.) nylon reinforced clear polyvinyl chloride tubing. I: Gate valve. Reprinted, by permission, from Jones and Gregory (11).

mitigating the levels of bacterial and fungal infections. Several variations of introducing fungicides have been developed and are being used; the California system (19), for example, uses nitrogen gas as the driving force (Fig. 4); others use compressed air in lieu of nitrogen. Several reports on tree injection reveal a variety of methods and uses (9,10,20,22).

EVALUATION CRITERIA, PROBLEMS, AND FUTURE NEEDS

The history of chemical tree treatments by the various methods reviewed herein is brief and fragmented, and their commercial aspect has really been developed only during the last two or three decades. Hence, we are in the infancy of research and development in this exciting and important aspect of tree medicine, and the potentials are vast.

Needs to be assessed and met might include the following: the prevention and cure of the killer wilt diseases, such as DED (24); control of the many foliar disorders caused by an array of biotic factors; management of insect and allied pests (such as gypsy moth); alleviation of nutrient deficiencies; and other problems.

From the review of the scientific literature, reports in trade journals, the vast unpublished data bank issued from various research meetings, and other sources, it is obvious that many needs must be met before any reliability or reputability can truly characterize a system touted to be effective. Some of these needs are annotated below:

1. Pesticide molecules, for example, need to be synthesized that can be introduced successfully into the correct tree tissues, be uniformly translocated, hit the target, and be adequately residual to effect pest and pathogen control at least at acceptable levels.
2. A standardized infusion/injection system must be developed and available commercially that will provide the maximum pest and disease control but will minimize tree injury.
3. Standardized treatments do not exist. Dosage levels (concentrations), volumes, and frequencies have never been scientifically established. For example, great disparity currently exists in the dosages of such compounds as thiabendazole (Arbotect 20S), methyl 2-benzimidazolecarbamate phosphate (Lignasan BLP), and Fungisol, all of which are close relatives chemically and purportedly have comparable antifungal effects on the DED fungus. Kolpak et al. (12) found greatly enhanced translocation of thiabendazole when a 5% concentration in low volume was used compared with the same amount of fungicide administered in higher volumes. In contrast, a distributor of methyl 2-benzimidazolecarbamate phosphate recommends the injection of large volumes of a dilute solution.
4. No one has scientifically evaluated the long-term effects on tree health of various tree treatments, although strong claims are made about their long-term safety.
5. Some who tout the management of DED with fungicide injection tend to minimize the great and overriding importance of sanitation. It has been well established by extensive and scientifically designed studies that sanitation is the critical and key component of DED control (2,5,24).
6. A vast array of problems that must be addressed and evaluated are the following phenomena that determine effectiveness of a systemic compound: mobility, phytotoxicity, compatibility, residual activity, absorption, and solubility (Fig. 5).
7. Since tree therapy, unlike human medicine, is not rigidly regulated, it has been, is now, and likely will be exploited by charlatans and those who prefer to ignore scientifically endorsed technology. In fact, there are currently marketed compounds that are not only worthless but injurious to trees.

In summary, it is evident that the potential for the use of tree injection is great. Even though there is commercial use of this strategy of disease and pest control, much of it has empirical base only, and some has none at all. Research and development must then become a combined priority of scientific laboratories and industries if tree injection technology is ever to become not only a respectable but a useful tool.

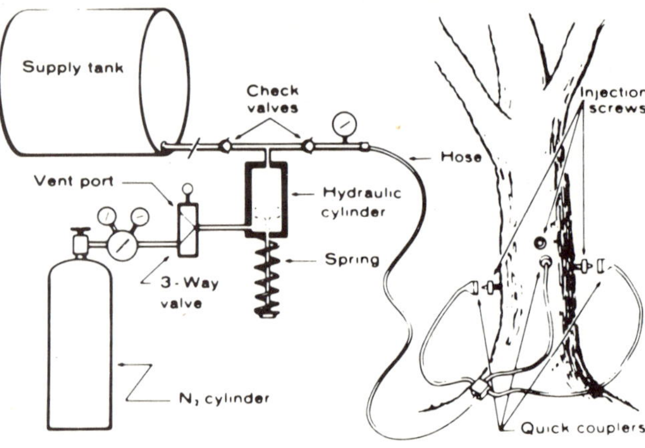

Fig. 4. Tree injection system developed by the University of California. Reprinted, by permission, from Reil (19).

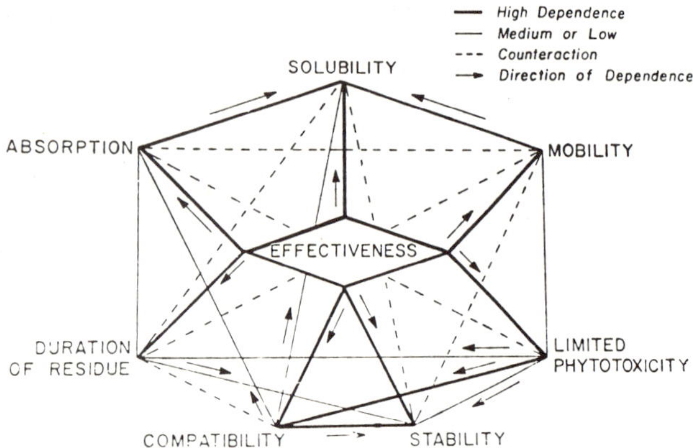

Fig. 5. Interrelation of characteristics of systemic chemicals at varied levels of dependence or counteraction. Reprinted, by permission, from Campana (4).

LITERATURE CITED

1. Banfield, W. M. 1941. Distribution by the sap stream of spores of three fungi that induce vascular wilt disease of elm. J. Agric. Res. 62:637-681.
2. Barger, J. H. 1977. Improved sanitation practice for control of Dutch elm disease. U.S. Dep. Agric. For. Serv. Res. Paper

NE-386. 4 pp.

3. Campana, R. J. 1970. Systemic fungicides for control of tree disease. Trees Mag. 30(5).
4. Campana, R. J. 1979. Characteristics of successful systemic chemicals. Pages 19-34 in: Proc. Symp. Systemic Chem. Treatments Tree Culture. J. J. Kielbaso, et al., eds. Michigan State University, East Lansing. 357 pp.
5. Cannon, W. N., Jr., Barger, J. H., and Worley, E. P. 1977. Dutch elm disease control: Intensive sanitation and survey economics. U.S. Dep. Agric. For. Serv. Res. Paper NE-387. 10 pp.
6. Erwin, D. C. 1973. Systemic fungicides: Disease control, translocation, and mode of action. Annu. Rev. Phytopathol. 11:389-422.
7. Gregory, G. F., and Jones, T. W. 1975. An improved apparatus for pressure injecting fluid into trees. U.S. For. Serv. Res. Note NE-214. 6 pp.
8. Gregory, G. F., Jones, T. W., and McWain, P. 1973. Pressure injection of methyl 2-benzimidazole carbamate hydrochloride solution as a control for Dutch elm disease. U.S. Dep. Agric. For. Serv. Res. Paper NE-176. 9 pp.
9. Himelick, E. B. 1972. High pressure injection of chemicals into trees. Arborist's News 37:97-103.
10. Hock, W. K., and Wilson, C. L. 1972. Pressure injection of chemicals into living trees. Am. Nurseryman 135:7-9.
11. Jones, T. W., and Gregory, G. F. 1971. An apparatus for pressure injection of solutions into trees. U.S. Dep. Agric. For. Serv. Res. Pap. NE-233. 7 pp.
12. Kolpak, M. X., Anderson, J., Stipes, R. J., Campana, R. J., and Landis, W. R. 1978. Distribution of gravity flow-administered 2-(4-thiazolyl)benzimidazole hypophosphite (Arbotect 20-S) in landscape American elms. (Abstr.) Phytopathol. News 12:70.
13. Kondo, E. S. 1972. A method for introducing water-soluble chemicals into mature elms. Can. For. Serv. Inf. Rep. O-X-171. 11 pp.
14. Kondo, E. S., and Huntley, G. D. 1973. Root-injection field trials of MBC-phosphate in 1972 for Dutch elm disease control. Can. For. Serv. Inf. Rep. O-X-182. 17 pp.
15. Kondo, E. S., Roy, D. N., and Jorgensen, E. 1973. Salts of methyl-2-benzimidazole carbamate (MBC) and assessment of their potential in Dutch elm disease control. Can. J. For. Res. 3:548-555.
16. Kozlowski, T. T., and Winget, C. H. 1963. Patterns of water movement in forest trees. Bot. Gaz. 124:300-311.
17. May, C. 1941. Methods of tree injection. Trees Mag. 4:7, 10–12, 14, 16.
18. Norris, D. M. 1967. Systemic insecticides in trees. Annu. Rev. Entomol. 12:127-148.
19. Reil, W. O. 1979. Pressure-injecting chemicals into trees. Calif. Agric. 33:16-19.
20. Sachs, R. M., Nyland, G., Hackett, W. P., Coffelt, J., Debie, J., and Giannini, G. 1977. Pressurized injection of aqueous solutions into tree trunks. Sci. Hortic. 6:297-310.
21. Smalley, E. B. 1978. Control tactics in research and practice. IV. Systemic chemical treatments of trees for protection and therapy. Pages 34-39 in: Dutch Elm Disease—Perspectives After 60 Years. W. A. Sinclair and R. J. Campana, eds. Search 8(5). Cornell University, Ithaca, NY. 52 pp.
22. Sterrett, J. P., and Creager, R. A. 1977. A miniature pressure injector for deciduous woody seedlings and branches. HortScience 12:156-158.
23. Stipes, R. J. 1975. Chemical control of *Ceratocystis ulmi*—an overview. Pages 1-15 in: Dutch Elm Disease. D. A. Burdekin and H. M. Heybroek, compilers. Proc. IUFRO Conf., Minneapolis-St. Paul, MN, September 1973. U.S. For. Serv., N.E. For. Exp. Stn. 94 pp.
24. Stipes, R. J., and Campana, R. J., eds. 1981. Compendium of Elm Diseases. American Phytopathological Society, St. Paul, MN. 96 pp.
25. Truax, P. A. 1978. Comparative distribution of Arbotect 20-S, CGA 64251, Lignasan BLP and nuarimol in *Ulmus americana* following administration with a Sterrett-Creager miniature pressure injector. M.S. thesis, Virginia Polytechnic Institute and State University, Blacksburg. 140 pp.
26. Wilson, C. L. 1979. Injection and infusion of trees. Pages 1-6 in: Proc. Symp. Systemic Chem. Treatments Tree Culture. J. J. Kielbaso, et al., eds. Michigan State University, East Lansing. 357 pp.
27. Zentmyer, G. A., Horsfall, J. G., and Wallace, P. P. 1946. Dutch elm disease and its chemotherapy. Conn. Agric. Exp. Stn. Bull. 498:1-70.
28. Zimmermann, M. H. 1983. Xylem Structure and the Ascent of Sap. Springer-Verlag, New York. 143 pp.
29. Zimmermann, M. H., and Brown, C. L. 1971. Trees: Structure and Function. Springer-Verlag, New York. 336 pp.

Evaluating Chemicals for Tree Infusion or Injection to Control Diseases Caused by Mycoplasmalike Organisms

GEORGE H. LACY, Department of Plant Pathology, Physiology and Weed Science, Virginia Polytechnic Institute and State University, Blacksburg 24061

Mycoplasmalike organisms (MLO), including *Spiroplasma* spp., cause diseases of many economically important trees (Table 1). Until recently, a number of these diseases were believed to be caused by viruses and, hence, to be refractory to conventional chemotherapeutic methods. Once the prokaryotic nature of the causal organisms was determined, tetracycline antibiotics were found effective for control of the diseases they cause. Presently, oxytetracycline (Terramycin) is used to control several MLO-caused diseases in the United States, including pear decline (PD), peach X-disease (XD), and lethal yellowing of palms. Methods for applying oxytetracycline follow three routes: foliar application, root application, and infusion or injection. Of these methods, root uptake of the antibiotic from solutions may be the most effective (14), but binding to soil limits its usefulness (3). Foliar application causes only transitory remission of symptoms; moreover, uptake and transport from leaves is poor (5). Placing oxytetracycline directly into plant tissues by infusion or injection is the method used most for control of MLO-caused diseases (11,12) and is the method discussed in this chapter. I will refer most to PD and XD because my experience with chemotherapy is more extensive with these MLO-caused diseases.

Infusion occurs by diffusion of antibiotics into the xylem and uptake by the transpirational stream from an antibiotic-filled hole drilled in woody parts of the tree. Injection with pressure forces antibiotics from the hole and into the surrounding tissues. Infusion techniques include simply filling a hole in the trunk with concentrated antibiotics (16) or infusion of more dilute solutions from Mauget cups (8) or from pipets (9). Injection is accomplished by pneumatic (8,11) or hydraulic pressure (5). Gravity flow applications of antibiotics from hanging bags or bottles (5,8,11) occur by a combination of hydraulic pressure, created by the height of the container above the application site, and infusion. The time required for application and uptake of antibiotics using these methods of infusion or injection varies greatly (Table 2). Chiefly the pipet method of infusion will be considered in the following sections.

Table 1. Economically important trees affected by diseases caused by mycoplasmalike organisms and use of oxytetracycline for control of these diseases

Host	Disease	Oxytetracycline chemotherapy
Prunus spp. (cherry, apricot, peach)	X-disease, eastern	Yes
	western	Yes
	Rosette	Yes
Pyrus communis (pear)	Decline (slow decline on domestic or French rootstock)	Yes
Cocos nucifera (coconut)	Lethal yellows	Yes
Citrus spp. (orange, grapefruit, tangelo, etc.)	Greening (caused by a phloem-limited bacterium rather than an MLO)	Yes
	Likubin (possibly an MLO-tristeza virus complex)	Yes
	Stubborn	Yes
Paulownia tomentosa (paulownia)	Witches'-broom	No
Morus alba (mulberry)	Dwarf	Yes
Euphoria longana (longan)	Witches'-broom	Yes
Musa paradisiaca (banana)	Bunchy top (may be caused by a virus rather than an MLO)	No
Castenada crenata (Japanese elm)	Yellows	No
Ulmus americana (American elm)	Phloem necrosis	Yes

Table 2. Time required to treat trees using several methods for infusing or injecting oxytetracycline

Method and reference	Application time per tree	Uptake, time/tree	Volume applied/tree (ml)
Infusion			
Sands and Walton concentrate method (8)	36 sec	30–90 min	1
Pipet method (8)	36 sec	30–90 min	10
Mauget cup[a] (8)	2.4–4.0 min	24 hr	50
Injection			
Pneumatic pressure (8)	20–30 min	24 hr	4,000
Mauget injection capsules[b] (11)	NR[c]	50–60 min	500
Hydraulic (5)	7.5–10 min[d]	7.5–10 min[d]	945
Gravity flow			
Plastic bottles (plastic milk bottles) (8)	2.4–4.0 min	24 hr	1,000
Injection bags (intravenous feeding bags) (11)	NR	Overnight	500

[a]Mauget cups (no longer available commercially) are open to atmospheric pressure.
[b]Mauget injection capsules use pneumatic pressure.
[c]Not reported.
[d]Includes trunk preparation, application, and uptake of oxytetracycline.

METHODS

Evaluation of the effectiveness of chemotherapy with oxytetracycline or other antibiotics for MLO-caused diseases is in its infancy compared with more conventional foliar or soil applications of chemicals for control of diseases caused by fungi and bacteria. However, enough literature now exists to address some of the problems that are encountered: evaluating the distribution of applied antibiotics in planta, rating remission of foliar symptoms, measuring yield, determining disease loss, evaluating tree vigor, analyzing for the presence of the causal organism, and measuring phytotoxicity.

Antibiotic distribution in plants. The most common method of evaluating the in planta distribution of applied antibiotics is to observe the distribution of transitory chlorosis and remission of symptoms. Chlorosis occurs because antimycoplasma antibiotics often affect 70S ribosome-mediated protein synthesis in chloroplasts. In using the pipet or gravity flow methods to apply oxytetracycline, it is obvious that neither transitory chlorosis nor remission of symptoms is complete in all parts of treated trees, especially in XD; therefore, distribution of the antibiotic is probably incomplete. Movement of antibiotics to a specific part of a tree is more effective if application sites are selected directly below the scaffolds supporting that part of the canopy (8). Pressure-injection techniques are reputed to give the most even distribution of antibiotics; however, quantitative comparisons with other techniques have not been reported. The trade-off between infusion and injection techniques for application is that the latter are labor, equipment, and time intensive (Table 2).

Although "cures" have been reported and probably occur, symptoms of MLO-caused diseases often reappear in treated trees. Whether these recurrences result from new infections or movement of sequestered pathogenic inocula from roots, trunks, or other scaffolds is not known. Careful observation of treated trees often reveals that some symptoms of XD persist even on treated scaffolds. These usually inconspicuous symptoms are distributed in a pattern suggesting that oxytetracycline provides an inverted cone of protection, radiating upward into the canopy from the application site (8). Outside this cone (i.e., above and behind the application site), mild symptoms of XD may persist. In PD, this pattern is less clear. The best studies of the distribution of an injected antibiotic used radioisotope-labeled oxytetracycline and dyes in pear trees (15). Similar studies have not been performed with other woody species or antibiotics, but they would be valuable to aid in understanding results of field use.

Infused or injected antibiotics must be applied after fruit harvest but before significant seasonal leaf abscission to prevent contamination of fruit and promote uptake via the transpirational flow (6). With pear and peach trees, this is best done on clear, warm autumn days prior to the first killing frost.

Antibiotics may be lost or exuded from as well as taken up by plants. Oxytetracycline activity may be found in naturally abscised leaves (9) or in those that develop during the growing season following treatment and winter dormancy (5). Finally, oxytetracycline may be exuded from roots of treated plants (1). These examples raise concerns that agricultural use of oxytetracycline, an antibiotic also used in human and animal chemotherapy, may lead to selection of tetracycline-resistant MLO or bacteria in the environment. Evaluation of antibiotics for control of MLO-caused diseases that are not useful for human or animal chemotherapy would eliminate this possibility.

After infusion or injection, oxytetracycline may be extracted from various plant tissues, concentrated, and detected in agar-well assays by inhibition of the growth of *Bacillus cereus* var. *mycoides* (8). Similar bioassays are available for most antibiotics.

Remission of foliar symptoms. Remission of foliar symptoms of MLO-caused diseases is the most striking feature of chemotherapy—and the most difficult to evaluate. Probably the best and the least expensive method for evaluation continues to be visual rating of symptom intensity or its remission. Photoelectric color-difference meters have been used for measuring light reflected from untreated and treated diseased plants (6). However, the data need to be interpreted to account for such irregularities as defoliation, distribution of symptoms, and leaf size. Therefore, the problem with visual rating is the same as with any other scientific method: that is, the need to establish its accuracy, precision, and limits. In the following paragraphs, I will mention problems with visual evaluations of symptom remission and suggest ways to deal with them.

Symptom intensity rating indexes. For PD, symptom intensity ratings may be based on the percentage of the canopy with reddened and curled leaves, provided the reduction in visible leaf area caused by the smaller curled leaves is taken into account (9). With XD, the problem is greater because defoliation and death of branches occur unevenly throughout the season and from year to year. With this disease, it is better to use stages of disease development as an index for symptom intensity (8). For both diseases, increases in percentage of reddened and curled leaves (PD) and the stages of pathogenesis (XD) are correlated with reductions in fruit production.

Observation technique. For both PD and XD, symptom intensity rating proved difficult since symptoms were distributed unevenly within the canopies of the trees. This problem was solved by rating the opposite sides of trees to obtain a 360° or "whole tree" view. To accomplish this, two or more people rated each tree independently from different perspectives (at least from both sides of a row of trees). The averaged rating approximates the rating of one person viewing both sides of each tree (Dodds, Lacy, McIntyre, and Walton, *unpublished data*). Having two or more persons rate each tree is faster than having one person rate each tree twice. Key to this method is a carefully constructed symptom intensity rating index and trained personnel.

Year-to-year evaluations. Memory, light quality, environmental conditions, and personnel turnover pose problems for consistent year-to-year evaluations of trees. These problems may be dealt with by making photographic records of each tree from the same perspective at each evaluation to substantiate field ratings, allow side-by-side comparisons of data from different years, and train new personnel or refresh the memories of experienced evaluators (8,9). For training purposes, standard photographs, on which the symptom intensity rating index is based, may be used for comparison with photographs of other trees. Prints of these standard photographs may be carried into the field for comparison with trees in the field.

Measuring yield. The most important result of disease control for the fruit grower or pathologist is the improved marketability of fruit on diseased trees. Fruit appearance, taste, and yield are components of marketability.

Appearance. For MLO diseases, problems may occur in grading fruit appearance. For instance, in XD fruit drop is accelerated so that thinning occurs at a rate greater than normal, resulting in some fruits in symptomless areas of diseased trees being large and having a good appearance while smaller "acorn" fruits have dropped prematurely before harvest from areas with symptoms in the same tree. Therefore, rating the fruit remaining at harvest gives an unrepresentative good impression of fruit appearance but an unrepresentative poor impression of the number of fruits produced. Further, any standard pruning scheme causes inequities in measuring appearance or yield from weak scaffolds with symptoms compared with strong or symptomless scaffolds on the same tree. Rather, it may be more accurate to measure fruit production and appearance on trees with MLO-caused diseases from comparable scaffolds without pruning (8).

Taste. Fruit taste and texture are major factors in marketability. Fruits from trees with MLO-caused diseases often mature early, are small, and are off-flavor. Although these characteristics have been mentioned often in the literature, they have rarely been confirmed in any quantitative manner. Organoleptic evaluations of fruit may be made by professional taste panels or by consumer preference analyses. The latter method is easily arranged by pathologists and is usually less expensive than taste panels. Consumer preference was used to determine that pears from trees with PD treated with oxytetracycline were preferred over fruit from trees with PD that were not treated (4).

Fruit yield. Fruit yield is the product of fruit size and number. Fruit size is simply the average weight of harvested fruit (grams per fruit). Fruit number is often determined as the number of fruits, bushels, etc., produced per tree. However, since the size of individual trees, even in a single planting, may be affected by many factors and compounded by the perennial nature of tree crops, the more dependable measurement is fruit numbers per cross-sectional area of the tree or scaffold (fruits per square centimeter). In the same manner, total fruit yield is most adequately expressed as grams per fruit × fruits per square centimeter = grams per square centimeter. For statistical purposes, these measurements are more comparable than those given as grams per tree or fruits per tree (8,9).

Disease loss determination. Losses from MLO-caused diseases are greatest over several seasons when trees either die (XD) or continue to grow less vigorously than apparently healthy trees (PD). For evaluation of chemotherapy, and on a short-term, year-to-year basis, losses are correlated most strongly with reduced fruit set and stage of pathogenesis (XD) or fruit weight (PD) and may be calculated as follows:

For XD, percentage of loss = $100\,[T_tY_1 - (T_1Y_1 + T_2Y_2 + \ldots + T_nY_n)]/T_tY_1$, where $T_t \ldots T_n$ = total trees (t) and number of trees in each symptom intensity rating index (1 to n), with 1 = apparently healthy and n = dead; and where $Y_1 \ldots Y_n$ = yield (grams per square centimeter) in the symptom intensity rating indexes 1 to n.

For PD, percentage of loss = $100\,\{T_tY_s - [Y_s(T_t - T_d) + T_dY_d]\}\,/T_tY_s$, where T_t = total number of trees; T_d = number of trees with symptoms; Y_s = yield (grams per square centimeter) of symptomless trees; and Y_d = yield of trees with symptoms.

Statistical analyses of data from trees with MLO-caused diseases are often difficult since trees with symptoms are distributed unevenly in orchards; the numbers of trees available for evaluations may limit the number of observations available and, hence, the number of treatments possible for statistical reasons. It may be possible to increase the number of observations and the power of the analyses by using more than one scaffold per tree (8).

Tree vigor. Tree vigor, or tree development apart from fruit production in response to chemotherapy of MLO-caused diseases, may be measured several ways: (i) by increase in plant weight, (ii) by increase in three-dimensional surface area, (iii) by increase in branching, (iv) by increase in girth, or (v) by elongation of shoots. Since the first two methods call for destructive sampling, they are not readily applicable to orchard settings. The third method is applicable to computer-digitized image analysis possibly using adaptations of the methods of Lindow and Webb (10). Girth measurements were less sensitive for measuring tree vigor than shoot elongation on a year-to-year basis for PD (9). At this time, therefore, shoot elongation measurements are recommended as a method for evaluating tree vigor or rate of recovery from MLO-caused diseases. Further, by measuring the distance between annular leaf scars on shoots, the vigor of a tree in response to chemotherapy can be compared with the vigor of that tree in previous seasons.

Analyses for the causal organism. Direct or indirect analyses for the presence of the mycoplasmalike causal agent of disease may be useful for assessing the effects of chemotherapy. With present technology, these methods, including direct microscopic observation, grafting, culture, and serology, are tedious and likely to produce false negatives, and they are difficult to quantitate. Direct observation by electron microscopy is costly and time-consuming. Fluorescent techniques are presently being developed (2) that may make quantitative detection of these organisms possible, but much developmental work remains before they can be put into routine use. Grafting is time-consuming, takes a large amount of greenhouse space, and suffers (especially with XD) in that a high percentage of diseased material fails to form successful grafts. Culture remains difficult or impossible with many fastidious pathogens and is presently a nonquantitative method. Finally, serology provides the best, most rapidly obtained quantitative information on the presence or absence of fastidious pathogens. However, serology is dependent on culture of the pathogens to provide adequate sources for antigens.

Phytotoxicity of oxytetracycline. Infusion or injection with antibiotics often causes application site damage and more or less severe foliar chlorosis. For oxytetracycline, the damage appears in a vertical, elongated pattern around the application site and results in cambial death, a depressed lesion, and cracking of the bark as adjacent parts of the trunk expand with growth. On peach trees, copious gum exudates may be present in these lesions. These wounds become sites for secondary invasion of the tree by wood-rotting fungi. The area of damaged tissue may be estimated as the product of its width across the application site and its vertical length (9). Some antimycoplasma antibiotics (such as tylosin) may cause less damage than oxytetracycline, and others (like spectinomycin) may cause more; but no compounds tested to date provided better in planta remission of symptoms than oxytetracycline (8). Phytotoxicity may be reduced by the method of application: infusion by gravity flow or pipet causes less damage than by cup infusion or pneumatic injection (8). In pear trees with PD, foliar chlorosis, unopened blossoms, necrosis of blossoms, and application site damage increased with the amount of oxytetracycline infused (9). Because of phytotoxicity, I do not recommend use of oxytetracycline for prophylactic treatment of MLO-caused diseases.

DISCUSSION

Development of scientific principles for application of oxytetracycline or other antimycoplasma antibiotics is in its infancy. This review and others (5,11,12) serve to encourage more and better research, to present the problems encountered, to suggest possible solutions to the problems, and to indicate directions for future research rather than to present absolute rules for evaluating chemotherapy of MLO-caused diseases. Much is still to be learned.

Future directions for research concerned with application technology and its evaluation must include development and testing of compounds that are less phytotoxic than oxytetracycline, more evenly distributable in planta, and more effective for controlling MLO-caused diseases; better or more refined methods for application; greater understanding of the way in which infused and injected compounds are distributed in planta; and new methods for controlling MLO-caused diseases. One aspect of disease control that needs special evaluation is the possibility of combining control of insect vectors for MLO with control of the MLO-caused diseases themselves. PD and XD are excellent models for these studies. PD incidence is reduced measurably in pear orchards with good psylla (*Psylla pyricola*) control. In XD, many leafhopper species may act as vectors; however, vector control may also be an important component of control (7). Several novel methods of combined insect and MLO controls should be considered, including controlling weeds that attract vector insects into orchards, using insecticide sprays, and infusing or injecting trees with compounds active against both MLO and insects (13). Improved horticultural techniques, such as rooting peach cuttings with XD (2), will aid immensely in designing greenhouse and orchard experiments with sound statistical bases for evaluating compounds, methods for application, and tactics useful for disease control.

ACKNOWLEDGMENTS

I thank V. K. and E. L. Stromberg and L. D. Moore for their critical reviews of this chapter.

LITERATURE CITED

1. Daniels, M. J. 1982. Possible side effects of antibiotic therapy in plants. Rev. Infect. Dis. 4 (Suppl.):167-170.
2. Douglas, S. M., and Rich, S. 1983. Production of naturally infected, rooted peach cuttings for studying X-disease. (Abstr.) Phytopathology 73:807.
3. Gonsalves, D., and Tucker, D. P. H. 1977. Behavior of oxytetracycline in Florida citrus and soils. Arch. Environ. Contam. Toxicol. 6:515-523.
4. Hankin, L., Lacy, G. H., and McIntyre, J. L. 1979. Organoleptic examination of pears from trees infused with oxytetracycline to remit symptoms of pear decline. J. Food Prot. 42:732-734.
5. Jones, A. L., and Rosenberger, D. A. 1979. Progress on the use of systemic injections and sprays of terramycin for control of peach X-disease. Pages 247-254 in: Proc. Symp. Systemic Chem. Treatments Tree Culture. J. J. Kielbaso et al., eds. Braun-Brumfield, Ann Arbor, MI. 357 pp.
6. Keil, H. L., and Civerolo, E. L. 1978. Effects of trunk injections of oxytetracycline on bacterial spot disease in peach trees. Plant Dis. Rep. 63:1-5.
7. Lacy, G. H. 1982. Occurrence and seasonal distribution of leafhopper vectors of the X-disease causal agent in methoxychlor-sprayed and unsprayed peach orchards. Crop Prot. 1:333-340.
8. Lacy, G. H. 1982. Peach X-disease: Treatment site damage and yield response following antibiotic infusion. Plant Dis. 66:1129-1133.
9. Lacy, G. H., McIntyre, J. L., Walton, G. S., and Dodds, J. A. 1980. Rapid method for and effects of infusing trees with concentrated oxytetracycline-HCl solutions for pear decline control. Can. J. Plant Pathol. 2:96-101.
10. Lindow, S. E., and Webb, R. R. 1983. Quantification of foliar plant disease symptoms by microcomputer-digitized video image analysis. Phytopathology 73:520-524.
11. McCoy, R. E. 1982. Use of tetracycline antibiotics to control yellows diseases. Plant Dis. 66:539-542.
12. Nienhaus, F., and Sikora, R. A. 1979. Mycoplasmas, spiroplasmas, and rickettsia-like organisms as plant pathogens. Annu. Rev. Phytopathol. 17:37-58.
13. Paddick, R. G., French, F. L., and Turner, P. L. 1971. Control of leafhopper-borne plant diseases possibly due to direct action of systemic biocides on mycoplasma. Plant Dis. Rep. 55:291-293.
14. Peterson, E. A., and Sinha, R. C. 1977. Uptake, distribution, and persistence of tetracycline antibiotics in various plant species susceptible to mycoplasma infection. Phytopathol. Z. 90:250-256.
15. Sachs, R. M., Hyland, G., Hackett, G., Coffelt, W. P., Debie, J., and Giannini, G. 1977. Pressurized injection of aqueous solutions into tree trunks. Sci. Hortic. 6:297-310.
16. Sands, D. C., and Walton, G. S. 1975. Tetracycline injections for control of eastern X-disease and bacterial spot of peach. Plant Dis. Rep. 59:573-576.

Soil Fumigation by Chemigation with Metham

PETER B. ADAMS, Soilborne Diseases Laboratory, Plant Protection Institute, ARS, USDA, Beltsville, MD 20705

Of the commonly used broad-spectrum soil fumigants, only metham (sodium methydithiocarbamate) is water soluble. Thus, metham is an ideal pesticide for application through sprinkler irrigation systems (chemigation). Metham, when applied as a drench or by chemigation, penetrates the soil profile to a particular depth depending on the soil type and the amount of water used. With this method of soil fumigation, an attempt should be made to treat the soil profile with the appropriate aqueous concentration of metham to the depth at which the target pest survives. In some situations, this may be as little as 2–3 cm or as much as 60 cm or more.

Metham applied by chemigation requires a uniform application of the irrigation treatment to the entire field and the metering of metham into the irrigation line at the appropriate rate for the entire irrigation period. When properly applied, metham can be used effectively at much less than the labeled rates and at less cost to the grower than conventional methods of soil fumigation (1).

Soil fumigation by chemigation has been shown to control lettuce drop (2), onion white rot (1), root and hypocotyl rot of seedlings caused by *Rhizoctonia solani* and *Pythium* spp. (6), Verticillium wilt and pod rot of peanuts (3), Verticillium wilt and root knot (*Meloidogyne* spp.) on potato (4,5), and several common weed species (7). This method of soil fumigation has great potential for control of numerous soilborne pests including fungi, nematodes, insects, and weeds. Research is needed not only to expand the list of pests controlled by this technology, but also on the minimum amount of metham and the optimum amount of irrigation water required for effective application of the chemical.

It is highly desirable for plant pathologists to try to involve nematologists, entomologists, and weed scientists in soil fumigation field tests. Much useful information can be obtained when other scientific disciplines are involved in the collection of data from such field tests.

METHODS AND PROCEDURES

Metham can be applied to fields by any sprinkler or overhead irrigation system ranging from solid-set irrigation lines covering less than 0.5 ha to traveling guns, wheel-move systems, or even center-pivot systems covering about 80 ha. Thus, when designing field tests to determine the efficacy of metham on a particular pest, each metham plot will be at least 0.04 ha in size. When 0.04 ha is multiplied by the number of treatments and replications, the field test is at least 0.3 ha in size. There are, however, some space-saving methods that can be used to reduce the field size required for a particular test.

When using solid-set irrigation lines to apply the metham treatments, untreated control plots can be established by plugging three successive sprinklers on two adjacent irrigation lines. Assuming that under this system the sprinklers are 12 m apart on each line and that adjacent lines are 12 m apart, each control plot would be 12 × 12 m (144 m^2). This method of establishing control plots works very well but requires large control plots and is restricted to solid-set irrigation systems.

In a small field (0.2 ha) with solid-set irrigation in which one rate of metham is to be compared with an untreated control, plastic sheets can be laid, with the edges buried, in the field just before treatment application (Fig. 1) and removed after the treatment is applied. The size of these untreated control plots depends on the type of crop and the management practices for it. In one field test for control of Sclerotinia lettuce drop, I successfully used plastic sheets that were 4 × 8 m (32 m^2) in size (2). In larger fields, larger control plots should be used. This method of establishing untreated control plots can be used with any sprinkler irrigation system.

The size of the field needed for a field test is greatly increased if more than one rate of metham is to be evaluated in a randomized block design. The only practical solution to this problem is to set up separate field tests for each rate of metham, with untreated control plots in each test.

The field selected for the test should be fairly level and have a population of the target pest that is sufficiently high to be quantitatively detected in each of the plots. Soil samples from each of the plots are collected before and 2–5 days after the treatments are applied. Analysis of these samples for the target pest populations gives an estimate of the efficacy of the metham treatments.

The metham treatments may be applied on top of a cover crop, after the field is plowed, after it is plowed and disked, or after the field is ready for planting. The choice of field preparation is dictated by the type of crop to be grown and the farmer's cultural practices. A field with a cover crop or a freshly plowed field is probably preferred because there would be less chance of the irrigation treatment puddling and running off the field.

To determine the rates of metham application to be evaluated in the field, a few simple laboratory experiments can be performed to determine the minimum concentration of metham required to kill the target pest (1,2). Field results in various locations in the United States and Israel indicated that 234–468 L of commercial product (32.7% metham) per hectare applied with 25 mm of irrigation water controls a wide variety of soilborne pests. This means that 234 or 468 L of product per hectare must be metered into the irrigation system over the period of time required to apply 25 mm of water. This period of time should, therefore, be determined. Assume that 234 L of product per hectare is to be applied to 2 ha with 25 mm of irrigation water and that it takes 2 hr to apply the water with the available irrigation system. This means that 468 L of product must be injected into the irrigation system over a 2-hr period, or 234 L/hr. A chemical injection pump will be required that can deliver 234 L/hr.

Several companies make injection pumps that are available directly from the companies or from distributors of irrigation equipment. I have used a John Blue injection pump (Model L-905, John Blue Co., Huntsville, AL 35807) and a Cheminjector Chemical Metering Pump (Model CV8047-0904, Hydroflo Corp., Plumsteadville, PA 18949) (Fig. 1) with excellent results. Regardless of the pump used, the capacity must be adjusted to that required for the particular field test. There must be a check valve at the point where the hose from the injection pump enters the irrigation pipe (Fig. 1) to prevent the irrigation water from backing up into the injection pump. Appropriate check valves, a vacuum breaker, and a low-pressure drain on the irrigation system are used to prevent possible contamination of the water source with metham.

Before the treatments are applied, at least 8–10 rain gauges are put throughout the area to be treated. The location of the rain gauges with respect to the sprinklers is noted on a field diagram. The data obtained from the rain gauges may help to explain any erratic results from the field test. If rain gauges are not available, cans, beakers, or whatever is available can be used to measure the amount and uniformity of the irrigation treatment.

During the application of the metham-irrigation treatment, the area should not be left unattended. If the irrigation pump on a solid-set irrigation system shuts down, the injection pump continues to pump metham into the irrigation lines, most likely causing the metham to leak out onto the field or to move backward in the irrigation main toward the source of water. Researchers must be on guard for clogged sprinklers and ruptured irrigation lines, especially if inexperienced people are conducting the field test or if an unfamiliar irrigation system is being used.

After the metham treatment has been applied and the field can be walked on, the amount of water in each rain gauge is determined. Also, the plastic sheets that were used for the untreated control plots are removed as soon as possible. The treated field can be worked within a week and planted in 2 weeks.

Data collected during the field test include disease incidence and severity at various intervals between planting and harvesting of the crop. With some plant diseases, such as vascular wilts, it is very important to compare incidence and severity of disease at various time periods in the untreated control plots with those in the treated plots. With diseases such as lettuce drop or onion white rot, this would not be necessary. The disease incidence and severity results are reported as well as being converted to percentage of disease control. One can then compare the efficacy of the treatments in

Fig. 1. The application of metham by chemigation. The portion of the field in the upper left corner is being treated with one rate of metham, whereas the portion in the foreground will be treated with a different rate.

Table 1. Relationship between rate of metham applied by chemigation, amount of irrigation water, and concentration of metham in the irrigation water

Rate of application of metham product[a]		Concentration of metham (μg/ml) when applied with the indicated amount of water			
L/ha	gal/A	6 mm	12 mm	25 mm	50 mm
47	5	296	148	70	35
94	10	592	296	140	70
141	15	888	444	210	105
187	20	1,184	592	280	140
234	25	1,480	740	350	175
281	30	1,776	888	420	210
374	40	2,368	1,184	560	280
468	50	2,960	1,480	700	350
701	75	4,440	2,220	1,050	525
935	100	5,920	2,960	1,400	700

[a]Product contains 32.7% metham.

several different field tests.

Crop yield data are probably more important than disease control data, and they are too often omitted when results of field tests are reported. With yield data, the economic benefit of a pest control system can easily be determined; the results thus look even more attractive, especially to farmers and administrators. This is especially important with soil fumigation field tests.

The soil texture, pH, and organic matter content of the soil, as well as the results of the soil analysis for the target pest, are reported. The amount and uniformity of the irrigation treatment as determined from the rain gauges are also mentioned. With this information the concentration of metham in the irrigation water (Table 1) can be calculated.

DISCUSSION

Several precautions should be taken when one applies metham by chemigation. The soil moisture and drainage characteristics of the field should be such that the field will absorb the amount of water that is applied with the treatment. Wind can raise havoc with the uniform distribution of water from an irrigation system. Thus, the treatments should not be applied when there is any significant wind (>10 kph). On solid-set irrigation lines, all the sprinklers should be of the same capacity. Frequently, farms have sprinklers of different capacities. Care should be taken to ensure that the sprinklers at the edge of the field do not treat adjacent fields, hedgerows, etc. Metham is a broad-spectrum pesticide that will kill established crops. It has been shown under laboratory conditions that metham can be moved down a soil column by the addition of water (1). Thus, the metham treatments should not be applied if significant rainfall is expected within 48 hr.

Metham is a rather unique broad-spectrum pesticide in that it is water soluble and relatively nontoxic. However, once in the soil it converts to methyl isothiocyanate that, at the concentrations used in these studies, is also water soluble and highly toxic. Metham can be moved to any depth in the soil profile by adjusting the amount of water used to apply the chemical (1). The important factor with this method of metham application is the concentration of the chemical in the irrigation water. Thus, there are many "games" that can be played with the method of application.

Assume that 234 L of product (32.7% metham) per hectare (25 gal/A) applied with 25 mm (1 in.) of water controls a particular plant pest. Under these conditions, the concentration of metham in the irrigation water is 350 μg/ml (Table 1). If the pest is restricted to the top 5 cm of the soil profile, then only about 6 mm of water is needed to carry the metham to that depth in the soil, and only 58 L of product is required to maintain the proper metham concentration of 350 μg/ml in the irrigation water. Conversely, to kill the pest to a depth of 50–60 cm, 50 mm of irrigation water is probably required. In this situation, one would have to double the rate of the product to 468 L/ha to maintain the proper metham concentration of 350 μg/ml in the irrigation water.

Many of these experiments can be conducted in the laboratory using columns containing soil infested with a particular pest. The soil columns can be treated with metham solutions in various ways and assayed for the viability of the pest (1,2). Successful laboratory experiments can then be confirmed under field conditions.

LITERATURE CITED

1. Adams, P. B., and Johnston, S. A. 1983. Factors affecting efficacy of metham applied through sprinkler irrigation for control of Allium white rot. Plant Dis. 67:978-980.
2. Adams, P. B., Johnston, S. A., Krikun, J., and Carpenter, H. E. 1983. Application of metham sodium by sprinkler irrigation to control lettuce drop caused by *Sclerotinia minor*. Plant Dis. 67:24-26.
3. Krikun, J., and Frank, Z. R. 1982. Metham sodium applied by sprinkler irrigation to control pod rot and Verticillium wilt of peanut. Plant Dis. 66:128-130.
4. Qualls, M. 1982. Disease control in Irish potatoes. Pages 61-63 in: Proc. Natl. Symp. Chemigation, 2d. J. R. Young and D. R. Sumner, eds. Tifton, GA, August 18–19.
5. Santo, G. S. 1982. Nemagation on Irish potato and grapes. Pages 64-68 in: Proc. Natl. Symp. Chemigation, 2d. J. R. Young and D. R. Sumner, eds. Tifton, GA, August 18–19.
6. Sumner, D. R. 1981. Application of foliar and soil fungicides and a soil fumigant through overhead irrigation water. Pages 82-88 in: Proc. Natl. Symp. Chemigation. J. R. Young, ed. Tifton, GA, August 20–21.
7. Teasdale, J. R., Adams, P. B., and Johnston, S. A. 1983. Weed control after chemigation with low rates of metham. Proc. Northeast Weed Sci. Soc. 37:258-262.

Evaluating Soil Treatments for Control of *Agrobacterium tumefaciens*

LARRY W. MOORE, Department of Botany and Plant Pathology, Oregon State University, Corvallis 97331

Crown gall disease is primarily a problem in the nursery production of woody plants grown for use as ornamentals, landscape plantings, and fruit production. Infected plants develop galls or tumorlike outgrowths on the roots and crowns. Occasional problems from crown gall also occur in orchards, as with infected almond trees that are debilitated and subject to blowover because of inferior development of the root system, and in the landscape with such plants as severely galled roses that gradually decline in vigor over several seasons of growth. In the nursery, emphasis is on prevention of infection to achieve disease control. Attempts to eradicate the galls from living plants have had mixed success, and the need remains for a soil treatment that will eradicate the pathogen from a landscape site so that a severely diseased plant can be removed and the site replanted.

The disease is caused by a bacterium, *Agrobacterium tumefaciens*, which occurs both as a soil inhabitant and an invader. Soil populations of *Agrobacterium* are typically several orders of magnitude lower than those populations associated with the rhizoplane-rhizosphere area. Strains of *Agrobacterium* isolated from the soil and from uninfected plants in the Pacific Northwest are almost exclusively saprophytic, and pathogenic strains are recovered only rarely from these habitats. Although galls are a good source of the pathogen, they are also often coinhabited by saprophytic strains of *Agrobacterium*. Paradoxically, epidemics of crown gall occasionally occur, which raises the question of where the primary inoculum comes from since the saprophytic agrobacteria predominate in the field. As an explanation for this paradox, my colleagues and I have hypothesized that the saprophytes can become pathogenic via some mechanism of virulence conversion.

These comments about the saprophytic and pathogenic *Agrobacterium* strains and their habitats are pertinent to this chapter because one must have some measure of the effectiveness of a given soil treatment. Typically, assessment of a treatment for control of crown gall is based on the number of uninfected plants harvested at the end of the growing season compared with the untreated controls. Alternatively, a large reduction or (ideally) elimination of the *Agrobacterium* population from the soil should correlate with a decline in the incidence of crown gall.

METHODS

Test site. Test sites can be greenhouse pots of soil, concrete tiles filled with soil, small plots of land, and commercial fields. The choice of sites may be influenced in part by the type of commercial operation under consideration, such as greenhouse operations, container production, seedbeds, or large field acreages. Smaller "plots" are used in preliminary studies to determine probable rates, efficiency of the treatment, potential cost, efficacy, modifications of the application, influence of soil edaphic factors on the treatment, etc. Obviously, extrapolation of data obtained from greenhouse tests to the field must be verified by field experiments.

Test plants. Several factors must be considered in choosing a test plant and a suitable pathogenic strain of *Agrobacterium*. Not only are there at least three biovars (13) of *Agrobacterium*, but pathogenic strains often exhibit a narrow host range (1). Grapevines are reportedly infected primarily by biovar-3 strains of *A. tumefaciens* (17). In addition, we have observed an interaction between pathogen, antagonist, and pesticide that was influenced by the particular host species (14; L. W. Moore, *unpublished data*). Even the incubation period between inoculation and the first sign of symptoms can vary extensively. Several strains of *A. tumefaciens* isolated from naturally galled incense cedar had incubation periods of 4–6 months, whereas other strains from our collection produced symptoms on incense cedar within 2 weeks (L. W. Moore, *unpublished data*). Incubation periods of 18 months have been reported for rose (15). Thus, preliminary inoculation tests should be conducted with multiple strains of the pathogen to find a suitable host-pathogen combination.

We often use *Prunus* species as host plants for experimentation because they are among the most susceptible and commonly grown plants in Pacific Northwest nurseries. Test procedures as outlined for use with *Prunus* could be adapted for other woody plants with possible minor modifications. Cultural and management practices that are normally followed in the nursery industry should be followed. Field test plots will then be compatible with routine nursery practices, and the data can be readily adapted to the industry.

Dormant seedlings of woody plants are used because they are more economical than budded or grafted trees, smaller and easier to handle, and relatively plentiful. Commercial procedures of root and top pruning are followed: lateral roots are pruned to about 1–1.3 cm from the main root and tops are pruned to about 45–50 cm in length. Because wounds are required for *A. tumefaciens* to infect the tree, the pruning wounds provide that point of entry. However, a higher incidence of infection can often be achieved by making a longitudinal slit 4–5 cm long at the crown with a razor blade or scalpel in addition to the root pruning. Wounds should be made only a few days before planting. Wounded plants are stored at temperatures below 10°C before planting to prevent wound callusing, which reduces wound susceptibility. Tree spacing in the test plot is the same (or less than) the spacing used in commercial practice.

Test plot management. Cultural and management practices that are recommended for commercial greenhouse and nursery production are used to maintain the test plots (media or land preparation, planting and harvesting methods, fertilization, cultivation, irrigation, pesticide application). Land recently cleared of galled trees or shrubs that may serve as a source of inoculum should be avoided (2). *Agrobacterium* can be spread by irrigation water (19) or runoff water draining onto lower land from a higher elevation where infected plants are located. Similarly, care should be taken to avoid cross-contamination of pots or containers from water splash.

Experimental design. The choice of experimental design is dictated in part by the experiment and test site. Experiments are more easily performed with greenhouse potted plants or containers than in the field. Recommendations such as those cited by L. A. Nelson (*this volume*) should be adapted to the specific problem, or a statistician may be consulted. It is best to use the simplest design that gives the precision needed.

Most of our test trees have been planted in rows in the field. Since "hot spots" of higher gall incidence can occur within a field, the treatment replications are randomized according to a pattern derived from a random numbers table. For small plots, we have utilized a Latin square design with border rows of untreated plants.

Although the title of this chapter refers to soil treatments, treatments to control the disease are most commonly applied to plant propagules, such as bare-rooted seedlings. Root-pruned seedlings are dipped in a chemical or biological control agent for a designated amount of time, making sure the aqueous suspension of test material covers the root zone to a height of 5–7 cm above the crown area. Again, the method of control chosen dictates the experimental design, especially if the treatments are located within a grower's field. For example, a soil fumigant that requires tarping would probably be applied through the entire length of the field, whereas seedlings from different treatments can be planted in the same row or randomized throughout the planting to reduce positional effects within the test site.

Control plots. Plots that have not been treated are included in each experiment to determine the effectiveness of a particular treatment. If a treatment of proven effectiveness is available, this is included as a standard reference point. Unfortunately, there are no satisfactory soil treatments to control crown gall in the field. Fumigation with a variety of fumigants has reduced but never eliminated the disease (18), and some growers (*personal communication*) and scientists (5) have even reported an increase in the incidence of crown gall following fumigation. For greenhouse or containerized production, autoclaved or gassed planting media may provide control, providing caution is taken to prevent reintroduction of the pathogen from other sources. I am unaware of any studies using aerated steam to control crown gall.

Pathogen. Like many diseases, crown gall usually occurs naturally at low incidence, with occasional severe outbreaks. Hence, plants or planting media usually need to be infested with the pathogen to ensure sufficient pressure of the pathogen to allow a meaningful comparison of the treatments. A strain or mixture (12) of strains that represent the pathogen in question for locality and plant flora is chosen. Preferably, the strains have been isolated recently from infected plants, purified by single-cell or single-colony procedures, tested for pathogenicity, and characterized as needed to confirm biovar identity (3,13). Strains that have been subcultured extensively should be avoided. Stock cultures of the pathogens are maintained in a lyophilized state or at −70°C to reduce the probability of selecting and using a mutant or "laboratory-tamed" strain that is less competent in the natural environment. The stock culture is used to obtain a new working strain for a new experiment.

The pathogen can be tested for sensitivity to the chemical or biological control treatment that is used. Sensitivity is typically determined by in vitro tests similar to that described by Stonier (20) for assaying bacteriocin activity and the standard cup or antibiotic disk assays. In vitro sensitivity is no guarantee that the compound or biological agent will be effective in the field or greenhouse and vice versa (4), but the test does give an indication of potential toxicity or competitiveness.

Preparation of inoculum. Inoculum concentrations of about 5×10^6 cfu/ml of *A. tumefaciens* are recommended for routine tests. The inoculum is grown about 7 days as a lawn on Difco potato-dextrose agar plus 0.5% calcium carbonate and harvested by washing the cells off the agar surface; the cells are suspended in sterile tap water or buffer, such as a 5-mM phosphate buffer at pH 6.5–7.0. Inoculum that has just entered the stationary phase is probably best for infestation of plants and planting media, but we have no data to support this proposition. There are data indicating that certain marine pseudomonads (8) survive longest in the stationary phase, probably because these cells are subject to less physiological shock than exponentially growing cells when they are introduced into a relatively harsh, foreign environment such as the soil or plant root. Experiments are needed to test this hypothesis.

Infection is generally proportional to the inoculum concentration. The correlation between inoculum density above 10^6 cfu/ml and number of tumors formed is quite good on bean leaves (9), but this is an idealized environment relative to the field or soil. Consequently, a higher inoculum density may be needed under some environmental conditions. The inoculum should be protected from high temperatures by being carried in ice chests during transport to the field. Water from wells that are free of chlorine can be used for resuspending the inoculum at the planting site.

Inoculum techniques and methods of applying test materials. The planting medium can be infested with *A. tumefaciens* when working with containers and relatively small plots, but this method becomes too costly with large field plots. We have used greenhouse pots and concrete tiles (1.2 m in diameter) that were filled with soil infested with *A. tumefaciens.* To ensure uniformity of the soil mix, moist soil with no detectable background population of *Agrobacterium* was placed in a cement mixer and tumbled while slowly spraying the tumbling soil or adding a prescribed amount of *A. tumefaciens* in liquid suspension. The mixer and transport equipment were washed thoroughly with water between strains. Alternatively, chemicals could be mixed in the potting mix in a similar fashion. Concrete boxes measuring 2 × 5 × 0.5 m containing infested soil have been used for tests in Spain (10). To study Verticillium wilt disease of potatoes (11), small tiles were buried in fields and filled with soil that was inoculated with *Verticillium* or with *Verticillium* plus a nematode, and this approach would seem adaptable to research with crown gall. The method is very labor intensive.

Small field plots of soil have been infested by spraying the soil surface with the pathogen and using a rotary tiller to mix the inoculum into the soil. Chemicals can be incorporated in a similar fashion. As the scale of plot size increases, soil infestation with laboratory-produced inoculum becomes less feasible, and treatment of the plant propagules is commonly used. However, small areas of carefully marked inoculated soil or nylon bags containing gall material (7) can be buried in the path of a fumigant applicator, thus exposing the material to a field application of the chemical. Survival of the pathogen can then be assayed using selective media (3,13). These same media and methods should be used to assay the populations of agrobacteria before any treatments are applied or the soil is infested. Combinations of selective media are required to assay for all the biovars.

In our experience, there is no correlation between the presence of natural *Agrobacterium* populations in the field and infections. Fairly high populations of *Agrobacterium* can be isolated from soil, but they are almost without exception avirulent. For example, natural populations of *Agrobacterium* strains in several Oregon nurseries were not affected or were reduced less than 10% in soils fumigated or treated with methyl bromide (Dowfume MC-2; 487 and 896 kg/ha [435 and 800 lb/A], tarped), dichloropropene (Telone II; 561 L/ha [60 gal/A], tarped, and the same rate untarped), sodium azide (112 and 224 kg of actual per hectare [100 and 200 lb of actual per acre]), and EPTC (Eptam; 5 and 8 L/ha [4 and 7 pt/A]). None of the agrobacteria isolated before soil treatment or after treatment were pathogenic, and none of the cherry and maple seedlings that were planted in the treated soil and harvested at the end of one growing season developed gall. Ideally, these plants would have been harvested as dormant seedlings, stored, planted, and observed for another growing season. Deep et al (5) did not report any galls on plants grown the first season in fumigated soil; but galls developed after planting out the second season, and the incidence of infection was highest in those seedlings that had been grown in fumigated soil. The effectiveness of a soil treatment is influenced by various soil edaphic factors (6,7,16), such as soil type, moisture content, pH, drainage, and crop residue, and these should be monitored carefully.

Timing of inoculation treatment. Plants are most susceptible to *A. tumefaciens* following any practice that wounds the plant. Planting or transplanting times are periods of high susceptibility. Dormant, bare-rooted seedlings are normally root pruned a few days to several weeks before they are planted in April or May (Pacific Northwest). It is critical to protect those wounds either by a direct chemical treatment before planting or to treat the planting medium to reduce the threat of infection after the seedlings are planted. These constraints indicate when the period of susceptibility is greatest and when the treatment is needed. However, soil treatments such as fumigation usually require high soil temperatures, which means a summer or early fall treatment. Thus, the timing of treatment must be scheduled carefully to fit within these constraints. Wounds can remain susceptible to infection for several months at reduced temperatures, but susceptibility decreases rapidly within 1 week after planting during the summer months (12); no wound healing occurred in cherry seedlings at 10°C (L. W. Moore, *unpublished data*). Inoculation of the trees with *A. tumefaciens* should be scheduled within these periods of susceptibility.

INTERPRETATION OF DATA

Data collection. Test trees or plants are harvested usually at the end of one growing season, but the time could be shortened to 3 weeks or extended to several years, depending upon the type of crop. If the plants are harvested too early, gall development may be difficult to read. Our primary focus is on treatments that reduce the incidence of galled plants. However, some treatments reduce the size and number of galls that develop on a plant, and these additional characteristics can be assayed if desired. A reduction in gall size and number is a positive result of the treatment, but the highest number of salable, gall-free plants is the ultimate goal.

Treated plants should be watched closely during the growing season to check for any chemically induced phytotoxicity. The treated plants are compared with those of the untreated control.

Data analysis. Statistical analysis of the data should be appropriate to the experimental design. A common problem with field data from plots planted with woody seedlings is unequal numbers of living trees among replications and treatments at the end of the experiment because of tree mortality during the growing season. It is advisable to consult a statistician if there is any question regarding experimental design and data analysis.

Reporting test results. The comments by Beer and Norelli (*this volume*) about reporting test results are equally pertinent here.

LITERATURE CITED

1. Anderson, A. R., and Moore, L. W. 1979. Host specificity in the genus *Agrobacterium*. Phytopathology 69:320-323.
2. Ark, P. A. 1954. Some important sources of crown gall bacteria in California orchards. Plant Dis. Rep. 38:207-208.
3. Brisbane, J. G., and Kerr, A. 1983. Selective media for three biovars of *Agrobacterium*. J. Appl. Bacteriol. 54:425-431.
4. Cooksey, D. A., and Moore, L. W. 1980. Biological control of crown gall with fungal and bacterial antagonists. Phytopathology 70:506-509.
5. Deep, I. W., McNeilan, R. A., and MacSwan, I. C. 1968. Soil fumigants tested for control of crown gall. Plant Dis. Rep. 52:102-105.
6. Dickey, R. S. 1961. Relation of some edaphic factors to *Agrobacterium tumefaciens*. Phytopathology 51:607-614.
7. Dickey, R. S. 1962. Efficacy of five fumigants for the control of *Agrobacterium tumefaciens* at various depths in the soil.

Plant Dis. Rep. 46:73-76.

8. Kurath, G., and Morita, R. Y. 1983. Starvation-survival physiological studies of a marine *Pseudomonas* spp. Appl. Environ. Microbiol. 45:1206-1211.

9. Lippincott, J. A., and Herberlein, G. T. 1965. The quantitative determination of the infectivity of *Agrobacterium tumefaciens*. Am. J. Bot. 52:856-863.

10. Lopez, M. M., Miro, M., Gorris, M. T., Salcedo, C. I., Temprano, F., and Orive, R. J. 1983. Comparative efficiency of inoculation treatments with *Agrobacterium tumefaciens* pv. *radiobacter* K84 against sensitive and resistant agrocin 84 strains of *Agrobacterium radiobacter* pv. *tumefaciens*. Pages 43-58 in: Proc. Int. Workshop Crown Gall. R. Grimm, ed. Swiss Fed. Res. Stn., Wadenswil, Switzerland.

11. Martin, M. J., Reidel, R. M., and Rowe, R. C. 1982. *Verticillium dahliae* and *Pratylenchus penetrans*: Interactions in the early dying complex of potato in Ohio. Phytopathology 72:640-644.

12. Moore, L. W. 1976. Latent infections and seasonal variability of crown gall development in seedlings of three *Prunus* species. Phytopathology 66:1097-1101.

13. Moore, L. W., Anderson, A., and Kado, C. I. 1980. *Agrobacterium*. Pages 17-25 in: Laboratory Guide for Identification of Plant Pathogenic Bacteria. N. W. Schaad, ed. American Phytopathological Society, St. Paul, MN. 72 pp.

14. Moore, L. W., and Warren, G. 1979. *Agrobacterium radiobacter* strain 84 and biological control of crown gall. Annu. Rev. Phytopathol. 17:163-179.

15. Munnecke, D. E., Chandler, P. A., and Starr, M. P. 1963. Hairy root (*Agrobacterium rhizogenes*) of field roses. Phytopathology 53:788-799.

16. Munnecke, D. E., and Ferguson, J. 1960. Effect of soil fungicides upon soil-borne plant pathogenic bacteria and soil nitrogen. Plant Dis. Rep. 44:552-555.

17. Panagopolus, C. G., Psallidas, P. G., and Alivizatos, A. S. 1978. Studies of biotype 3 of *Agrobacterium radiobacter* var. *tumefaciens*. Proc. Int. Conf. Plant Pathog. Bact., 4th, 1:221-228. Angers, France.

18. Ross, N., Schroth, M. N., Sanborn, R., O'Reilly, H. J., and Thompson, J. P. 1970. Reducing loss from crown gall disease. Calif. Agric. Exp. Stn. Bull. 845:1-10.

19. Smith, C. O., and Cochran, L. C. 1944. Crown gall and irrigation water. Plant Dis. Rep. 28:160-162.

20. Stonier, T. 1960. *Agrobacterium tumefaciens* Conn. II. Production of an antibiotic substance. J. Bacteriol. 79:889-898.

Assaying Populations and Evaluating Fungicides for Control of Soilborne Pathogenic Fungi

DONALD R. SUMNER and ALEX S. CSINOS, Department of Plant Pathology, University of Georgia, Coastal Plain Station, Tifton 31793

Many studies have been conducted on control of root, hypocotyl, and crown diseases with chemicals without determining the soilborne pathogens causing the diseases. Many root and crown diseases have common symptoms, and in most instances the causal organism must be identified to identify the disease. Since many chemicals being developed, or recently developed, have activity only against specific groups of microorganisms, it is especially important for pathologists to identify the fungi associated consistently with the disease being studied. It is beyond the scope of this chapter to discuss detailed methods of isolating and identifying species and races of fungi, but methods, and references to general methods, are given that can be used as guides to specific methods or media. Sometimes it may be necessary to develop selective media when those described in the literature are deemed to be unsatisfactory.

ASSAYING SOILBORNE FUNGI

***Rhizoctonia solani* Kühn and *Rhizoctonia*-like fungi.** Populations of *R. solani* and related fungi commonly occur in fields in populations of 0–100 propagules per 100 g of soil. Therefore, populations cannot be readily determined by the usual methods involving soil dilutions.

R. solani is usually present in soil in colonized particles of organic matter or as sclerotia (5). These particles and sclerotia can be removed by wet-sieving through 0.25- or 0.35-mm mesh sieves and incubated on petri plates of water agar or selective media (7,20), or soil can be placed directly into petri plates of selective media with a multiple-pellet soil sampler (4,17). With practice, the latter method provides a quick, convenient method of assaying numerous soil samples and identifying a wide range of basidiomycetes that exist in soil. *R. solani* and *Rhizoctonia*-like fungi can also be isolated from soil by incubating internodal stem pieces of mature, dry stems of buckwheat, cotton, or bean in soil for 3–4 days. After incubation, stem segments are removed, washed under tap water, and incubated on petri dishes of selective media (5). With all assay methods, hyphal tips of *R. solani* and *Rhizoctonia*-like fungi are transferred to potato-dextrose agar or potato-dextrose yeast extract casein hydrolosate agar (PDYCA) (17) and identified.

***Fusarium* spp.** The fusaria are common inhabitants of all soils (100–10,000 propagules per gram of soil), and most are saprophytes. However, numerous pathogenic *Fusarium* spp. survive in soil as chlamydospores or in colonized plant residues. Several media have been developed to assist in separating the soilborne pathogens from indigenous soil saprophytes. Soil samples are diluted 1:100 to 1:400 into tap water or 0.3–0.5% water agar. Samples in tap water may be shaken by hand or placed into flasks or beakers with a stirring bar and mixed with a magnetic stirrer. Dilute water agar is desirable for sandy soils to help prevent sand from settling from the suspension before dilutions are completed. The agar-gel suspension with soil can be shaken briefly by hand in bottles, and the soil will remain suspended in the gel until the dilutions are completed. One-milliliter aliquants are pipetted from the tap water suspension while it is being stirred, or from the gel suspension in bottles, onto each of five petri dishes of the desired selective media. Calibrated 1-ml, wide-tip pipettes are preferable to prevent sand particles from plugging the opening. For *F. solani* (Mart.) Appel & Wr., pentachloronitrobenzene (PCNB) or modified-PCNB medium is satisfactory (5). For *F. oxysporum* Schlecht., Komada's medium may be more satisfactory (8). If several *Fusarium* spp. or cultivars of *F. roseum* (Lk.) em. Snyd. & Hansen are to be assayed, it may be necessary to use combinations of two or more media (9). The soil suspension is evenly distributed over each petri dish with a spoon or L-shaped glass rod, or by gently swirling immediately after pipetting. Petri dishes are incubated at 20–30°C under fluorescent lights for 5–7 days and the colonies identified.

***Thielaviopsis basicola* (Berk. & Br.) Ferraris.** Soil dilutions of 1:1,000 to 1:10,000 may be assayed on a variety of selective media (5,10). However, a carrot disk technique may be the most sensitive assay in detecting low concentrations of the pathogen in naturally infested soils (18). Soil is spread over the surface of carrot disks in petri dishes, moistened by being sprayed with an atomizer, and incubated 2–4 days at room temperature. Soil is washed from the disks, the disks are reincubated for approximately 6 days in a moist chamber, and the gray colonies of the pathogen are identified.

***Sclerotium rolfsii* Sacc. and *S. cepivorum* Berk.** Sclerotia of *S. rolfsii* can be separated by washing soil on sieves or by a combination of flotation of sclerotia in a blackstrap molasses solution containing a flocculating agent, followed by sieving (13). Sclerotia are then incubated on selective media and the number of viable sclerotia determined. Similar methods have been used for *S. cepivorum* (5). A recent method for determining the number of viable sclerotia of *S. rolfsii* in field soil uses methanol to stimulate germination. The method is as accurate and is simpler than soil-sieving and flotation methods (13).

***Pythium* spp. and *Phytophthora* spp.** For soil dilutions, 1- to 5-g samples are placed into 50–100 ml of water or 0.1–0.3% water agar and serially diluted on benomyl pentachloronitrobenzene rifampicin ampicillin agar (BNPRA) (11), pimaricin ampicillin rifampicin pentachloronitrobenzene agar (PARP) (6), or other selective media (6,11,19). Numerous techniques have also been developed using various baits to isolate and identify pathogens (5,19). Chlamydospores of *Phytophthora cinnamomi* Rands may be separated from soil by sieving and placing the washings collected from nested sieves (149, 61, 44, and 38 mesh) onto selective media (5).

Other fungi. Populations of *Verticillium albo-atrum* Reinke & Berth. may be determined by soil dilutions in water agar (2) or by using an Anderson air sampler (5). Wet-sieving techniques or an Anderson air sampler may be used to assay *V. dahliae* Kleb on selective media (1). *Cylindrocladium crotalariae* (Loos) Bell & Sobers microsclerotia can be separated from soil with a semiautomatic elutriator (12) or wet-sieving (3), and debris suspensions or soil suspensions can then be diluted onto selective media (3). Similar methods have been used for other *Cylindrocladium* spp. (5). Techniques for assaying soils for populations of *Aspergillus* spp. and *Fomes annosus* (Fr.) Cke. by soil dilutions on selective media have been described (5).

A flotation method using mineral oil may be used for counting conidia of *Helminthosporium sativum* P.K. & B, other *Helminthosporium* spp., and large spores of other soilborne pathogens, such as *Alternaria* and *Trichothecium* spp. (5). Sclerotia of *Phymatotrichum omnivorum* Shear (Dug.) can be separated from soil with a sieving technique (5). Sclerotia of *Macrophomina phaseolina* (Tassi) Goid. can be removed from soil by suspending soil in 0.5% NaOCl for 10 min to kill mycelial tissue, wet-sieving, and washing residues from sieves into molten (45°C) selective media (14).

EXPERIMENTAL PROCEDURES

Location and design of experiments. Plots should be established in fields with a history of the specific diseases to be studied. The influence of such cultural practices as tillage, soil fertility, irrigation, row spacing, plant spacing, and depth of seeding on disease severity should be considered. For example, turning soil 20–30 cm deep with a moldboard plow usually reduces populations of *R. solani* AG-4 in topsoil and leads to decreased root and hypocotyl rot in snap bean, lima bean, and other vegetables and reduces disease from *S. rolfsii* on peanut. In soils with a hardpan, root diseases may be more severe on the restricted root systems. Therefore, the selection of cultural practices may determine the kind and amount of diseases to be controlled and whether sufficient disease occurs for a fungicide evaluation to be successful.

Cultivars used may determine the disease severity and the kinds of evaluations that can be obtained. In many crops there is a range of susceptibility to different pathogens among cultivars, and the cultivars selected may influence the range of disease control, yield, and quality differences among fungicide treatments.

Application of fungicides. Several different application methods and types of equipment may be used to apply fungicides in field plots. A common method with beans, peas, and cotton is an in-furrow spray with two nozzles attached to the planter, one flat-fan nozzle directed downward into the furrow onto the seed and another at a 45° angle to the first spraying into the covering soil in front of the press wheels. In-furrow granules may be applied in a 15- to 25-cm band behind the seed drop tubes and in front of the press wheels at a right angle to the row. For small-seeded crops such as crucifers, cucurbits, celery, carrot, lettuce, and spinach, a surface spray with one nozzle or a band application behind the planter may be more feasible. In plant beds seeded for transplant production, broadcast spray or granule treatments are probably more practical than individual row treatments.

Transplant treatments are usually added as solutions or suspensions into the water used during transplanting. Tobacco, cabbage, tomato, collard, pepper, and some other crops are commonly transplanted because of short growing seasons in cooler climates, or to achieve the desired plant spacing.

Postemergence or posttransplant treatments may be applied as directed sprays or granules at the base of the stem, as sprays or granules over the row, or as broadcast treatments at some point during the growing season to provide middle and late-season control of root, pod, peg, or tuber rot.

Preplant incorporated treatments may be applied either as sprays or granules on a broadcast basis. Fungicides may be incorporated to the required depth by double disking or with a rotary tiller. Soil fumigants are usually injected with chisels under or adjacent to the row or applied on a broadcast basis. Because of the toxicity of the vapors to plants, fumigants are usually applied 2–8 weeks before planting or transplanting, and the treated area may have to be covered with plastic or irrigated immediately after application for maximum efficacy.

A method developed recently involves application of soil fungicides through irrigation water by using overhead sprinklers (16). It is possible that chemicals also can be applied with trickle, drip, and other kinds of irrigation systems. With this method it is essential to have approved safety devices to prevent backflow of chemicals into irrigation water sources, to have uniform mixing of the fungicide in the water, and to monitor the distribution of the water per unit area or per plant.

Rates of application are expressed as active ingredient per unit area (kg/ha, lb/A) on a broadcast basis, or per linear length of row (g/m, lb/1,000 ft), and sprays are applied in a specified amount of water per area or per row. In-furrow treatments should specify the amount per linear length of row and the width of the row to be treated. In transplanting, the amount of chemical in a certain amount of water (g/L, lb/gal) and the rate of solution or suspension per plant or field unit (liters, cups, pints) are determined. Postemergence and posttransplanting treatments may be on a broadcast rate per unit area or per linear length of row. Application through irrigation water is by unit of chemical per volume of water per unit area, or with trickle irrigation per unit area or per plant.

Test chemicals may be applied to small plots by hand with sprinkler cans or sprayers, or by hand-powered equipment for spreading granules. Chemicals and water should be measured separately for each plot, as small errors in calculations are magnified when converted to a per-acre or per-hectare basis.

Sampling soil. Soil samples are collected from areas in the soil profile where root diseases are to be controlled. For shallow-rooted crops, the soil 0–10 cm or 0–15 cm under the row or adjacent to the row is collected. For deep-rooted crops, it may be desirable to sample a meter or more in depth. An easy, fast way to collect samples is with a soil probe that removes a core 2.5 cm in diameter. The probe can be inserted as deeply as it can be pushed and the cores divided by depth, if desired. Moisture content of the soil may be the single most important factor in taking a good representative soil sample. Water-logged soil or dry, sandy soil is extremely difficult to sample with a soil probe. Power-driven soil probes or augers may be used to collect large cores 3–10 cm in diameter and 1–3 m deep. Several (5–10) cores are taken in each plot, composited, and mixed thoroughly, and the amount of soil needed for assays is removed and placed into a plastic bag, soil can, or glass jar as soon as possible. Samples are placed into a dry refrigerator or storage room at 2–5°C until processing. Samples should be processed as soon as possible and not be stored more than a few weeks, or the assays may not accurately represent the microflora in the soil at the date of sampling.

It may be desirable to sample soils immediately before and after treatment with fungicides and at intervals of one or more weeks after application, depending on the crop and the pathogen being studied. Specific stages of crop development such as bloom, fruit set, and maturation or harvest may also be used as time increments for soil sampling.

Root disease evaluations. Emergence, plant stands, and postemergence damping-off are recorded 1–4 weeks after planting, depending on the crop and weather conditions. With many vegetables it is advantageous to take two or three stand counts 7–10 days apart, as small seedlings may emerge, die, and disintegrate within a few days. Seedlings are dug from a predetermined area of row 1–4 weeks after planting, the roots washed in cool (10–20°C) water, and the roots and hypocotyls rated for disease severity. A rating system commonly used is a 1–5 scale with 1 = <2, 2 = 2–10, 3 = 11–50, and 4 = >50% discoloration and decay; 5 = plant dead or dying. The scale could be expanded to 1–10 with finer divisions of disease severity, if desired. Other data that could be recorded include the kinds and colors of lesions on specific roots or hypocotyls (i.e., the number of plants with hypocotyl cankers in legumes and many other plants) and the number and kind of roots (seminal, crown, tap, fibrous) with lesions or terminal decay.

Isolations are made from lesions and causal organisms identified to confirm the presence of the intended pathogens in the experiment. An empirical scale of 1–5 can be used for root growth, with 1 = very poor growth and 5 = excellent growth. Wet and dry weight of roots, foliage, or roots plus foliage can be taken for additional data for evaluating treatments. Data on phytotoxicity to roots and foliage at one or more stages of crop growth is recorded by comparisons to plants grown in untreated soil free of pathogens in greenhouse or field tests.

Yield. Data on the quality and quantity of the plant part (grain, green foliage, roots, tubers, ears, pods, fruits, transplants) normally harvested for yield is desirable for each treatment, as root disease severity and root growth may not be related to yield. Visual observations may not detect differences in fibrous root injury or rhizosphere ecology related to fungicide treatments. For edible crops, the plant part that is eaten may have to be sampled at specific periods of growth for chemical residues and the influence of fungicides on flavor and nutrition determined. When roots, tubers, bulbs, or pods are the usually harvested plant part (carrot, Irish potato, sweet potato, turnip, peanut, radish, beet, sugar beet, onion, taro, rutabaga), a disease severity rating is also taken at harvest as part of the quality evaluation.

Greenhouse and environmental chamber studies. Soil naturally infested with a specific pathogen or a mixture of pathogens can be used. Soil treated with aerated steam, steam, or dry heat at 65–75°C for 30 min and infested artificially with cultures may also be used. Heating soil to higher temperatures may cause undesirable changes in chemical and physical properties. Cultures of pathogens can be grown on 3% cornmeal-sand (w/w) or natural or artificial media (5) and blended with soil on a weight-per-weight basis. Fungicides are applied on a weight-per-volume (milligram-per-liter) basis so that the amount of fungicide per volume of soil is constant regardless of the size of container used (15). The soil, fungicide, cultures, and fertilizer can be blended in a concrete mixer, or small volumes of soil and chemicals can be poured back and forth from one container to another at least five times. Then the mixed soil is placed in the desired container, the seeds are planted or plants are transplanted, and the plants are grown under the desired environmental conditions. If the fungicide is to be mixed with soil only at a certain depth or placement, soil without the fungicide can be placed in the containers separately from soil with the fungicide. However, all soil within a container should be from the same source and have the same fertility. If fungicides are to be drenched onto soil after planting or transplanting, rates are still calculated on a weight-per-volume basis, rather than by surface area of the container.

Statistical analysis. Data are analyzed according to the experimental design by using statistical techniques discussed in more detail in other chapters (see especially L. A. Nelson, *this volume*). If possible, correlation and regression analyses are desirable to show the relationship of disease symptoms to yield and the influence of emergence, heading, pollination, and other plant growth characteristics to fungicide treatments and root disease control.

LITERATURE CITED

1. Butterfield, E. J., and Devay, J. E. 1977. Reassessment of soil assays for *Verticillium dahliae*. Phytopathology 67:1073-1078.
2. Christen, A. A. 1982. A selective medium for isolating *Verticillium albo-atrum* from soil. Phytopathology 72:47-49.
3. Griffin, G. J. 1977. Improved selective media for isolating *Cylindrocladium crotalariae* microsclerotia from naturally infested soils. Can. J. Microbiol. 23:680-683.
4. Henis, Y., Ghaffar, A., Baker, R., and Gillespie, S. L. 1978. A new pellet soil-sampler and its use for the study of population dynamics of *Rhizoctonia solani* in soil. Phytopathology 68:371-376.

5. Johnson, L. F., and Curl, E. A. 1972. Methods for research on the ecology of soil-borne plant pathogens. Burgess Publishing Co., Minneapolis. 247 pp.
6. Kannwischer, M. E., and Mitchell, D. J. 1981. Relationships of numbers of spores of *Phytophthora parasitica* var. *nicotianae* to infection and mortality of tobacco. Phytopathology 71:69-73.
7. Ko, W., and Hora, F. K. 1971. A selective medium for the quantitative determination of *Rhizoctonia solani* in soil. Phytopathology 61:707-710.
8. Komada, H. 1975. Development of a selective medium for quantitative isolation of *Fusarium oxysporum* from natural soil. Rev. Plant Prot. Res. 8:114-125.
9. McMullen, M. P., and Stack, R. W. 1983. Effects of isolation techniques and media on the differential isolation of *Fusarium* species. Phytopathology 73:458-462.
10. Maduewesi, J. N. C., Sneh, B., and Lockwood, J. L. 1976. Improved selective media for estimating populations of *Thielaviopsis basicola* in soil on dilution plates. Phytopathology 66:526-530.
11. Masago, H., Yoshikawa, M., Fukada, M., and Nakanishi, H. 1977. Selective inhibition of *Pythium* spp. on a medium for direct isolation of *Phytophthora* spp. from soils and plants. Phytopathology 67:425-428.
12. Phipps, P. M., Beute, M. K., and Barker, K. R. 1976. An elutriation method for quantitative isolation of *Cylindrocladium crotalariae* microsclerotia from peanut field soil. Phytopathology 66:1255-1259.
13. Rodriguez-Kabana, R., Beute, M. K., and Backman, P. A. 1980. A method for estimating numbers of viable sclerotia of *Sclerotium rolfsii* in soil. Phytopathology 70:917-919.
14. Short, G. E., Wyllie, T. D., and Bristow, P. R. 1980. Survival of *Macrophomina phaseolina* in soil and in residue of soybean. Phytopathology 70:13-17.
15. Smith, F. F. 1952. Conversion of per-acre dosages of soil insecticide to equivalents for small units. J. Econ. Entomol. 45:339-340.
16. Sumner, D. R. 1981. Application of foliar and soil fungicides and a soil fumigant through overhead irrigation water. Pages 82-88 in: Proc. Natl. Symp. Chemigation. J. R. Young, ed. Rural Development Center, Tifton, GA.
17. Sumner, D. R., and Bell, D. K. 1982. Root diseases induced in corn by *Rhizoctonia solani* and *Rhizoctonia zeae*. Phytopathology 72:86-91.
18. Tabachnik, M., Devay, J. E., Garber, R. H., and Wakeman, R. J. 1979. Influence of soil inoculum concentrations on host range and disease reactions caused by isolates of *Thielaviopsis basicola* and comparison of soil assay methods. Phytopathology 69:974-977.
19. Tsao, P. H. 1970. Selective media for isolation of pathogenic fungi. Annu. Rev. Phytopathol. 8:157-186.
20. Weinhold, A. R. 1977. Population of *Rhizoctonia solani* in agricultural soils determined by a screening procedure. Phytopathology 67:566-569.

Test Materials and Environmental Conditions for Field Evaluation of Nematicides

A. W. JOHNSON, U.S. Department of Agriculture, Agricultural Research Service, Coastal Plain Experiment Station, Tifton, GA 31793

Test materials and environmental and cultural conditions are important factors that influence the results of secondary and field evaluation of nematicides. This chapter describes the information on these factors that should be recorded during field tests of experimental nematicides to achieve meaningful evaluation.

TEST MATERIALS

All test materials should be compared with an untreated control and with a known standard, usually one of the materials currently recommended. To evaluate field trials of experimental nematicides, researchers should know as much as possible about the biological activity and the chemical and physical properties of the test material. Pertinent literature or technical reports should be reviewed before field trials are designed.

Formulation

The recorded data should include formulation type—emulsifiable concentrate, wettable powder, flowable, water-soluble, or granular (mesh size). The names and percentage of every ingredient in the formulation, the lot number on the package label, and dates sent and received should be recorded. If a nematicide is diluted before application, the amount of diluent used should be recorded and the diluent material should be specified by common and chemical names.

Application Rates

Rates should be clearly and precisely stated as formulation and active ingredient in one or more of the following terms: the quantity per unit of area if treated overall (broadcast), the quantity per linear distance and row spacing if row-treated, and the width of band and row spacing if band-treated.

Number and Timing of Applications

Dates (month, day, year) of preplant or postplant applications or both should be recorded. Proper timing of application is critical. Information should specify the time of application in terms of crop planting date, emergence date, growth stage, preharvest interval, and intervals between applications. Target pest population levels and incidence should be specified. For postplant applications (after seeds or seedlings are planted or on established plantings), plant size, stage of growth, number of days since emergence, or a combination of these statistics should be recorded.

Method of Application

Method of application, including specialized equipment, should be specified. Such descriptive terms as spraying, injecting, spacing, soaking, rinsing, and flooding should be used when appropriate. Soil applications should include such information as band width, row spacing, chisel spacing, depth of application, and time interval between application and incorporation. If applied on the surface, the method and depth of incorporation, if any, should be stated. Descriptions of row applications should include whether they were in-furrow, band-over-row, or side-dressed (preplant, at planting, postplant, postemergence). For side-dressed applications, placement in relation to seed or plant should be given.

Treatment may be broadcast, strip, row, site, root-dip, or foliar spray, and the test material may be injected into the soil or applied to the soil surface as a drench, spray, granule, or solution in irrigation water. Granules may be incorporated into the top few centimeters of soil, or the active ingredient may be washed from the granules by irrigation or rainfall. Care should always be taken to prevent recontamination of treated areas by cultivation or other means whereby soil from untreated areas is blended with soil from treated areas. Methods of application in irrigation water include overhead sprinkler, flood, trickle, row or furrow, and basin.

ENVIRONMENTAL AND CULTURAL CONDITIONS

Information about environmental and cultural factors before treatment, at the time of treatment, and after treatment that could affect nematicidal efficacy should be recorded. Erratic results of incomplete experimentation may be caused, at least in part, by the effects of such factors as relative humidity, wind, rainfall, and air temperature during the test period. The relationship of all environmental and cultural factors to the crop, pathogen, and nematicide should be considered and explained in any evaluation of results.

Soil factors include the identity of target and nontarget nematodes and their relative density before, during, and after testing; temperature; soil types,

Revision of a 1978 paper prepared by the Joint SON/ASTM E35.16 Task Force on Test Material and Environmental Conditions in Field Evaluation of Nematode Control Agents, A. W. Johnson, chairman.

including textural variations with depth; pH; field capacity; nutrient levels; percentage of organic matter; presence or absence of crop refuse or trash; percentage of soil moisture and, if possible, an estimate of drainage; amounts and frequency of rainfall, irrigation, or both, and type of irrigation (e.g., flood, sprinkler, row, basin); and other data that may affect the application or performance of the nematicide being tested. Occurrence and quantity of other organisms that affect crop growth and nematode populations should also be noted. Applications of fertilizers, lime, and other soil amendments such as herbicides, fungicides, and insecticides should be recorded. Previous cropping history and pesticide usage are also important.

Determining Nematode Population Responses to Control Agents

K. R. BARKER, Department of Plant Pathology, North Carolina State University, Raleigh 27695-7616; J. L. TOWNSHEND, Agriculture Canada, Vineland Station, Ontario L0R 2E0; G. W. BIRD, Department of Entomology, Michigan State University, East Lansing 48823; I. J. THOMASON, Department of Plant Nematology, University of California, Riverside 92521; and D. W. DICKSON, Department of Entomology and Nematology, Nematology Laboratory, University of Florida, Gainesville 32611

Chemical control agents are necessary for the efficient production of many crops. If nematicides are to be tested and used properly, there must be an informed consensus on reliable, standardized sampling procedures and extraction methods for the target nematodes. Regardless of the methods used to evaluate nematode responses to nematicides, any weaknesses in the experiments will be reflected in the results.

In evaluating the efficacy of nematode control agents, characterizing nematode population responses is as important as determining plant responses. Most nematicides exert their maximum effect on nematodes shortly after application, but long-term changes in soil biology often occur. For example, the proportions of species in nematode communities often shift after chemical treatments of soil. In addition, some nematicides may stimulate plant growth whether or not nematodes are present; others affect insect, fungal, or bacterial populations or a combination of these.

Revision of a 1978 paper prepared by the Joint SON/ASTM E35.16 Task Force on Nematode Population Response, K. R. Barker, chairman.

The most important considerations in determining nematode population responses to control agents are collection of representative soil or root samples or both from each plot, adequate mixing of these samples, and extraction of nematodes by the most appropriate procedure for the target species and soil type. Such procedures depend on the kinds and numbers of nematodes present, their characteristics, and the nature and condition of the samples, including soil texture and time of collection (Tables 1 and 2). If any of these procedures is to be used reliably to evaluate responses to nematode control agents, investigators must understand the population dynamics of the target nematode on the particular crop and the specific geographic region involved. Where nematode numbers are below a detectable level, an appropriate bioassay should be used.

Because no single method of collecting nematode samples for assay or extracting nematodes to evaluate

Table 1. Extraction procedures for combinations of nematodes and soil types[a]

	Type of nematode (and soil type)											
	Meloidogyne spp.[b]			*Globodera* and *Heterodera* spp.[c]			Endoparasites[d]			Ectoparasites[e]		
Extraction technique	Sandy soil	Clay soil	Organic soil	Sandy soil	Clay soil	Organic soil	Sandy soil	Clay soil	Organic soil	Sandy soil	Clay soil	Organic soil
Baermann trays (31)	X	X	X	X	X	X	X	X	X	X	X	X
Blender and Baermann trays (31)							X	X	X			
Cobb's decanting-sieving and Baermann trays (31)	X			X			X	X		X	X	
Centrifugal flotation (16)	X	X	X	X	X	X	X	X	X	X	X	X
Centrifugal flotation with heavy sugar (9)				X	X	X						
Elutriation and Baermann trays (4)	X	X		X	X		X	X		X	X	
Elutriation and centrifugal flotation (4)	X	X		X	X		X	X		X	X	
Shaker extraction (3)							X	X	X			
Elutriation and dissolution egg masses (4,5)	X	X	X									
Elutriation (flotation) and dissolution cysts (1,4,5)				X	X							
Fenwick or flotation can (1,27)				X	X	X						
Mist chamber (24)							X	X	X			
Sugar-flotation sieving (7)	X						X			X		

[a]See Table 2 for selection of specific techniques for given sampling times.
[b]Also for *Nacobbus*, *Rotylenchulus*, and *Tylenchulus* spp.
[c]If hatching factor is required, use Fenwick or flotation can only.
[d]*Pratylenchus, Radopholus, Hoplolaimus*; for some plants, *Helicotylenchus, Tylenchorhynchus, Ditylenchus*, and *Aphelenchoides*.
[e]*Belonolaimus, Dolichodorus, Helicotylenchus, Hemicycliophora, Longidorus, Paratylenchus, Rotylenchus, Scutellonema, Paratrichodorus, Tylenchorhynchus*, and *Xiphinema*. Sugar-flotation sieving is most efficient for large forms such as *Belonolaimus* and *Xiphinema*. Researchers should use centrifugal flotation or elutriation and centrifugation for *Criconemella* spp.

nematicides can be used in all situations, the procedures described in this chapter are presented only as guidelines. However, specific procedures for sampling and extracting nematodes are recommended for various combinations of crops, soil textures, sampling periods, and nematodes. The last section is a glossary of some useful descriptive terms.

SAMPLING

Sampling procedures vary, depending on the crop, nematode species, host-parasite relationships, and soil type. Ideal times for sampling (when differences in population densities between treated and untreated plots are greatest) also vary with the crop, nematode, and geographic location.

The primary objective in sampling is to collect a sample that truly represents the population in a given plot or field at a given time. The spatial distribution of nematode populations varies greatly both horizontally and vertically. Most species occur as aggregates (egg masses or cysts) in a clustered (contagious) distribution (2,13) rather than in either a uniform or a random pattern (Fig. 1). The populations also may be skewed in distribution because of plant or soil influences. For these reasons, sampling is frequently the weakest link in field evaluations of nematicides (Fig. 2).

Important considerations in soil sampling include the number, diameter, and depth of cores needed to provide

Table 2. Extraction procedures for use at various sampling times for nematode parasites of annual and perennial crops[a]

	Type of nematode (and sampling time)[b]																			
	Meloidogyne spp.[c]				*Globodera* and *Heterodera* spp.[d]				*Criconemella* spp.[e]				Endoparasites[f]				Ectoparasites[g]			
Extraction technique	Pi	Pie	Pm	Pf	Pi	Pie	Pm	Pf	Pi	Pie	Pm	Pf	Pi	Pie	Pm	Pf	Pi	Pie	Pm	Pf
Annuals																				
Baermann trays (31)	X	X	X		X	X	X	X					X	X	X		X	X	X	X
Bioassay-gall index (19)			X(?)	X																
Blender and Baermann trays (31)													X		X	X				
Cobb's decanting-sieving and Baermann trays (31)	X	X	X	X	X	X	X	X					X	X	X	X	X	X	X	X
Centrifugal flotation (16)	X		X(?)	X			X(?)	X	X	X[e]	X[e]	X	X		X(?)	X	X		X(?)	X
Centrifugal flotation with heavy sugar (9)					X			X												
Elutriation and Baermann trays (4)	X	X	X	X	X	X	X	X					X	X	X	X	X	X	X	X
Elutriation and centrifugal flotation (4)	X		X		X			X	X	X[e]	X[e]	X	X			X	X		X(?)	X
Elutriation and dissolution egg masses (4,5)			X	X																
Elutriation (flotation) and dissolution cysts (1,4,5)					X		X(?)	X												
Fenwick or flotation can (1,27)					X		X(?)	X												
Mist chamber (24)															X	X				
Shaker extraction (3)															X	X				
Sugar-flotation sieving (7)	X												X		X	X	X		X	X
Vital stain (22)										X[e]	X[e]									
Perennials																				
Baermann trays	X	X	X	X									X	X	X	X	X	X	X	X
Bioassay-gall index	X		X	X																
Blender and Baermann trays													X	X	X	X				
Cobb's decanting-sieving and Baermann trays	X	X		X									X	X	X	X	X	X	X	X
Centrifugal flotation	X		X(?)	X					X	X[e]	X[e]	X	X	X	X	X	X	X	X	X
Centrifugal flotation with heavy sugar					X		X	X												
Elutriation and Baermann trays	X	X	X	X	X	X	X	X					X	X	X	X	X	X	X	X
Elutriation and centrifugal flotation	X		X(?)	X	X		X	X	X	X[e]	X[e]	X	X		X(?)	X	X	X	X	X
Elutriation and dissolution egg masses	X		X	X																
Elutriation and dissolution cysts					X		X	X												
Fenwick or flotation can					X		X	X												
Mist chamber													X		X	X				
Shaker extraction													X		X	X				
Sugar-flotation sieving	X		X	X									X		X	X	X		X	X
Vital stain										X[e]	X[e]									

[a]See Table 1 for consideration of soil type, Table 3 for data required for sampling times.

[b]Sampling times: Pi = initial nematode density, Pie = posttreatment nematode density, Pm = midseason nematode density, and Pf = final nematode density.

[c]Also for *Nacobbus*, *Rotylenchulus*, and *Tylenchulus* spp.

[d]If hatching factor is required, use Fenwick or flotation can only.

[e]Need vital stain to identify living specimens for Pie and sometimes Pm sampling.

[f]*Pratylenchus*, *Radopholus*, *Hoplolaimus*; for some plants, *Helicotylenchus*, *Tylenchorhynchus*, *Ditylenchus*, and *Aphelenchoides*.

[g]*Belonolaimus*, *Dolichodorus*, *Helicotylenchus*, *Hemicycliophora*, *Longidorus*, *Paratylenchus*, *Rotylenchus*, *Scutellonema*, *Paratrichodorus*, *Tylenchorhynchus*, and *Xiphinema*. (Sugar-flotation sieving is most efficient for large forms such as *Belonolaimus* and *Xiphinema*.)

A

50	100	0	0	0	0
100	400	0	0	0	0
0	0	0	125	100	0
0	0	0	0	50	0
0	50	0	0	0	0
125	275	0	0	0	335

B

40	45	50	55	45	50
45	55	45	40	40	50
50	45	45	45	50	55
45	50	55	45	50	45
55	50	40	50	45	45
50	40	50	50	55	40

C

30	30	47	38	30	43
27	45	60	52	101	64
36	30	60	50	25	53
24	21	68	61	50	0
71	52	76	45	74	67
50	43	30	50	60	47

Fig. 1. Theoretical distribution patterns of nematode populations. **A,** Aggregate or contagious ($\sigma^2 > \bar{x}$; $\bar{x} = 47.5$, $\sigma^2 = 9{,}316$, $\Sigma = 1{,}710$). **B,** Uniform ($\sigma^2 < \bar{x}$; $\bar{x} = 47.5$, $\sigma^2 = 22.9$, $\Sigma = 1{,}710$). **C,** Random ($\sigma^2 = \bar{x}$; $\bar{x} = 47.5$, $\sigma^2 = 360$, $\Sigma = 1{,}710$; although $\sigma^2 > \bar{x}$ in this case, the population distribution is more random than the example in **A**).

an adequate sample; the choice of a sampling pattern that will give reliable, representative data on population densities; the choice of sampling times that will reflect population density at critical stages during the growing season; the condition of the soil; and the proper handling and storage of samples.

Soil should be moist but not wet, preferably below 60-cm tension (percentage) of water suction (field capacity) for most sampling. The percentage of moisture should be determined from three to five representative samples per sampling time. Before extraction for cysts of *Heterodera* or *Globodera* spp., certain clay or silt soils (which become hard when dry) may be dried slowly and crumbled daily during the drying process.

Four samplings should be made to evaluate fully the effects of chemical soil treatments on nematode populations and plant growth: a pretreatment sampling (within 1 week before treatment), an intermediate posttreatment sampling (2–4 weeks after treatment), a midseason sampling, and a final sampling near or at harvest or the end of the growing season. In each case, the nematode population density should be determined per 100–500 cm^3 of soil. Nematicide treatments are often applied in the fall to avoid excessive delays in spring planting in northern areas. In such situations, sampling schedules need to be modified accordingly (the intermediate posttreatment sampling would be in the spring). Posttreatment samples collected in late fall and early spring are also necessary to determine nematicide effects on perennials.

The ideal sampling times may be identified more precisely by experimentally determining the degree-hours or -days (11) or the cumulative index of activity (17) required for the maximum population increase. The application of degree-hours (or -days) is helpful except where extremes of temperatures and/or moisture are encountered. A comparable model that relates temperature to the development (D) of root endoparasites, offered by Jones (17), is as follows:

$$D = k \sum_{1}^{n} (T_n - T_b) \ ,$$

where k is a temperature coefficient, T_n is the mean daily temperature, T_b is the basal temperature below which

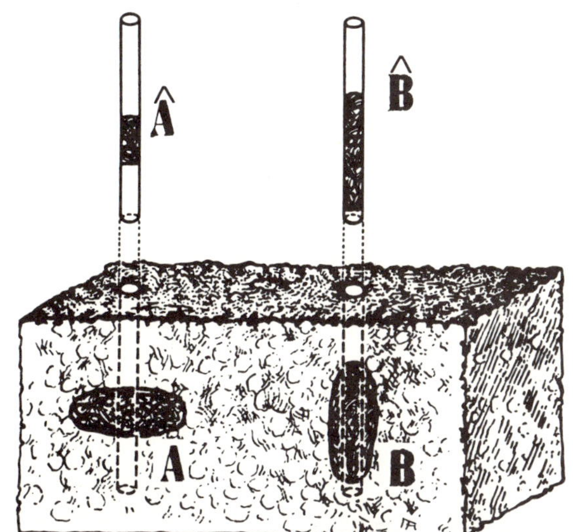

Fig. 2. Soil profile containing two aggregates of nematodes of equal population density (A = B/100 cm^3 of soil). Because of the difference in the axis orientation, the population densities estimated from soil cores are vastly different ($\hat{A} \sim 0.3\,\hat{B}/100\ cm^3$ of soil).

growth is essentially zero, and $T_n - T_b$ is the effective temperature that is survived for n days. Jones developed a similar model, based on rainfall, for ectoparasites. His concept of combining cumulative temperature and rainfall for calculating an "index of activity" for nematodes may eventually be helpful in developing sampling schedules.

To standardize sampling procedures, researchers should use a sampling tube 2.0 cm in inner diameter (ID) or a cone-shaped tube (14) and sample to a depth of 20 cm unless otherwise indicated. Cores should be taken to a depth of 30–45 cm for the pretreatment sampling in regions with hot, dry summers; in such situations, a hydraulic sampling device may be necessary. Each core (2.0 cm in diameter by 20 cm deep) provides about 62 cm^3 of soil. Enough soil cores should be collected in a systematic procedure (Fig. 3A) to cover each plot according to its size as follows: small plots (1–5 m^2), 10 cores; medium-size plots (5–100 m^2), 20 or more cores, and large plots (greater than 100 m^2), 30 or more cores.

The pretreatment distribution pattern of target nematodes should be determined on a quadrat basis when a high degree of precision is required. The actual distribution patterns can be characterized by a number of statistical procedures and parameters. The negative binomial distribution, described by two parameters, the mean ($\bar{x}$) and the dispersion parameter (k), can be fitted to nematode sample data from clustered distributions (2,13). The distribution approaches that of the Poisson as k approaches infinity (for practical purposes, $k \geqslant 8$ indicates a population with a near-random distribution). Values of k less than 1.0 indicate that the distribution is approaching the logarithmic series that occurs when $k = 0$ (10,28).

A relatively simple equation for estimating k is:

$$k = \frac{\bar{x}^2}{s^2 - \bar{x}} ,$$

where $\bar{x}$ is the population mean and s^2 is the sample population variance. Fitting the negative binomial distribution to population data and estimating k can be done by available computer programs and more complex models (10,18).

The mean and k parameters may be used as a basis for estimating the number of samples needed for given degrees of precision (18). The number of cores needed for clustered populations, which can be described by the negative binomal distribution, may be determined as follows (18):

$$n = (Z\alpha/2)^2 \frac{(1/\bar{x} + 1/k)}{E^2} ,$$

where n is the number of samples, Z is the upper $\alpha/2$ point of the standard normal distribution, $\bar{x}$ is the population mean, k is the experimentally derived dispersion parameter of the negative binomial distribution, and E is the predetermined standard error as a decimal of the mean. This model, described by Karandinos (18), is more rigorous than those given by other authors (2,28) but is useful for single-composite sample estimations of populations. The simple model

$$n = \frac{(1/\bar{x} + 1/k)}{E^2}$$

given by Southwood (28) should be adequate for determining nematicide efficacy in small-plot tests.

Fig. 3. Schemes for collecting soil samples. **A,** Recommended pattern for collecting a minimum of 20–30 cores for pretreatment nematode assay of large test field or plot. **B,** Sampling pattern for collecting soil from two center rows in four-row plot (performance is determined only from the two center rows). **C,** Procedure for sampling single-plant plot. **D,** Pattern for sampling in feeder-root zone of established perennials (for small plots, additional cores should be collected from each plant).

Careful collection of soil samples by well-trained, responsible individuals is of utmost importance for reliable nematode assays. The sampling procedure, however, must be modified for different kinds of crops. With established annual row crops, sampling should be done in the row, with cores coming from the root zone, 5–12 cm from the stems (Fig. 3B). Special care must be given to single-plant plots (Fig. 3C). For deep-rooted perennials such as grape and citrus, samples should be collected at two or three depths (15, 30, and 60–100 cm). Sampling to a depth of 8–12 cm is sufficient for shallow-rooted perennials such as turfgrasses. For ornamentals and other perennials such as fruit trees, borings should be collected in the drip line to a depth of 20 cm (Fig. 3D). Samples should be taken at several depths down to 90 cm for some fruit trees on deep alluvial soils.

Extracting nematodes from roots and soil at mid-season and final samplings is often necessary for certain endoparasites such as *Pratylenchus* spp. (Tables 2 and 3). A larger sampling tube (5.0 cm ID) should be used to obtain roots with such soil samples. Alternatively, the researcher may dig and collect root samples from a minimum of 10 plants per plot.

Initial Nematode Density

Each plot should be sampled in a systematic manner within 1 week before treatment. This sampling is particularly important to establish a base population density for each individual plot, because nematode populations may vary greatly from plot to plot. These base population densities allow more precise estimation of the effects of various nematicides in each plot.

Posttreatment Nematode Density

The intermediate posttreatment sampling can be used to relate numbers of nematodes to crop performance for fast-acting nematicides (fumigants).

Caution: Do not collect soil treated with highly toxic, slow-acting materials such as organic phosphates or carbamates within 14 days after application.

Fumigant nematicides. With fumigant nematicides, plots should be sampled 2–3 weeks after application in the same pattern as the initial sampling if the treatment is broadcast. For row treatment, the sampling should be done in the row.

Table 3. Nematode population data required from standardized-volume soil samples from annual and perennial crops[a]

	Type of nematode				
	Sedentary forms			Migratory forms	
Sampling time[b]	*Meloidogyne* spp.	*Globodera* and *Heterodera* spp.	*Criconemella* spp.	Endoparasites	Ectoparasites
			Annuals		
Pi	Juveniles in soil	Cysts, eggs, and juveniles in soil	Nemas in soil	Nemas in soil	Nemas in soil
Pie	Living juveniles in soil	Living juveniles	Living nemas in soil	Living nemas in soil	Living nemas in soil
Pm	Juveniles and eggs (soil and roots)	Eggs, juveniles, and cysts	Living nemas in soil	Living nemas in soil and roots	Nemas living in soil
Pf	Above plus gall index	As above	Nemas in soil	Nemas in soil and roots	Nemas in soil
			Perennials		
Pi	Juveniles and eggs (soil and roots)	—	Nemas in soil	Nemas in soil and roots	Nemas in soil
Pie	Living juveniles	—	Living nemas in soil	Living nemas in soil and roots	Living nemas in soil
Pm and Pf	Same as for annuals				

[a]See Tables 1 and 2 for specific extraction procedures for specific combinations of types of nematodes and soil.
[b]Sampling times: Pi = initial nematode density, Pie = posttreatment nematode density, Pm = midseason nematode density, and Pf = final nematode density.

Table 4. Nomograph of root-knot galling indexes for *Meloidogyne* spp.

Percentage of root system galled	Galling index system[a] 0–4	0–5	1–6[b]	0–10
0	0	0	1	0
10		1	2	1
20		2	3	2
	1			
30				3
40				4
50	2		4	5
		3		
60				6
70				7
	3			
80		4	5	8
90				9
100	4	5	6	10

[a]Roots are scored for degree of galling using one of several root-galling indexes, all of which are comparable and relatively interchangeable. A minimum of 10 root systems should be evaluated per treatment.
[b]We recommend this scheme for use in evaluating nematicides (see Fig. 5 for examples similar to those of Zeck [34]).

Nonfumigant nematicides. Sampling should be done 4–6 weeks after application, because most of these compounds are slow-acting. Extraction methods that depend on nematode motility may be used, or a vital stain (22) may be used to recognize living nematodes at this sampling time.

Midseason Nematode Density

The optimum time for the critical midseason sampling for determining the efficacy of nematicides varies with the life cycle of the target nematode species, host-parasite relationships, crop, climate, and type of nematicide. Normally it is best done when the population density in untreated controls is high (6–12 weeks after treatment). If too much time elapses after treatment, nematode numbers in treated plots often exceed those in controls because of the larger root system and lack of competitors or predators or both in the treated plots. As mentioned earlier, sampling may be based on the physiologic time (cumulative heat units above the activity threshold [11]) required to complete a life cycle. When this information is available, the midseason sampling particularly should be scheduled accordingly. For many perennials, e.g. turfgrasses, or when little is known about the population dynamics of the target species, sampling monthly for 5 or 6 months is desirable. Final population densities are often more useful than midseason densities in northern geographic regions with short growing seasons.

Final Nematode Density

The final population is usually determined at the time of harvest for annual crops. This sampling is most useful for determining residual effects of nematicides. Nematode numbers in treated plots often exceed those in untreated plots at this sampling. For *Meloidogyne* spp., root-knot indexes should be obtained at this time (Table 4).

CARE AND CONDITIONING OF SOIL SAMPLES

Soil samples for nematode assays should be regarded as perishable and handled accordingly. Exposure to temperatures of 40°C or above, even for a short time, kills some species. All soil and root samples should be placed in plastic bags to prevent drying, kept out of the sun, and transported in an insulated container.

Ideally, nematodes should be extracted no later than 2 days after soil samples are collected; however, longer storage is often necessary. Samples should be stored at 10–15°C to keep the nematodes physiologically young and active. Lower storage temperatures may be desirable in northern regions. High temperatures allow hatching of eggs and rapid aging of motile forms; low temperatures may cause chilling injury.

Adequate mixing of composite soil samples before an aliquant (usually 100 cm^3) is removed for extracting nematodes is also important when the entire sample (500 cm^3 or more) cannot be processed. Such samples should be screened first and then mixed thoroughly by a suitable procedure, e.g., running them through a soil sample splitter two to four times. Also, soil samples may be mixed effectively by coning and quartering. Mixing the composite sample avoids undue variation caused by nematode aggregation. Root or other plant tissues should be cut and mixed when an aliquant is selected for extraction. Chopping tissues in blenders may be satisfactory in some instances but is not recommended for general use (because of toxic substances released by some plants, such as peach).

Caution: Although mixing soil is necessary, special care (gentle mixing or use of flotation procedures) must be given to samples containing *Xiphinema, Longidorus*, or *Paratrichodorus* spp., which are often injured and lose their motility by mechanical mixing of soil.

CALIBRATION AND STANDARDIZATION

Concentrations of all chemicals used for extracting nematodes should be expressed on a molar basis—micrograms per milliliter or gram (see specific procedures). In procedures using the centrifuge, the relative force times gravity (*g*) should be determined rather than simply the number of revolutions per minute (rpm). When stirrers are used, rpm is sufficient. Subsamples should be based on volume of soil rather than weight, because weight varies greatly with soil moisture. The water content should be determined and reported as a percentage of oven-dry weight of soil.

SAFETY PRECAUTIONS

Certain hazards are associated with some of the methods described. When using the NaOCl procedure for extracting eggs of *Meloidogyne* spp. (5) and other nematodes that produce external egg masses or for dissolving cysts to free eggs of *Heterodera* or *Globodera* spp., workers should use a fume hood to avoid inhaling the vapors. Soil samples should not be collected within 14 days after application of highly toxic nematicides (organic phosphates or carbamates). When soil assays are made within 2–4 weeks after applying chemicals at any residual concentration, appropriate precautions (protective gloves) should be taken in handling and mixing soil.

EXTRACTION METHODS

Knowledge of the biology and population dynamics of various nematode species is essential for selecting the most appropriate extraction method for each sampling time (Tables 1–3). Root samples are best used for endoparasites (such as *Meloidogyne* and *Pratylenchus*) and semiendoparasites (such as *Hoplolaimus* and *Helicotylenchus*) in evaluating chemical soil treatments, because a significant portion of these nematode populations may exist in the roots. Root sampling may not be necessary shortly after a chemical treatment, because the fractions of the populations in the soil reflect the relative efficacy of the test material. Many sedentary (immobile) ectoparasites can be extracted only by flotation procedures. For soil samples collected within 1–3 weeks after chemical soil treatments, methods that yield only motile nematodes are best. Vital stains (Phloxine B and new Blue R) may be used with procedures such as the centrifugal flotation method, which yields dead as well as live specimens (22). Many nonfumigant nematicides act over a period of 6 weeks or more, causing nematode starvation and slow disappearance from the soil (15). The timing of early posttreatment sampling must be adjusted for such materials.

Several factors may affect the efficiency of specific extraction procedures. Certain problems, such as losing nematodes through sieve openings, occur with numerous techniques. Procedures and potential major difficulties are listed in Table 5.

Selection of assay procedures depends on the kinds and numbers of nematodes present, host, nature and condition of the samples (including soil texture), time of collection, and chemical soil treatments used. Nematode populations may consist of relatively motile to quiescent forms—cysts, individual eggs, eggs in masses, or various combinations of these. Ectoparasitic forms occur primarily in the soil, whereas high percentages of endoparasites are in roots, root fragments, or other plant parts during certain periods of the year. Actively moving juveniles or adults (in roots or soil) can be

Table 5. Extraction procedures and potential major difficulties

Procedure	Difficulties
Baermann methods, sieving-Baermann trays	Excessive soil or debris; inappropriate temperature used for incubation; nematode motility; excessive microbial activity; excessive water over soil
Centrifugal flotation	Inappropriate type or use of centrifuge or sieves; high amount of organic matter; nematodes primarily in roots; incorrect sugar concentration
Sugar-flotation sieving	High moisture, especially with clay soils; high clay or organic content of soil; incorrect sugar or Separan concentration; incorrect mesh of sieves; dark color of molasses, which, if used in lieu of sugar, interferes with decanting
Elutriators	Incorrect rate of water flow; wrong sieve size; soil texture (high organic matter or clay)
Mist techniques	Evaporation in dry climates, which may reduce water temperature below optimum (22–25°C at root level) for some nematodes; toxic substances in roots (e.g., peach if chopped in blender)
Fenwick or flotation can	Excessive moisture in cysts; high organic matter
Shaker incubation	Excessive microbial activity; inappropriate incubation temperature (22–25°C optimum for nematode species)
Semiautomatic elutriator	Incorrect water, airflow, sieves, or all three; high organic matter
Elutriator-NaOCl egg extraction	Erratic root distribution and egg hatching before assay; incorrect concentration of NaOCl for dissolving cysts of *Heterodera* or *Globodera* spp.

extracted by Baermann trays or modifications thereof, a combination of flotation or sieving or both and Baermann methods, or flotation (or elutriation) and cotton wool methods. Methods adapted to both motile and nonmotile forms include sieving, centrifugal flotation, sugar-flotation sieving, and Seinhorst's two-Erlenmeyer-flask sedimentation apparatus. The flotation methods are of limited use for endoparasites when most are inside the roots or other plant parts, but they are superior for ectoparasites. For the extraction of migratory nematodes from roots, the Seinhorst mist apparatus is ideal, but incubation on a gyratory shaker (3) is also satisfactory. Methods for extracting free eggs and egg masses are now being developed. Extraction of cysts of *Heterodera* and *Globodera* spp. also requires special techniques, such as the Fenwick or flotation can or "heavy sugar" centrifugation. If population densities are below a detectable level, an appropriate bioassay may be used (e.g., Rutgers tomato for most *Meloidogyne* spp., Lee soybean for *Heterodera glycines*).

Specific outlines for each recommended method of extracting nematodes follow. Several methods that require special equipment, excessive labor inputs, or both are not described. As indicated earlier, many nematodes are lost during sieving in any procedure. Such losses often can be minimized by allowing the nematodes to settle out of suspension and decanting the excess water instead of sieving. A second passage of the suspension may be necessary when population densities are low. Required equipment and chemicals are listed for each of the following methods.

EXTRACTION OF NEMATODES IN SOIL ONLY

Centrifugal Flotation

Centrifugal flotation (16, modified) is an excellent method for routine assays and is the best procedure when *Criconemella* is the target genus.

Equipment: Centrifuge with horizontal (swinging bucket) head with 50-ml or larger tubes and operation to 420 *g*; mechanical stirrers; 35-, 325-, and 400-mesh sieves; 100-, 150-, and 1,000-ml beakers; soil sample splitters (W. S. Tyler Co., Mentor, OH 44060) or coarse sieves (for mixing).

Chemicals: Sucrose solution, 454 g, in sufficient water to make 1 L of solution (specific gravity 1.18 or 38.5% by weight).

Note: Commercial detergents may enhance recovery by this method (33).

Procedure:

1. Mix the soil.
2. Place a 100-cm^3 aliquant in a 1,000-ml beaker and add sufficient water to bring the total volume to 600 ml.
3. Stir for 20 sec and allow the soil to settle for 60 sec (maximum time of 20–30 sec is best for *Criconemella* spp.).
4. Decant onto a 40-mesh sieve over a 325-mesh sieve. Hold sieves at an angle of about 35–40° during all decanting processes to minimize the chance of small nematodes passing directly through the sieve.
5. Using a wash bottle, rinse the 40-mesh sieve while still over the 325-mesh sieve (excessive rinsing washes small nematodes through both sieves). For *Heterodera* and *Globodera* spp., add a 60-mesh sieve between the 40- and 325-mesh sieves to collect cysts.
6. Wash the debris and nematodes from the 325-mesh sieve into a 150-ml beaker.
7. Pour the washings (step 6) into 50-ml centrifuge tubes.
8. Place the tubes in the centrifuge (be sure to balance the tubes).
9. Centrifuge at 420 *g* for 5 min.
10. Decant water from the tubes (nematodes are in the soil pellet in the bottom of the tubes).
11. Refill centrifuge tubes with sucrose solution and mix with stirring rod or vibrator mixer.
12. Centrifuge for 30 sec at 420 *g* (nematodes remain suspended in sugar solution). Do not use the brake on certain centrifuges, because it may cause enough vibration to dislodge the pellet.
13. Decant the sugar solution-nematode suspension onto a 400-mesh sieve (pour slowly when recovering small nematodes).
14. Rinse the residue and nematodes from the 400-mesh sieve into a 150-ml beaker (about 20 ml of water, suitable for making counts).

Flotation-Sieving Method

This method (7) is well suited for extracting large nematodes such as *Xiphinema* spp. from sandy soils. It generally is less efficient than centrifugal flotation but is acceptable if centrifuges are not available.

Equipment: Mechanical stirrers; 35-, 325-, and 400-mesh sieves; 100-, 150-, and 1,000-ml beakers; wash bottles or mist hoses; soil sample splitter or sieves (for mixing).

Chemicals: Sucrose solution (0.7*M* or 22.0% by weight); Separan NP10 (Dow Chemical Company, Midland, MI 48640), final concentration of 12.5 μg/ml.

Note: Molasses can be used in lieu of sucrose (specific gravity 1.1) (23).

Procedure:

1. Mix soil.
2. Place a 100-cm^3 subsample in a 1,000-ml beaker and add sufficient 0.7*M* sucrose and Separan (12.5 μg/ml) solution to bring total volume to 500 ml.
3. Stir with a motorized stirrer for 20 sec.
4. Allow the soil to settle for about 2 min.
5. Decant the liquid onto a 40-mesh sieve over a 325-mesh sieve.
6. Rinse the 40-mesh sieve while it is still over the 325-mesh sieve.
7. Using a wash bottle, rinse the nematodes and debris from the 325-mesh sieve into a 150-ml beaker (about 50 ml of water).
8. Swirl the 150-ml beaker and allow the contents to settle for 5–10 sec).
9. Decant the nematode suspension onto a 400-mesh sieve and rinse the residue into a 150-ml beaker (about 20 ml of water).

Semiautomatic Elutriator

This approach (4) to nematode extraction includes an elutriator similar to Oostenbrink's (20) plus a sample splitter and sieve shaker. It may be used in combination with Baermann trays or centrifugal flotation.

Equipment: Simple elutriator, Oostenbrink type; aqueous sample splitter; water and air supplies; motorized sieve shaker; 10-, 24-, 40-, 60-, 400-, and 500-mesh sieves. (This method may be semiautomated with time clocks, etc. [4].)

Procedure:

1. Add 500 cm^3 of unmixed soil to the elutriator (with air and water flowing at desired rates).

2. Run the elutriator for 3 min, catching roots on the 40-mesh sieve over a sample splitter and "free" nematodes on the 400-mesh sieve on the motorized shaker. A 10- or 24-mesh sieve should be used rather than the 40-mesh sieve for large nematodes such as *Xiphinema* that might be trapped on a 40-mesh sieve.

3. Rinse the sieves.

4. For eggs of *Meloidogyne* spp., process the roots from the 40-mesh sieve by the NaOCl method (5). For *Pratylenchus* spp. and other migratory endoparasites, roots are trapped on the 40-mesh sieve and incubated in the mist chamber. Cysts of *Heterodera* or *Globodera* spp. may be collected on a 60-mesh sieve under 10- and 24-mesh sieves. The eggs from cysts may be extracted by the NaOCl method (5).

Any fraction (1/15, 1/5, etc.) of nematodes in the soil is collected on sieves on the shaker. *Criconemella* and related genera may be cleaned by centrifugation flotation and other species by Baermann methods, sugar-flotation sieving, or centrifugal flotation with 500-mesh sieves.

Cobb's Decanting and Sieving

The simple, modified version described below of Cobb's (31) sifting and gravity method is useful in extracting nematodes for inoculation purposes and routine assays when combined with Baermann methods. Addition of Separan eliminates the need for numerous sievings. Thorne (30) describes the original method, and Townshend (31) gives illustrations.

Equipment: 10-L pails; 10-, 25-, 50-, 100-, 200-, 325-, and 400-mesh sieves; 50-, 150-, 250-, and 500-ml beakers.

Chemicals: Separan with sucrose can be used to reduce the steps involved. A stirrer and 600-ml suspension with 100 cm^3 of soil should be used as in the description of sugar-flotation sieving.

Procedure (short, modified method):

1. Place 500 cm^3 of soil in a large pan or pail and cover well with water containing Separan (12.5 μg/ml). Thoroughly break all lumps, mix by hand, and allow to settle about 2 min.

2. Decant onto a 40-mesh sieve over a 325-mesh sieve.

3. Resuspend the original soil in water and repeat step 2 if maximum recovery is desired.

4. Combine the washings from steps 2 and 3, stir, and allow them to settle for 10 sec.

5. Decant through a 400-mesh sieve and rinse the residue (nematodes) into a clean beaker for counting (may place on Baermann tray for cleaner samples).

EXTRACTION OF MOTILE SPECIES

Baermann Trays

The Baermann trays procedure (32) is useful for extracting nematodes from small soil samples, root fragments, and debris coming from elutriators and for extracting juveniles of *Heterodera* and *Globodera* spp. (in the spring).

Equipment: Plastic screen (sieve-type) 17.5 cm in diameter (20-mesh) supported with legs 4 mm high or similar metal unit, wet-strength facial tissues, epoxy resin-coated aluminum pie pans 20 cm in diameter or small plastic salad bowls.

Procedure:

1. Mix the soil.

2. Place a 100-cm^3 aliquant of soil uniformly over the tissue, which is superimposed on a plastic or stainless-steel screen (do not use copper screens).

3. Place the screen with soil in a 20-cm pie pan and add water just to cover the soil.

4. Incubate the samples at 21–24°C (cover or stack to reduce evaporation). Add water as needed.

5. Collect the nematodes from the pans after 3 days. If dirty, clean by pouring through a 500-mesh sieve. For maximum recoveries, nematodes should be collected over a period of 1–14 days.

Baermann Trays Plus Elutriation or Sieving

The combination of Baermann trays with elutriation (or sieving), a modification of the Christie-Perry procedure (8), is still invaluable for extracting nematodes from soil and root fragments.

Equipment: 10-, 24-, and 400-mesh sieves; elutriator; screen supports; pie pans (stainless-steel or resin-treated) 15–20 cm in diameter; wet-strength facial tissues; milk filter or muslin filter.

Procedure:

1. Add an unmixed 500-cm^3 soil sample to the elutriator.

2. Collect the root and soil fractions from 10- (or 24-) and 400-mesh sieves, respectively (see section on the semiautomatic elutriator), or use the modified Cobb's decanting-sieving method.

3. Place these fractions on facial tissues supported on screen.

4. Carefully add enough water to just cover the residue. Add more water as needed with time.

5. Collect the nematodes after incubation at 21–24°C for 3 days (for maximum recoveries, collect them over a period of 10–14 days).

Others

The semiautomatic elutriator approach is described herein. The Seinhorst elutriator (25) or Oostenbrink elutriator (20) procedures may also be used to extract motile nematodes.

EXTRACTION OF CYSTS

The use of a reliable method for determining numbers of eggs in cysts, and egg masses if present, is essential for evaluating the efficacy of nematicides on *Heterodera* and *Globodera* spp. A Ten-Broeck homogenizer (27) is satisfactory for this purpose. The NaOCl (sodium hypochlorite) method described in the section Extraction of Nematode Eggs also gives good results in dissolving cysts and freeing the eggs for counts (although the researcher may need to use two to three times the concentration indicated for egg masses of *Meloidogyne* spp.).

Flotation Can

The flotation can method (1) is an alternative to the Fenwick can (see next subsection). The use of 9:1 ethanol-glycerin in lieu of acetone-carbon tetrachloride offers many advantages. In addition to being very safe, the ethanol-glycerin solution can be filtered and reused.

Equipment: Flotation can and device for separating cysts from organic debris (Fig. 4), sieves (20- and 100-mesh), soil crusher (for "brick-like" soils), stirrers, Whatman no. 4 filter paper, general glassware.

Chemicals: Ethanol, glycerin, materials as described in the section Extraction of Nematode Eggs for extraction of eggs by NaOCl.

Procedure for extracting cysts from soil:

1. Place 500 cm^3 of well-mixed soil in flotation can. Add water to within 10 cm of spout and mix by hand.
2. Add water at the bottom of the flotation can so that the flow rate is sufficient to wash over floating and suspended material but not silt or sand particles. Continue for 2 min, catching the overflow on 20- and 100-mesh Tyler sieves (clay particles that are washed over will pass through both sieves).
3. Turn off water and stir sediment thoroughly.
4. Add water at the correct rate for a further 2 min, catching the overflow on the 20- and 100-mesh sieves.
5. Turn off water and pour the contents of the can through the sieves until just before the sediment starts to flow out. Do not pour the sediment onto the sieves.
6. Wash the debris on the 20-mesh sieve with a jet of water so that cysts are flushed onto the 100-mesh sieve.
7. Wash the contents of the 100-mesh sieve (the "flotsam") onto tissue paper supported by a wire gauze sieve 10 cm in diameter.
8. Leave to dry.

Procedure for separating cysts from organic debris:

1. Fold Whatman no. 4 filter paper (24 cm in diameter) and place in funnel in flask with side-arm attachment (Fig. 4). Close funnel tap and fill with ethanol-glycerin mixture (9:1, v/v) to 2 cm from surface.
2. If the "flotsam" has aggregated, gently crumble it to free the cysts from adhering debris without crushing them.
3. Pour the finely divided "flotsam" onto the ethanol-glycerin mixture in the funnel. Cysts move outward to the filter surface. The major portion of the debris sinks to the bottom of the funnel and should be stirred with a spatula to release trapped cysts.
4. When all of the floating portion has reached the filter paper, apply a partial vacuum to the flask through the side arm and open the funnel tap cautiously. The ethanol-glycerin mixture passes through into the flask and can be reused. The floating portion containing the cysts remains as a thin line around the filter paper.
5. Unfold the filter paper and place on a shallow watch glass or glass plate. Cysts are almost all on the outermost edge of the line and can be counted under a dissecting microscope.
6. After cysts have been counted, wash the line through the funnel into a beaker.

Procedure for releasing and counting eggs from cysts (basic method as outlined in the section Extraction of Nematode Eggs for eggs of *Meloidogyne* spp.):

1. Blend cysts in water-sodium hypochlorite (NaOCl 5.25%) mixture (1:1, v/v), using a tissue homogenizer for 30 sec (time will vary with type of homogenizer and condition of cysts).
2. Transfer blended material onto 500-mesh sieve and rinse with gentle flow to remove NaOCl.
3. Transfer material from the 500-mesh sieve into a

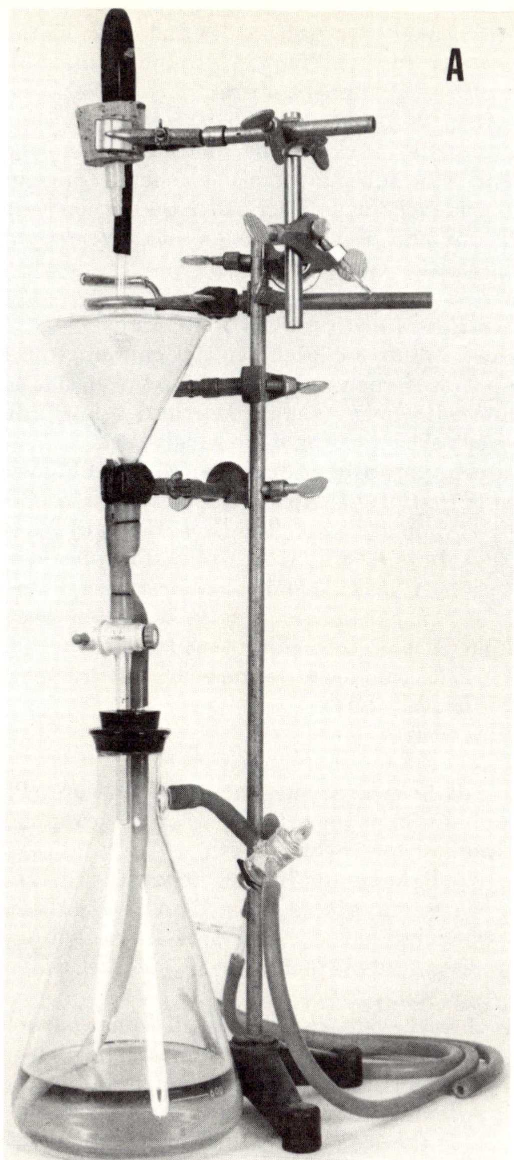

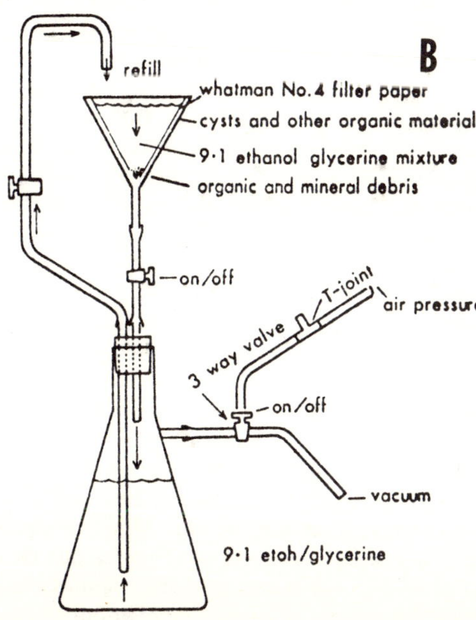

Fig. 4. A, Device for separating cysts of *Heterodera* or *Globodera* from organic debris. **B,** Diagram of the apparatus. (Courtesy Herb Quick)

150-ml beaker with fluted sides and make up to 100 ml with water (to 50 ml if low egg count is anticipated).

4. Stir with magnetic stirrer.

5. Remove 1-ml sample (5–10 ml for low egg count) with pipette into counting dish and count eggs and juveniles (do not count empty egg shells). (Staining eggs with acid fuchsin, as described in the section Extraction of Nematode Eggs for eggs of *Meloidogyne* spp. facilitates the egg counts.)

Flotation-Modified Fenwick Can

The flotation-modified Fenwick can procedure (27) is useful for extracting cysts from dry soil, but centrifugation with heavy sugar (described below) and other methods are becoming more widely used.

Equipment and chemicals: Modified Fenwick can, sieve (60–80 mesh) with bowl, camel's-hair brush (no. 00, 0, or 1), homogenizer (Ten-Broeck), plain glass slide, 100-ml bottle, pipette, aquarium pump, 400-ml Erlenmeyer flask, 250-ml volumetric flask, funnel, petri dish, filter papers 18.5 cm in diameter, acetone or acetone-carbon tetrachloride 3:1 (may use ethanol-glycerin [9:1] in lieu of acetone-carbon tetrachloride; see previous procedure).

Procedure:

1. Mix the soil thoroughly.

2. Fill the modified Fenwick can with water. Place the sample of 100 cm^3 of well-mixed soil in the top sieve (20.5 cm in diameter, 18- or 24-mesh).

3. Wash the sample into the apparatus via the funnel. The coarse material is retained on the top sieve, heavy soil particles such as sand sink to the bottom of the apparatus, and the floating cysts are carried off over the overflow collar.

4. Collect cysts, root debris, and other particles on a sieve 20 cm in diameter (60–80 mesh). Particles 175 μm and smaller pass with water through the sieve.

5. After washing, dry the debris at room temperature. Transfer the somewhat dried debris retained on the sieve to a 250-ml flask.

6. Pour technical acetone (or a mixture of 3 parts acetone and 1 part carbon tetrachloride) into a volumetric flask up to level 1 (neck of flask). Shake the flask and fill it completely. Be sure to use the exhaust hood.

7. After 1 min, decant the floating cysts and debris through a filter paper (18.5 cm in diameter) in a glass funnel into a second or Erlenmeyer flask while rotating the original flask. The acetone passes through the filter.

8. Place the filter in a petri dish and view it through a dissecting microscope (magnification ×50) with overhead light. Pick up the cysts with a camel's-hair brush (no. 00, 0, or 1) and transfer them to a small watch glass with moist filter paper. Identify the cysts under the dissecting microscope using an overhead light. Use a camel's-hair brush to transfer the cysts of the desired species into a small drop of water in the glass tube of the homogenizer. Place the piston in the tube and very carefully rotate it by hand. Pour the eggs and juveniles that were released from the cysts into a bottle. Fill the bottle with water up to 100 ml. Mix the suspension carefully using compressed air. Pipette out two 10-ml aliquants and place them in 10-ml Perspex dishes for counting. As indicated earlier, eggs should be freed by NaOCl or Ten-Broeck homogenizer and numbers determined as outlined in the subsection on the flotation can method.

Centrifugation with Heavy Sugar

This procedure (9, modified) is useful for isolating cysts and juveniles of *Heterodera* and *Globodera* spp., but problems may be encountered with some fine clay soils.

Equipment: Same as for centrifugal flotation for routine use, except that 24- and 100-mesh sieves are needed.

Chemicals: 1.8*M* (50% by weight) sucrose solution (615 g of sucrose) dissolved in enough warm water to make 1 L of solution (specific gravity 1.23).

Procedure:

1. Wash 100 cm^3 of soil through a 24-mesh sieve and collect it in a beaker (use about 1 L of water).

2. Mix the suspension thoroughly and allow it to settle for 10–15 sec.

3. Pour the supernatant through a 100-mesh screen (add a 325-mesh sieve for juveniles).

4. Wash any residue from the screen into a centrifuge tube or tubes with 1.8*M* sucrose solution.

5. Centrifuge at 420 *g* for 2.5 min.

6. Collect the supernatant on a 100-mesh screen (add a 325-mesh sieve for juveniles).

7. Rinse thoroughly.

8. Wash the sample into a beaker, using about 20 ml of water.

9. Crush the cysts with a Ten-Broeck homogenizer or dissolve them with NaOCl, as described for the flotation can method, and count the eggs and juveniles.

Others

The semiautomatic elutriator method (4) is useful for recovering cysts of *Globodera* and *Heterodera* (see previous description). Seinhorst's extraction procedure (25) may also be used for *Heterodera* and *Globodera* cysts from moist soil.

SEPARATION OF NEMATODES FROM PLANT TISSUES

Modified Seinhorst Mist Apparatus

This mist chamber method (24) is the most widely used method for obtaining nematodes from plant tissues.

Equipment: Time clock, water mixer-warmer, water regulator and filter, solenoid switch, cover (fiberglass, Plexiglas, stainless steel, or other suitable material) with doors to funnel racks or supports, superfine nozzles, plastic petri dishes, glass funnels 10 cm in diameter with rubber tubing, 1-L plastic cups with holes in the bottom, clamps, 500-mesh sieve (specifications available from K. R. Barker).

Procedure:

1. Place a representative sample of roots or other plant tissues in plastic cups superimposed over an open Baermann funnel. This funnel is supported over a second Baermann funnel by a plastic petri dish with a 2.5-cm hole in the center. Use wet-strength facial tissues in the cups to reduce debris.

2. Set the time clock to regulate desired mist (on 1 min, off 2 min). Adjust the water mixer to the flow rate that gives a temperature of 24°C.

3. Collect the nematodes from the funnels every 3–5 days through a 14-day period (nematodes such as *Pratylenchus* spp. will continue to emerge for weeks).

4. Concentrate and clean the nematode suspensions if necessary with a 500-mesh sieve.

Shaker

Several researchers have found that this shaker procedure (3) yields numbers of *Pratylenchus* spp. and other genera from roots similar to those obtained from mist chambers.

Equipment: Gyratory shaker, 125-ml flasks.

Chemicals: Ethoxyethyl mercuric chloride (Aretan), streptomycin sulfate or other suitable antibiotics.

Procedure:

1. Wash roots and cut into segments 1–2 cm long.
2. Place a representative sample of root tissue (0.5–5.0 g) in a 125-ml flask.
3. Cover the tissue with a mixture of 10 μg/ml of ethoxyethyl mercuric chloride and 50 μg/ml of streptomycin sulfate.
4. Incubate the mixture at 100 rpm for 48 hr.
5. Collect nematodes on a 325- or 400-mesh sieve; rinse them into a 150-ml beaker and count them.

Blender-Baermann Tray

Although many problems are encountered with this procedure (29), it is useful for limited situations.

Equipment: Blender, beakers, 325-mesh sieve, Baermann funnels or pans.

Chemicals: Ethoxyethyl mercuric chloride in a 10-μg/ml solution (other materials such as antibiotics or certain fungicides [captan] may be used instead).

Procedure:

1. Rinse the plant tissues until they are free of soil.
2. Weigh the tissues to be processed and place them in the blender. Use not more than 50 g of tissue (fresh-weight basis) with 200 ml of water in 1.9-L (2-qt) blender.
3. Homogenize the mixture for 15 sec.
4. Decant the suspension from the blender, including the rinse water, over a 325-mesh sieve.
5. Using a wash bottle containing antibiotic solution (see shaker method), wash the debris from the sieve into a beaker.
6. Pour the liquid that passed through the sieve over the sieve again.
7. Combine the material collected on the sieve in the second passage with that from the first passage.
8. Gently decant the suspension of material collected on the sieve over the filter of the Baermann apparatus.
9. Fill the Baermann apparatus with enough antibiotic solution to barely submerge the debris on the filter.
10. Replenish the incubation solution with water as needed.
11. Collect the nematodes after incubation at 21–24° C for 2–3 days.

In addition to the combined procedure just described, blending followed by wet sieving is useful for certain nematodes such as *Radopholus similis* on banana.

EXTRACTION OF NEMATODE EGGS

Elutriation-Dissolution

Variation of egg numbers in the field sometimes causes problems with this assay method of elutriation, dissolution of gelatinous matrices of egg masses, and staining (5). The method is useful, however, for midseason to late-season assays of *Meloidogyne* spp. and other nematodes that form egg masses and/or cysts.

Equipment: Sample splitter or semiautomatic elutriator, 150- and 600-ml beakers, stirrers, 15-cm household sieve, 40- and 500-mesh sieves, exhaust hood, 5-ml dipper, compressed air (for cleaning sample splitter and use with elutriator).

Chemicals: Sodium hypochlorite (NaOCl), antifoam spray, acid fuchsin, lactic acid.

Procedure:

1. Mix the soil.
2. With water flow adjusted to 350 ml/sec, or 60–80 ml/sec if air-water mixture is used, place a 500-cm^3 aliquant of soil in the elutriator. Turn the water on for 2–3 min, trapping root fragments on a 40-mesh sieve.
3. With a spray nozzle, wash the residue off the sieve into a 600-ml beaker and add water to 200 ml.
4. Add 20 ml of 5.25% NaOCl and spray with an antifoam agent.
5. Stir the mixture under an exhaust hood for 10 min.
6. Using a household sieve to retain debris, take a 5-ml sample with a dipper and rinse into a 150-ml beaker.
7. Pour this subsample onto a 500-mesh sieve and wash the eggs from the sieve into a clean 150-ml beaker (should have 20–25 ml of suspension).
8. Add two drops of 0.35% acid fuchsin in 25% lactic acid and boil for 1 min under the exhaust hood (a microwave oven is useful).
9. Allow the mixture to cool before counting.

Centrifugation

Nematode eggs may be extracted by centrifugation (12,22).

BIOASSAYS OF NEMATODE POPULATIONS

With low natural populations of nematodes such as *Meloidogyne, Heterodera,* and *Ditylenchus* spp., bioassays (19,22,26) are the most reliable procedures. The following is a typical bioassay for *Meloidogyne.*

Equipment: Fumigated sandy loam soil, 10-cm clay pots, 15-cm plastic pots, pot labels, 3-week-old Rutgers or other susceptible tomato seedlings, nutrient solution, greenhouse space. Other recommended tomato cultivars are Person A1, Heinz 1350, Marglobe, Bonny Best, and Manapal.

Procedure:

1. Fill the bottom 2 cm of 10-cm clay pots with sterile soil.
2. Add 250 cm^3 of test soil to the clay pots. The field soil used for a bioassay should represent an entire plot or a portion of a given field. Large fields should be marked off into units of about 1 ha with two composite (20-core) samples used for bioassays from each unit (19). This sampling-bioassay scheme gives a ratio of standard error to mean of about 25%. More bioassays are needed if greater precision is desired (19).
3. Transplant a tomato seedling to each pot, and fill the remainder of each pot with sterile soil (place each 10-cm pot in a 15-cm plastic pot to minimize contamination).
4. Grow plants for 5–6 weeks at 24–28° C, providing nutrients and water as needed. Do not overwater.
5. Harvest the plants by washing the roots out carefully. Rate the nematode development by using the gall index (Table 4) or by determining the numbers of

eggs (see NaOCl method).

The third root-knot index (1–6) in Table 4 (nomograph) is recommended for evaluating the effects of nematicides on *Meloidogyne* spp. treated in the field or greenhouse. Figure 5 illustrates examples of each class. The other schemes in the nomograph (Table 4) are also acceptable.

Rating galling caused by different *Meloidogyne* spp. may be confusing, because the number and size of galls incited by a given number of juveniles vary with species and host plant. Concentrating primarily on the proportion of roots galled rather than the size of galls can minimize this problem.

When the cause of galls is uncertain, roots should be stained with 0.05% acid fuchsin and cleared in glycerin or lactophenol (6). The contents of galls (numbers of juveniles, eggs, and adults) can then be determined.

Gall indexes for *Meloidogyne* spp. should be limited to galling. To rate necrosis associated with root knot, a separate rating system such as the one Powell et al (21) developed is suitable. In their classification system, 0 = no necrosis, 1 = less than 10% of the root system necrotic, 2 = 11–25% necrotic, 3 = 25–50% necrotic, 4 = 51–75% necrotic, and 5 = 76–100% necrotic. Based on all root systems per treatment, the disease or necrosis index is computed as follows:

$$DI = \frac{(n_1 \times 1) + (n_2 \times 2) + \ldots + (n_5 \times 5)\ \times 100}{5N},$$

where DI = disease index; n_i = number of plants in class $i, i = 1, 2, \ldots, 5$; and N = number of plants in treatment ($N = n_1 + n_2 + n_3 + n_4 + n_5$). A similar lesion or necrosis index is also useful for rating banana roots for infection by *Radopholus similis*.

Numbers of cysts may be counted on bioassay plants for *Heterodera* spp. (cereal cyst nematode, *H. avenae*, on wheat [26] or soybean cyst nematode, *H. glycines*, on soybean).

REPORTING RESULTS

The following results should be determined for each replicate and averaged for each treatment:

1. Mean numbers of nematodes per 100 cm^3 or greater volume of soil (use of a mechanical stage on the microscope to control counting dish increases efficiency in making nematode counts).

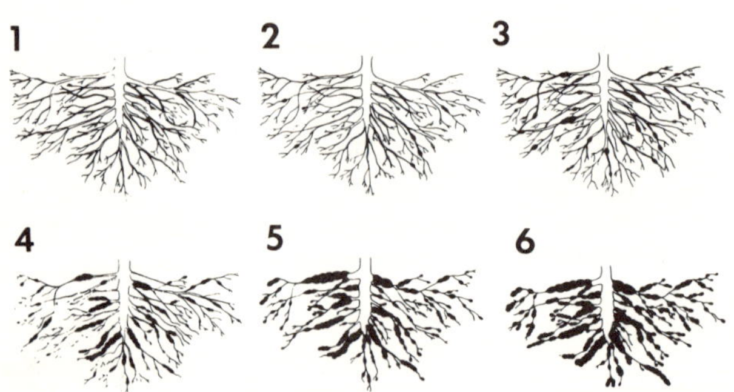

Fig. 5. Scheme for rating field and greenhouse infestation and bioassay evaluation of *Meloidogyne* spp. See Table 4 (galling index system 1–6) for descriptions of each root-knot nematode gall index class.

2. Numbers of nematodes per gram of root or other plant part (fresh or dry weight) when endoparasites are extracted from same plots. Total nematodes per plant should be included for greenhouse experiments.
3. For *Heterodera* spp., numbers of juveniles and cysts as well as numbers of eggs per cyst or per gram of soil.
4. For *Meloidogyne* spp., numbers of eggs and juveniles per unit volume of soil or weight of roots (fresh or dry weight); also record root-knot indexes based on the index given under bioassay procedure. Table 3 summarizes the data required for various types of nematodes and host plants.
5. Plant response.

Precision in determining nematode population responses to nematicides depends on how the samples are collected, handled, stored, extracted, and counted. With proper execution of each of these activities, variation usually ranges from 10 to 30%. Oostenbrink found that the usual coefficient of variation for this error is about 25%. This error varies, however, with nematode species and population densities. In estimating egg numbers for *Meloidogyne* spp., this coefficient of variation is considerably greater. Thus, proper statistical analyses of all data must be used to estimate the variance for a given experiment, including coefficient of variation and standard error. Because some zero readings and considerable variation are normally encountered, the $\log_{10}(P + 1)$ or another transformation should be used for nematode population data. Standardization of reporting nematode population data is needed to facilitate communication regarding nematicide efficacy, especially because most assay procedures vary in efficiency (22). The extraction technique and rates of recovery for each target species for the soils involved should be noted in all nematicide evaluations.

GLOSSARY OF DESCRIPTIVE TERMS

Bioassay. Use of appropriate hosts to determine relative population levels (has been used primarily for endoparasites such as root-knot and cyst nematodes, but can be used for any nematode under suitable propagation conditions; usually done in greenhouse).

Detection level. Minimum population density of a nematode species that can be identified routinely from a soil or plant tissue sample (varies with each sampling and extraction technique).

Economic threshold density. Minimum number of nematodes per unit volume of soil (or weight of plant tissue) required to cause an economically significant loss in crop production.

Fumigant nematicide. Gas, volatile liquid, or solid that diffuses through the soil pore spaces as a vapor and is water-soluble.

Grid or systematic sampling (stratified). Collection of specified number of borings over entire plot area (one for each square meter or other area measure) for pretreatment sampling. If crops are established, collection of borings from root zone (in the rows for row crops).

Horizontal head. For centrifuge, head is so structured that individual tubes (swinging bucket) are in horizontal position when centrifuge is operating.

Molar solution. 1 g mol wt in sufficient water to make 1 L of solution.

Multipurpose chemicals. Of three basic types:

Fumigant. Volatile chemical that disperses throughout the soil and controls nematodes, fungi, bacteria, insects, weeds, or a combination of these.

Nonfumigant nematicide (contact nematicide). Nonvolatile chemical that does not disperse readily except through soil water but controls nematodes (and in some cases, soil insects) that come in contact with it.

Systemic nematicide. Chemical that is absorbed by a plant and translocated throughout the plant and controls nematodes.

Nonfumigant nematicide. Nematicide that has little or no volatility and that must be mixed thoroughly in soil as a granular formulation or applied in water (water-soluble formulation).

Population density. Number of nematodes per unit volume of soil or weight of plant tissue.

ppm. Parts per million; 1 ppm = 1 mg/L (1 lb/120,000 gal) of water. With water, 1 ppm is equivalent to 1 μg/ml (w/v) or μg/g (w/w). We recommend using μg/ml or μg/g instead of ppm. The solvent should be specified when solutions other than water are used.

Random sampling. Points for collecting borings determined by random chance (not usually suitable for nematodes such as *Meloidogyne* spp., which occur in aggregates).

Relative *g* force (relative centrifugal force). For centrifuge, force (minimum, average, maximum) times that of gravity: relative g force = $0.00001118 \times r \times \text{rpm}^2$, where r = radius in centimeters.

Sampling tube. Metal (all or part) tool for collecting soil cores of constant diameter and volume. Major types are the Oakfield soil sampling tube (2.0-cm ID, open side, 38 cm long; good for routine sampling to a depth of 20 cm); the Viemeyer soil sampling tube (2.5-cm ID, 120 cm long, with reinforced head; tube can be driven into soil to a depth of 90 cm); the can or bucket auger (steel auger with bucket 5–7.5 cm in diameter by 15 cm high, fitted with cutting edges at the base; useful for obtaining large soil samples and roots to depths of 90–100 cm); and the cone-shaped sampling tube (useful for routine sampling of loose, sandy soils) (22).

Sieve mesh number. Number of wires (filaments running in each direction) per inch. We recommend that the size of openings be given in micrometers and that mesh number be given in parentheses.

Soil sterilant. A chemical (e.g., methyl bromide) that kills all soil organisms.

Tolerance limit. Minimum number of nematodes that inhibit plant growth (number per unit volume of soil or weight of plant tissue). May not have economic significance.

LITERATURE CITED

1. Andersson, S. 1970. A method for the separation of *Heterodera* cysts from organic debris. Nematologica 16:222-226.
2. Barker, K. R., and Campbell, C. L. 1981. Sampling nematode populations. Pages 451-474 in: Plant Parasitic Nematodes, Vol. III. B. M. Zuckerman and R. A. Rohde, eds. Academic Press, New York.
3. Bird, G. W. 1971. Influence of incubation solution on the rate of recovery of *Pratylenchus brachyurus* from cotton roots. J. Nematol. 3:378-385.
4. Byrd, D. W., Jr., Barker, K. R., Ferris, H., Nusbaum, C. J., Griffin, W. E., Small, R. H., and Stone, C. A. 1976. Two semi-automatic elutriators for extracting nematodes and certain fungi from soil. J. Nematol. 8:206-212.
5. Byrd, D. W. Jr., Ferris, H., and Nusbaum, C. J. 1972. A method for estimating numbers of eggs of *Meloidogyne* spp. in soil. J. Nematol. 4:266-269.
6. Byrd, D. W., Jr., Kirkpatrick, T., and Barker, K. R. 1983. An improved technique for clearing and staining plant tissues for detection of nematodes. J. Nematol. 15:142-143.
7. Byrd, D. W., Jr., Nusbaum, C. J., and Barker, K. R. 1966. A rapid flotation-sieving technique for extracting nematodes from soil. Plant Dis. Rep. 50:954-957.
8. Christie, J. R., and Perry, V. G. 1951. Removing nematodes from soil. Proc. Helminthol. Soc. Wash. 18:106-108.
9. Dunn, R. A. 1969. Extraction of cysts of *Heterodera* species from soils by centrifugation in high density solutions. (Abstr.) J. Nematol. 1:7.
10. Elliott, J. M. 1977. Some Methods for Statistical Analysis of Samples of Benthic Invertebrates. Sci. Pub. No. 25. Freshwater Biological Association, Ambleside, Cumbria. 144 pp.
11. Ferris, H. 1978. Modification of a computer simulation model for a plant-nematode system. J. Nematol. 10:198-201.
12. Flegg, J. J. M., and McNamara, D. G. 1968. A direct sugar-centrifugation method for the recovery of eggs of *Xiphinema, Longidorus* and *Trichodorus* from soil. Nematologica 14:156-157.
13. Goodell, P., and Ferris, H. 1980. Plant parasitic nematode distributions in an alfalfa field. J. Nematol. 12:136-141.
14. Goodey, J. B. 1963. Laboratory Methods for Work with Plant and Soil Nematodes. 4th ed. Min. Agric. Fish. Food Tech. Bull. 2. Her Majesty's Stationery Office, London.
15. Hough, A., and Thomason, I. J. 1975. Effects of aldicarb on the behavior of *Heterodera schachtii* and *Meloidogyne javanica.* J. Nematol. 7:221-229.
16. Jenkins, W. R. 1964. A rapid centrifugal-flotation technique for separating nematodes from soil. Plant Dis. Rep. 48:692.
17. Jones, F. G. W. 1975. Accumulated temperature and rainfall as measures of nematode development and activity. Nematologica 21:62-70.
18. Karandinos, M. C. 1976. Optimum sample size and comments on some published formulas. Bull. Entomol. Soc. Am. 22:417-421.
19. McSorley, R., and Parrado, J. L. 1983. A bioassay sampling plan for *Meloidogyne incognita.* Plant Dis. 67:182-184.
20. Oostenbrink, M. 1960. Estimating nematode populations by some selected methods. Pages 85-102 in: Nematology—Fundamentals and Recent Advances with Emphasis on Plant Parasitic and Soil Forms. J. N. Sasser and W. R. Jenkins, eds. University of North Carolina Press, Chapel Hill.
21. Powell, N. T., Meléndez, P. L., and Batten, C. K. 1971. Disease complexes in tobacco involving *Meloidogyne incognita* and certain soil-borne fungi. Phytopathology 61:1332-1337.
22. Rickard, D. A., and Barker, K. R. 1982. Nematode assays and advisory services. Pages 8-20 in: Nematology in the Southeastern Region of the United States. R. D. Riggs, ed. Southern Coop. Series Bull. 276. Agric. Exp. Stn., Univ. Arkansas, Fayetteville.
23. Rodriguez-Kabana, R., and King, P. S. 1972. The use of sugarcane molasses as an economical substitute for sugar in the extraction of nematodes from soil by the flotation-sieving technique. Plant Dis. Rep. 56:1093-1096.
24. Seinhorst, J. W. 1950. De betekenis van de toestand van de grond voor het optreden van aantasting door het stengelaaltje (*Ditylenchus dipsaci* [Kuhn] Filipjev). Tijdschr. Plantenziekten 56:289-348.
25. Seinhorst, J. W. 1964. Methods for the extraction of *Heterodera* cysts from not previously dried soil samples. Nematologica 10:87-94.
26. Simon, A. 1980. A plant assay of soil to assess potential damage to wheat by *Heterodera avenae.* Plant Dis. 64:917-919.
27. Southey, J. F., ed. 1970. Laboratory Methods for Work with

Plant and Soil Nematodes. 5th ed. Min. Agric. Fish. Food Tech. Bull. 2. Her Majesty's Stationery Office, London.
28. Southwood, T. R. E. 1978. Ecological Methods with Particular References to the Study of Insect Populations. Chapman and Hall, London; John Wiley & Sons, New York.
29. Taylor, A. L., and Loegering, W. Q. 1953. Nematodes associated with root lesions in abaca. Turrialba 3:8-13.
30. Thorne, G. 1961. Principles of Nematology. McGraw-Hill Inc., New York.
31. Townshend, J. L. 1962. An examination of the efficiency of the Cobb decanting and sieving method. Nematologica 8:293-300.
32. Townshend, J. L. 1963. A modification and evaluation of the apparatus for the Oostenbrink direct cottonwool filter extraction method. Nematologica 9:106-110.
33. Wehunt, E. J. 1973. Sodium-containing detergents enhance the extraction of nematodes. J. Nematol. 5:79-80.
34. Zeck, W. M. 1971. A rating scheme for field evaluation of root-knot nematode infestations. Pflanzenschutz-Nachr. (Am. Ed.) 24:141-144.

Plant Responses in the Evaluation of Nematode Control Agents

C. C. ORR, Agricultural Research Service, U.S. Department of Agriculture, Texas Agricultural Experiment Station, Lubbock 79401, and C. M. HEALD, Agricultural Research Service, U.S. Department of Agriculture, Weslaco, TX 78596

Nematicide registration must be supported by efficacy data that demonstrate control of the target nematode species and absence of adverse effects on the plant being treated. When symptoms of nematode feeding are definitive and diagnostic, reduction of such symptoms in treated plants demonstrates control of the target nematode.

The reason for using a nematode control agent is to prevent, arrest, or mitigate adverse plant reactions attributable to plant-parasitic nematodes. Use of a nematicide usually results in a beneficial plant response. Such responses may be due to control of the target nematode, control of one or more nontarget pests, or a stimulative effect on the plant. Chemicals can induce adverse as well as beneficial plant responses. Plant response to a control agent may range from accelerated growth and vigor because of nematode control to no response to chlorosis, stunting, or death.

This chapter describes the major types of plant responses found in efficacy and phytotoxicity tests associated with use and development of nematode control agents and some of the factors that affect these responses.

GROSS REACTIONS

Plant Symptoms

Plant parasitic nematodes possess a piercing structure (stylet) in the anterior portion of the body. The stylet is used to puncture plant cells for feeding and to aid in tunneling through and between cells as the nematode moves within plant tissue. Secretory glands located in the nematode's neck region produce enzymes and other substances that are injected through the hollow stylet into plant tissue. These substances facilitate penetration of plant tissue and transform cell contents into an ingestible form. In addition, these secretions may cause characteristic changes in the tissues and cells of plants.

Underground symptoms. Root galls are an obvious symptom of nematode infestation. Galling of root tissue may be observed as discrete swellings, or when susceptible plants are heavily infested, the entire root system may appear as a mass of abnormal fleshy tissue. Stimulation by the nematode causes susceptible plants to develop abnormal multinucleate cells (giant cells) upon which the nematode feeds, surrounded by abnormal xylem. Giant cells are essential for the survival of many sedentary parasites. Nematode stimulation may also cause the formation of numerous lateral roots near the infested area. This type of plant injury is typical of nematodes in the genera *Meloidogyne* and *Heterodera.*

Feeding of ectoparasitic and endoectoparasitic nematode genera (e.g., *Helicotylenchus, Criconemoides, Pratylenchus,* and *Radopholus*) may first appear as tiny necrotic root lesions. The lesions may enlarge from a few cells to cause extensive injury or death of a root. Larger lesions are often complicated by the action of secondary pathogenic organisms.

Trichodorus and *Xiphinema* are examples of nematodes that attack root tips. Apical growth is retarded or stopped, causing the root system to appear dwarfed and thickened.

Dry rot of bulbs, corms, or tubers of certain crops resulting from tissue infested with species of *Ditylenchus* is yet another symptom expressed by plants in response to nematode attack.

Aboveground symptoms. Severe nematode infestations are recognized by greatly reduced growth of aboveground plant parts. The standard condition is characteristically accompanied by thin stands of field crops resulting from plants killed by nematodes, other pathogenic organisms in a disease complex, and environmental extremes that are more lethal to weakened plants. Infested plants may also be chlorotic.

Certain nematode species feed on aerial plant parts, causing thickening, twisting, or galling of stems, leaves, and fruiting structures. Symptoms of *D. dipsaci* infestation are thickening of the base of shoots and accelerated tillering, distortion, and twisting of stems and leaves of many plant species. Similar symptoms but without basal thickening and accelerated tillering are observed in wheat plants infested by *Anguina tritici.* Wheat kernels are "chaffy," because the seed contents are replaced with infective larvae of the nematode. *Aphelenchoides* spp. cause distortion, wrinkling, and dwarfing of leaves and stems of chrysanthemum and strawberry. Leaves of silverleaf nightshade infested with *Orrina phyllobia* develop massive thickening, which increases their biomass many times. Invasion of emerging shoots severely deforms and kills plants (9).

Estimates of plant response include stand count, yield, weight, height, and visual ratings of growth and development (2,4–7,10–12). Root reactions are measured by infection sites, lesions, galls, nematodes or eggs per unit of plant tissue, root pruning, root proliferation, or

Revision of a 1978 paper prepared by the Joint SON/ASTM E35.16 Task Force on Plant Responses in Evaluation of Nematode Control Agents, C. C. Orr, chairman.

other malformations (1,3,13). The survival of sedentary nematodes depends on the long life of surrounding host cells. In spite of this accommodation of the host at the infection site, the overall growth of the plant may be severely reduced, and crop failure is a common result.

Disease Complexes and Synergism

Nematodes cause serious injury to plants quite independently of other microorganisms that may be present in the plant's environment. Also, species of fungi, bacteria, and viruses are important pathogens that injure plants without influence from other biotic agents. Plants are constantly exposed to complexes of these pathogenic organisms that are natural components of the soil biosphere. Nematodes contribute to disease complexes mostly by modifying the physiology of the host plant. However, the metabolic activities of one pathogen in a disease complex must influence those of the other components and may result in severe damage to the host plant. A well-documented example of a nematode-fungus interaction is the increased severity of Fusarium wilt in cotton plants infested with root-knot nematodes.

Soil Properties

Inherent properties of the soil influence plant responses to nematode control agents. The type, texture, moisture content, temperature, and organic matter content of soil control the movement and fate of nematicides, thereby influencing the degree of nematicidal activity and the vigor of the host plant. Insufficient fertilizer and trace elements result in stunting or chlorosis similar to symptoms of nematode injury. Other edaphic factors of mycorrhizae, microflora, microfauna, and insects also influence plant response.

Growth and Development

Plant symptom expression is influenced by the nematode species as well as the growth habit of the host plant. Seedlings of annual crop plants exposed to excessive nematode pressure emerge more slowly and may wilt during the heat of the day. Stunting is observed as early as the 4–6-leaf stage, and frequently plants die. Young plants may show small galls on the roots, lesions on the root surface, damaged root tips, excessive root proliferation, or nematodes embedded on or in the root tissue.

As the season progresses, plant injury symptoms from nematodes become more pronounced. Stunting and thin stands are more evident, and roots are often darkened from invasion by secondary microorganisms. At maturity the yield, quality, or appearance of the crop may be much reduced.

Perennial crops heavily infested with nematodes display many of the symptoms of annual plants but with additions. Trees attacked by nematodes are often chlorotic and decline, with dieback of twigs and branches. Root crops, bulbs, and ornamentals may have reduced underground plant parts where cracking or dry rot may be present. Distortion, galling, or twisting of stems and leaves is observed where leaf-feeding nematodes are present.

Chemical Control Agents

The ability of a nematicide to kill or suppress nematodes is influenced by the mode of action and concentration of the chemical and by how long the nematode is exposed to the chemical. Temperature, moisture, soil type, and tilth are important to the diffusion of fumigant nematicides as well as movement of nonfumigants from point of placement. Correct timing and method of application of nematicides influence the degree of control obtained. Nematode genera react differently to nematicides. For example, root-knot nematode juveniles survived four times longer than the citrus nematode at the same concentration of the soil fumigant ethylene dibromide. Immature forms and juveniles in molt are more easily killed than adults. Nematode eggs are highly resistant to nematicides.

YIELD AND ECONOMIC RESPONSE

Crop losses to nematodes are known throughout the world. The magnitude of the loss depends on the nematode species, degree of infestation, plant species, and conditions of cultivation. Losses range from insignificant in some crops and areas of production to total crop failure. An estimate of cotton losses in West Texas, based on extensive research data, suggests that 10% of cotton production in the area is lost to nematodes (8).

Apart from yield reduction, further crop losses from nematodes occur in marketing and storage. The quality of tuber and root crops infested with nematodes may be unacceptable in the marketplace. Moreover, nematode-infested produce is more susceptible to invasion from secondary fungi that cause rapid deterioration in storage. Ornamental and nursery crops infested with nematodes may lose part or all of their aesthetic value or marketability.

VARIETAL RESPONSES

The host-parasite physiology of resistant and susceptible plants differs appreciably. Three general resistance reactions have been described for root-knot nematodes:

Failure to initiate giant cells. Equal numbers of juveniles may penetrate resistant roots and susceptible roots, but few giant cells are initiated in the resistant roots.

Arrested giant cell development. Nuclear and nucleolar enlargement and perhaps some nuclear division occur, but only slight cell enlargement takes place and the giant cell usually collapses before the walls thicken.

Limited susceptible host response. Few juveniles manage to initiate giant cells successfully and maintain their structure long enough for the nematode to complete its life cycle.

The factor(s) responsible for resistance in plants to nematodes is elusive. However, changes in the concentrations of secondary metabolites after infection have been implicated in resistance to root-knot nematodes. Few examples can be cited where nematode resistance approaching immunity can be found within the gene pool of a plant species.

LITERATURE CITED

1. Beingefors, S. 1971. Resistance to nematodes and the possible value of induced mutations. Pages 209–235 in:

Mutation Breeding for Disease Resistance. IAEA-PL-412/21. International Atomic Energy Agency, Vienna.
2. Bergeson, G. B. 1968. Evaluation of factors contributing to the pathogenicity of *Meloidogyne incognita*. Phytopathology 58:49-53.
3. Endo, B. Y. 1975. Pathogenesis of nematode-infected plants. Annu. Rev. Phytopathol. 13:213-238.
4. Heald, C. M. 1967. Pathogenicity of five root-knot nematode species on *Ilex crenata* Helleri. Plant Dis. Rep. 51:581-585.
5. Hoestra, H., and Oostenbrink, M. 1962. Nematodes in relation to plant growth. IV. *Pratylenchus penetrans* (Cobb) on orchard trees. Neth. J. Agric. Sci. 10(4):286-296.
6. Jones, J. E., Newsom, L. D., and Finley, E. L. 1959. Effect of reniform nematode on yield, plant characters, and fiber properties of upland cotton. Agron. J. 51:353-356.
7. Oostenbrink, M. 1966. Major characteristics of the relation between nematodes and plants. Meded. Landbouwhogesch. Wageningen 66:1-46.
8. Orr, C. C. 1983. Cotton losses to nematodes on the Texas southern high plains. J. Nematol. 4:487.
9. Orr, C. C., Abernathy, J. R., and Hudspeth, E. B. 1975. *Nothanguina phyllobia*, a nematode parasite of silverleaf nightshade. Plant Dis. Rep. 59:416-418.
10. Seinhorst, J. W. 1973. Principles and possibilities of determining degrees of nematode control leading to maximum returns. I. Protection of one crop sown or planted soon after treatment. Nematol. Med. 1:93-105.
11. Taylor, A. L., and Golden, A. M. 1954. Preliminary trials of D-D Hi-Sil as a soil fumigant. Plant Dis. Rep. 38:63-64.
12. Thorne, G. 1961. Principles of Nematology. McGraw-Hill Book Co., New York. Page 553.
13. Townshend, J. L. 1958. The effect of *Pratylenchus penetrans* on a clone of *Fragaria vesca*. Can. J. Bot. 36:683-685.

Evaluating Nematicides Applied Through Sprinkler Irrigation Systems

A. W. JOHNSON, Agricultural Research Service, U.S. Department of Agriculture, Coastal Plain Experiment Station, Tifton, GA 31793

The need for and interest in using sprinkler irrigation systems effectively and economically in the United States have resulted in cooperative research among universities, the U.S. Department of Agriculture, and industry that has developed new technology for applying nematicides through sprinkler irrigation systems (nemagation). Soil types on which nemagation research has been conducted include sand, sandy loam, and loamy sand containing less than 1% organic matter; these soil types are common in the Southeast and Pacific Northwest areas of the United States. Most of the research on nemagation has been conducted at the Coastal Plain Experiment Station in Tifton, Georgia, and the Irrigated Agriculture Research and Extension Center in Prosser, Washington. This chapter reviews and discusses methods of evaluating nematicides applied through sprinkler irrigation systems.

TEST MATERIALS

All test materials should be compared with an untreated control, a standard, and a conventional method of application currently recommended for control of nematodes. Researchers should know as much as possible about the biological activity and chemical and physical properties of the nematicide being evaluated. Pertinent literature or technical reports should be reviewed before field tests are designed.

Formulation

Test materials that have been applied through sprinkler irrigation systems for control of nematodes include emulsifiable concentrations of ethoprop (*O*-ethyl *S,S*-dipropyl phosphorodithioate), fenamiphos (ethyl 3-methyl-4-(methylthio) phenyl (1-methylethyl) phosphoramidate), oxamyl (methyl *N′,N′*-dimethyl-*N*-[(methyl carbamoyl)oxy]-1-thiooxamimidate), carbofuran (2,3-dihydro-2,2-dimethyl-7-benzofuranyl methylcarbamate), and metam-sodium (sodium *N*-methyldithiocarbamate) (6–8,12,13). Data records should include formulation type (e.g., emulsifiable concentrate, wettable powder, flowable), water solubility, name and percentage of ingredients in the formulation, lot number on the package label, and dates sent and received. If a nematicide is diluted before application, the amount of diluent used should be specified and the common and chemical names of the diluent should be given.

Application of Water and Nematicide

Uniform application. Uniform application of the irrigation water is a must if the nematicide is to be spread uniformly over the field. This requires good engineering assembly and installation of the irrigation system, precise calibration, and good irrigation management. Center-pivot sprinklers give uniformity coefficients of 85 (E. D. Threadgill, *personal communication*) when engineered and managed properly on most types of soil. On steep slopes and hilly terrains, pressure regulators on the sprinkler heads and higher pressures may be required to get satisfactory uniformity of water and nematicide application.

Rates of nematicide application. Rates should be clearly and precisely stated in terms of formulation and active ingredient applied per unit area of land.

Rate of water application. The rate of movement of a nematicide in the soil is influenced by the solubility of the nematicide, the amount of water applied during and after application, soil moisture and temperature at the time of application, soil type, and other factors such as soil adsorption capacity. It is important to record data related to these parameters at the time of application.

Studies conducted in 1981 at the Coastal Plain Experiment Station on Bonifay sand (93.5% sand, 2.9% silt, and 3.6% clay, pH 6.0–6.7, organic matter less than 1%) provided data on the amount of water to use as a carrier for applying fenamiphos via sprinkler irrigation. Based on root-gall indexes, number of nematodes in the soil, and crop response, the nematicide should be applied with 0.25 ha-cm (0.1 acre-in.) of water (25,385 L/ha; 2,715 gal/acre) when the Bonifay sand is at or near field capacity (7). Under other conditions, the water application rate must be adjusted, based on soil type and soil moisture, to transport the nematicide to the rhizosphere.

Number and timing of applications. The principal use of nematicides, regardless of how they are applied, is to control nematode populations in the soil before crops are planted (16). In plots treated with fenamiphos (6.7 kg a.i./ha) via sprinkler irrigation, root-gall indexes of squash were lower and yield was greater from plants in plots treated at planting compared with those treated 2 weeks after planting (7). Similar studies on corn indicated that root-gall indexes were lower and yield was greater from plants in plots treated at planting than in those treated 1 week after planting. Data available in the literature indicate that it is more beneficial to apply the nematicide before or immediately after planting than later (7). Information is not available on the effects of multiple postplant applications of nematicides via sprinkler irrigation for controlling nematodes on crops. However, studies in Washington (1981–1983) showed that four monthly applications of oxamyl (1.1 and 2.2 kg a.i./ha per application) via sprinkler irrigation to control nematodes on Concord grapes increased yields (G. S. Santo, *personal communication*).

EVALUATING CHEMICAL EFFECTS

The methods used to evaluate the effects of nematicides applied via sprinkler irrigation do not differ from those used to evaluate nematicides applied by conventional methods.

Nematode Species

Soil samples should be collected from the test site before and after treatment application, handled, and assayed for nematodes by one or more of the extraction procedures described by Barker et al (*this volume*). All plant-parasitic nematodes should be identified to species, and to races (16) if desirable.

Soil treated with highly toxic, slow-acting materials such as organic phosphates or carbamates should *not* be collected within 14 days of the application. Nonfumigant nematicides may not kill nematodes directly but may alter their physiology to prevent feeding and reproduction. Sampling should be done 4–6 weeks after application of nonfumigant nematicides, because most of these compounds are slow-acting.

Researchers may use methods of extraction that depend on nematode motility, or they may use a vital stain (4,15) to recognize living nematodes. In farm fields, application of nematicides is followed by a decrease in infectivity of nematodes in the soil. Because the decrease is not necessarily correlated with the number of living nematodes in the soil, it is best measured by comparing the infection of indicator plants from treated and untreated soil.

In addition, nematodes may be extracted from root tissue, nematode eggs may be extracted from soil and examined in roots, and root-gall indexes may be determined. Details of these methods are also presented by Barker et al (*this volume*).

Plant Responses

Responses of plants to nematicides applied via sprinkler irrigation are determined by examining seedlings, roots, stems, and leaves. Symptoms on aboveground parts may include stunting, erratic stand, chlorosis, or malformations from nematode injury; delayed seedling emergence, poor stand, stunting, chlorosis, scorch, abnormal growth, or death from nematicide injury; accelerated vigor and growth; or no symptoms due to treatment.

Estimates of plant response include stand count, visual ratings of growth and development, height, weight, and yield (2,5,9,10,14). Root reactions are measured by galls, lesions, root pruning, root proliferation, nematodes per gram of root, or malformation (1,3).

Emergence. Seedling response to nematicides applied via sprinkler irrigation may be measured by the rate of plant emergence and final stand counts. Greater stand counts or earlier emergence may indicate nematode control. Delayed emergence and growth may indicate inactivity of the nematicide or phytotoxicity, while rapid emergence and good stands resulting from nematode control frequently are accompanied by less seedling disease.

Early growth. As plants grow and mature, nematode control may be expressed as larger plants, more expanded leaves, longer internodes, or other evidence of increased vigor and growth. Conversely, injury symptoms—stunting, root pruning, discoloration, or morphological abnormalities—can be observed and rated.

YIELD AND ECONOMIC RESPONSE

Yield is perhaps the most common measure of the benefits derived from control of nematodes. Yield can be measured by weight, size, quality, or aesthetic value of plants or plant parts. Yield of field crops is usually expressed in kilograms per hectare or pounds per acre, while other crop responses may be expressed in terms of grade of fruit, leaf, stem, or root commodity; quality; smoothness; color; date of blossom; or eye appeal.

Nematicides are relatively expensive, but their application through sprinkler irrigation requires minimal labor and specialized equipment. The current costs of nematicide application with a tractor-powered rototiller vary from $4.79 to $25.27 per hectare ($1.94–$10.23/acre), based on whether the farmer owns or must purchase a rototiller and the time required to till 1 ha (acre) of land (11). For example, assuming an interest rate of 10%, gasoline cost of $0.955/gal, and diesel cost of $0.855/gal and pricing the farmer's labor at $3.50/hr, it costs the farmer $1.94/acre to apply a nematicide if the farmer owns a rototiller and rototills 1 acre in 0.31 hr, $3.18/acre if the farmer buys a rototiller and rototills 1 acre in 0.31 hr, $6.24/acre if the farmer owns a rototiller and rototills 1 acre in 1 hr, and $10.23/acre if the farmer buys a rototiller and rototills 1 acre in 1 hr. The current cost of nematicide application with 0.1 acre-in. of water (2,715 gal/acre) via sprinkler irrigation is $0.45/acre (11). Therefore, applying nematicides via sprinkler irrigation saves 77–96% of the costs for conventional application. When growers use nematicides, they expect an increase in crop value of at least three or four times the investment.

SPECIALIZED EQUIPMENT

The basic system for delivering a diluted or undiluted nematicide into pressurized irrigation water facilities includes a chemical supply tank, a positive-displacement pump capable of operating at pressures in excess of the water system, and an injection port located between the irrigation system and the appropriate safety and antisiphon devices that prevent potential contamination of the water source. Commercial equipment is available for injecting nematicides into sprinkler irrigation systems. Details of delivery systems, injection pumps, supply tanks, equipment construction, safety features and design, and injector calibration are presented by Threadgill (*this volume*).

The advantages of applying nematicides through sprinkler irrigation systems are uniform application; greater control of nematodes with liquid formulations of nematicides than when the nematicides are applied by conventional methods; possible reduction in the need for applying nematicide before planting each crop in intensive cropping systems; reduced human exposure and risk; reduced field traffic and compaction; and lower application costs.

Many growers are using their irrigation systems to apply nitrogen fertilizers to crops. The same equipment can be used to apply pesticides. The practicality of the commercial use of nematicides applied through sprinkler irrigation systems for nematode control needs further investigation. More research is needed on a number of crops, soil types, and costs to determine whether application of nematicides through sprinkler irrigation systems can be a feasible alternative to conventional methods of applying nematicides.

LITERATURE CITED

1. Beingefors, S. 1971. Resistance to nematodes and the possible value of induced mutations. Pages 209–235 in: Mutation Breeding for Disease Resistance. IAEA-PL-412/21. International Atomic Energy Agency, Vienna.
2. Bergeson, G. B. 1968. Evaluation of factors contributing to the pathogenicity of *Meloidogyne incognita*. Phytopathology 58:49-53.
3. Endo, B. Y. 1975. Pathogenesis of nematode-infected plants. Annu. Rev. Phytopathol. 13:213-238.
4. Fenner, L. M. 1962. Determination of nematode mortality. Plant Dis. Rep. 46:383.
5. Hoestra, H., and Oostenbrink, M. 1962. Nematodes in relation to plant growth. IV. *Pratylenchus penetrans* (Cobb) on orchard trees. Neth. J. Agric. Sci. 10(4):286-296.
6. Johnson, A. W. 1978. Effect of nematicides applied through overhead irrigation on control of root-knot nematodes on tomato transplants. Plant Dis. Rep. 62:48-51.
7. Johnson, A. W. 1982. Application of nematicides through an overhead sprinkler irrigation system for control of nematodes on crops. Proc. Natl. Symp. Chemigation, Univ. Ga. (Tifton) 2:69-73.
8. Johnson, A. W., Young, J. R., and Mullinix, B. G. 1981. Applying nematicides through an overhead sprinkler irrigation system for control of nematodes. J. Nematol. 13(2):154-159.
9. Jones, J. E., Newsom, L. D., and Finley, E. L. 1959. Effect of reniform nematode on yield, plant characters, and fiber properties of upland cotton. Agron. J. 51:343-346.
10. Oostenbrink, M. 1966. Major characteristics of the relation between nematodes and plants. Meded. Landbouwhogesch. Wageningen 66:1-46.
11. Osteen, C., Johnson, A. W., and Dowler, C. C. 1982. Applying the economic threshold concept to control lesion nematodes on corn. U.S. Dep. Agric. Econ. Res. Serv. Tech. Bull. 1670.
12. Qualls, M. 1982. Disease control in Irish potatoes. Proc. Natl. Symp. Chemigation, Univ. Ga. (Tifton) 2:61-63.
13. Santo, G. S. 1982. Nemagation on Irish potato and grapes. Proc. Natl. Symp. Chemigation, Univ. Ga. (Tifton) 2:64-67.
14. Seinhorst, J. W. 1973. Principles and possibilities of determining degrees of nematode control leading to maximum returns. I. Protection of one crop sown or planted soon after treatment. Nematol. Med. 1:93-105.
15. Shepherd, A. M. 1962. New Blue R, a stain that differentiates between living and dead nematodes. Nematologica 8:201-208.
16. Taylor, A. L., and Sasser, J. N. 1978. Biology, identification and control of root-knot nematodes (*Meloidogyne* species). Dep. Plant Pathol. N.C. State Univ., Raleigh. 111 pp.

Testing Nematicidal Efficacy Using Eggs, Juveniles, or Cysts of *Heterodera schachtii*

ARNOLD E. STEELE, Agricultural Research Service, U.S. Department of Agriculture, Salinas, CA 93915

The methods discussed in this chapter use cysts, eggs, or hatched second-stage juveniles of *Heterodera schachtii* Schmidt, 1871 to screen chemicals. They should be used to evaluate the effects of chemical treatments on the viability of cyst contents and to obtain eggs or juveniles for in vitro evaluation of chemicals or for inoculation to evaluate chemical foliar sprays, drenches, or soil treatments.

APPARATUS AND REAGENTS

The following items are required for the methods suggested (in each case, equivalent materials may be substituted): collection cups (small sieves) (Hykro Pet Industries, Esbjerg, Denmark); plastic portion cups (about 12-ml capacity) (Thunderbird Container Corporation, El Paso, TX 79912); a sieve made of rigid plastic tubing and nylon mesh (40 per centimeter); long-stemmed glass funnels of convenient size; a refrigerator with countertop cabinet measuring about 0.04 m^3, with holes large enough to accept the funnel stems drilled through the top of the cabinet; an electronic thermoregulator equipped with thermistor probe; flexible heating tape 11 × 2 mm, 35 W; pinch clamps; pipettes; camel's-hair brushes; and a twin shell blender.

Hatching agents, such as aqueous solutions of 4×10^{-6}M zinc chloride or sugar beet root diffusate leached from the soil of pot-grown sugar beet plants, are required.

PROCEDURES

To evaluate chemical effects on juvenile hatching and emergence from cysts or penetration and development in host plants. Newly formed cysts are separated from sugar beet roots and soil by washing, floating, and decanting suspended debris into screens. Newly formed cysts with eggs and juveniles are selected, manually separated from washed debris, and stored in water at 8°C until needed (1).

Two to 5 ml of test solutions are transferred to the portion cups using a precalibrated pipette. Groups of 20 nematode cysts are transferred with a fine-bristle brush to a small wedge-shaped piece of filter paper before going to the collection cups (sieves). This step prevents dilution of the test solution during transfer of the cysts. Treatments, including controls with diluent but no added chemicals, are replicated at least four times. The hatching vessels and their contents are incubated at 24°C during the entire test period.

The cysts are treated with the chemical solutions for 1–7 days and then transferred to tap water, which is changed daily for 4 days to remove the test materials. If the objective is to determine the effect of chemicals on the ability of juveniles to penetrate and develop in host plant roots, the cysts are broken open and the contents inoculated as described below in the section on evaluating the effects of chemical treatments on the hatching and emergence of juveniles. Whole, undamaged cysts are placed in a hatching agent for 2–4 weeks, after which cysts should be broken open and the numbers of eggs with unhatched juveniles and juveniles that have hatched but not emerged from cysts should be counted. Data on numbers of emerged larvae may then be converted to percentage hatch.

An alternative to the above procedure is to remove embryonated eggs judged to be viable from cysts before initiating the chemical treatments. The advantage of this method is that fewer nematodes are required and far more replications can be managed with less effort required for counting. However, groups of eggs are more difficult to manipulate than cyst lots.

To hatch eggs. Washed and screened root debris containing cysts are added to a large screen with openings of about 250 μm and 10 cm in diameter, which then is placed in a funnel containing a hatching solution. The level of the solution is adjusted so that the debris is wet but not completely covered.

The stem of the funnel is inserted through the top of the refrigerator cabinet. The cabinet interior is maintained at 8°C. The temperature of the solution bathing the cysts is maintained at 24°C using an electronic thermoregulator equipped with a thermistor and heating tape. If ambient temperatures are above 24°C, supplementary heating of the solution may not be required (2).

To evaluate chemical effects on hatched second-stage juveniles. Newly hatched second-stage juveniles are treated for 24 hr at 24°C with aqueous solutions of 1, 5, 10, 25, 50, and 100 μg/ml of test chemical. If the researcher suspects that sublethal rates of nematicides may increase parasitism over that of untreated controls, rates ranging from 0.01 to 1.0 μg/ml of chemical may be included.

Effects of chemical treatments on mobility may be estimated by placing the juveniles on tissue paper supported by collection cups (sieves) that are in turn placed in the chemical solution. Juveniles that remain

Revision of a 1978 paper prepared in connection with the effort of the Joint SON/ASTM E35.16 Subcommittee on Nematode Control Agents to develop guidelines on assessing nematicide efficacy.

on the tissue paper are assumed to be either immobilized or incapable of purposeful movement (disoriented).

After the initial treatment, the juveniles are washed with several liters of tap water. The nematodes are easily concentrated in a small volume of water using a Büchner-type funnel with fritted disk.

Roots of sugar beet (or other suitable host such as cabbage, broccoli, Brussels sprouts, or cauliflower) grown in a steam-sterilized sand-soil mixture are inoculated with treated and untreated juveniles. At 18 and 30–35 days after inoculation, the plants are harvested and the roots and soil examined for adult sugar beet nematodes.

Bioassay of nematicidal efficacy. The numbers of juveniles added per plant or unit weight of soil should be specified; however, not less than 2,000 juveniles, or eggs and juveniles, from 20 selected viable cysts should be added per plant. If cysts are used, the numbers of eggs and juveniles may be estimated by counting them or by hatching the eggs in a solution containing a hatching agent. At least four, and preferably six to 10, plant replicates should be used for each treatment or untreated check.

If plants are inoculated at the time of transplanting, the inoculum (cysts, eggs, or juveniles) is placed in a hole just large enough to accept the transplant roots. If established plants are inoculated, the inoculum should be evenly distributed in three holes large enough to accept the pipette tip (1.5–3.0 cm deep) and located 1.5–3.0 cm from the plant stem.

To test efficacy on pot-grown plants. Experiments to evaluate the efficacy of chemicals for controlling *H. schachtii* on pot-grown plants may be conducted in a growth chamber or greenhouse. Chemicals may be applied as aqueous foliar sprays, drenched on the soil surface, or incorporated in soil before potting. Granular formulations may be applied to the surface of potted soil either before or after seeding or transplanting seedlings, or the material may be incorporated into the soil using a twin shell blender before potting. Chemical rates should be reported as amount of active ingredient per gram of solute and the volume of formulated material applied per plant or per pot. If granular formulations are used, the formulation and amount of active ingredient per unit volume of soil should be reported.

To assess nematode populations. Nematicide efficacy may be evaluated one or more times during plant growth. At least one count should be obtained before the second nematode generation is produced.

Counts may be made on nematodes extracted from soil or plant tissues and may include any or all stages. Adult males may be counted by harvesting plants 18 days after inoculation and placing roots on funnels in a moist chamber for 5–10 days. To count adult females, plants are harvested 30–35 days after inoculation and the roots and potting soil are washed and examined for nematodes. Barker et al (*this volume*) have described preferred methods of population assessment.

Counts of accumulated juvenile hatches should be subjected to analysis of variance, and the statistical procedures used should be reported.

LITERATURE CITED

1. Steele, A. E. 1972. Evaluation of cyst selection as a means of reducing variation in sugarbeet nematode inocula. J. Am. Soc. Sugar Beet Technol. 17:22-29.
2. Steele, A. E. 1976. Improved methods of hatching *Heterodera schachtii* larvae for screening chemicals. J. Nematol. 8:23-25.

Nematicide Evaluation as Affected by Disease Complexes and Nematicide Effects on Nontarget Organisms

J. L. STARR and C. M. KENERLEY, Department of Plant Sciences, Texas A&M University, College Station 77843

Agricultural soils are a complex matrix of interacting physical, chemical, and biological components. Treatment of soil in a specific manner to achieve one desired effect may result in additional unanticipated changes in the soil complex. Application of nematicides to the soil to reduce populations of plant-parasitic nematodes and alleviate crop damage is known to affect other components of the soil complex, especially the biological components. These side effects may be beneficial, harmful, or innocuous and, therefore, can affect the evaluation of nematicide activity. This chapter is not intended to be a complete review of the data available on the effects of nematicides on nontarget organisms; rather, it illustrates the types of side effects that may be expected when nematicides are applied to the soil. We emphasize those nematicides generally considered to have a relatively narrow spectrum of activity compared to the broad-spectrum, multipurpose soil fumigants (e.g., methyl bromide, chloropicrin, and methyl isothiocyanate). The insecticidal properties of the organophosphate and carbamate nonfumigant nematicides are not discussed.

NEMATODE INVOLVEMENT IN DISEASE COMPLEXES

Frequently, diseases caused by nematodes are actually disease complexes that involve not only the nematodes but also various soilborne fungi and bacteria. The most frequently observed disease complexes are the Fusarium wilt–root-knot complexes and those involving *Meloidogyne* spp. and seedling disease pathogens (e.g., *Pythium, Rhizoctonia,* and *Thielaviopsis* spp.). Numerous other disease complexes have also been described (for an extensive listing of interactions of nematodes with other pathogens in disease etiology, see Western Regional Research Project [20]).

In most of the disease complexes that have been described, the amount of plant disease caused by the "secondary pathogen" depends on the effects of the "primary pathogen" (the nematode). In the absence of the nematode pathogen, the secondary pathogens may cause little or no disease. Under these conditions, controlling plant disease caused by nematodes also controls the disease caused by the secondary pathogens. Thus, improved growth of the test plant as a result of nematicide treatment is not entirely the result of alleviation of damage caused directly by the nematodes. Testing nematicides under field conditions, therefore, requires a careful assessment of the potential involvement of disease complexes in the etiology of the nematode-incited disease at the time of the test.

FUNGICIDAL ACTIVITY OF NEMATICIDES

The broad-spectrum soil fumigants such as methyl bromide, chloropicrin, and methyl isothiocyanate have fungicidal activity and are effective in suppressing disease caused by soilborne fungi (19). Fumigant nematicides with narrower spectra of activity also exhibit varying degrees of fungicidal activity. Dibromochloropropane (DBCP) at concentrations of 0.02% in cornmeal agar inhibited mycelial growth of several pythiaceous fungi by 60–75% (3). In the same study, DBCP also suppressed soil populations of these fungi in field tests when applied at 26.6 L/ha. In a separate study, DBCP decreased damping-off of snap beans caused by *P. irregulare* or *R. solani* but increased damping-off caused by *P. myriotylum* (14). Soil treatment with dichloropropene and dichloropropane (D-D) suppressed populations of *Phytophthora* spp. but at three times the rate required to suppress nematode population densities (2). In contrast to these reports, D-D increased the severity of lettuce drop, caused by *Sclerotinia sclerotiorum*, by increasing the production of primary inoculum (ascospores) for the disease (9). Vanachter (19) provides a more extensive review of the effects of soil fumigants on soilborne fungi.

Some nonfumigant nematicides are also reported to have fungicidal activity. Fenamiphos had ED_{50} values of less than 10 μg/ml against 24 of 28 common soilborne fungi. Oxamyl, on the other hand, had ED_{50} values of more than 100 μg/ml against 27 of these same 28 fungi (4). Both fensulfothion and ethoprop inhibited the maturation of sclerotia of *Sclerotium rolfsii* in soil plate assays (11,12) and provided some early season control of the pathogen in field tests. Fensulfothion and ethoprop, however, had little or no effect on the mycoparasite *Trichoderma harzianum* or on species of *Rhizopus* and *Aspergillus* (11,12). In other tests, ethoprop at 9 kg a.i./ha increased damping-off of snap beans and turnips in fields infested with *R. solani, Pythium* spp., and *Fusarium* spp. (14,15). Similarly, in two separate studies, aldicarb increased the severity of diseases of sugar beet and potato caused by *R. solani* (6,17). These data illustrate the variety of effects some nematicides can have on soilborne pathogens other than nematodes.

VESICULAR-ARBUSCULAR MYCORRHIZAE AND RHIZOBIA-LEGUME ASSOCIATIONS

The beneficial effects that vesicular-arbuscular (VA) mycorrhizal fungi have on plant health and vigor by enhancing plant absorption of various soil minerals have been demonstrated with numerous plant species (5,7). A substantial number of studies have documented that nematicides (fumigants and nonfumigants) can affect root infection and chlamydospore development by VA mycorrhizal fungi (8). In general, fumigants reduce root infection and chlamydospore development. The response of VA mycorrhizal fungi to the nonfumigants tested has been varied. This variation may be attributed to sensitivity of the fungus, plant, or specific fungus-plant combination to the nematicide, the timing of the chemical application (before or after planting), the reduction of organisms that feed on or compete with VA mycorrhizal fungi for substrates, or stimulation of microorganisms that may enhance root infection (1,8).

Unless the role of VA mycorrhizal fungi in promoting plant health and yield is taken into consideration, erroneous conclusions may be reached in nematicide evaluations. When applications of a nematicide increase root infection by VA mycorrhizal fungi, any subsequent increase in plant vigor or yield compared to control plants cannot be attributed solely to the action of the nematicide in reducing populations of parasitic nematodes. Also, if plant stunting occurs after an experimental nematicide is applied, plant roots should be assayed to ensure that roots are infected with VA mycorrhizal fungi. What may appear to be phytotoxicity could in fact be a lack of VA mycorrhizal fungi in root tissue.

These examples illustrate the need to conduct experiments to determine the effect of nematicides on VA mycorrhizal fungi. The results from such experiments may indicate whether reinoculation with VA mycorrhizal fungi is necessary or would assist in reducing variability among plots. Methods for making such determinations have been published elsewhere (13).

Data on the effects of nematicides on nodulation by *Rhizobium* spp. are rather sparse, and those that are available vary with the strain and nematicide (18). Tewfik et al (16) transplanted broad bean seedlings into tubes of agar, then inoculated each seedling with *Rhizobium leguminosarum* and subsequently treated seedlings with aldicarb or DBCP. They reported that no nodules developed on seedlings treated with the nematicides at 20–2,000 μg/ml. However, Reddy and Kumar Rao (10) reported no effect of aldicarb on nodulation of soybeans by *Rhizobium japonicum* in field plots. These differences may reflect the nature of the test conditions and illustrate the need for uniformity in test procedures (cultivar and strain identification, nitrogen efficiency, field or greenhouse assay) to properly identify what effects nematicides may have on nodulation.

SUMMARY

The data presented provide ample evidence that soil treatment with nematicides can affect plant growth and yield by means other than controlling plant-parasitic nematodes. During routine testing of nematicidal chemicals, it is not uncommon to use application rates two to three times higher than those normally recommended. These conditions increase the probability of encountering side effects of the nematicide which may affect plant growth. Therefore, when evaluating nematicide efficacy, the scientist must be alert for effects on nontarget organisms that may influence the evaluation of the nematicide. Although it is not possible to adequately test for all such side effects during routine nematicide evaluation, careful observation and plant response can signal the possibility of nontarget effects.

It is particularly important to note any differences among treatments in the incidence or severity of other diseases. With legume crops, any differences in rhizobia nodules among treatments should be noted. Adequate control of insect pests is also necessary, especially when evaluating chemicals known to have both insecticidal and nematicidal activity. As stated previously, the soil is a complex environment with many interacting factors, and it should not be surprising that attempts to alter the activity of one set of factors (e.g., plant-parasitic nematodes) also affect the activity of other factors in the environment.

LITERATURE CITED

1. Atilano, R. A., and Van Gundy, S. D. 1979. Effects of some systemic, nonfumigant nematicides on grape mycorrhizal fungi and citrus nematode. Plant Dis. Rep. 63:729-733.
2. Baines, R. C., Klotz, L. J., DeWolfe, T. A., Small, R. H., and Turner, G. O. 1966. Nematocidal and fungicidal properties of some soil fumigants. Phytopathology 56:691-698.
3. Bumbieris, M. 1970. Effect of DBCP on pythiaceous fungi. Plant Dis. Rep. 54:622-624.
4. Bunt, J. A. 1975. Effect and mode of action of some systemic nematicides. Meded. Landbouwhogesch. Wageningen 75:1-127.
5. Hayman, D. S. 1978. Endomycorrhizae. Pages 401-442 in: Interactions Between Non-pathogenic Soil Microorganisms and Plants. Y. R. Dommergues and S. V. Krupa, eds. Elsevier Scientific Publishing Co., Amsterdam. 475 pp.
6. Leach, S. S., and Frank, J. A. 1982. Influence of three systemic insecticides on Verticillium wilt and Rhizoctonia disease complex of potato. Plant Dis. 66:1180-1182.
7. Maronek, D. M., Hendrix, J. W., and Kiernan, J. 1981. Mycorrhizal fungi and their importance in horticultural crop production. Hortic. Rev. 3:172-213.
8. Menge, J. A. 1982. Effect of soil fumigants and fungicides on vesicular-arbuscular fungi. Phytopathology 72:1125-1132.
9. Partyka, R. E., and Mai, W. F. 1958. Nematocides in relation to sclerotial germination in *Sclerotinia sclerotiorum*. Phytopathology 48:519-520.
10. Reddy, D. D. R., and Kumar Rao, J. V. D. K. 1975. Effect of selected nonvolatile nematicides and benomyl on nodulation, root-knot nematode control, and yield of soybeans. Plant Dis. Rep. 59:592-595.
11. Rodriguez-Kabana, R., Backman, P. A., Kerr, G. W., Jr., and King, P. S. 1976. Effects of the nematicide fensulfothion on soilborne pathogens. Plant Dis. Rep. 60:521-524.
12. Rodriguez-Kabana, R., Backman, P. A., and King, P. S. 1976. Antifungal activity of the nematicide ethoprop. Plant Dis. Rep. 60:255-259.
13. Schenck, N. C., ed. 1982. Methods and Principles of Mycorrhizal Research. Am. Phytopathol. Soc., St. Paul, MN. 244 pp.
14. Sumner, D. R. 1974. Interactions of herbicides and nematicides with root diseases of snapbean and southern pea. Phytopathology 64:1353-1358.
15. Sumner, D. R., and Glaze, N. C. 1978. Interactions of

herbicides and nematicides with root diseases of turnip grown for leafy greens. Phytopathology 68:123-129.

16. Tewfik, M. S., Embabi, M. S., and Hambi, Y. A. 1975. Efficiency of *Rhizobium leguminosarum* as affected by certain herbicides and nematicides. Zbl. Bakt. II. 130:725-731.
17. Tisserat, N., Altman, J., and Campbell, C. L. 1977. Pesticide-plant disease interactions: The influence of aldicarb on growth of *Rhizoctonia solani* and damping-off of sugar beet seedlings. Phytopathology 67:791-793.
18. Trinick, M. J. 1982. Biology. Pages 76–146 in: Nitrogen Fixation, Vol. 2: Rhizobium. W. J. Broughton, ed. Clarendon Press, Oxford. 353 pp.
19. Vanachter, A. 1979. Fumigation against fungi. Pages 163-183 in: Soil Disinfestation. D. Mulder, ed. Elsevier Scientific Publishing Co., New York. 368 pp.
20. Western Regional Research Project (W-56). 1976. Bibliography of nematode interactions with other organisms in plant disease complexes. Agric. Exp. Stn. Oreg. State Univ. Corvallis Stn. Bull. 623.

Conversion Factors for U.S. Customary and SI (Metric) Units of Measurement

An act of Congress in 1866 made the International System of Units**, SI (Metric System), "lawful throughout the United States of America," and the government is actively promoting its use. We are encouraged to report agricultural data in SI units. U.S. Customary Units are converted to SI units by multiplication by a factor having five (5) significant digits. Reciprocal conversion factors and symbols are given in the middle column. The spellings meter and liter are preferred.

U.S. customary to SI (metric) units	SI to U.S. units and symbols (abbreviations)[1]	Comments[1]
	Length	
1 mile(statute) × 1.6093 = kilometers	1 km × 0.62137 = mi	mile;spell out
1 foot × 0.30480* = meters	1 m × 3.2808 = ft	
1 inch × 2.5400* = centimeters	1 cm × 0.39370 = in	inch:",spell out
	Area	
1 square mile × 2.5900 = square kilometer	1 km^2 × 0.38610 = mi^2	square:(sq);mile:spell out
1 acre × 0.40469 = hectares	1 ha × 2.4710 = acre	acre:spell out,A,a
1 square foot × 0.092903 = square meters	1 m^2 × 10.764 = ft^2	square:sq
1 square inch × 6.4516 = square centimeters	1 cm^2 × 0.15500 = in^2	square:sq;inch:spell out
	Volume-Capacity	
1 barrel(31.5 gal) × 119.24 = liters	1 L × 0.0083864 = bbl	other bbl(bl) = 31,40,42 (oil) gal
1 bushel × 0.35239 = hectoliters	1 hl × 2.8378 = bu	cubic:cu
1 cubic inch × 16.387 = cubic centimeters	1 cm^3 × 0.061024 = $in.^3$	cm^3:cc
1 cubic foot × 0.028317 = cubic meters	1 m^3 × 35.315 = ft^3	cubic:cu
1 quart(liquid) × 0.94635 = liters	1 L × 1.0567 = qt	liter:spell out
1 quart(dry) × 1.1012 = liters	1 L × 0.90808 = qt	liter:spell out
1 gallon(liquid) × 3.784 = liters	1 L × 0.26417 = gal	liter:spell out
1 fluid ounce × 29.574 = milliliters	1 ml × 0.033814 = fl oz	
	Weight-Mass	
1 ton(short 2000 lb) × 0.90718 = tons(metric)	1 t(m) × 1.1023 = t(s)	ton:T,spell out
1 pound × 0.00045359 = tons(metric)	1 t(m) × 2204.6 = lb	ton:T,spell out
1 ton(short 2000 lb) × 907.18 = kilograms	1 kg × 0.0011023 = t(s)	ton:T,spell out
1 pound × 0.45359 = kilograms	1 kg × 2.2046 = lb	
1 pound × 453.59 = grams	1 g × 0.0022046 = lb	gram:gm
1 avoirdupois ounce × 28.350	1 g × 0.035274 = oz av	gram:gm
	Rates	
1 mile per hour × 1.6093 = kilometers per hour	1 km/hr × 0.62137 = mph	
1 mile per hour × 26.822 = meters per minute	1 km/hr × 54.681 = ft/min	
1 ton(short) per acre × 2.2417 = tons(metric) per hectare	1 t/ha × 0.44609 = t(s)/a	acre:(A),ton(T):spell out
1 pound per acre × 1.1209 = kilograms per hectare	1 kg/ha × 0.89218 = lb/a	acre:A,spell out
1 pound per square foot × 4.8824 = kilograms per square meter	1 kg/m^2 × 0.20481 = lb/ft^2	square:sq
1 pound per square inch × 0.070307 = kilograms per sq cm	1 kg/cm^2 × 14.223 = lb/in^2	lb/in^2:psi;Pa pressure
1 pound per cubic foot × 16.018 = kilograms per cubic meter	1 kg/m^3 × 0.062428 = lb/ft^3	cubic:cu
1 ounce(avdp) per gallon × 7.4892 = grams per liter	1 g/L × 0.13353 = oz/gal	liter:spell out
1 ounce (avdp) per acre × 70.054 = grams per hectare	1 g/ha × 0.014275 = oz/a	acre:A,spell out
1 pound per gallon × 0.11983 = kilograms per liter	1 kg/L × 9.3454 = lb/gal	liter:spell out
1 gallon per acre × 9.3538 = liters per hectare	1 L/ha × 0.10691 = gal/a	liter,acre(A):spell out
1 ounce(fl) per acre × 73.079 = milliliters per hectare	1 ml/ha × 0.013684 = oz/a	acre:A,spell out
1 ounce(fl) per gallon × 7.8125 = milliliters per liter	1 ml/L × 0.12800 = fl oz/gal	fluid,liter:spell out
1 bushel per acre × 0.87077 = hectoliters per hectare	1 hl/ha × 1.1484 = bu/a	acre:A,spell out
1 cubic foot per acre × 0.069971 = cubic meters per hectare	1 m^3/ha × 14.291 = ft^3/a	acre:(A),cubic:spell out
	Other	
1 pound per square inch × 6894.8 = pascal	1 Pa × 0.00014504 = psi	Use Pa or kPa for pressure
1 atmosphere × 101.325 = kilopascals	1 kPa × 0.0098692 = atm	Normal atmosphere
1 bar × 100,000* = pascals	1 Pa × 0.00001* = bar	
1 part per million × 0.00083305 = pounds per 100 gal of water	1 ppm × 0.0010000 = g/L water	at 20 C
degrees Celsius × 1.8 plus 32 = degrees Fahrenheit	C = (°F − 32)/1.8	(centigrade)
1 British thermal unit × 252.12 = calories	1 cal × 0.0039663 = Btu	at 20 C
1 horsepower × 1.0139 = metric horsepower	1 (m)hp × 0.98632 = hp	
1 horsepower × 0.74570 = kilowatt	1 kw × 1.3410 = hp	
1 foot-pound per second × 1.3558 = watts	1 W × 0.73756 = ft-lb/sec	
1 foot-candle × 10.764 = lux	1 lux × 0.092902 = ft-c	

**American Society for Testing and Materials, 1977. Standard for Metric Practice. E380-76. 1916 Race Street, Philadelphia, Pennsylvania 19103. 34 pp.
*Exact conversion factor.
[1] Symbols for SI units have no periods or plurals and are not capitalized unless the unit is derived from a proper name. To avoid confusion, "L" is often used for liter, similarly "in." is used for inch.

Note: Reprinted, by permission, from Charles W. Averre and Marvin K. Beute from *Fungicide and Nematicide Tests.*

Index